The Autobiography of
Nikola Tesla
and Other Works

The Autobiography of Nikola Tesla and Other Works

INTRODUCTION BY KEN MONDSCHEIN, PhD

Canterbury Classics

San Diego

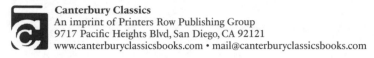

Canterbury Classics
An imprint of Printers Row Publishing Group
9717 Pacific Heights Blvd, San Diego, CA 92121
www.canterburyclassicsbooks.com • mail@canterburyclassicsbooks.com

Printers Row Publishing Group is a division of Readerlink Distribution Services, LLC.
Canterbury Classics is a registered trademark of Readerlink Distribution Services, LLC.

Correspondence concerning the content of this book should be sent to Canterbury Classics, Editorial Department, at the above address.

Publisher: Peter Norton • Associate Publisher: Ana Parker
Art Director: Charles McStravisk
Senior Developmental Editor: April Graham Farr
Editor: Traci Douglas
Editorial Team: Dan Mansfield
Production Team: Julie Greene, Rusty von Dyl

Cover Designer: Ray Caramanna
Interior Designer: Riann Bender

Image Credits: http://www.classictesla.com/photos/tesla/tesla.html via Wikimedia Commons; *Electrical Experimenter Magazine*

Library of Congress Cataloging-in-Publication Data

Names: Tesla, Nikola, 1856-1943, author. | Martin, Thomas Commerford,
     1856-1924, author. | Tesla, Nikola, 1856-1943. My inventions. | Martin,
     Thomas Commerford, 1856-1924. inventions, researches, and writings of
     Nikola Tesla.
Title: The autobiography of Nikola Tesla and other works / Nikola Tesla,
     Thomas Martin ; introduction by Ken Mondschein, PhD.
Description: San Diego : Canterbury Classics, [2021] | Series:
     Leather-bound classics | Summary: "Part of our Canterbury Classics
     leather-bound editions, this volume contains Tesla's autobiography,
     originally published as a collection of articles in Electrical
     Experimenter magazine during the early 1900s. Also included are several
     other articles written by the man himself, and the complete volume of
     Thomas Commerford Martin's The Inventions, Researches, and Writings of
     Nikola Tesla"-- Provided by publisher.
Identifiers: LCCN 2020048522 | ISBN 9781645174387
Subjects: LCSH: Tesla, Nikola, 1856-1943. | Electrical engineers--United
     States--Biography. | Serbian American inventors--Biography. |
     Inventors--United States--Biography. | Electrical engineering. |
     Electric currents, Alternating.
Classification: LCC TK140.T4 A3 2021 | DDC 621.3092 [B]--dc23
LC record available at https://lccn.loc.gov/2020048522

PRINTED IN CHINA

25 24 23 22 21   1 2 3 4 5

Editor's Note: These works have been published in their original form to preserve the authors' intent and style.

CONTENTS

PART II: The Tesla Effects with High Frequency and
High Potential Currents

PART III: Miscellaneous Inventions and Writings

Part IV: Appendix: Early Phase Motors and the Tesla Mechanical
and Electrical Oscillator

Introduction

THE AUTOBIOGRAPHY OF
NIKOLA TESLA AND OTHER WORKS

More than 125 years after his birth—which was, fittingly enough, during a lightning storm—the Serbian-American electrical engineer Nikola Tesla has become synonymous with scientific genius. His name has been lent not only to the well-known electric car company and the SI unit of magnetic flux density, but also to an asteroid, a crater on the moon, and even a rock band from the 1980s. Likewise, his life story has been made into books, movies, and even an opera. After achieving the pinnacle of wealth and success, he fell into genteel poverty and embodied a new archetype— the mad scientist, prone to making grandiose claims about new inventions that never materialized. However, in the past few decades, the image of Tesla the crackpot has faded into the background, leaving his considerable accomplishments to enjoy the limelight.

To be sure, Tesla is a fascinating subject: He was a man of considerable talents and even more considerable flaws, combining an inventive brain with a poetic and even mystical turn of mind. As Arthur Edwin Kennelly of Harvard University said upon awarding Tesla his Edison medal, "Tesla set wheels going round all over the world. . . . What he showed was a revelation to science and art unto all time." Chief and most lasting among Tesla's innovations was the alternating current, or AC, motor. The value of this device is that it can take an electrical current that is both easy to generate and can be transmitted over long distances and exploit it to do useful work. AC motors are also quiet, long lasting, powerful, and efficient. They are used in everything from water pumps to vacuum cleaners to spacecraft to the electric cars that bear Tesla's name. In other words, the AC motor is one of the fundamental inventions of the modern age.

In order to appreciate the significance of Tesla's innovation, it's necessary to understand the difference between alternating current and direct current. Picture electric charge flowing through a wire as water going through a hose. A direct current (DC) is what you are used to seeing in your backyard: the water flows one direction. However, in an alternating current (AC), the water flows back and forth, as if you had a section of hose with some water in it and alternately raised and lowered each end. Because electrical power is dependent on the strength (potentiality) of the stream and not the direction of the flow, both AC and DC are useful for doing different kinds of work—though using AC takes some clever engineering tricks. Tesla was a brilliant engineer who discovered many applications for AC power.

Tesla had other accomplishments, as well. He invented generators more efficient than any ever seen before, and, with the backing of George Westinghouse, he was one of the chief people responsible for the adoption of AC power grids in the United States. In this campaign, Tesla and Westinghouse were opposed by Thomas Edison, who championed DC current. The "current war" ended with victory for Westinghouse and Tesla. Among other inventions, Tesla also improved street lighting and invented ways to meter electricity and convert AC to DC. His "magnifying transmitter"—the famous Tesla coil—was used in radio and medical applications for decades and is still used for entertainment. In all, Tesla is estimated to have had about 300 patents to his name, though not all of them have been accounted for by historians.

We must not, however, invest too heavily in the idea of the "lone genius." While a man of undoubted talents and considerable charisma, as well as an indisputable flair for self-promotion, Tesla, like all inventors, built on the accomplishments of others. He did not invent the electric dynamo; rather, he built on work by men from many countries and many walks of life, such as the British inventor Michael Faraday, the Italian professor Antonio Pacinotti, the German industrialist Ernst Siemens, the Hungarian priest Ányos Jedlik, and of course, Thomas Alva Edison. Similarly, the labor and financial backing of others was required to bring his ideas to fruition. Edison and Westinghouse helped to develop the market and infrastructure that allowed Tesla's inventions to be deployed on a large scale. Finally, countless nameless men and women labored to construct, install, and maintain his innovations.

EARLY LIFE

Nikola Tesla was born to Milutin Tesla, an Eastern Orthodox priest, and his wife, Djuka, in 1857. His maternal grandfather was also a priest. As for his mother, though she had no formal schooling, Nikola would attribute his prodigious mental talents to his inheritance from her. The town where he was born, Smiljan, is today in Croatia, but at the time it was part of the Austro-Hungarian Empire. Because of this, Tesla's schooling was conducted in German, the language of the empire.

Tesla attended a prestigious Gymnasium in Karlovac, now in Croatia—that is, a college-preparatory high school. There, he would have studied languages, science, and mathematics. He first became fascinated by electricity when his physics teacher demonstrated a dynamo, a device that produces electricity by rotating a metal piece (appropriately called the rotor) in a magnetic field: "It is impossible for me to convey an adequate idea of the intensity of feeling I experienced in witnessing his exhibitions of these mysterious phenomena. Every impression produced a thousand echoes in my mind. I wanted to know more of this wonderful force; I longed for experiment and investigation."

Unfortunately, Nikola was his father's sole surviving son (his elder brother Dane died at the age of 12 in a horse-riding accident) and had been earmarked for the "family business"—that is, the priesthood. A career as an electrical engineer and inventor was thus out of the question. However, shortly thereafter, Tesla had the bad—or good—fortune to fall gravely ill with cholera, which was ravaging the district at the time, and was bedridden for nine months. The elder Tesla promised his son that if he recovered, he could study engineering at university.

After taking a year to recuperate by roaming the mountains near his home (and incidentally avoiding conscription into the army), Tesla began his studies at Austrian Polytechnic in the city of Graz. Tesla recounts in his memoirs how a professor at the Polytechnic named Poeschl publicly mocked him for suggesting a modification of a Gramme dynamo. This was an early form of a generator that turned a magnetized wire-wrapped iron ring through a ring-shaped armature; the electric current was then "picked" off by copper brushes, which only allowed the electricity to flow in one direction (a device called a commutator). At the time, most dynamos (electrical generators) were made to produce direct current, since alternating current was not seen as very useful. Tesla dared to suggest that the metal-brush commutator was inefficient and unnecessary. This presaged his later interest in AC generators—the sort most commonly used today. The Gramme dynamo also presaged Tesla's interest in motors: supplied with power and run in reverse, the shaft would turn on its own— the first practical electromagnetic motor.

Ironically, though he was a fanatical and talented student—he had graduated high school one year early and was known for being able to do calculus in his head—Tesla never graduated from the

Polytechnic. His grades for his first year were outstanding, and he later claimed to have worked 20 hours a day, seven days a week—so much so that his professors wrote to his father, claiming that he risked dying of overwork. However, at the end of his second year, Tesla lost his scholarship and developed a gambling problem; by his third year, he had stopped going to class entirely. In fact, he had some sort of mental breakdown, hiding from his family and friends and moving to Maribor, Slovenia, where he worked as a draftsman, whiling away his spare time playing cards and chess. His father begged him to come home, but Nikola only returned to Gospić in police custody, since he did not have a residence permit to stay in Maribor. One week after his homecoming, Tesla's father died.

Tesla tried his luck at formal schooling once more: his uncles Petar and Pavle put together enough money for him to escape Gospić. He went to Prague, that ancient city of culture and learning, to study at the prestigious Charles-Ferdinand University, but as he had never studied ancient Greek and could neither speak nor write Czech, he was only able to audit lectures. (However, in addition to his native Serbo-Croatian, Tesla also studied Latin, Italian, French, German, and English.)

After a year, he moved to the rapidly modernizing city of Budapest, where he worked at the telephone company and met the famous Hungarian engineer Tivadar Puskás, who had invented the telephone exchange. Always ambitious, Tesla resumed his usual exhausting work schedule in Budapest. Unsurprisingly, he fell ill, with symptoms more mental than physical: he became hypersensitive to any sound, no matter how small or distant. One of his college friends, Anthony Szigety, persuaded him to go for a walk on a February afternoon, hoping exercise would restore his strength. While watching the sunset and quoting a passage from Goethe's Faust, a sudden inspiration struck him. To Szigety's amazement, he seized a stick and began to draw in the sand. Tesla had hit upon the design for an alternating-current motor in which rotating magnetic field produces the electrical charge in the motor, which in turn produces rotational torque that can be used to do useful work. "No more will men be slaves to hard tasks," he declared. "My motor will set them free, it will do the work of the world."

Edison, America, and First Business Failure

After Tesla's tenure at the telephone company, Puskás found him employment working for the Continental Edison Company in Paris. His advanced knowledge of physics and engineering skill made him an extraordinarily valuable employee, and he traveled around Europe installing new lighting systems and troubleshooting problems. In 1884, the company moved Tesla's boss, the British-born inventor Charles Batchelor, back to the United States. Batchelor brought the promising young engineer with him. It was a momentous move, and he would spend the rest of his life in the United States.

Tesla was not happy in Edison's machine works in Manhattan. He never worked directly with Edison, and apparently only met him on a couple of occasions. Rather, he was assigned to work on the problem of installing electric arc-lighting—a difficult problem, as arc lamps, which produce light from electricity jumping an air gap between two electrodes, required high-voltage alternating current, while the Edison systems were low-voltage and direct current. (If we liken electricity to water flowing through a hose, then voltage is the water pressure. The more voltage, the further the current travels and the more work it can do.) Tesla quit after just six months; in his autobiography, he claims that the manager had offered him $50,000 (almost $13 million in 2020) for his designs for new sorts of generators and other devices and then reneged. In any case, Tesla had learned a harsh lesson about American capitalism: employers work to enrich themselves, not their employees.

He was soon to learn another harsh lesson: investors care about profits, not innovation for its own sake. Tesla took his arc-lighting ideas to Edison's patent attorney, Lemuel W. Serrell, who in turn introduced him to two businessmen, Robert Lane and Benjamin Vail. Tesla, together with Lane and Vail, founded the short-lived Tesla Electric Light & Manufacturing company. Their one accomplishment was installing an advanced arc-lighting generator in Rahway, New Jersey. In the end, Lane and Vail decided that the "manufacturing" part of the company was too much work and that they would prefer to simply run an electric utility. They pulled out their money, leaving Tesla penniless. He didn't even have his patents to his name, as he had assigned them to the company in exchange for now-worthless stock. He spent the rest of 1886 doing menial labor, even digging ditches for $2 a day.

The Miraculous Motor

In late 1886, Tesla had a fortuitous meeting with Alfred S. Brown and Charles F. Peck, two experienced businessmen and investors. Unlike Lane and Vail, they were interested in speculating on innovation. Tesla set up a laboratory in lower Manhattan and quickly filed for patents on new types of AC generators and motors. The problem with previous attempts at AC motors was that if the current fluctuates between positive and negative, the net charge is zero. If we want to imagine an AC current as a wave, then while the "crest" can do useful work, rotating the shaft of the rotor forwards, the "trough" cancels it out. Tesla's great innovation was to take alternating currents that were a quarter-cycle out of phase and feed them into two different coils set around a rotor. The magnetic field thus produced would move the rotor around the axis.

In May of 1888, Tesla demonstrated his motor before the American Institute of Electrical Engineers. The novelty of his innovation, as well as his showmanship, impressed the onlookers. Tesla had done what no one else had been able to do. The inventor and entrepreneur George Westinghouse, who was constructing AC power systems throughout the United States, took notice. Westinghouse licensed Tesla's designs for $60,000—$10,000 more than his manager at Edison had promised him for his earlier inventions—plus a minimum royalty of $15,000 a year. It seemed that Tesla's financial independence and ability to work on his projects was assured.

The Current War

Because of the way electricity is produced in a generator—a magnet moving within a larger magnetic field—a device that produces an alternating current is easier to construct than one that produces a direct current. Remember Tesla's observation of the Gramme dynamo: it required commutators to make the current flow in a single direction. The commutators, however, were inefficient and required maintenance. Another advantage of alternating current is that it can be transmitted much greater distances and at higher voltages than direct current. At the receiving end, a transformer can "step down" the current for a user to power whatever device they might wish. Conversely, Edison's DC system only worked in dense cities, since the current could not be transmitted more than a mile.

On the downside, high-powered AC transmission lines are quite dangerous, and led to some highly publicized deaths of utility workers who accidentally electrocuted themselves or of bystanders who merely came into contact with New York City's unregulated maze of poorly maintained suspended wires. In fact, Edison felt that alternating current was so harmful that he backed demonstrations that used AC to euthanize stray dogs to drive home the hazard. Of course, Edison had an ulterior motive: his incandescent bulbs ran on low-voltage and relatively safe DC systems, and AC posed a threat to the system in which he had so heavily invested. For a while, it seemed that electrical transmission would be limited to cities, and that power would be generated by dynamos at specific sites for different machines, rather than coming from one more efficient centralized utility that could benefit from economies of scale.

George Westinghouse changed all that—with Nikola Tesla's help. The Westinghouse Electric Company was formed in 1886, and by the end of the year had installed a demonstration system in Great Barrington, Massachusetts, and a commercial system in Buffalo, New York. Furious, Edison publicly charged that Westinghouse was infringing on his patents while at the same time secretly aiding New York State authorities in devising a new means of executing criminals—the electric chair. When it came time to construct the murderous device, Edison made sure that the state bought used Westinghouse AC generators. However, when his backroom dealing was exposed in 1889, it backfired with the bad publicity falling on Edison. Another series of well-publicized linemen deaths led to electric main lines finally being buried in New York City, making AC transmission much safer. Edison, having decisively lost the current war, was rapidly forced out of his own company. The coup de grace was in 1892, when Westinghouse outbid General Electric (formerly Edison General Electric) for the opportunity to electrify the World's Columbian Exhibition in Chicago, at which numerous modern marvels made their debut.

LATER CAREER

Westinghouse's licensing of Tesla's motor had given the inventor financial freedom. But a workable, practical design was still being developed when Westinghouse came to call with some bad news: The financial crisis of 1890 had put him in a difficult position. His investors were demanding he cut back on research and development. Tesla agreed to forgo the guaranteed $15,000 a year that Westinghouse promised him.

The two continued to collaborate in business ventures, however, and when Tesla was asked to consult on building a power plant at Niagara Falls in 1893, he recommended the Westinghouse company. Three years later, Westinghouse would buy the patent for Tesla's motor for a one-time payment of $216,000.

Despite his success, and though he had close female friends and was both handsome and personable, Tesla never married. Since doing his own domestic chores was unthinkable, he instead chose to live in a series of hotels, as did many well-off bachelors of the time. Beyond that, his personal life is something of a mystery. The thing we know for certain is that he claimed his work left him no time for romantic relationships and that his chastity contributed to his creativity. Ever the mystic, he saw his entire purpose as furthering the human condition through harnessing ever-more-powerful forces.

The 1890s were indeed a productive time for Tesla. He experimented with radio waves (two years before Marconi was credited with inventing practical long-distance radio transmission), remote control for automated boats (that is, the first drones), and X-rays. He also invented a steam-powered oscillating AC generator that never found a practical application—though he did claim that it caused an earthquake. At the World's Columbian Exhibition, celebrating the 400th anniversary of Columbus's arrival in the Americas, Tesla demonstrated his induction motor and wireless induction lamps, which presaged his fascination with transmitting power through the air and the earth. He also devised a "Columbus egg" for the exhibition in which he used the magnetic field of an AC motor to spin a copper egg around until it stood on its end. (The name referred to an apocryphal story about Christopher Columbus, who challenged the wise men of Spain to stand an egg on one end; when they failed, he crushed one end, so the egg stood up.)

Perhaps the most iconic of Tesla's inventions was his magnifying transmitter—the "Tesla coil"—which he first tested in 1891, the year he became an American citizen. By adding a second circuit to a preexisting design for a spark-gap circuit, which spectacularly discharged electricity across an air gap, he created a resonance that allowed high frequency, high-voltage electric charges to build up. These found practical application in wireless telegraphy and also in various medical uses that, while popular in the nineteenth and early twentieth centuries, have since been proven ineffective. Using principles similar to Tesla's, Guglielmo Marconi demonstrated workable long-distance radio in 1895. While

Marconi is generally accepted to have had primacy and received a Nobel Prize for his work, the Supreme Court granted Tesla's patent primacy in 1943—not without ulterior motives, since the U.S. was at war with Italy at the time and didn't want to pay royalties for the use of radio.

While today Tesla coils are mainly used for entertainment and spectacular displays of indoor lightning, at the time, Tesla saw his invention as key to a new way of transmitting power. By permeating the air with electromagnetism, he could power any attuned device—for instance, in one demonstration, lighting a bulb held in his hand. He even claimed to have been able to wirelessly light a field of light bulbs a kilometer away.

On March 13, 1895, a fire ravaged Tesla's New York City lab, destroying $50,000 worth of notes, materials, photographs, and prototypes. "I am in too much grief to talk," he told the New York Times. He relocated to another shop in Manhattan and continued his work. Having noticed unusual effects when attempting to take a photograph of Mark Twain using a lighting device called a Crookes tube, he performed further experiments following Wilhelm Röntgen's discovery of X-rays. In 1898, during the height of the Spanish–American War, he demonstrated a radio-controlled boat in Madison Square Garden. Such ships, he said, would enable battles to be fought without the loss of human life.

In 1899, Tesla moved to Colorado Springs where he set up a high-altitude testing station in order to continue his distance-transmission experiments. Tesla claimed that it would be possible to send electromagnetic energy wirelessly around the Earth by setting up harmonies with the planet's natural resonances. This was not as insane as it sounds—it was known that the earth has telluric currents that travel through its surface, and today, we know that the space between the Earth's surface and the upper atmosphere also has its own electromagnetic frequency and can be used to track lightning storms. However, despite a $100,000 investment from millionaire John Jacob Astor, Tesla was never able to make a workable demonstration of these principles—though he did discuss the possibility of using such resonances to control the weather. Even more fantastically, he received radio signals that he claimed must be from intelligent beings from other planets. Though he was met with disbelief, most likely, the signals were from Marconi's early experiments or possibly from a natural source such as Jupiter's moon Io passing through the planet's magnetic field.

Returning to New York, he moved into Astor's hotel, the Waldorf-Astoria, and feted investors with fantastic promises. The financier J.P. Morgan provided him with $150,000, and he began planning his greatest project yet: the Wardenclyffe Tower in Shoreham, New York. Looking eastwards over the sea to England, it was intended to transmit not only Morse code, as Marconi did, but also telephony and images. Tesla was ahead of his time with his ambitions, but hopelessly wrong in his methods: He disbelieved Hertz' theories of airborne radio waves, instead maintaining that all energy, like light, travels in straight lines and dissipates at distance. At the same time, Tesla's writing, such as "The Problem of Increasing Human Energy" published in Century magazine in June 1900, became more eccentric and philosophical.

Wardenclyffe, which was apparently only activated once, in 1903, was to be Tesla's undoing. The project fell apart by 1906—among other tragedies Morgan would not advance any further funds, and the architect in charge of construction, Stanford White, was murdered by his secret lover's husband. His patents had all expired, his money was all spent, and he had no workable inventions with which to follow up. Even the Navy passed on his remote-control device. The financial problems led to Tesla having something of a nervous breakdown. He mortgaged the Wardenclyffe property to cover the debt from his high living at the Waldorf-Astoria; his inability to pay led to the site later being foreclosed on. The tower was finally dismantled and sold for scrap in 1917—his Colorado Springs lab had already been torn down in 1904 when the electric company foreclosed on the property after he failed to pay the power bill.

This is not to say that Tesla was out of ideas—notably bladeless turbines and air-friction speedometers. However, he failed to monetize them effectively. Other revolutionary ideas, such as aircraft that could take off and land vertically, proved impossible. His old backers were rapidly disappearing. Astor died on the Titanic in 1912, and Westinghouse died in 1914. Rumors flew that Tesla and Edison would split the Nobel Prize in 1915, but neither received it. The reason why—perhaps the two men refused to share it—was never revealed. In 1916, court proceedings forced him to publicly acknowledge the shabby state of his finances. His mental health failed, too: When the American Institute of Electrical Engineers gave him the Edison medal, its highest award in 1917, he disappeared during the ceremony and was found feeding the pigeons in the park. Despite his deteriorating circumstances, editor Hugo Gernsback managed to persuade Tesla to write his autobiography in 1919.

Tesla's final years were spent trying to recapture his fading glory, receiving a never-ending stream of honors from learned societies and foreign governments, feeding (and, he claimed, communicating with) his beloved pigeons, obsessively washing, and obsessing over the number three. He never accepted Einstein's theory of relativity or even that electrons are one of the primal components of atoms. Beginning on his seventy-fifth birthday in 1931, he gave annual press conferences at which he made outlandish claims of new discoveries. In 1932, he claimed to have an engine that worked on cosmic rays. In 1933, he claimed to have found a new form of energy. Even more fantastic was his claim in 1934 to have invented "teleforce"—though he denied it was a science-fiction "death ray," he did see it as a device to end war:

> My apparatus projects particles which may be relatively large or of microscopic dimensions, enabling us to convey to a small area at a great distance trillions of times more energy than is possible with rays of any kind. Many thousands of horsepower can thus be transmitted by a stream thinner than a hair, so that nothing can resist. The nozzle would send concentrated beams of particles through the free air, of such tremendous energy that they will bring down a fleet of 10,000 enemy airplanes at a distance of 200 miles from a defending nation's border and will cause armies to drop dead in their tracks.

Always tall and thin, the aging Tesla became emaciated, especially after he adopted a vegetarian diet in later years. Always surviving on little sleep, he now hardly slept at all. He was kicked out of (and sued by) the hotels he lived in as his debt accumulated. On more than one occasion, he paid debts with models of, or parts for, his teleforce machine. He finally achieved some measure of financial stability in 1934 when the Westinghouse Corporation agreed to pay him $125 per month, as well as cover his rent at the Hotel New Yorker.

In August of 1937, Tesla was hit by a taxi while crossing the street in the middle of the night to go feed the pigeons in the park. He claimed not to have been seriously injured, but he likely broke several ribs. Nonetheless, as was his habit, he refused medical attention. His health never recovered, and he was found dead a little more than five years later in his room, 3327, on the 33rd floor of the Hotel New Yorker. He was 86 years old. Tesla was given a state funeral at St. John the Divine and a eulogy by Mayor Fiorello La Guardia, while the Federal government briefly seized his equipment to ensure there were, in fact, no secret death rays. Tesla's personal effects and his ashes were finally shipped to Belgrade in 1952, where they today reside in the Nikola Tesla Museum.

Ken Mondschein, PhD

The Autobiography of Nikola Tesla

MY INVENTIONS
as First Serialized in *Electrical Experimenter* Magazine

PART I
FEBRUARY 1919

MY EARLY LIFE

How does the world's greatest inventor invent? How does he carry out an invention? What sort of mentality has Nikola Tesla? Was his early life as commonplace as most of ours? What was the early training of one of the World's Chosen? These, and many other very interesting questions are answered in an incomparable manner by Nikola Tesla himself in this, his first article.

In his autobiography, treating mainly on his early youth, we obtain a good insight into the wonderful life this man has led. It reads like a fairy tale, which has the advantage of being true. For Tesla is no common mortal. He has led a charmed life—struck down by the pest, the cholera and what not—given up by doctors at least three times as dead—we find him at sixty, younger than ever. But—read his own words. You have never read the like before.

— EDITOR

The progressive development of man is vitally dependent on invention. It is the most important product of his creative brain. Its ultimate purpose is the complete mastery of mind over the material world, the harnessing of the forces of nature to human needs. This is the difficult task of the inventor who is often misunderstood and unrewarded. But he finds ample compensation in the pleasing exercises of his powers and in the knowledge of being one of that exceptionally privileged class without whom the race would have long ago perished in the bitter struggle against pitiless elements.

Speaking for myself, I have already had more than my full measure of this exquisite enjoyment, so much that for many years my life was little short of continuous rapture. I am credited with being one of the hardest workers and perhaps I am, if thought is the equivalent of

Nikola Tesla at the age of 23, from an unpublished photograph.

labor, for I have devoted to it almost all of my waking hours. But if work is interpreted to be a definite performance in a specified time according to a rigid rule, then I may be the worst of idlers. Every effort under compulsion demands a sacrifice of life-energy. I never paid such a price. On the contrary, I have thrived on my thoughts.

In attempting to give a connected and faithful account of my activities in this series of articles which will be presented with the assistance of the Editors of the *Electrical Experimenter* and are chiefly addrest to our young men readers. I must dwell, however reluctantly, on the impressions of my youth and the circumstances and events which have been instrumental in determining my career.

Our first endeavors are purely instinctive, promptings of an imagination vivid and undisciplined. As we grow older reason asserts itself and we become more and more systematic and designing. But those early impulses, tho not immediately productive, are of the greatest moment and may shape our very destinies. Indeed, I feel now that had I understood and cultivated instead of suppressing them, I would have added substantial value to my bequest to the world. But not until I had attained manhood did I realize that I was an inventor.

This was due to a number of causes. In the first place I had a brother who was gifted to an extraordinary degree—one of those rare phenomena

Mr. Tesla at the age of 29.

of mentality which biological investigation has failed to explain. His premature death left my parents disconsolate. We owned a horse which had been presented to us by a dear friend. It was a magnificent animal of Arabian breed, possest of almost human intelligence, and was cared for and petted by the whole family, having on one occasion saved my father's life under remarkable circumstances. My father had been called one winter night to perform an urgent duty and while crossing the mountains, infested by wolves, the horse became frightened and ran away, throwing him violently to the ground. It arrived home bleeding and exhausted, but after the alarm was sounded immediately dashed off again, returning to the spot, and before the searching party were far on the way they were met by my father, who had recovered consciousness and remounted, not realizing that he had been lying in the snow for several hours. This horse was responsible for my brother's injuries from which he died. I witnessed the tragic scene and altho fifty-six years have elapsed since, my visual impression of it has lost none of its force. The recollection of his attainments made every effort of mine seem dull in comparison.

Anything I did that was creditable merely caused my parents to feel their loss more keenly. So I grew up with little confidence in myself. But

I was far from being considered a stupid boy, if I am to judge from an incident of which I have still a strong remembrance. One day the Aldermen were passing thru a street where I was at play with other boys. The oldest of these venerable gentlemen—a wealthy citizen—paused to give a silver piece to each of us. Coming to me he suddenly stopt and commanded, "Look in my eyes." I met his gaze, my hand outstretched to receive the much valued coin, when, to my dismay, he said, "No, not much, you can get nothing from me, you are too smart." They used to tell a funny story about me. I had two old aunts with wrinkled faces, one of them having two teeth protruding like the tusks of an elephant which she buried in my check every time she kist me. Nothing would scare me more than the prospect of being hugged by these as affectionate as unattractive relatives. It happened that while being carried in my mother's arms they asked me who was the prettier of the two. After examining their faces intently, I answered thoughtfully, pointing to one of them, "This here is not as ugly as the other."

Then again, I was intended from my very birth for the clerical profession and this thought constantly opprest me. I longed to be an engineer but my father was inflexible. He was the son of an officer who served in the army of the Great Napoleon and, in common with his brother, professor of mathematics in a prominent institution, had received a military education but, singularly enough, later embraced the clergy in which vocation he achieved eminence. He was a very erudite man, a veritable natural philosopher, poet and writer and his sermons were said to be as eloquent as those of Abraham a Sancta-Clara. He had a prodigious memory and frequently recited at length from works in several languages. He often remarked playfully that if some of the classics were lost he could restore them. His style of writing was much admired. He penned sentences short and terse and was full of wit and satire. The humorous remarks he made were always peculiar and characteristic. Just to illustrate, I may mention one or two instances. Among the help there was a cross-eyed man called Mane, employed to do work around the farm. He was chopping wood one day. As he swung the axe my father, who stood nearby and felt very uncomfortable, cautioned him, "For God's sake, Mane, do not strike at what you are looking but at what you intend to hit." On another occasion he was taking out for a drive a friend who carelessly permitted his costly fur coat to rub on the carriage wheel. My father reminded him of it saying, "Pull in your coat, you are ruining my tire." He had the odd habit of talking to himself and would often carry on an animated conversation and indulge in heated argument, changing the tone of his voice. A casual listener might have sworn that several people were in the room.

Altho I must trace to my mother's influence whatever inventiveness I possess, the training he gave me must have been helpful. It comprised all sorts of exercises—as, guessing one another's thoughts, discovering the defects of some form or expression, repeating long sentences or performing mental calculations. These daily lessons were intended to strengthen memory and reason and especially to develop the critical sense, and were undoubtedly very beneficial.

* * * * *

NIKOLA TESLA: THE MAN
BY H. GERNSBACK

The door opens and out steps a tall figure—over six feet high—gaunt but erect. It approaches slowly, stately. You become conscious at once that you are face to face with a personality of a high order. Nikola Tesla advances and shakes your hand with a powerful grip, surprising for a man over sixty. A winning smile from piercing light blue-gray eyes, set in extraordinarily deep sockets, fascinates you and makes you feel at once at home.

You are guided into an office immaculate in its orderliness. Not a speck of dust is to be seen. No papers litter the desk, everything just so. It reflects the man himself, immaculate in attire, orderly and precise in his every movement. Drest in a dark frock coat, he is entirely devoid of all jewelry. No ring, stickpin, or even watch-chain can be seen.

Tesla speaks—a very high almost falsetto voice. He speaks quickly and very convincingly. It is the man's voice chiefly which fascinates you.

As he speaks you find it difficult to take your eyes off his own. Only when he speaks to others do you have a chance to study his head, predominant of which is a very high forehead with a bulge between the eyes—the never-failing sign of an exceptional intelligence. Then the long, well-shaped nose, proclaiming the scientist.

How does this man, who has accomplished such a tremendous work, keep young and manage to surprise the world with more and more new inventions as he grows older? How does this youth of sixty, who is a professor of mathematics, a great mechanical and electrical engineer and the greatest inventor of all times, keep his physical as well as remarkable mental freshness?

To begin with, Tesla, who is by birth a Serbian, comes from a long-lived hardy race. His family tree abounds with centenarians. Accordingly, Tesla—barring accidents—fully expects to be still inventing in AD 1960.

But the chief reason for his perpetual youth is found in his gastronomical frugality. Tesla has learned the great fundamental truth that most people not only eat all of their bodily ills, but actually eat themselves to death by either eating too much or else by food that does not agree with them.

When Tesla found out that tobacco and black coffee interfered with his physical well-being, he quit both. This is the simple daily menu of the great inventor:

Breakfast: One to two pints of warm milk and a few eggs, prepared by himself—yes, he is a bachelor!

Lunch: None whatsoever, as a rule.

Dinner: Celery or the like, soup, a single piece of meat or fowl, potatoes and one other vegetable; a glass of light wine. For dessert, perhaps a slice of cheese, and invariably a big raw apple. And that's all.

Tesla is very fussy and particular about his food: he eats very little, but what he does eat must be of the very best. And he knows, for outside of being a great inventor in science he is an accomplished cook who has invented all sorts of savory dishes. His only vice is his generosity. The man who, by the ignorant onlooker has often been called an idle dreamer, has made over a million dollars out of his inventions—and spent them as quickly on new ones. But Tesla is an idealist of the highest order and to such men money itself means but little.

* * * * *

My mother descended from one of the oldest families in the country and a line of inventors. Both her father and grandfather originated numerous implements for household, agricultural and other uses. She was a truly great woman, of rare skill, courage and fortitude, who had braved the storms of life and past thru many a trying experience. When she was sixteen a virulent pestilence swept the country. Her father was called away to administer the last sacraments to the dying and during his absence she went alone to the assistance of a neighboring family who were stricken by the dread disease. All of the members, five in number, succumbed in rapid succession. She bathed, clothed and laid out the bodies, decorating them with flowers according to the custom of the country and when her father returned he found everything ready for a Christian burial. My mother was an inventor of the first order and would, I believe, have achieved great things had she not been so remote from modern life and its multifold opportunities. She invented and constructed all kinds of tools and devices and wove the finest designs from thread which was spun by her. She even planted the seeds, raised the plants and separated the fibers herself. She worked indefatigably, from break of day till late at night, and most of the wearing apparel and furnishings of the home was the product of her hands. When she was past sixty, her fingers were still nimble enough to tie three knots in an eyelash.

There was another and still more important reason for my late awakening. In my boyhood I suffered from a peculiar affliction due to the appearance of images, often accompanied by strong flashes of light, which marred the sight of real objects and interfered with my thought and action. They were pictures of things and scenes which I had really seen, never of those I imagined. When a word was spoken to me the image of the object it designated would present itself vividly to my vision and sometimes I was quite unable to distinguish whether what I saw was tangible or not. This caused me great discomfort and anxiety. None of the students of psychology or physiology whom I have consulted could ever explain satisfactorily these phenomena. They seem to have been unique although I was probably predisposed as I know that my brother experienced a similar trouble. The theory I have formulated is that the images were the result of a reflex action from the brain on the retina

Mr. Tesla at the age of 39.

under great excitation. They certainly were not hallucinations such as are produced in diseased and anguished minds, for in other respects I was normal and composed.

To give an idea of my distress, suppose that I had witnessed a funeral or some such nerve-racking spectacle. Then, inevitably, in the stillness of night, a vivid picture of the scene would thrust itself before my eyes and persist despite all my efforts to banish it. Sometimes it would even remain fixt in space tho I pushed my hand thru it. If my explanation is correct, it should be possible to project on a screen the image of any object one conceives and make it visible. Such an advance would revolutionize all human relations. I am convinced that this wonder can and will be accomplished in time to come; I may add that I have devoted much thought to the solution of the problem.

To free myself of these tormenting appearances, I tried to concentrate my mind on something else I had seen, and in this way I would often obtain temporary relief; but in order to get it I had to conjure continuously new images. It was not long before I found that I had exhausted all of those at my command; my "reel" had run out, as it were, because I had seen little of the world—only objects in my home and the immediate surroundings. As I performed these mental operations for the second or third time, in order to chase the appearances from my vision, the remedy gradually lost all its force. Then I instinctively commenced to make excursions beyond the limits of the small world of which I had knowledge, and I saw new scenes. These were at first very blurred and indistinct, and would flit away when I tried to concentrate my attention upon them, but by and by I succeeded in fixing them; they gained in strength and distinctness and finally assumed the concreteness of real things. I soon discovered that my best comfort was attained if I simply went on in my vision farther and farther, getting new impressions all the time, and so I began to travel—of course, in my mind. Every night (and sometimes during the day), when alone, I would start on my journeys—see new places, cities and countries—live there, meet people and make friendships and acquaintances and, however unbelievable, it is a fact that they were just as dear to me as those in actual life and not a bit less intense in their manifestations.

This I did constantly until I was about seventeen when my thoughts turned seriously to invention. Then I observed to my delight that I could visualize with the greatest facility. I needed no models, drawings or experiments. I could picture them all as real in my mind. Thus I have been led unconsciously to evolve what I consider a new method of materializing inventive concepts and ideas, which is radically opposite to the purely experimental and is in my opinion ever so much more expeditious—and efficient. The moment one constructs a device to carry into practise a crude idea he finds himself unavoidably engrost with the details and defects of the apparatus. As he goes on improving and reconstructing, his force of concentration diminishes and he loses sight of the great underlying principle. Results may be obtained but always at the sacrifice of quality.

My method is different. I do not rush into actual work. When I get an idea I start at once building it up in my imagination. I change the construction, make improvements and operate the device in my mind. It is absolutely immaterial to me whether I run my turbine in thought or test it in my shop. *I even note if it is out of balance.* There is no difference whatever, the results are the same. In this way I am able to rapidly develop and perfect a conception without touching anything. When I have gone so far as to embody in the invention every possible improvement I can think of and see no fault anywhere, I put into concrete form this final product of my brain. Invariably my device works as I conceived that it should, and the experiment comes out exactly as I planned it. In twenty years there has not been a single exception. Why should it be otherwise? Engineering, electrical and mechanical, is positive in results. There is scarcely a subject that cannot be mathematically treated and the effects calculated or the results determined beforehand from the available theoretical and practical data. The carrying out into practise of a crude idea as is being generally done is, I hold, nothing but a waste of energy, money and time.

My early affliction had, however, another compensation. The incessant mental exertion developed my powers of observation and enabled me to discover a truth of great importance. I had noted that the appearance of images was always preceded by actual vision of scenes under peculiar and generally very exceptional conditions and I was impelled on each occasion to locate the original impulse. After a while this effort grew to be almost automatic and I gained great facility in connecting cause and effect. Soon I became aware, to my surprise, that every thought I conceived was suggested by an external impression. Not only this but all my actions were prompted in a similar way. In the course of time it became perfectly evident to me that I was merely an *automaton* endowed with power of movement, responding to the stimuli of the sense organs and thinking and acting accordingly. The practical result of this was the art of *telautomatics* which has been so far carried out only in an imperfect manner. Its latent possibilities will, however, be eventually shown. I have been since years planning self-controlled automata and believe that mechanisms can be produced which will act as if possest of reason, to a limited degree, and will create a revolution in many commercial and industrial departments. I was about twelve years old when I first succeeded in banishing an image from my vision by wilful effort, but I never had any control over the flashes of light to which I have referred. They were, perhaps, my strangest experience and inexplicable.

They usually occurred when I found myself in a dangerous or distressing situation or when I was greatly exhilarated. In some instances I have seen all the air around me filled with tongues of living flame. Their intensity, instead of diminishing, increased with time and seemingly attained a maximum when I was about twenty-five years old. While in Paris, in 1883, a prominent French manufacturer sent me an invitation to a shooting expedition which I accepted. I had been long confined to the factory, and the fresh air had a wonderfully invigorating effect on me. On my return to the city that night I felt a positive sensation that my brain had caught fire.

I saw a light as tho a small sun was located in it and I past the whole night applying cold compressions to my tortured head. Finally the flashes diminished in frequency and force but it took more than three weeks before they wholly subsided. When a second invitation was extended to me my answer was an emphatic *No!*

These luminous phenomena still manifest themselves from time to time, as when a new idea opening up possibilities strikes me, but they are no longer exciting, being of relatively small itensity. When I close my eyes I invariably observe first, a background of very dark and uniform blue, not unlike the sky on a clear but starless night. In a few seconds this field becomes animated with innumerable scintillating flakes of green, arranged in several layers and advancing towards me. Then there appears, to the right, a beautiful pattern of two systems of parallel and closely spaced lines, at right angles to one another, in all sorts of colors with yellow-green and gold predominating. Immediately thereafter the lines grow brighter and the whole is thickly sprinkled with dots of twinkling light. This picture moves slowly across the field of vision and in about ten seconds vanishes to the left, leaving behind a ground of rather unpleasant and inert grey which quickly gives way to a billowy sea of clouds, seemingly trying to mould themselves in living shapes. It is curious that I cannot project a form into this grey until the second phase is reached. Every time, before falling asleep, images of persons or objects flit before my view. When I see them I know that I am about to lose consciousness. If they are absent and refuse to come it means a sleepless night.

To what an extent imagination played a part in my early life I may illustrate by another odd experience. Like most children I was fond of jumping and developed an intense desire to support myself in the air. Occasionally a strong wind richly charged with oxygen blew from the mountains rendering my body as light as cork and then I would leap and float in space for a long time. It was a delightful sensation and my disappointment was keen when later I undeceived myself.

During that period I contracted many strange likes, dislikes and habits, some of which I can trace to external impressions while others are unaccountable. I had a violent aversion against the earrings of women but other ornaments, as bracelets, pleased me more or less according to design. The sight of a pearl would almost give me a fit but I was fascinated with the glitter of crystals or objects with sharp edges and plane surfaces. I would not touch the hair of other people except, perhaps, at the point of a revolver. I would get a fever by looking at a peach and if a piece of camphor was anywhere in the house it caused me the keenest discomfort. Even now I am not insensible to some of these upsetting impulses. When I drop little squares of paper in a dish filled with liquid, I always sense a peculiar and awful taste in my mouth. I counted the steps in my walks and calculated the cubical contents of soup plates, coffee cups and pieces of food—otherwise my meal was unenjoyable. All repeated acts or operations I performed had to be divisible by three and if I mist I felt impelled to do it all over again, even if it took hours.

Up to the age of eight years, my character was weak and vacillating. I had neither courage or strength to form a firm resolve. My feelings came in waves and surges and vibrated unceasingly between extremes. My wishes were of consuming force and like the heads of the hydra, they multiplied. I was opprest by thoughts of pain in life and death and religious fear. I was swayed by superstitious belief and lived in constant dread of the spirit of evil, of ghosts and ogres and other unholy monsters of the dark. Then, all at once, there came a tremendous change which altered the course of my whole existence.

Of all things I liked books the best. My father had a large library and whenever I could manage I tried to satisfy my passion for reading. He did not permit it and would fly into a rage when he caught me in the act. He hid the candles when he found that I was reading in secret. He did not want me to spoil my eyes. But I obtained tallow, made the wicking and cast the sticks into tin forms, and every night I would bush the keyhole and the cracks and read, often till dawn, when all others slept and my mother started on her arduous daily task. On one occasion I came across a novel entitled "Abafi" (the Son of Aba), a Serbian translation of a well known Hungarian writer, Josika. This work somehow awakened my dormant powers of will and I began to practise self-control. At first my resolutions faded like snow in April, but in a little while I conquered my weakness and felt a pleasure I never knew before—that of doing as I willed. In the course of time this vigorous mental exercise became second nature. At the outset my wishes had to be subdued but gradually desire and will grew to be identical. After years of such discipline I gained so complete a mastery over myself that I toyed with passions which have meant destruction to some of the strongest men. At a certain age I contracted a mania for gambling which greatly worried my parents. To sit down to a game of cards was for me the quintessence of pleasure.

My father led an exemplary life and could not excuse the senseless waste of time and money in which I indulged. I had a strong resolve but my philosophy was bad. I would say to him, "I can stop whenever I please but is it worth while to give up that which I would purchase with the joys of Paradise?" On frequent occasions he gave vent to his anger and contempt but my mother was different. She understood the character of men and knew that one's salvation could only be brought about thru his own efforts. One afternoon, I remember, when I had lost all my money and was craving for a game, she came to me with a roll of bills and said, "Go and enjoy yourself. The sooner you lose all we possess the better it will be. I know that you will get over it." She was right. I conquered my passion then and there and only regretted that it had not been a hundred times as strong. I not only vanquished but tore it from my heart so as not to leave even a trace of desire. Ever since that time I have been as indifferent to any form of gambling as to picking teeth.

During another period I smoked excessively, threatening to ruin my health. Then my will asserted itself and I not only stopt but destroyed all inclination. Long ago I suffered from heart trouble until I discovered that it was due to the innocent cup of coffee I consumed every morning. I discontinued at once, tho I confess it was not an easy task. In this way I checked and bridled other habits and passions and have not only preserved my life but derived an immense amount of satisfaction from what most men would consider privation and sacrifice. After finishing the studies at the Polytechnic Institute and University I had a complete nervous breakdown and while the malady lasted I observed many phenomena strange and unbelievable.

MY FIRST EFFORTS IN INVENTION

Boys will be boys, the world over. The Boy Tesla was no exception to the universal rule, as this, his second autobiographical article clearly proves.

Mr. Tesla in his own inimitable, delightful way, here paints with a literary artist's brush his own intimate boyhood in charming as well as vivid colors.

We have often heard of Tesla, the dreamer. But if he is entitled to the epithet, his early boyhood certainly fails to reveal it. Tesla did not allow much grass to grow under his feet while a boy, for he assuredly was a strenuous, red-blooded youngster.

You will wish to read all about the greatest inventor's early boyhood. It is doubly valuable because it comes from his own pen. We promise you an interesting twenty-minutes' entertainment.

—EDITOR

I shall dwell briefly on these extraordinary experiences, on account of their possible interest to students of psychology and physiology and also because this period of agony was of the greatest consequence on my mental development and subsequent labors. But it is indispensable to first relate the circumstances and conditions which preceded them and in which might be found their partial explanation.

From childhood I was compelled to concentrate attention upon myself. This caused me much suffering but, to my present view, it was a blessing in disguise for it has taught me to appreciate the inestimable value of introspection in the preservation of life, as well as a means of achievement. The pressure of occupation and the incessant stream of impressions pouring into our consciousness thru all the gateways of knowledge make modern existence hazardous in many ways. Most persons are so absorbed in the contemplation of the outside world that they are wholly oblivious to what is passing on within themselves. The premature death of millions is primarily traceable to this cause. Even among those who exercise care it is a common mistake to avoid imaginary, and ignore the real dangers. And what is true of an individual also applies, more or less, to a people as a whole. Witness, in illustration, the prohibition movement. A drastic, if not unconstitutional, measure is now being put thru in this country to prevent the consumption of alcohol and yet it is a positive fact that coffee, tea, tobacco, chewing gum and other stimulants, which are freely indulged in even at the tender age, are vastly more injurious to the national body, judging from the number of those who succumb. So, for instance, during my student years I gathered from the published necrologues in Vienna, the home of coffee drinkers, that deaths from heart trouble sometimes reached *sixty-seven percent* of the total.

Similar observations might probably be made in cities where the consumption of tea is excessive. These delicious beverages super excite and gradually exhaust the fine fibers of the brain. They also interfere seriously with arterial circulation and should be enjoyed all the more sparingly as their deleterious effects are slow and imperceptible. Tobacco, on the other hand, is conducive to easy and pleasant thinking and detracts from the intensity and concentration necessary to all original and vigorous effort of the intellect. Chewing gum is helpful for a short while but soon drains the glandular system and inflicts irreparable damage, not to speak of the revulsion it creates. Alcohol in small quantities is an excellent tonic, but is toxic in its action when absorbed in larger amounts, quite immaterial as to whether it is taken in as whiskey or produced in the stomach from sugar. But it should not be overlooked that all these are great eliminators assisting Nature, as they do, in upholding her stern but just law of the survival of the fittest. Eager reformers should also be mindful of the eternal perversity of mankind which makes the indifferent "laissez-faire" by far preferable to enforced restraint. The truth about this is that we need stimulants to do our best work under present living conditions, and that we must exercise moderation and control our appetites and inclinations in every direction. That is what I have been doing for many years, in this way maintaining myself young in body and mind. Abstinence was not always to my liking but I find ample reward in the agreeable experiences I am now making. Just in the hope of converting some to my precepts and convictions I will recall one or two. A short time ago I was returning to my hotel. It was a bitter cold night, the ground slippery, and no taxi to be had. Half a block behind me followed another man, evidently as anxious as myself to get under cover. Suddenly my legs went up in the air. In the same instant there was a flash in my brain, the nerves responded, the muscles contracted, I swung thru 180 degrees and landed on my hands. I resumed my walk as tho nothing had happened when the stranger caught up with me. "How old are you?" he asked, surveying me critically. "Oh, about fifty-nine," I replied. "What of it?" "Well," said he, "I have seen a cat do this but never a man." About a month since I wanted to order new eyeglasses and went to an oculist who put me thru the usual tests. He lookt at me incredulously as I read off with ease the smallest print at considerable distance. But when I told him that I was past sixty he gasped in astonishment. Friends of mine often remark that my suits fit me like gloves but they do not know that all my clothing is made to measurements which were taken nearly 35 years ago and never changed. During this same period my weight has not varied one pound. In this connection I may tell a funny story. One evening, in the winter of 1885, Mr. Edison, Edward H. Johnson, the President of the Edison Illuminating Company, Mr. Batchellor, Manager of the works, and myself entered a little place opposite 65 Fifth Avenue where the offices of the company were located. Someone suggested guessing weights and I was induced to step on a scale. Edison felt me all over and said: "Tesla weighs 152 lbs. to an ounce," and he guest it exactly. Stript I weighed 142 lbs. and that is still my weight. I whispered to Mr. Johnson: "How is it possible that Edison could guess my weight so closely?" "Well," he said, lowering his voice, "I will tell you, confidentially, but you must not say anything. He was employed for a long time in a Chicago slaughterhouse where he weighed thousands of hogs every day! That's why." My friend, the Hon. Chauncey M. Depew, tells of an Englishman on whom he sprung one of his original anecdotes and who listened with a puzzled expression but—a year later—laughed out loud. I will frankly confess it took me longer than that to appreciate Johnson's joke.

Now, my well-being is simply the result of a careful and measured mode of living and perhaps the most astonishing thing is that three times in my youth I was rendered by illness a hopeless physical wreck and given up by physicians. More than this, thru ignorance and lightheartedness, I got into all sorts of difficulties, dangers and scrapes from which I extricated myself as by enchantment. I was almost drowned a dozen times; was nearly boiled alive and just mist being cremated. I was entombed, lost and frozen. I had hair-breadth escapes from mad dogs, hogs, and other wild animals. I past thru dreadful diseases and met with all kinds of odd mishaps and that I am hale and hearty today seems like a miracle. But as I recall these incidents to my mind, I feel convinced that my preservation was not altogether accidental.

An inventor's endeavor is essentially lifesaving. Whether he harnesses forces, improves devices, or provides new comforts and conveniences, he is adding to the safety of our existence. He is also better qualified than the average individual to protect himself in peril, for he is observant and resourceful. If I had no other evidence that I was, in a measure, possest of such qualities I would find it in these personal experiences. The reader will be able to judge for himself if I mention one or two instances. On one occasion, when about 14 years old, I wanted to scare some friends who were bathing with me. My plan was to dive under a long floating structure and slip out quietly at the other end. Swimming and diving came to me as naturally as to a duck and I was confident that I could perform the feat. Accordingly, I plunged into the water and, when out of view, turned around and proceeded rapidly towards the opposite side. Thinking that I was safely beyond the structure, I rose to the surface but to my dismay struck a beam. Of course, I quickly dived and forged ahead with rapid strokes until my breath was beginning to give out. Rising for the second time, my head came again in contact with a beam.

Now I was becoming desperate. However, summoning all my energy, I made a third frantic attempt, but the result was the same. The torture of supprest breathing was getting unendurable, my brain was reeling and I felt myself sinking. At that moment, when my situation seemed absolutely hopeless, I experienced one of those flashes of light and the structure above me appeared before my vision. I either discerned or guest that there was a little space between the surface of the water and the boards resting on the beams and, with consciousness nearly gone, I floated up, prest my mouth close to the planks and managed to inhale a little air, unfortunately mingled with a spray of water which nearly choked me. Several times I repeated this procedure as in a dream until my heart, which was racing at a terrible rate, quieted down and I gained composure. After that I made a number of unsuccessful dives, having completely lost the sense of direction, but finally succeeded in getting out of the trap when my friends had already given me up and were fishing for my body.

An interesting study of the great inventor, contemplating the glass bulb of his famous wireless light. A full description of the invention will appear shortly in the *Electrical Experimenter*. This is the only profile photograph of Mr. Tesla in existence. It was taken specially for the *Electrical Experimenter*.

That bathing season was spoiled for me thru recklessness, but I soon forgot the lesson and only two years later I fell into a worse predicament. There was a large flour mill with a dam across the river near the city where I was studying at that time. As a rule the height of the water was only two or three inches above the dam and to swim out to it was a sport not very dangerous in which I often indulged. One day I went alone to the river to enjoy myself as usual. When I was a short distance from the masonry, however, I was horrified to observe that the water had risen and was carrying me along swiftly. I tried to get away but it was too late. Luckily, tho, I saved myself from being swept over by taking hold of the wall with both hands. The pressure against my chest was great and I was barely able to keep my head above the surface. Not a soul was in sight and my voice was lost in the roar of the fall. Slowly and gradually I became exhausted and unable to withstand the strain longer. Just as I was about to let go, to be dashed against the rocks below, I saw in a flash of light a familiar diagram illustrating the hydraulic principle that the pressure of a fluid in motion is proportionate to the area exposed, and automatically I turned on my left side. As if by magic the pressure was reduced and I found it comparatively easy in that position to resist the force of the stream. But the danger still confronted me. I knew that sooner or later I would be carried down, as it was not possible for any help to reach me in time, even if I attracted attention. I am ambidextrous now but then I was left-handed and had comparatively little strength in my right arm. For this reason I did not dare to turn on the other side to rest and nothing remained but to slowly push my body along the dam. I had to get away from the mill towards which my face was turned as the current there was much swifter and deeper. It was a long and painful ordeal and I came near to failing at its very end for I was confronted with a depression in the masonry. I managed to get over with the last ounce of my force and fell in a swoon when I reached the bank, where I was found. I had torn virtually all the skin from my left side and it took several weeks before the fever subsided and I was well. These are only two of many instances but they may be sufficient to show that had it not been for the inventor's instinct I would not have lived to tell this tale.

Interested people have often asked me how and when I began to invent. This I can only answer from my present recollection in the light of which the first attempt I recall was rather ambitious for it involved the invention of an *apparatus* and a *method*. In the former I was anticipated but the latter was original. It happened in this way. One of my playmates had come into the possession of a hook and fishing-tackle which created quite an excitement in the village, and the next morning all started out to catch frogs. I was left alone and deserted owing to a quarrel with this boy. I had never seen a real hook and pictured it as something wonderful, endowed with peculiar qualities, and was despairing not to be one of the party. Urged by necessity, I somehow got hold of a piece of soft iron wire, hammered the end to a sharp point between two stones, bent it into shape, and fastened it to a strong string. I then cut a rod, gathered some bait, and went down to the brook where there were frogs in abundance. But I could not catch any and was almost discouraged when it occurred to me to dangle the empty hook in front of a frog sitting on a stump. At first, he collapsed but by and by his eyes bulged out and became bloodshot, he swelled to twice his normal size and made a vicious snap at the hook. Immediately I pulled him up. I tried the same thing again and again and the method proved infallible. When my comrades, who in spite of their fine outfit had caught nothing, came to me they were green with envy. For a long time, I kept my secret and enjoyed the monopoly but finally yielded to the spirit of Christmas. Every boy could then do the same and the following summer brought disaster to the frogs.

In my next attempt I seem to have acted under the first instinctive impulse which later dominated me—to harness the energies of nature to the service of man. I did this thru the medium of May-bugs—or June-bugs as they are called in America—which were a veritable pest in that country and sometimes broke the branches of trees by the sheer weight of their bodies. The bushes were black with them. I would attach as many as four of them to a crosspiece, notably arranged on a thin spindle, and transmit the motion of the same to a large disc and so derive considerable "power." These creatures were remarkably efficient, for once they were started they had no sense to stop and continued whirling for hours and hours and the hotter it was the harder they worked. All went well until a strange boy came to the place. He was the son of a retired officer in the Austrian Army. That urchin ate May-bugs alive and enjoyed them as tho they were the finest blue-point oysters. That disgusting sight terminated my endeavors in this promising field and I have never since been able to touch a May-bug or any other insect for that matter.

After that, I believe, I undertook to take apart and assemble the clocks of my grandfather. In the former operation I was always successful but often failed in the latter. So it came that he brought my work to a sudden halt in a manner not too delicate and it took thirty years before I tackled another clockwork again. Shortly thereafter I went into the manufacture of a kind of pop-gun which comprised a hollow tube, a piston, and two plugs of hemp. When firing the gun, the piston was prest against the stomach and the tube was pushed back quickly with both hands. The air between the plugs was comprest and raised to high temperature and one of them was expelled with a loud report. The art consisted in selecting a tube of the proper taper from the hollow stalks which did very well with that gun but my activities interfered with the windowpanes in our house and met with painful discouragement. If I remember rightly, I then took to carving swords from pieces of furniture which I could conveniently obtain. At that time I was under the sway of the Serbian national party and full of admiration for the feats of the heroes. I used to spend hours in mowing down my enemies in the form of corn-stalks which ruined the crops and netted me several spankings from my mother. Moreover these were not of the formal kind but the genuine article.

I had all this and more behind me before I was six years old and had past thru one year of elementary school in the village of Smiljan where I was born. At this juncture we moved to the little city of Gospic nearby. This change of residence was like a calamity to me. It almost broke my heart to part from our pigeons, chickens and sheep, and our magnificent flock of geese which used to rise to the clouds in the morning and return from the feeding grounds at sundown in battle formation, so perfect that it would have put a squadron of the best aviators of the present day to shame. In our new house I was but a prisoner, watching the strange people I saw thru the window blinds. My bashfulness was such that I would rather have faced a roaring lion than one of the city dudes who strolled about. But my hardest trial came on Sunday when I had to dress up and attend the service. There I met with an accident, the mere thought of which made my blood curdle like sour milk for years afterwards. It was my second adventure in a church. Not long before, I was entombed for a night in an old chapel on an inaccessible mountain which was visited only once a year. It was an awful experience, but this one was worse. There was a wealthy lady in town, a good but pompous woman, who used to come to the church gorgeously painted up and attired with an enormous train and attendants. One Sunday I had just finished ringing the bell in the belfry and rushed downstairs when this grand dame was sweeping out and I jumped on her train. It tore off with a ripping noise which sounded like a salvo of musketry fired by raw recruits. My father was livid with rage. He gave me a gentle slap on the cheek, the only corporal punishment he ever administered to me but I almost feel it now. The embarrassment and confusion that followed are indescribable. I was practically ostracized until something else happened which redeemed me in the estimation of the community.

An enterprising young merchant had organized a fire department. A new fire engine was purchased, uniforms provided and the men drilled for service and parade. The engine was, in reality, a pump to be worked by sixteen men and was beautifully painted red and black. One afternoon the official trial was prepared for and the machine was transported to the river. The entire population turned out to witness the great spectacle. When all the speeches and ceremonies were concluded, the command was given to pump, but not a drop of water came from the nozzle. The professors and experts tried in vain to locate the trouble. The fizzle was complete when I arrived at the scene. My knowledge of the mechanism was nil and I knew next to nothing of air pressure, but instinctively I felt for the suction hose in the water and found that it had collapsed. When I waded in the river and opened it up the water rushed forth and not a few Sunday clothes were spoiled. Archimedes running naked thru the streets of Syracuse and shouting Eureka at the top of his voice did not make a greater impression than myself. I was carried on the shoulders and was the hero of the day.

This photograph shows in the background the house in which Mr. Tesla's family resided. The edifice at the right is the "Real Gymnasium" where he studied. The ecclesiastic gentleman is his uncle, the metropolitan of Bosnia, who was a great statesman and who thwarted the designs of Austria upon Serbia at a critical period.

Upon settling in the city I began a four-years' course in the so-called Normal School preparatory to my studies at the College or *Real-Gymnasium*. During this period my boyish efforts and exploits, as well as troubles, continued. Among other things I attained the unique distinction of champion crow catcher in the country. My method of procedure was extremely simple. I would go in the forest, hide in the bushes, and imitate the call of the bird. Usually I would get several answers and in a short while a crow would flutter down intopthe shrubbery near me. After that all I needed to do was to throw a piece of cardboard to detract its attention, jump up and grab it before it could extricate itself from the undergrowth. In this way I would capture as many as I desired. But on one occasion something occurred which made me respect them. I had caught a fine pair of birds and was returning home with a friend. When we left the forest, thousands of crows had gathered making a frightful racket. In a few minutes they rose in pursuit and soon enveloped us. The fun lasted until all of a sudden I received a blow on thc back of my head which knocked me down. Then they attacked me viciously. I was compelled to release the two birds and was glad to join my friend who had taken refuge in a cave.

In the schoolroom there were a few mechanical models which interested me and turned my attention to water turbines. I constructed many of these and found great pleasure in operating them. How extraordinary was my life an incident may illustrate. My uncle had no use for this kind of pastime and more than once rebuked me. I was fascinated by a description of Niagara Falls I had perused, and pictured in my imagination a big wheel run by the Falls. I told my uncle that I would go to America and carry out this scheme. Thirty years later I saw my ideas carried out at Niagara and marveled at the unfathomable mystery of the mind.

I made all kinds of other contrivances and contraptions but among these the arbalists I produced were the best. My arrows, when shot, disappeared from sight and at close range traversed a plank of pine one inch thick. Thru the continuous tightening of the bows I developed skin on my stomach very much like that of a crocodile and I am often wondering whether it is due to this exercise that I am able even now to digest cobble-stones! Nor can I pass in silence my performances with the sling which would have enabled me to give a stunning exhibit at the

Hippodrome. And now I will tell of one of my feats with this antique implement of war which will strain to the utmost the credulity of the reader. I was practicing while walking with my uncle along the river. The sun was setting, the trout were playful and from time to time one would shoot up into the air, its glistening body sharply defined against a projecting rock beyond. Of course any boy might have hit a fish under these propitious conditions but I undertook a much more difficult task and I foretold to my uncle, to the minutest detail, what I intended doing. I was to hurl a stone to meet the fish, press its body against the rock, and cut it in two. It was no sooner said than done. My uncle looked at me almost scared out of his wits and exclaimed *"Iade retro Satanas!"* and it was a few days before he spoke to me again. Other records, however great, will be eclipsed but I feel that I could peacefully rest on my laurels for a thousand years.

* * * * *

(*In the April issue Mr. Tesla will describe in detail how he conceived one of his most important and far reaching inventions: "The Rotary Magnetic Field."* —EDITOR.)

PART III
APRIL 1919

MY LATER ENDEAVORS

This installment, no doubt the most interesting of the three published so far, reveals many extraordinary occurrences and experiences in the world's greatest inventor's life—experiences such as do not fall to the lot of ordinary mortals. And Tesla, the many sided, aside of inventing, knows the rare art of painting word-pictures. He does so here in a masterly fashion. He tells us how he finally conceived the induction motor—perhaps his greatest discovery—the invention which changed the face of the globe, the invention which made possible the street car, the subway, the electric train, power transmission, the harnessing of water falls and countless others. But let Tesla tell you himself how it all came about. It is a classic worth reading.

—EDITOR

THE DISCOVERY OF THE ROTATING MAGNETIC FIELD

At the age of ten I entered the Real Gymnasium which was a new and fairly well equipt institution. In the department of physics were various models of classical scientific apparatus, electrical and mechanical. The demonstrations and experiments performed from time to time by the instructors fascinated me and were undoubtedly a powerful incentive to invention. I was also passionately fond of mathematical studies and often won the professor's praise for rapid calculation. This was due to my acquired facility of visualizing the figures and performing the operations, not in the usual intuitive manner, but as in actual life. Up to a certain degree of complexity it was absolutely the same to me whether I wrote the symbols on the board or conjured them before my mental vision. But freehand drawing, to which many hours of the course were devoted, was an annoyance I could not endure. This was rather remarkable as most of the members of the family excelled in it. Perhaps my aversion was simply due to the predilection I found in undisturbed thought. Had it not been for a few exceptionally stupid boys, who could not do anything at all, my record would have been the worst. It was a serious handicap as under the then existing educational regime, drawing being obligatory, this deficiency threatened to spoil my whole career and my father had considerable trouble in railroading me from one class to another.

In the second year at that institution I became obsest with the idea of producing continuous motion thru steady air pressure. The pump incident, of which I have told, had set afire my youthful imagination and imprest me with the boundless possibilities of a vacuum. I grew frantic in my desire to harness this inexhaustible energy but for a long time I was groping in the dark. Finally, however, my endeavors crystallized in an invention which was to enable me to achieve what no other mortal ever attempted.

Nikola Tesla at 60. A very recent portrait
of the great inventor. An excellent likeness.

Imagine a cylinder freely rotatable on two bearings and partly surrounded by a rectangular trough which fits it perfectly. The open side of the trough is closed by a partition so that the cylindrical segment within the enclosure divides the latter into two compartments entirely separated from each other by air-tight sliding joints. One of these compartments being sealed and once for all exhausted, the other remaining open, a perpetual rotation of the cylinder would result, at least, I thought so. A wooden model was constructed and fitted with infinite care and when I applied the pump on one side and actually observed that there was a tendency to turning, I was delirious with joy. Mechanical flight was the one thing I wanted to accomplish altho still under the discouraging recollection of a bad fall I sustained by jumping with an umbrella from the top of a building. Every day I used to transport myself thru the air to distant regions but could not understand just how I managed to do it. Now I had something concrete—a flying machine with nothing more than a rotating shaft, flapping wings, and a vacuum of unlimited power! From that time on I made my daily aerial excursions in a vehicle of comfort and luxury as might have befitted King Solomon. It took years before I understood that the atmospheric pressure acted at right angles to the surface of the cylinder and that the slight rotary effort I observed was due to a leak. Tho this knowledge came gradually it gave me a painful shock. I had hardly completed my course at the Real Gymnasium when I was prostrated with a dangerous illness or rather, a score of them, and my condition became so desperate that I was given up by physicians. During this period I was permitted to read constantly, obtaining books from the Public Library which had been neglected and entrusted to me for classification of the works and preparation of the catalogues. One day I was handed a few volumes of new literature unlike anything I had ever read before and so captivating as to make me utterly forget my hopeless state. They were the earlier works of Mark Twain and to them might have been due the miraculous recovery which followed. Twenty-five years later, when I met Mr. Clements and we formed a friendship between us, I told him of the experience and was amazed to see that great man of laughter burst into tears.

My studies were continued at the higher Real Gymnasium in Carlstadt, Croatia, where one of my aunts resided. She was a distinguished lady, the wife of a Colonel who was an old war-horse having participated in many battles. I never can forget the three years I past at their home. No fortress in time of war was under a more rigid discipline. I was fed like a canary bird. All the meals were of the highest quality and deliciously prepared but short in quantity by a thousand percent. The slices of ham cut by my aunt were like tissue paper. When the Colonel would put something substantial on my plate she would snatch it away and say excitedly to him: "Be careful, Niko is very delicate." I had a voracious appetite and suffered like Tantalus.

But I lived in an atmosphere of refinement and artistic taste quite unusual for those times and conditions. The land was low and marshy and malaria fever never left me while there despite of the enormous amounts of quinin I consumed. Occasionally the river would rise and drive an army of rats into the buildings, devouring everything even to the bundles of the fierce paprika. These pests were to me a welcome diversion. I thinned their ranks by all sorts of means, which won me the unenviable distinction of rat-catcher in the community. At last, however, my course was completed, the misery ended, and I obtained the certificate of maturity which brought me to the cross-roads.

During all those years my parents never wavered in their resolve to make me embrace the clergy, the mere thought of which filled me with dread. I had become intensely interested in electricity under the stimulating influence of my Professor of Physics, who was an ingenious man and often demonstrated the principles by apparatus of his own invention. Among these I recall a device in the shape of a freely rotatable bulb, with tinfoil coatings, which was made to spin rapidly when connected to a static machine. It is impossible for me to convey an adequate idea of the intensity of feeling I experienced in witnessing his exhibitions of these mysterious phenomena. Every impression produced a thousand echoes in my mind. I wanted to know more of this wonderful force; I longed for experiment and investigation and resigned myself to the inevitable with aching heart.

Just as I was making ready for the long journey home I received word that my father wished me to go on a shooting expedition. It was a strange request as he had been always strenuously opposed to this kind of sport. But a few days later I learned that the cholera was raging in that district and, taking advantage of an opportunity, I returned to Gospic in disregard of my parents' wishes. It is incredible how absolutely ignorant people were as to the causes of this scourge which visited the country in intervals of fifteen to twenty years. They thought that the deadly agents were transmitted thru the air and filled it with pungent odors and smoke. In the meantime they drank the infected water and died in heaps. I contracted the awful disease on the very day of my arrival and altho surviving the crisis, I was confined to bed for nine months with scarcely any ability to move.

My energy was completely exhausted and for the second time I found myself at death's door. In one of the sinking spells which was thought to be the last, my father rushed into the room. I still see his pallid face as he tried to cheer me in tones belying his assurance. "Perhaps," I said, "I may get well if you will let me study engineering." "You will go to the best technical institution in the world," he solemnly replied, and I knew that he meant it. A heavy weight was lifted from my mind but the relief would have come too late had it not been for a marvelous cure brought about thru a bitter decoction of a peculiar bean. I came to life like another Lazarus to the utter amazement of everybody. My father insisted that I spend a year in healthful physical outdoor exercises to which I reluctantly consented. For most of this term I roamed in the mountains, loaded with a hunter's outfit and a bundle of books, and this contact with nature made me stronger in body as well as in mind. I thought and planned, and conceived many ideas almost as a rule delusive. The vision was clear enough but the knowledge of principles was very limited. In one of my inventions I proposed to convey letters and packages across the seas, thru a submarine tube, in spherical containers of sufficient strength to resist the hydraulic pressure. The pumping plant, intended to force the water thru the tube, was accurately figured and designed and all other particulars carefully worked out. Only one trifling detail, of no consequence, was lightly dismist. I assumed an arbitrary velocity of the water and, what is more, took pleasure in making it high, thus arriving at a stupendous performance supported by faultless calculations. Subsequent reflections, however, on the resistance of pipes to fluid flow determined me to make this invention public property.

Another one of my projects was to construct a ring around the equator which would, of course, float freely and could be arrested in its spinning motion by reactionary forces, thus enabling travel at a rate of about one thousand miles an hour, impracticable by rail. The reader will smile. The plan was difficult of execution, I will admit, but not nearly so bad as that of a well-known New York professor, who wanted to pump the air from the torrid to the temperate zones, entirely forgetful of the fact that the Lord had provided a gigantic machine for this very purpose.

Still another scheme, far more important and attractive, was to derive power from the rotational energy of terrestrial bodies. I had discovered that objects on the earth's surface, owing to the diurnal rotation of the globe, are carried by the same alternately in and against the direction of translatory movement. From this results a great change in momentum which could be utilized in the simplest imaginable manner to furnish motive effort in any habitable region of the world. I cannot find words to describe my disappointment when later I realized that I was in the predicament of Archimedes, who vainly sought for a fixt point in the universe.

At the termination of my vacation I was sent to the Polytechnic School in Gratz, Styria, which my father had chosen as one of the oldest and best reputed institutions. That was the moment I had eagerly awaited and I began my studies under good auspices and firmly resolved to succeed. My previous training was above the average, due to my father's teaching and opportunities afforded. I had acquired the knowledge of a number of languages and waded thru the books of several libraries, picking up information more or less useful. Then again, for the first time, I could choose my subjects as I liked, and freehand drawing was to bother me no more. I had made up my mind to give my parents a surprise, and during the whole first year I regularly started my work at three o'clock in the morning and continued until eleven at night, no Sundays or holidays excepted. As most of my fellow-students took things easily, naturally enough I eclipsed all records. In the course of that year I past thru nine exams and the professors thought I deserved more than the highest qualifications. Armed with their battering certificates, I went home for a short rest, expecting a triumph, and was mortified when my father made light of these hard-won honors. That almost killed my ambition; but later, after he had died, I was pained to find a package of letters which the professors had written him to the effect that unless he took me away from the Institution I would be killed thru overwork. Thereafter I devoted myself chiefly to physics, mechanics and mathematical studies, spending the hours of leisure in the libraries. I had a veritable mania for finishing whatever I began, which often got me into difficulties. On one occasion I started to read the works of Voltaire when I learned, to my dismay, that there were close on one hundred large volumes in small print which that monster had written while drinking seventy-two cups of black coffee per diem. It had to be done, but when I laid aside the last book I was very glad, and said, "Never more!"

My first year's showing had won me the appreciation and friendship of several professors. Among these were Prof. Rogner, who was teaching arithmetical subjects and geometry; Prof. Poeschl, who held the chair of theoretical and experimental physics; and Dr. Allé, who taught integral calculus and specialized in differential equations. This scientist was the most brilliant lecturer to whom I ever listened. He took a special interest in my progress and would frequently remain for an hour or two in the lecture room, giving me problems to solve, in which I delighted. To him I explained a flying machine I had conceived, not an illusionary invention, but one based on sound, scientific principles, which has become realizable thru my turbine and will soon be given to the world. Both Professors Rogner and Poeschl were curious men. The former had peculiar ways of expressing himself and whenever he did so there was a riot, followed by a long and embarrassing pause. Prof. Poeschl was a methodical and thoroly grounded German. He had enormous feet and hands like the paws of a bear, but all of his experiments were skillfully performed with clock-like precision and without a miss.

It was in the second year of my studies that we received a Gramme dynamo from Paris, having the horseshoe form of a laminated field magnet, and a wire-wound armature with a commutator. It was connected up and various effects of the currents were shown. While Prof. Poeschl was making demonstrations, running the machine as a motor, the brushes gave trouble, sparking badly, and I observed that it might be possible to operate a motor without these appliances. But he declared that it could not be done and did me the honor of delivering a lecture on the subject, at the conclusion of which he remarked: "Mr. Tesla may accomplish great things, but he certainly never will do this. It would be equivalent to converting a steadily pulling force, like that of gravity, into a rotary effort. It is a perpetual motion scheme, an impossible idea." But instinct is something which transcends knowledge. We have, undoubtedly, certain finer fibers that enable us to perceive truths when logical deduction, or any other willful effort of the brain, is futile. For a time I wavered, imprest by the professor's authority, but soon became convinced I was right and undertook the task with all the fire and boundless confidence of youth.

Tesla's first induction motor. This historic model is one of the two first presented before the American Institute of Electrical Engineers.

WHAT IS THE INDUCTION MOTOR?

The induction motor operates on alternating current. It has no commutator like a direct current motor, nor slip rings like an alternating current motor. Contrary to the two types just cited the "field" current is not steady, but the current itself rotates constantly pulling around with it—by induction—the only moving part of the motor—the rotor or armature. Having no armature nor slip rings, the induction motor never sparks. It consequently knows no "brush" trouble. It needs no attention because of its ruggedness. Only the bearings wear out. Its efficiency too is higher. On account of all this the induction motor is used in a prepondering proportion in street cars, electric trains, factories, etc.

I started by first picturing in my mind a direct-current machine, running it and following the changing flow of the currents in the armature. Then I would imagine an alternator and investigate the processes taking place in a similar manner. Next I would visualize systems comprising motors and generators and operate them in various ways. The images I saw were to me perfectly real and tangible. All my remaining term in Gratz was past in intense but fruitless efforts of this kind, and I almost came to the conclusion that the problem was insolvable.

In 1880 I went to Prague, Bohemia, carrying out my father's wish to complete my education at the University there. It was in that city that I made a decided advance, which consisted in detaching the commutator from the machine and studying the phenomena in this new aspect, but still without result. In the year following there was a sudden change in my views of life. I realized that my parents had been making too great sacrifices on my account and resolved to relieve them of the burden. The wave of the American telephone had just reached the European continent and the system was to be installed in Budapest, Hungary. It appeared an ideal opportunity, all the more as a friend of our family was at the head of the enterprise. It was here that I suffered the complete breakdown of the nerves to which I have referred. What I experienced during the period of that illness surpasses all belief. My sight and hearing were always extraordinary. I could clearly discern objects in the distance when others saw no trace of them. Several times in my boyhood I saved the houses of our neighbors from fire by hearing the faint crackling sounds which did not disturb their sleep, and calling for help.

In 1899, when I was past forty and carrying on my experiments in Colorado, I could hear very distinctly thunderclaps at a distance of 550 miles. The limit of audition for my young assistants was scarcely more than 150 miles. My ear was thus over thirteen times more sensitive. Yet at that time I was, so to speak, stone deaf in comparison with the acuteness of my hearing while under the nervous strain. In Budapest I could hear the ticking of a watch with three rooms between me and the time-piece. A fly alighting on a table in the room would cause a dull thud in my ear. A carriage passing at a distance of a few miles fairly shook my whole body. The whistle of a locomotive twenty or thirty miles away made the bench or chair on which I sat vibrate so strongly that the pain was unbearable. The ground under my feet trembled continuously. I had to support my bed on rubber cushions to get any rest at all. The roaring noises from near and far often produced the effect of spoken words which would have frightened me had I not been able to resolve them into their accidental components. The sun's rays, when periodically intercepted, would cause blows of such force on my brain that they would stun me. I had to summon all my will power to pass under a bridge or other structure as I experienced a crushing pressure on the skull. In the dark I had the sense of a bat and could detect the presence of an object at a distance of twelve feet by a peculiar creepy sensation on the forehead. My pulse varied from a few to two hundred and sixty beats and all the tissues of the body with twitchings and tremors which was perhaps the hardest to bear. A renowned physician who gave me daily large doses of Bromid of Potassium pronounced my malady unique and incurable. It is my eternal regret that I was not under the observation of experts in physiology and psychology at that time. I clung desperately to life, but never expected to recover.

Can anyone believe that so hopeless a physical wreck could ever be transformed into a man of astonishing strength and tenacity, able to work thirty-eight years almost without a day's interruption, and find himself still strong and fresh in body and mind? Such is my case. A powerful desire to live and to continue the work, and the assistance of a devoted friend and athlete accomplished the wonder. My health returned and with it the vigor of mind. In attacking the problem again I almost regretted that the struggle was soon to end. I had so much energy to spare. When I undertook the task it was not with a resolve such as men often make. With me it was a sacred vow, a question of life and death. I knew that I would perish if I failed. Now I felt that the battle was won.

Back in the deep recesses of the brain was the solution, but I could not yet give it outward expression. One afternoon, which is ever present in my recollection, I was enjoying a walk with my friend in the City Park and reciting poetry. At that age I knew entire books by heart, word for word. One of these was Goethe's "Faust." The sun was just setting and reminded me of the glorious passage:

"Sic rückt and weicht, der Tag ist iiberlebt,
Dort eilt sic hin und fördert neues Leben.
Oh, dass kein Flügel mich vom Boden hebt
Ihr nach und immer nach zu streben!**

Ein schöner Traum indessen sic entweicht.
Ach, zit des Geistes Flügeln wird so leicht
Kein körperlicher Flügel sich gesellen!"†

As I uttered these inspiring words the idea came like a flash of lightning and in an instant the truth was revealed. I drew with a stick on the sand the diagrams shown six years later in my address before the American Institute of Electrical Engineers, and my companion understood them perfectly. The images I saw were wonderfully sharp and clear and had the solidity of metal and stone, so much so that I told him: "See my motor here; watch me reverse it." I cannot begin to describe my emotions. Pygmalion seeing his statue come to life could not have been more deeply moved. A thousand secrets of nature which I might have stumbled upon accidentally I would have given for that one which I had wrested from her against all odds and at the peril of my existence.

* * * * *

Tesla on High Frequency Generators

Editor, Electrical Experimenter:

It is to be regretted that a letter addrest to me by Mr. J. Harris Rogers, in your care, was published in the March number of the Electrical Experimenter, *altho the concurrence of our views in some wireless features might have made this desirable to so wide-awake and enterprising a periodical as yours.*

---

* "The glow retreats, done is the day of toil; / It yonder hastes, new fields of life exploring; / Ah, that no wing can lift me from the soil, / Upon its track to follow, follow soaring!"

† "A glorious dream! though now the glories fade. / Alas! the wings that lift the mind no aid / Of wings to lift the body can bequeath me."

The purpose of this analog is to portray, as faithfully as possible, the phenomena of Tesla's rotating magnetic field so as to make them easily understandable to the average reader. The two alternating fluxes are represented by streams of water having the same relation as to phase, amplitude and direction. The magnetic polarity of the rotor's imitated by the employment of a body so shaped as to behave, with respect to the streams, exactly as the rotor with respect to the poles. Moreover, the corresponding rotating and stationary parts are given a similar appearance and are disposed in like manner. To make the analog complete, it may further be assumed that the liquid is compressible so that there willbe a phase displacement between pressure and flow as that existing between electro-motive Force and current.

Mr. Rogers seems to be a very appreciative gentleman and nothing would be farther from my thoughts than to detract anything from his merit, but in a separate contribution, which I expect to prepare for your next issue, I shall express myself on this subject without prejudice and in the interest of truth. However, the article by your Mr. H. Winfield Secor on "America's Greatest War Invention: The Rogers Underground Wireless" contains a reference to "a novel and original high frequency generator" of Mr. Rogers' invention. May I not—to use the President's elegant expression—call attention to the fact that this device was described by me years ago, as will be evident from the following excerpt of a communication which appeared in the Electrical Review *of March 15, 1899. In speaking of circuit controllers, I said: "I may mention here, based on a different principle, which is incomparably more effective, more efficient, and also simpler on the whole. It comprises a fine stream of conducting fluid which is made to issue, with any desired speed, from an orifice connected with one pole of a generator, thru the primary of the induction coil, against the other terminal of the generator placed at a small distance. This device gives discharges of a remarkable suddenness, and the frequency may be brought within reasonable limits, almost to anything desired. I have used this device for a long time in connection with ordinary coils and in a form of my own coil with results greatly superior in every respect to those obtainable with the form of your letter, make a few statements referring to such make-and-break devices in general, and various forms based on this new principle."*

I may add that a great many forms of this apparatus were constructed and employed by me for a long time, proving very convenient and useful. Water does not give particularly good results, being incapable of causing very abrupt changes, but electrolytes have the property of diminishing enormously in resistance when they are heated and the effects are much more intense. Salts of lithium are especially efficient.

—Nikola Tesla
New York, February 20, 1919

* * * * *

PART IV
MAY 1919

THE DISCOVERY OF THE
TESLA COIL AND TRANSFORMER

The proverbial trials and tribulations known to every inventor were not spared Tesla, the world's greatest inventor of all times. In this article we see him, arrived at young manhood, struggling along in a cold world. Already his fame has spread far and wide and his genius is recognized. But converting genius and fame into dollars and cents is quite a different matter, and the world is full of unappreciative and unscrupulous men. Tesla, the idealist, cared little for money and thus was promptly taken advantage of. But let Tesla himself tell you in his own inimitable style. It is a wonderful story.

In this month's installment Tesla also tells us how he made one of his most important as well as sensational discoveries—the Tesla Coil. Few inventions have caused such a sensation as this one which culminated in the only man-made lightning ever produced. The Tesla Coil has so many uses and has been built in so many styles that it would take a catalog to list them all. From the spectacular high frequency stunts on the stage down to the "violet" ray machine in your home: all are Tesla Coils in one form or another. Wireless without the Tesla Coil would not be possible today. Without an oscillation transformer, spark gap and condenser— which is a Tesla Coil—the sending station would be crippled.

But it is for industrial purposes where the Tesla Coil will shine brightest in the future. The production of Ozone, the extraction of Nitrogen from the air in huge quantities—all are children of Tesla's fertile brain. His coil is the key to them all.

—EDITOR

For a while I gave myself up entirely to the intense enjoyment of picturing machines and devising new forms. It was a mental state of happiness about as complete as I have ever known in life. Ideas came in an uninterrupted stream and the only difficulty I had was to hold them fast. The pieces of apparatus I conceived were to me absolutely real and tangible in every detail, even to the minutest marks and signs of wear. I delighted in imagining the motors constantly running, for in this way they presented to the mind's eye a more fascinating sight. When natural inclination develops into a passionate desire, one advances towards his goal in seven-league boots. In less than two months I evolved virtually all the types of motors and modifications of the system which are now identified with my name. It was, perhaps, providential that the necessities of existence commanded a temporary halt to this consuming activity of the mind. I came to Budapest prompted by a premature report concerning the telephone enterprise and, as irony of fate willed it, I had to accept a position as draftsman in the Central Telegraph Office of the Hungarian Government at a salary which I deem it my privilege not to disclose! Fortunately, I soon won the interest of the Inspector-in-Chief and was thereafter employed on calculations, designs and estimates in connection with new installations, until the Telephone Exchange was started, when I took charge of the same. The knowledge and practical experience I gained in the course of this work was most valuable and the employment gave me ample opportunities for the exercise of my inventive faculties. I made several improvements in the Central Station apparatus and perfected a telephone repeater or amplifier which was never patented or publicly described but could be creditable to me even today. In recognition of my efficient assistance the organizer of the undertaking, Mr. Puskas, upon disposing of his business in Budapest, offered me a position in Paris which I gladly accepted.

I never can forget the deep impression that magic city produced on my mind. For several days after my arrival I roamed thru the streets in utter bewilderment of the new spectacle. The attractions were many and irresistible, but, alas, the income was spent as soon as received. When Mr. Puskas asked me how I was getting along in the new sphere, I described the situation accurately in the statement that "the last twenty-nine days of the month are the toughest!" I led a rather strenuous life in what would now be termed "Rooseveltian fashion." Every morning, regardless of weather, I would go from the Boulevard St. Marcel, where I resided, to a bathing house on the Seine, plunge into the water, loop the circuit twenty-seven times and then walk an hour to reach Ivry, where the Company's factory was located. There I would have a woodchopper's breakfast at half-past seven o'clock and then eagerly await the lunch hour, in the meanwhile cracking hard nuts for the Manager of the Works, Mr. Charles Batchellor, who was an intimate friend and assistant of Edison. Here I was thrown in contact with a few Americans who fairly fell in love with me because of my proficiency in billiards.

FIG. 1. Tesla oscillation transformer (Tesla Coil) presented by Lord Kelvin before the British Association in August, 1897. This small and compact instrument, only 8 inches high, developed two square feet of streamers with twenty-five watts from the 110 volt D. C. supply circuit. The instrument contains a Tesla primary and secondary, condenser, and a circuit controller.

To these men I explained my invention and one of them, Mr. D. Cunningham, Foreman of the Mechanical Department, offered to form a stock company. The proposal seemed to me comical in the extreme. I did not have the faintest conception of what that meant except that it was an American way of doing things. Nothing came of it, however, and during the next few months I had to travel from one to

Mechanical Analog of Tesla Oscillation Transformer *(Tesla Coil)*

This revolutionary improvement was exhibited and explained by Tesla for the first time in his lecture before the American Institute of Electrical Engineers May 20, 1891. It has made possible to generate automatically damped or undamped oscillations of any desired frequency and, what is equally important, of perfectly constant period. It has been instrumental in many great achievements and its use has become universal. The underlying principle may be briefly stated as follows: A source of electricity is made to charge a condenser and when the difference of potential at the terminals of the latter has reached a predetermined value, an air-gap is bridged, permitting the accumulated energy to be discharged through a circuit under resonant conditions, this resulting in a long series of isochronous impulses. These are either directly used or converted to any desired volume or pressure by means of a second circuit inductively linked with the first and tuned to the same. The above diagram is taken from Tesla's lecture before the Franklin Institute and National Electric Light Association in 1893 and shows more elaborate arrangements of circuits, now quite familiar, for the conversion of ordinary direct or alternating currents into high frequency oscillations by this general method. In the mechanical apparatus illustrated, an attempt is made to convey an idea of the electrical operations as closely as practicable. The reciprocating and centrifugal pumps, respectively, represent an alternating and a direct current generator. The water takes the place of the electric fluid. The cylinder with its elastically restrained piston represents the condenser. The inertia of the moving parts corresponds to the self-induction of the electric circuit and the wide ports around the cylinder, through which the fluid can escape, perform the function of the air-gap. The operation of this apparatus will now be readily understood. Suppose first that the water is admitted to the cylinder from the centrifugal pump, this corresponding to the action of a continuous current generator. As the fluid is forced into the cylinder, the piston moves upward until the ports are uncovered, when a great quantity of the fluid rushes out, suddenly reducing the pressure so that the force of the compressed spring asserts itself and sends the piston down, closing the ports, whereupon these operations are repeated in as rapid succession as it may be desired. Each time the system, comprising the piston, rod, weights and adjustable spring, receives a blow, it quivers at its own rate which is determined by the inertia of the moving parts and the reliability of the spring exactly as in the electrical system the period of the circuit is determined by the self-induction and capacity. Under the best conditions the natural period of the elastic system will be the same as that of the primarily impressed oscillations, and then the energy of the movement will be greatest. If, instead of the centrifugal, the reciprocating pump is employed, the operation is the same in principle except that the periodic impulses of the pump impose certain limitations. The best results are again obtained when synchronism is maintained between these and the natural oscillations of the system.

another place in France and Germany to cure the ills of the power plants. On my return to Paris I submitted to one of the administrators of the Company, Mr. Rau, a plan for improving their dynamos and was given an opportunity.

My success was complete and the delighted directors accorded me the privilege of developing automatic regulators which were much desired. Shortly after there was some trouble with the lighting plant which has been installed at the new railroad station in Strassburg, Alsace. The wiring was defective and on the occasion of the opening ceremonies a large part of a wall was blown out thru a short-circuit right in the presence of old Emperor William I. The German Government refused to take the plant and the French Company was facing a serious loss. On account of my knowledge of the German language and past experience, I was entrusted with the difficult task of straightening out matters and early in 1883 I went to Strassburg on that mission.

THE FIRST INDUCTION MOTOR IS BUILT

Some of the incidents in that city have left an indelible record on my memory. By a curious coincidence, a number of men who subsequently achieved fame, lived there about that time. In later life I used to say, "There were bacteria of greatness in that old town. Others caught the disease but I escaped!" The practical work, correspondence, and conferences with officials kept me preoccupied day and night, but as soon as I was able to manage I undertook the construction of a simple motor in a mechanical shop opposite the railroad station, having brought with me from Paris some material for that purpose. The consummation of the experiment was, however, delayed until the summer of that year when I finally had the satisfaction of *seeing rotation effected by alternating currents of different phase, and without sliding contacts or commutator,* as I had conceived a year before. It was an exquisite pleasure but not to compare with the delirium of joy following the first revelation.

Among my new friends was the former Mayor of the city, Mr. Bauzin, whom I had already in a measure acquainted with this and other inventions of mine and whose support I endeavored to enlist. He was sincerely devoted to me and put my project before several wealthy persons but, to my mortification, found no response. He wanted to help me in every way possible and the approach of the first of July, 1919, happens to remind me of a form of "assistance" I received from that charming man, which was not financial but none the less appreciated. In 1870, when the Germans invaded the country, Mr. Bauzin had buried a good sized allotment of St. Estèphe of 1801 and he came to the conclusion that he knew no worthier person than myself to consume that precious beverage. This, I may say, is one of the unforgettable incidents to which I have referred. My friend urged me to return to Paris as soon as possible and seek support there. This I was anxious to do but my work and negotiations were protracted owing to all sort of petty obstacles I encountered so that at times the situation seemed hopeless.

GERMAN "EFFICIENCY"

Just to give an idea of German thoroness and "efficiency," I may mention here a rather funny experience. An incandescent lamp of 16 c.p. was to be placed in a hallway and upon selecting the proper location I ordered the *monteur* to run the wires. After working for a while he concluded that the engineer had to be consulted and this was done. The latter made several objections but ultimately agreed that the lamp should be placed two inches from the spot I had assigned, whereupon the work proceeded.

Then the engineer became worried and told me that inspector Averdeck should be notified. That important person called, investigated, debated, and decided that the lamp should be shifted back two inches, which was the place I had marked. It was not long, however, before Averdeck got cold feet himself and advised me that he had informed *Ober-Inspector* Hieronimus of the matter and that I should await his decision. It was several days before the *Ober-Inspector* was able to free himself of other pressing duties but at last he arrived and a two-hour debate followed, when he decided to move the lamp two inches farther. My hopes that this was the final act were shattered when the *Ober-Inspector* returned and said to me: "*Regierungsrath* Funke is so particular that I would not dare to give an order for placing this lamp without his explicit approval." Accordingly arrangements for a visit from that great man were made. We started cleaning up and polishing early in the morning. Everybody brushed up. I put on my gloves and when Funke came with his retinue he was ceremoniously received. After two hours' deliberation he suddenly exclaimed "I must be going," and pointing to a place on the ceiling, he ordered me to put the lamp there. It was the exact spot which I had originally chosen.

So it went day after day with variations, but I was determined to achieve at whatever cost and in the end my efforts were rewarded. By the spring of 1884 all the differences were adjusted, the plant formally accepted, and I returned to Paris with pleasing anticipations. One of the administrators had promised me a liberal compensation in case I succeeded, as well as a fair consideration of the improvements I had made in their dynamos and I hoped to realize a substantial sum. There were three administrators whom I shall designate as A, B and C for convenience. When I called on A he told me that B had the say. This gentleman thought that only C could decide and the latter was quite sure that A alone had the power to act. After several laps of this *circulous viciosus*, it dawned upon me that my reward was a castle in Spain. The utter failure of my attempts to raise capital for development was another disappointment and when Mr. Batchellor prest me to go to America with a view of redesigning the Edison machines, I determined to try my fortunes in the Land of Golden Promise. But the chance was nearly mist. I liquefied my modest assets, secured accommodations and found myself at the railroad station as the train was pulling out. At that moment I discovered that my money and tickets were gone. What to do was the question. Hercules had plenty of time to deliberate but I had to decide while running alongside the train with opposite feelings surging in my brain like condenser oscillations. Resolve, helped by dexterity, won out in the nick of time and upon passing thru the usual experiences, as trivial as unpleasant, I managed to embark for New York with the remnants of my belongings, some poems and articles I had written, and a package of calculations relating to solutions of an unsolvable integral and to my flying machine. During the voyage I sat most of the time at the stern of the ship watching for an opportunity to save somebody from a watery grave, without the slightest thought of danger. Later when I had absorbed some of the practical American sense I shivered at the recollection and marvelled at my former folly.

FIG. 2. This illustrates tests with spark discharges from a ball of forty centimeters radius in Tesla's wireless plant erected at Colorado Springs in 1899. The ball is connected to the free end of a grounded resonant circuit seventeen meters in diameter. The disruptive potential of a ball, is, according to Tesla, in volts approximately V=75,400r (r being in centimeters), That is, in this case 75,400 x 40 = 3,016,000 volts. The gigantic Tesla Coil which produced these bolts of Thor was capable of furnishing a current of 1,100 amperes in the high tension secondary. The primary coil had a diameter of 51 feet! This Tesla Coil produced discharges which were the nearest approach to lightning ever made by man.

Tesla in America

I wish that I could put in words my first impressions of this country. In the Arabian Tales I read how genii transported people into a land of dreams to live thru delightful adventures. My case was just the reverse. The genii had carried me from a world of dreams into one of realities. What I had left was beautiful, artistic and fascinating in every way; what I saw here was machined, rough and unattractive. A burly policeman was twirling his stick which looked to me as big as a log. I approached him politely with the request to direct me. "Six blocks down, then to the left," he said, with murder in his eyes. "Is this America?" I asked myself in painful surprise. "It is a century behind Europe in civilization."

When I went abroad in 1889—five years having elapsed since my arrival here—I became convinced that it was more than one hundred years *ahead* of Europe and nothing has happened to this day to change my opinion.

Tesla Meets Edison

The meeting with Edison was a memorable event in my life. I was amazed at this wonderful man who, without early advantages and scientific training, had accomplished so much. I had studied a dozen languages, delved in literature and art, and had spent my best years in libraries reading all sorts of stuff that fell into my hands, from Newton's *Principia* to the novels of Paul de Kock, and felt that most of my life had been squandered. But it did not take long before I recognized that it was the best thing I could have done. Within a few weeks I had won Edison's confidence and it came about in this way.

The S. S. *Oregon*, the fastest passenger steamer at that time, had both of its lighting machines disabled and its sailing was delayed. As the superstructure had been built after their installation it was impossible to remove them from the hold. The predicament was a serious one and Edison was much annoyed. In the evening I took the necessary instruments with me and went aboard the vessel where I stayed for the night. The dynamos were in bad condition, having several short-circuits and breaks, but with the assistance of the crew I succeeded in putting them in good shape. At five o'clock in the morning, when passing along Fifth Avenue on my way to the shop, I met Edison with Batchellor and a few others as they were returning home to retire. "Here is our Parisian running around at night," he said. When I told him that I was coming from the *Oregon* and had repaired both machines, he looked at me in silence and walked away without another word. But when he had gone some distance I heard him remark: "Batchellor, this is a d—n good man," and from that time on I had full freedom in directing the work.

For nearly a year my regular hours were from 10:30 A.M. until 5 o'clock the next morning without a day's exception. Edison said to me: "I have had many hard-working assistants but you take the cake." During this period I designed twenty-four different types of standard machines with short cores and of uniform pattern which replaced the old ones. The manager had promised me fifty thousand dollars on the completion of this task but it turned out to be a practical joke. This gave me a painful shock and I resigned my position.

Immediately thereafter some people approached me with the proposal of forming an arc light company under my name, to which I agreed. Here finally was an opportunity to develop the motor, but when I broached the subject to my new associates they said: "No, we want the arc lamp. We don't care for this alternating current of yours." In 1886 my system of arc lighting was perfected and adapted for factory and municipal lighting, and I was free, but with no other possession than a beautifully engraved certificate of stock of hypothetical value. Then followed a period of struggle in the new medium for which I was not fitted, but the reward came in the end and in April, 1887, the Tesla Electric Company was organized, providing a laboratory and facilities. The motors I built there were exactly as I had imagined them. I made no attempt to improve the design, but merely reproduced the pictures as they appeared to my vision and the operation was always as I expected.

In the early part of 1888 an arrangement was made with the Westinghouse Company for the manufacture of the motors on a large scale. But great difficulties had still to be overcome. My system was based on the use of low frequency currents and the Westinghouse experts had adopted 133 cycles with the object of securing advantages in the transformation. They did not want to depart from their standard forms of apparatus and my efforts had to be concentrated upon adapting the motor to these conditions. Another necessity was to produce a motor capable of running efficiently at this frequency on two wires which was not easy of accomplishment.

Fig. 3. Scheme of circuit connections in Tesla's Oscillation Transformer shown in Fig. 1. The secondary circuit which slips into the primary is omitted.

At the close of 1889, however, my services in Pittsburg being no longer essential, I returned to New York and resumed experimental work in a laboratory on Grand Street, where I began immediately the design of high frequency machines. The problems of construction in this unexplored field were novel and quite peculiar and I encountered many difficulties. I rejected the inductor type, fearing that it might not yield perfect sine waves which were so important to resonant action. Had it not been for this I could have saved myself a great deal of labor. Another discouraging feature of the high frequency alternator seemed to be the inconstancy of speed which threatened to impose serious limitations to its use. I had already noted in my demonstrations before the American Institution of Electrical Engineers that several times the tune was lost, necessitating readjustment, and did not yet foresee, what I discovered long afterwards, a means of operating a machine of this kind at a speed constant to such a degree as not to vary more than a small fraction of one revolution between the extremes of load.

The Invention of the Tesla Coil

From many other considerations it appeared desirable to invent a simpler device for the production of electric oscillations. In 1856 Lord Kelvin had exposed the theory of the condenser discharge, but no practical application of that important knowledge was made. I saw the possibilities and undertook the development of induction apparatus on this principle. My progress was so rapid as to enable me to exhibit at my lecture in 1891 a coil giving sparks of *five inches*. On that occasion I frankly told the engineers of a defect involved in the transformation by the new method, namely, the loss in the spark gap. Subsequent investigation showed that no matter what medium is employed, be it air, hydrogen, mercury vapor, oil or a stream of electrons, the efficiency is the same. It is a law very much like that governing the conversion of mechanical energy. We may drop a weight from a certain height vertically down or carry it to the lower level along any devious path, it is immaterial insofar as the amount of work is concerned. Fortunately however, this drawback is not fatal as by proper proportioning of the resonant circuits *an efficiency of 85 percent* is attainable. Since my early announcement of the invention it has come into a universal use and wrought a revolution in many departments. But a still greater future awaits it. When in 1900 I obtained powerful discharges of 100 feet and flashed a current around the globe, I was reminded of the first tiny spark I observed in my Grand Street laboratory and was thrilled by sensations akin to those I felt when I discovered the rotating magnetic field.

This photograph shows the famous Tesla tower erected at Shoreham, L. I., N. Y. The tower was dismantled at the outbreak of the War. It was 187 feet high. The spherical top was 68 feet in diameter. Note the huge size of the structure by comparing the two-story power plant in the rear. The tower which was to be used by Tesla in his "World Wireless," was never finished.

THE MAGNIFYING TRANSMITTER

Imagine a man a century ago, bold enough to design and actually build a huge tower with which to transmit the human voice, music, pictures, press news and even power, thru the earth to any distance whatever without wires! He probably would have been hung or burnt at the stake. So when Tesla built his famous tower on Long Island he was a hundred years ahead of his time. And foolish ridicule by our latter day armchair "savants," does not in the least near Tesla's greatness. The titanic brain of Tesla has hardly produced a more amazing wonder than this "magnifying transmitter." Contrary to popular belief his tower was not built to radiate Hertzian waves into the ether. Tesla's system sends out thousands of horsepower thru the earth; he has shown experimentally how power can be sent without wires over distances from a central point. Nor is there any mystery about it how he accomplishes the result. His historic U.S. patents and articles describe the method used. Tesla's Magnifying Transmitter is truly a modern lamp of Aladdin.

—EDITOR

As I review the events of my past life I realize how subtle are the influences that shape our destinies. An incident of my youth may serve to illustrate. One winter's day I managed to climb a steep mountain, in company with other boys. The snow was quite deep and a warm southerly wind made it just suitable for our purpose. We amused ourselves by throwing balls which would roll down a certain distance, gathering more or less snow, and we tried to outdo one another in this exciting sport. Suddenly a ball was seen to go beyond the limit, swelling to enormous proportions until it became as big as a house and plunged thundering into the valley below with a force that made the ground tremble. I looked on spellbound, incapable of understanding what had happened. For weeks afterward the picture of the avalanche was before

This photograph of a model shows how the Tesla tower built on Long Island, eighteen years ago, would have looked completed. From its appearance nobody would infer that it was to be used for the great purposes which are set forth in his accompanying article.

my eyes and I wondered how anything so small could grow to such an immense size. Ever since that time the magnification of feeble actions fascinated me, and when, years later, I took up the experimental study of mechanical and electrical resonance, I was keenly interested from the very start. Possibly, had it not been for that early powerful impression, I might not have followed up the little spark I obtained with my coil and never developed my best invention, the true history of which I will tell here for the first time.

Scrapping the World's Engines

"Lionhunters" have often asked me which of my discoveries I prize most. This depends on the point of view. Not a few technical men, very able in their special departments, but dominated by a pedantic spirit nearsighted, have asserted that excepting the induction motor I have given to the world little of practical use. This is a grievous mistake. A new idea must not be judged by its immediate results. My alternating system of power transmission came at a psychological moment, as a long-sought answer to pressing industrial questions, and altho considerable resistance had to be overcome and opposing interests reconciled, as usual, the commercial introduction could not be long delayed. Now, compare this situation with that confronting my turbine, for example. One should think that so simple and beautiful an invention, possessing many features of an ideal motor, should be adopted at once and, undoubtedly, it would under similar conditions. But the prospective effect of the rotating field was not to render worthless existing machinery; on the contrary, it was to give it additional value. The system lent itself to new enterprise as well as to improvement of the old. My turbine is an advance of a character entirely different. It is a radical departure in the sense that its success would mean the abandonment of the antiquated types of prime movers on which billions of dollars have been spent. Under such circumstances the progress must needs be slow and perhaps the greatest impediment is encountered in the prejudicial opinions created in the minds of experts by organized opposition. Only the other day I had a disheartening experience when I met my friend and former assistant, Charles F. Scott, now professor of Electrical Engineering at Yale. I had not seen him for a long time and was glad to have an opportunity for a little chat at my office. Our conversation naturally enough drifted on my turbine and I became heated to a high degree. "Scott," I exclaimed, carried away by the vision of a glorious future, "my turbine will scrap all the heat-engines in the world." Scott stroked his chin and looked away thoughtfully as though making a mental calculation. "That will make quite a pile of scrap," he said, and left without another word!

"ALADDIN'S LAMP"

These and other inventions of mine, however, were nothing more than steps forward in certain directions. In evolving them I simply followed the inborn instinct to improve the present devices without any special thought of our far more imperative necessities. The "Magnifying Transmitter" was the product of labors extending through years, having for their chief object the solution of problems which are infinitely more important to mankind than mere industrial development.

If my memory serves me right, it was in November, 1891, that I performed a laboratory experiment which was one of the most extraordinary and spectacular ever recorded in the annals of science. In investigating the behaviour of high frequency currents I had satisfied myself that an electric field of sufficient intensity could be produced in a room to light up electrodeless vacuum tubes. Accordingly, a transformer was built to test the theory and the test trial proved a marvelous success. It is difficult to appreciate what those strange phenomena meant at that time. We crave for new sensations but soon become indifferent to them. The wonders of yesterday are today common occurrences. When my tubes were first publicly exhibited they were viewed with amazement impossible to describe. From all parts of the world I received urgent invitations and numerous honors and other flattering inducements were offered to me, which I declined.

IN FARADAY'S CHAIR

But in 1892 the demands became irresistible and I went to London where I delivered a lecture before the institution of Electrical Engineers. It had been my intention to leave immediately for Paris in compliance with a similar obligation, but Sir James Dewar insisted on my appearing before the Royal Institution. I was a man of firm resolve but succumbed easily to the forceful arguments of the great Scotchman. He pushed me into a chair and poured out half a glass of a wonderful brown fluid which sparkled in all sorts of iridescent colors and tasted like nectar. "Now," said he, "you are sitting in Faraday's chair and you are enjoying whiskey he used to drink." In both aspects it was an enviable experience. The next evening I gave a demonstration before that institution, at the termination of which Lord Rayleigh addressed the audience and his generous accords gave me the first start in these endeavors. I fled from London and later from Paris to escape favors showered upon me, and journeyed to my home where I passed through a most painful ordeal and illness. Upon regaining my health I began to formulate plans for the resumption of work in America. Up to that time I never realized that I possessed any particular gift of discovery but Lord Rayleigh, whom I always considered as an ideal man of science, had said so and if that was the case I felt that I should concentrate on some big idea.

NATURE'S TRIGGER

One day, as I was roaming in the mountains, I sought shelter from an approaching storm. The sky became overhung with heavy clouds but somehow the rain was delayed until, all of a sudden, there was a lightning flash and a few moments after a deluge. This observation set me thinking. It was manifest that the two phenomena were closely related, as cause and effect, and a little reflection led me to the conclusion that the electrical energy involved in the precipitation of the water was inconsiderable, the function of lightning being much like that of a sensitive trigger. Here was a stupendous possibility of achievement. If we could produce electric effects of the required quality, this whole planet and the conditions of existence on it could be transformed. The sun raises the water of the oceans and winds drive it to distant regions where it remains in a state of most delicate balance. If it were in our power to upset it when and wherever desired, this mighty life-sustaining stream could be at will controlled. We could irrigate arid deserts; create lakes and rivers and provide motive power in unlimited amounts. This would be the most efficient way of harnessing the sun to the uses of man. The consummation depended on our ability to develop electric forces of the order of those in nature. It seemed a hopeless undertaking, but I made up my mind to try it and immediately on my return to the United States, in the summer of 1892, work was begun which was to me all the more attractive, because a means of the same kind was necessary for the successful transmission of energy without wires.

Four Million Volts

The first gratifying result was obtained in the spring of the succeeding year when I reached tensions of about 1,000,000 volts with my conical coil. That was not much in the light of the present art, but it was then considered a feat. Steady progress was made until the destruction of my laboratory by fire in 1895, as may be judged from an article by T. C. Martin which appeared in the April number of the *Century Magazine*. This calamity set me back in many ways and most of that year had to be devoted to planning and reconstruction.

However, as soon as circumstances permitted, I returned to the task. Although I knew that higher electro-motive forces were attainable with apparatus of larger dimensions, I had an instinctive perception that the object could be accomplished by the proper design of a comparatively small and compact transformer. In carrying on tests with a secondary in the form of a flat spiral, as illustrated in my patents, the absence of streamers surprised me, and it was not long before I discovered that this was due to the position of the turns and their mutual action. Profiting from this observation I resorted to the use of a high tension conductor with turns of considerable diameter sufficiently separated to keep down the distributed capacity, while at the same time preventing undue accumulation of the charge at any point. The application of this principle enabled me to produce pressures of 4,000,000 volts, which was about the limit obtainable in my new laboratory at Houston Street, as the discharges extended through a distance of 10 feet. A photograph of this transmitter was published in the *Electrical Review* of November 1898. In order to advance further along this line I had to go into the open, and in the spring of 1899, having completed preparations for the erection of a wireless plant, I went to Colorado where I remained for more than one year. Here I introduced other improvements and refinements which made it possible to generate currents of any tension that may be desired. Those who are interested will find some information in regard to the experiment I conducted there in my article, "The Problem of Increasing Human Energy" in the *Century Magazine* of June 1900, to which I have referred on a previous occasion.

THE MAGNIFYING TRANSMITTER

I have been asked by the *Electrical Experimenter* to be quite explicit on this subject so that my young friends among the readers of the magazine will clearly understand the construction and operation of my "Magnifying Transmitter" and the purposes for which it is intended. Well, then, in the first place, it is a *resonant transformer* with a secondary in which the parts, charged to a high potential, are of considerable area and arranged in space along ideal enveloping surfaces of very large radii of curvature, and at proper distances from one another thereby insuring a *small electric surface density everywhere* so that *no leak can occur even if the conductor is bare*. It is suitable for any frequency, from a few to many thousands of cycles per second, and can be used in the production of currents of tremendous volume and moderate pressure, or of smaller amperage and immense electromotive force. The maximum electric tension is *merely dependent on the curvature of the surfaces* on which the charged elements are situated and the area of the latter.

100 Million Volts Possible

Judging from my past experience, as much as 100,000,000 volts are perfectly practicable. On the other hand currents of many thousands of amperes may be obtained in the antenna. A plant of but very moderate dimensions is required for such performances. Theoretically, a terminal of less than 90 feet in diameter is sufficient to develop an electro-motive force of that magnitude while for antenna currents of from 2,000–4,000 amperes at the usual frequencies it need not be larger than 30 feet in diameter.

In a more restricted meaning this wireless transmitter is one in which the Hertz-wave radiation is an entirely negligible quantity as compared with the whole energy, under which condition the damping factor is extremely small and an enormous charge is stored in the elevated capacity. Such a circuit may then be excited with impulses of any kind, even of low frequency and it will yield sinusoidal and continuous oscillations like those of an alternator.

Taken in the narrowest significance of the term, however, it is a resonant transformer which, besides possessing these qualities, is accurately proportioned to fit the globe and its electrical constants and properties, by virtue of which design it becomes highly efficient and effective in the wireless transmission of energy. Instance is then absolutely eliminated, there being *no diminution in the intensity of the transmitted impulses.* It is even possible to make the actions *increase with the distance from the plant* according to an exact mathematical law.

This invention was one of a number comprised in my "World-System" of wireless transmission which I undertook to commercialize on my return to New York in 1900. As to the immediate purposes of my enterprise, they were clearly outlined in a technical statement of that period from which I quote:

> The "World-System" has resulted from a combination of several original discoveries made by the inventor in the course of long continued research and experimentation. It makes possible not only the instantaneous and precise wireless transmission of any kind of signals, messages or characters. to all parts of the world, but also the inter-connection of the existing telegraph, telephone, and other signal stations without any

change in their present equipment. By its means, for instance, a telephone subscriber here may call up and talk to any other subscriber on the Globe. An inexpensive receiver, not bigger than a watch, will enable him to listen anywhere, on land or sea, to a speech delivered or music played in some other place, however distant. These examples are cited merely to give an idea of the possibilities of this great scientific advance which annihilates distance and makes that perfect natural conductor, the Earth, available for all the innumerable purposes which human ingenuity has found for a line-wire. One far-reaching result of this is that any device capable of being operated thru one or more wires (at a distance obviously restricted) can likewise be actuated, without artificial conductors and with the same facility and accuracy, at distances to which there are no limits other than those imposed by the physical dimensions of the Globe. Thus, not only will entirely new fields for commercial exploitation be opened up by this ideal method of transmission but the old ones vastly extended. "The World-System" is based on the application of the following important inventions and discoveries:

1. *The "Tesla Transformer."* This apparatus is in the production of electrical vibrations as revolutionary as gunpowder was in warfare. Currents many times stronger than any ever generated in the usual ways, and sparks over one hundred feet long, have been produced by the inventor with an instrument of this kind.

2. *The "Magnifying Transmitter."* This is Tesla's best invention—a peculiar transformer specially adapted to excite the Earth, which is in the transmission of electrical energy what the telescope is in astronomical observation. By the use of this marvelous device he has already set up electrical movements of greater intensity than those of lightning and passed a current, sufficient to light more than two hundred incandescent lamps, around the Globe.

3. *The "Tesla Wireless System."* This system comprises a number of improvements and is the only means known for transmitting economically electrical energy to a distance without wires. Careful tests and measurements in connection with an experimental station of great activity, erected by the inventor in Colorado, have demonstrated that power in any desired amount can be conveyed, clear across the Globe if necessary, with a loss not exceeding a few percent.

4. *"The Art of Individualization."* This invention of Tesla is to primitive "tuning" what refined language is to unarticulated expression. It makes possible the transmission of signals or messages absolutely secret and exclusive both in the active and passive aspect, that is, non-interfering as well as non-interferable. Each signal is like an individual of unmistakable identity and there is virtually no limit to the number of stations or instruments which can be simultaneously operated without the slightest mutual disturbance.

5. *"The Terrestrial Stationary Waves."* This wonderful discovery, popularly explained, means that the Earth is responsive to electrical vibrations of definite pitch just as a tuning fork to certain waves of sound. These particular electrical vibrations, capable of powerfully exciting the Globe, lend themselves to innumerable uses of great importance commercially and in many other respects.

The first "World-System" power plant can be put in operation in nine months. With this power plant it will be practicable to attain electrical activities up to ten million horsepower and it is designed to serve for as many technical achievements as are possible without due expense. Among these the following may be mentioned:

(1) The inter-connection of the existing telegraph
exchanges or offices all over the world;

(2) The establishment of a secret and non-interferable
government telegraph service;

(3) The inter-connection of all the present telephone
exchanges or offices on the Globe;

(4) The universal distribution of general news, by
telegraph or telephone, in connection with the Press;

(5) The establishment of such a "World-System" of
intelligence transmission for exclusive private use;

(6) The inter-connection and operation of all stock tickers
of the world;

(7) The establishment of a World-System of musical
distribution, etc.;

(8) The universal registration of time by cheap clocks
indicating the hour with astronomical precision and
requiring no attention whatever;

(9) The world transmission of typed or handwritten characters, letters, checks, etc.;

(10) The establishment of a universal marine service enabling the navigators of all ships to steer perfectly without compass, to determine the exact location, hour and speed, to prevent collisions and disasters, etc.;

(11) The inauguration of a system of world-printing on land and sea;

(12) The world reproduction of photographic pictures and all kinds of drawings or records.

I also proposed to make demonstrations in the wireless transmission of power on a small scale but sufficient to carry conviction. Besides these I referred to other and incomparably more important applications of my discoveries which will be disclosed at some future date.

A plant was built on Long Island with a tower 187 feet high, having a spherical terminal about 68 feet in diameter. These dimensions were adequate for the transmission of virtually any amount of energy. Originally only from 200 to 300 K.W. were provided but I intended to employ later several thousand horsepower. The transmitter was to emit a wave-complex of special characteristics and I had devised a unique method of telephonic control of any amount of energy.

The tower was destroyed two years ago but my projects are being developed and another one, improved in some features, will be constructed. On this occasion I would contradict the widely circulated report that the structure was demolished by the Government which owing to war conditions, might have created prejudice in the minds of those who may not know that the papers, which thirty years ago conferred upon me the honor of American citizenship, are always kept in a safe, while my orders, diplomas, degrees, gold medals and other distinctions are packed away in old trunks. If this report had a foundation I would have been refunded a large sum of money which I expended in the construction of the tower. On the contrary it was in the interest of the Government to preserve it, particularly as it would have made possible—to mention just one valuable result—the location of a submarine in any part of the world. My plant, services, and all my improvements have always been at the disposal of the officials and ever since the outbreak of the European conflict I have been working at a sacrifice on several inventions of mine relating to aerial navigation, ship propulsion and wireless transmission which are of the greatest importance to the country. Those who are well informed know that my ideas have revolutionized the industries of the United States and I am not aware that there lives an inventor who has been, in this respect, as fortunate as myself especially as regards the use of his improvements in the war. I have refrained from publicly expressing myself on this subject before as it seemed improper to dwell on personal matters while all the world was in dire trouble. I would add further, in view of various rumors which have reached me, that Mr. J. Pierpont Morgan did not interest himself with me in a business way but in the same large spirit in which he has assisted many other pioneers. He carried out his generous promise to the letter and it would have been most unreasonable to expect from him anything more. He had the highest regard for my attainments and gave me every evidence of his complete faith in my ability to ultimately achieve what I had set out to do. I am unwilling to accord to some small-minded and jealous individuals the satisfaction of having thwarted my efforts. These men are to me nothing more than microbes of a nasty disease. My project was retarded by laws of nature. The world was not prepared for it. It was too far ahead of time. But the same laws will prevail in the end and make it a triumphal success.

THE ART OF TELAUTOMATICS

In this article, Dr. Tesla dwells on the future possibilities of his magnifying transmitter, especially in connection with the art of Telautomatics, which was first conceived by him and doubtless constitutes one of his most brilliant gifts to the world.

Tesla was the first to build and successfully operate Automata in the form of boats steered and otherwise controlled by tuned wireless circuits and agents ensuring reliable action despite of all attempts to interfere.

But this was only the first step in the evolution of his invention. What he wanted, was to produce machines capable of acting as though possessed of intelligence. It will be readily perceived that if Dr. Tesla has practically realized his conception, the world will witness a revolution in every field of endeavor. In particular will his inventions affect the art of warfare and the peace of the world.

Dr. Tesla dwells eloquently on a number of topics agitating the public mind, and this article of his is perhaps the most brilliant and absorbing he has written.

HOW TESLA'S MIND RECUPERATES

No subject to which I have ever devoted myself has called for such concentration of mind and strained to so dangerous a degree the finest fibers of my brain as the system of which the Magnifying Transmitter is the foundation. I put all the intensity and vigor of youth in the development of the rotating field discoveries, but those early labors were of a different character. Although strenuous in the extreme, they did not involve that keen and exhausting discernment which had to be exercised in attacking the many puzzling problems of the wireless. Despite my rare physical endurance at that period the abused nerves finally rebelled and I suffered a complete collapse, just as the consummation of the long and difficult task was almost in sight. Without doubt I would have paid a greater penalty later, and very likely my career would have been prematurely terminated, had not providence equipt me with a safety device, which has seemed to improve with advancing years and unfailingly comes into play when my forces are at an end. So long as it operates I am safe from danger, due to overwork, which threatens other inventors and, incidentally, I need no vacations which are indispensable to most people. When I am all but used up I simply do as the darkies, who "naturally fall asleep while white folks worry." To venture a theory out of my sphere—the body probably accumulates little by little a definite quantity of some toxic agent and I sink into a nearly lethargic state which lasts half an hour to the minute. Upon awakening I have the sensation as though the events immediately preceding had occurred very long ago, and if I attempt to continue the interrupted train of thought I feel a veritable mental nausea. Involuntarily I then turn to other work and am surprised at the freshness of the mind and ease with which I overcome obstacles that had baffled me before.

After weeks or months my passion for the temporarily abandoned invention returns and I invariably find answers to all the vexing questions with scarcely any effort. In this connection I will tell of an extraordinary experience which may be of interest to students of psychology. I had produced a striking phenomenon with my grounded transmitter and was endeavoring to ascertain its true significance in relation to the currents propagated through the earth. It seemed a hopeless undertaking, and for more than a year I worked unremittingly, but in vain. This profound study so entirely absorbed me that I became forgetful of everything else, even of my undermined health. At last, as I was at the point of breaking down, nature applied the preservative inducing lethal sleep. Regaining my senses, I realized with consternation that I was unable to visualize scenes from my life except those of infancy, the very first ones that had entered my consciousness. Curiously enough, these appeared before my vision with startling distinctness and afforded me welcome relief. Night after night, when retiring, I would think of them and more and more of my previous existence was revealed. The image of my mother was always the principal figure in the spectacle that slowly unfolded, and a consuming desire to see her again gradually took possession of me.

This feeling grew so strong that I resolved to drop all work and satisfy my longing. But I found it too hard to break away from the laboratory, and several months elapsed during which I had succeeded in reviving all the impressions of my past life up to the spring of 1892. In the next picture that came out of the mist of oblivion, I saw myself at the *Hotel de la Paix* in Paris just coming to from one of my peculiar sleeping spells, which had been caused by prolonged exertion of the brain. Imagine the pain and distress I felt when it flashed upon my mind that a dispatch was handed to me at that very moment bearing the sad news that my mother was dying; I remembered how I made the long journey home without an hour of rest and how she passed away after weeks of agony! It was especially remarkable that during all this period of partially obliterated memory I was fully alive to everything touching on the subject of my research. I could recall the smallest details and the least insignificant observations in my experiments and even recite pages of text and complex mathematical formulae.

My belief is firm in a law of compensation. The true rewards are ever in proportion to the labor and sacrifices made. This is one of the reasons why I feel certain that of all my inventions, the Magnifying Transmitter will prove most important and valuable to future generations. I am prompted to this prediction not so much by thoughts of the commercial and industrial revolution which it will surely bring about, but of the humanitarian consequences of the many achievements it makes possible. Considerations of mere utility weigh little in the balance against the higher benefits of civilization. We are confronted with portentous problems which cannot be solved just by providing for our material existence, however abundantly. On the contrary, progress in this direction is fraught with hazards and perils not less menacing than those born from want and suffering. If we were to release the energy of atoms or discover some other way of developing cheap and unlimited power at any point of the globe this accomplishment, instead of being a blessing, might bring disaster to mankind in giving rise to dissension and anarchy which would ultimately result in the enthronement of the hated regime of force. The greatest good will come from technical improvements tending to unification and harmony, and my wireless transmitter is preeminently such. By its means the human voice and likeness will be reproduced everywhere and factories driven thousands of miles from waterfalls furnishing the power; aerial machines will be propelled around the earth without a stop and the sun's energy controlled to create lakes and rivers for motive purposes and transformation of arid deserts into fertile land. Its introduction for telegraphic, telephonic and similar uses will automatically cut out the statics and all other interferences which at present impose narrow limits to the application of the wireless. This is a timely topic on which a few words might not be amiss.

One of the Telautomatic Boats (Submersible) constructed by Tesla and exhibited by him in 1898. Controlled by wireless without aerials.

Tesla Raps "Static" Men Vigorously

During the past decade a number of people have arrogantly claimed that they had succeeded in doing away with this impediment. I have carefully examined all of the arrangements described and tested most of them long before they were publicly disclosed, but the finding was uniformly negative. A recent official statement from the U.S. Navy may, perhaps, have taught some beguilable news editors how to appraise these announcements at their real worth. As a rule the attempts are based on theories so fallacious that whenever they come to my notice I cannot help thinking in a lighter vein. Quite recently a new discovery was heralded, with a deafening flourish of trumpets, but it proved another case of a mountain bringing forth a mouse. This reminds me of an exciting incident which took place years ago when I was conducting my experiments with currents of high frequency. Steve Brodie had just jumped off the Brooklyn Bridge. The feat has been vulgarized since by imitators, but the first report electrified New York. I was very impressionable then and frequently spoke of the daring printer. On a hot afternoon I felt the necessity of refreshing myself and stepped into one of the popular thirty thousand institutions of this great city where a delicious twelve per cent beverage was served which can now be had only by making a trip to the poor and devastated countries of Europe. The attendance was large and not over-distinguished and a matter was discussed which gave me an admirable opening for the careless remark: "This is what I said when I jumped off the bridge." No sooner had I uttered these words than I felt like the companion of Timotheus in the poem of Schiller. In an instant there was a pandemonium and a dozen voices cried: "It is Brodie!" I threw a quarter on the counter and bolted for the door but the crowd was at my heels with yells: "Stop, Steve!" which must have been misunderstood for many persons tried to hold me up as I ran frantically for my haven of refuge. By darting around corners I fortunately managed—through the medium of a fire-escape—to reach the laboratory, where I threw off my coat, camouflaged myself as a hard-working blacksmith, and started the forge. But these precautions proved unnecessary; I had eluded my pursuers. For many years afterward, at night, when imagination turns into spectres the trifling troubles of the day, I often thought, as I tossed on the bed, what my fate would have been had that mob caught me and found out that I was not Steve Brodie!

Now the engineer, who lately gave an account before a technical body of a novel remedy against statics based on a "heretofore unknown law of nature," seems to have been as reckless as myself when he contended that these disturbances propagate up and down, while those of a transmitter proceed along the earth. It would mean that a condenser, as this globe, with its gaseous envelope, could be charged and discharged in a manner quite contrary to the fundamental teachings propounded in every elemental text-book of physics. Such a supposition would have been condemned as erroneous, even in

Franklin's time, for the facts bearing on this were then well known and the identity between atmospheric electricity and that developed by machines was fully established. Obviously, natural and artificial disturbances propagate through the earth and the air in exactly the same way, and both set up electro-motive forces in the horizontal, as well as vertical, sense. Interference cannot be overcome by any such methods as were proposed. The truth is this: In the air the potential increases at the rate of about fifty volts per foot of elevation, owing to which there may be a difference of pressure amounting to twenty, or even forty thousand volts between the upper and lower ends of the antenna. The masses of the charged atmosphere are constantly in motion and give up electricity to the conductor, not continuously but rather disruptively, this producing a grinding noise in a sensitive telephonic receiver. The higher the terminal and the greater the space encompast by the wires, the more pronounced is the effect, but it must be understood that it is purely local and has little to do with the real trouble. In 1900, while perfecting my wireless system, one form of apparatus comprised four antennae. These were carefully calibrated to the same frequency and connected in multiple with the object of magnifying the action, in receiving from any direction.

When I desired to ascertain the origin of the transmitted impulses, each diagonally situated pair was put in series with a primary coil energizing the detector circuit. In the former case the sound was loud in the telephone; in the latter it ceased, as expected, the two antennae neutralizing each other, but the true statics manifested themselves in both instances and I had to devise special preventives embodying different principles.

The Remedy for Static

By employing receivers connected to two points of the ground, as suggested by me long ago, this trouble caused by the charged air, which is very serious in the structures as now built, is nullified and besides, the liability of all kinds of interference is reduced to about one-half, because of the directional character of the circuit. This was perfectly self-evident, but came as a revelation to some simple-minded wireless

Tesla's new self-propelled aerial tel-automaton. Devoid of propeller, sustaining wings and all other means of external control. Can attain a speed of 350 miles per hour, and will reach a predetermined point a thousand miles away accurately within a few feet.

folks whose experience was confined to forms of apparatus that could have been improved with an axe, and they have been disposing of the bear's skin before killing him. If it were true that strays performed such antics, it would be easy to get rid of them by receiving without aerials. But, as a matter of fact, a wire buried in the ground which, conforming to this view, should be absolutely immune, is more susceptible to certain extraneous impulses than one placed vertically in the air. To state it fairly, a slight progress has been made, but not by virtue of any particular method or device. It was achieved simply by discarding the enormous structures, which are bad enough for transmission but wholly unsuitable for reception, and adopting a more appropriate type of receiver. As I pointed out in a previous article, to dispose of this difficulty for good, a radical change must be made in the system, and the sooner this is done the better.

Radio Government Control Not Wanted

It would be calamitous, indeed, if at this time when the art is in its infancy and the vast majority, not excepting even experts, have no conception of its ultimate possibilities, a measure would be rushed through the legislature making it a government monopoly. This was proposed a few weeks ago by Secretary Daniels, and no doubt that distinguished official has made his appeal to the Senate and House of Representatives with sincere conviction. But universal evidence unmistakably shows that the best results are always obtained in healthful commercial competition. There are, however, exceptional reasons why wireless should be given the fullest freedom of development. In the first place it offers prospects immeasurably greater and more vital to betterment of human life than any other invention or discovery in the history of man. Then again, it must be understood that this wonderful art has been, in its entirety, evolved here and can be called "American" with more right and propriety than the telephone, the incandescent lamp or the aeroplane. Enterprising press agents and stock jobbers have been so successful in spreading misinformation that even so excellent a periodical as the *Scientific American* accords the chief credit to a foreign country. The Germans, of course, gave us the Hertz-waves and the Russian, English, French and Italian experts were quick in using them for signaling purposes. It was an obvious application of the new agent and accomplished with the old classical and unimproved induction coil—scarcely anything more than another kind of heliography. The radius of transmission was very limited, the results attained of little value, and the Hertz oscillations, as a means for conveying intelligence, could have been advantageously replaced by sound-waves, which I advocated in 1891.

Moreover, all of these attempts were made three years after the basic principles of the wireless system, which is universally employed to-day, and its potent instrumentalities had been clearly described and developed in America. No trace of those Hertzian appliances and methods remains today. We have proceeded in the very opposite direction and what has been done is the product of the brains and efforts of citizens of this country. The fundamental patents have expired and the opportunities are open to all. The chief argument of the Secretary is based on interference. According to his statement, reported in the *New York Herald* of July 29th, signals from a powerful station can be intercepted in every village of the world. In view of this fact, which was demonstrated in my experiments of 1900, it would be of little use to impose restrictions in the United States.

America First

As throwing light on this point, I may mention that only recently an odd looking gentleman called on me with the object of enlisting my services in the construction of world transmitters in some distant land. "We have no money," he said, "but carloads of solid gold and we will give you a liberal amount." I told him that I wanted to see first what will be done with my inventions in America, and this ended the interview. But I am satisfied that some dark forces are at work, and as time goes on the maintenance of continuous communication will be rendered more difficult. The only remedy is a system immune against interruption. It has been perfected, it exists, and all that is necessary is to put it in operation.

The terrible conflict is still uppermost in the minds and perhaps the greatest importance will be attached to the Magnifying

Transmitter as a machine for attack and defense, more particularly in connection with Telautomatics. This invention is a logical outcome of observations begun in my boyhood and continued thruout my life. When the first results were publisht the *Electrical Review* stated editorially that it would become one of the "most potent factors in the advance and civilization of mankind." The time is not distant when this prediction will be fulfilled. In 1898 and 1900 it was offered to the Government and might have been adopted were I one of those who would go to Alexander's shepherd when they want a favor from Alexander. At that time I really thought that it would abolish war, because of its unlimited destructiveness and exclusion of the personal element of combat. But while I have not lost faith in its potentialities, my views have changed since.

The Road to Permanent Peace

War cannot be avoided until the physical cause for its recurrence is removed and this, in the last analysis, is the vast extent of the planet on which we live. Only thru annihilation of distance in every respect as, the conveyance of intelligence, transport of passengers and supplies and transmission of energy will conditions be brought about some day, insuring permanency of friendly relations. What we now want most is closer contact and better understanding between individuals and communities all over the earth, and the elimination of that fanatic devotion to exalted ideals of national egoism and pride which is always prone to plunge the world into primeval barbarism and strife. No league or parliamentary act of any kind will ever prevent such a calamity. These are only new devices for putting the weak at the mercy of the strong. I have exprest myself in this regard fourteen years ago, when a combination of a few leading governments—a sort of Holy Alliance—was advocated by the late Andrew Carnegie, who may be fairly considered as the father of this idea, having given to it more publicity and impetus than anybody else prior to the efforts of the President. While it cannot be denied that such a pact might be of material advantage to some less fortunate peoples, it cannot attain the chief object sought. Peace can only come as a natural consequence of universal enlightenment and merging of races, and we are still far from this blissful realization. As I view the world of today, in the light of the gigantic struggle we have witnest, I am filled with conviction that the interests of humanity would be best served if the United States remained true to its traditions and kept out of "entangling alliances." Situated as it is, geographically, remote from the theaters of impending conflicts, without incentive to territorial aggrandizement, with inexhaustible resources and immense population thoroly imbued with the spirit of liberty and right, this country is placed in a unique and privileged position. It is thus able to exert, independently, its colossal strength and moral force to the benefit of all, more judiciously and effectively, than as member of a league.

THE MECHANISTIC THEORY OF LIFE

In one of these biographical sketches, published in the *Electrical Experimenter*, I have dwelt on the circumstances of my early life and told of an affliction which compelled me to unremitting exercise of imagination and self-observation. This mental activity, at first involuntary under the pressure of illness and suffering, gradually became second nature and led me finally to recognize that I was but an automaton devoid of free will in thought and action and merely responsive to the forces of the environment. Our bodies are of such complexity of structure, the motions we perform are so numerous and involved, and the external impressions on our sense organs to such a degree delicate and elusive that it is hard for the average person to grasp this fact. And yet nothing is more convincing to the trained investigator than the mechanistic theory of life which had been, in a measure, understood and propounded by Descartes three hundred years ago. But in his time many important functions of our organism were unknown and, especially with respect to the nature of light and the construction and operation of the eye, philosophers were in the dark. In recent years the progress of scientific research in these fields has been such as to leave no room for a doubt in regard to this view on which many works have been published. One of its ablest and most eloquent exponents is, perhaps, Felix Le Dantec, formerly assistant of

Pasteur. Prof. Jacques Loeb has performed remarkable experiments in heliotropism, clearly establishing the controlling power of light in lower forms of organisms, and his latest book, *Forced Movements,* is revelatory. But while men of science accept this theory simply as any other that is recognized, to me it is a truth which I hourly demonstrate by every act and thought of mine. The consciousness of the external impression prompting me to any kind of exertion, physical or mental, is ever present in my mind. Only on very rare occasions, when I was in a state of exceptional concentration, have I found difficulty in locating the original impulses.

LACK OF OBSERVATION A FORM OF IGNORANCE

The by far greater number of human beings are never aware of what is passing around and within them, and millions fall victims of disease and die prematurely just on this account. The commonest, every-day occurrences appear to them mysterious and inexplicable. One may feel a sudden wave of sadness and rake his brain for an explanation when he might have noticed that it was caused by a cloud cutting off the rays of the sun. He may see the image of a friend dear to him under conditions which he construes as very peculiar, when only shortly before he has passed him in the street or seen his photograph somewhere. When he loses a collar button he fusses and swears for an hour, being unable to visualize his previous actions and locate the object directly. Deficient observation is merely a form of ignorance and responsible for the many morbid notions and foolish ideas prevailing.

There is not more than one out of every ten persons who does not believe in telepathy and other psychic manifestations, spiritualism and communion with the dead, and who would refuse to listen to willing or unwilling deceivers. Just to illustrate how deeply rooted this tendency has become even among the clear-headed American population, I may mention a comical incident.

Dr. Tesla is rapidly becoming younger.
Judge for yourself from his latest photograph.

Psychic Phenomena
in the Manufacture of Flivvers

Shortly before the war, when the exhibition of my turbines in this city elicited widespread comment in the technical papers, I anticipated that there would be a scramble among manufacturers to get hold of the invention, and I had particular designs on that man from Detroit who has an uncanny faculty for accumulating millions. So confident was I that he would turn up some day, that I declared this as certain to my secretary and assistants. Sure enough, one fine morning a body of engineers from the Ford Motor Company presented themselves with the request of discussing with me an important project. "Didn't I tell you?" I remarked triumphantly to my employees, and one of them said, "You are amazing, Mr. Tesla; everything comes out exactly as you predict." As soon as these hard-headed men were seated I, of course, immediately began to extol the wonderful features of my turbine, when the spokesmen interrupted me and said, "We know all about this, but we are on a special errand. We have formed a psychological society for the investigation of psychic phenomena and we want you to join us in this undertaking." I suppose those engineers never knew how near they came to being fired out of my office.

Confuting Spiritism

Ever since I was told by some of the greatest men of the time, leaders in science whose names are immortal, that I am possest of an unusual mind, I bent all my thinking faculties on the solution of great problems regardless of sacrifice. For many years I endeavored to solve the enigma of death, and watched eagerly for every kind of spiritual indication. But only once in the course of my existence have I had an experience which momentarily impressed me as supernatural. It was at the time of my mother's death. I had become completely exhausted by pain and long vigilance, and one night was carried to a building about two blocks from our home. As I lay helpless there, I thought that if my mother died while I was away from her bedside she would surely give me a sign. Two or three months before I was in London in company with my late friend, Sir William Crookes, when spiritualism was discussed, and I was under the full sway of these thoughts. I might not have paid attention to other men, but was susceptible to his arguments as it was his epochal work on radiant matter, which I had read as a student, that made me embrace the electrical career. I reflected that the conditions for a look into the beyond were most favorable, for my mother was a woman of genius and particularly excelling in the powers of intuition. During the whole night every fiber in my brain was strained in expectancy, but nothing happened until early in the morning, when I fell in a sleep, or perhaps a swoon, and saw a cloud carrying angelic figures of marvelous beauty, one of whom gazed upon me lovingly and gradually assumed the features of my mother. The appearance slowly floated across the room and vanished, and I was awakened by an indescribably sweet song of many voices. In that instant a certitude, which no words can express, came upon me that my mother had just died. And that was true. I was unable to understand the tremendous weight of the painful knowledge I received in advance, and wrote a letter to Sir William Crookes while still under the domination of these impressions and in poor bodily health. When I recovered I sought for a long time the external cause of this strange manifestation and, to my great relief, I succeeded after many months of fruitless effort. I had seen the painting of a celebrated artist, representing allegorically one of the seasons in the form of a cloud with a group of angels which seemed to actually float in the air, and this had struck me forcefully. It was exactly the same that appeared in my dream, with the exception of my mother's likeness. The music came from the choir in the church nearby at the early mass of Easter morning, explaining everything satisfactorily in conformity with scientific facts.

This occurred long ago, and I have never had the faintest reason since to change my views on psychical and spiritual phenomena, for which there is absolutely no foundation. The belief in these is the natural outgrowth of intellectual development. Religious dogmas are no longer accepted in their orthodox meaning, but every individual clings to faith in a supreme power of some kind. We all must have an ideal to govern our conduct and insure contentment, but it is immaterial whether it be one of creed, art, science or anything else, so long as it fulfills the function of a dematerializing force. It is essential to the peaceful existence of humanity as a whole that one common conception should prevail.

Tesla's Astounding Discovery

While I have failed to obtain any evidence in support of the contentions of psychologists and spiritualists, I have proved to my complete satisfaction the automatism of life, not only through continuous observations of individual actions, but even more conclusively through certain generalizations. These amount to a discovery which I consider of the greatest moment to human society, and on which I shall briefly dwell. I got the first inkling of this astounding truth when I was still a very young man, but for many years I interpreted what I noted simply as coincidences. Namely, whenever either myself or a person to whom I was attached, or a cause to which I was devoted, was hurt by others in a particular way, which might be best popularly characterized as the most unfair imaginable, I experienced a singular and undefinable pain which, for want of a better term, I have qualified as "cosmic," and shortly thereafter, and invariably, those who had inflicted it came to grief. After many such cases I confided this to a number of friends, who had the opportunity to convince themselves of the truth of the theory which I have gradually formulated and which may be stated in the following few words: Our bodies are of similar construction and exposed to the same external influences.

This results in likeness of response and concordance of the general activities on which all our social and other rules and laws are based. We are automata entirely controlled by the forces of the medium, being tossed about like corks on the surface of the water, but mistaking the resultant of the impulses from the outside for free will. The movements and other actions we perform are always life preservative and tho seemingly quite independent from one another, we are connected by invisible links. So long as the organism is in perfect order it responds accurately to the agents that prompt it, but the moment that there is some derangement in any individual, his self-preservative power is impaired. Everybody understands, of course, that if one becomes deaf, has his eyesight weakened, or his limbs injured, the chances for his continued existence are lessened. But this is also true, and perhaps more so, of certain defects in the brain which deprive the automaton, more or less, of that vital quality and cause it to rush into destruction. A very sensitive and observant being, with his highly developed mechanism all intact, and acting with precision in obedience to the changing conditions of the environment, is endowed with a transcending mechanical sense, enabling him to evade perils too subtle to be directly perceived. When he comes in contact with others whose controlling organs are radically faulty, that sense asserts itself and he feels the "cosmic" pain. The truth of this has been borne out in hundreds of instances and I am inviting other students of nature to devote attention to this subject, believing that thru combined and systematic effort results of incalculable value to the world will be attained.

DR. TESLA'S FIRST AUTOMATON

The idea of constructing an automaton, to bear out my theory, presented itself to me early but I did not begin active work until 1893, when I started my wireless investigations. During the succeeding two or three years a number of automatic mechanisms, to be actuated from a distance, were constructed by me and exhibited to visitors in my laboratory. In 1896, however, I designed a complete machine capable of a multitude of operations, but the consummation of my labors was delayed until late in 1897. This machine was illustrated and described in my article in the *Century Magazine* of June, 1900, and other periodicals of that time and, when first shown in the beginning of 1898, it created a sensation such as no other invention of mine has ever produced. In November, 1898, a basic patent on the novel art was granted to me, but only after the Examiner-in-Chief had come to New York and witnesst the performance, for what I claimed seemed unbelievable. I remember that when later I called on an official in Washington, with a view of offering the invention to the Government, he burst out in laughter upon my telling him what I had accomplished. Nobody thought then that there was the faintest prospect of perfecting such a device. It is unfortunate that in this patent, following the advice of my attorneys, I indicated the control as being effected thru the medium of a single circuit and a well-known form of detector, for the reason that I had not yet secured protection on my methods and apparatus for individualization. As a matter of fact, my boats were controlled thru the joint action of several circuits and interference of every kind was excluded. Most generally I employed receiving circuits in the form of loops, including condensers, because the discharges of my high-tension transmitter ionized the air in the hall so that even a very small aerial would draw electricity from the surrounding atmosphere for hours. Just to give an idea, I found, for instance, that a bulb 12" in diameter, highly exhausted, and with one single terminal to which a short wire was attached, would deliver well on to one thousand successive flashes before all charge of the air in the laboratory was neutralized. The loop form of receiver was not sensitive to such a disturbance and it is curious to note that it is becoming popular at this late date. In reality it collects much less energy than the aerials or a long grounded wire, but it so happens that it does away with a number of defects inherent to the present wireless devices. In demonstrating my invention before audiences, the visitors were requested to ask any questions, however involved, and the automaton would answer them by signs. This was considered magic at that time but was extremely simple, for it was myself who gave the replies by means of the device.

At the same period another larger telautomatic boat was constructed, a photograph of which is shown in this number of the *Electrical Experimenter*. It was controlled by loops, having several turns placed in the hull, which was made entirely water-tight and capable of submergence. The apparatus was similar to that used in the first with the exception of certain special features I introduced as, for example, incandescent lamps which afforded a visible evidence of the proper functioning of the machine.

TELAUTOMATICS OF THE FUTURE

These automata, controlled within the range of vision of the operator, were, however, the first and rather crude steps in the evolution of the Art of Telautomatics as I had conceived it. The next logical improvement was its application to automatic mechanisms beyond the limits of vision and at great distance from the center of control, and I have ever since advocated their employment as instruments of warfare in preference to guns. The importance of this now seems to be recognized, if I am to judge from casual announcements thru the press of achievements which are said to be extraordinary but contain no merit of novelty, whatever. In an imperfect manner it is practicable, with the existing wireless plants, to launch an aeroplane, have it follow a certain approximate course, and perform some operation at a distance of many hundreds of miles. A machine of this kind can also be mechanically controlled in several ways and I have no doubt that it may prove of some usefulness in war. But there are, to my best knowledge, no instrumentalities in existence today with which such an object could be accomplished in a precise manner. I have devoted years of study to this matter and have evolved means, making such and greater wonders easily realizable. As stated on a previous occasion, when I was a student at college I conceived a flying machine quite unlike the present ones. The underlying principle was sound but could not be carried into practice for want of a prime-mover of sufficiently great activity. In recent years I have successfully solved this problem and am now planning aerial machines devoid of sustaining planes, ailerons, propellers and other external attachments, which will be capable of immense speeds and are very likely to furnish powerful arguments for peace in the near future. Such a machine, sustained and propelled entirely by reaction, is shown on one of the pages and is supposed to be controlled either mechanically or by wireless energy. By installing proper plants it will be practicable to project a missile of this kind into the air and drop it almost on the very spot designated, which may be thousands of miles away. But we are not going to stop at this. Teleautomata will be ultimately produced, capable of acting as if possesst of their own intelligence, and their advent will create a revolution. As early as 1898 I proposed to representatives of a large manufacturing concern the construction and public exhibition of an automobile carriage which, left to itself, would perform a great variety of operations involving something akin to judgment. But my proposal was deemed chimerical at that time and nothing came from it.

At present many of the ablest minds are trying to devise expedients for preventing a repetition of the awful conflict which is only theoretically ended and the duration and main issues of which I have correctly predicted in an article printed in the *Sun* of December 20, 1914. The proposed League is not a remedy but, on the contrary, in the opinion of a number of competent men, may bring about results just the opposite. It is particularly regrettable that a punitive policy was adopted in framing the terms of peace, because a few years hence it will be possible for nations to fight without armies, ships or guns, by weapons far more terrible, to the destructive action and range of which there is virtually no limit. Any city, at a distance, whatsoever, from the enemy, can be destroyed by him and no power on earth can stop him from doing so. If we want to avert an impending calamity and a state of things which may transform this globe into an inferno, we should push the development of flying machines and wireless transmission of energy without an instant's delay and with all the power and resources of the nation.

Other Articles

FROM *ELECTRICAL EXPERIMENTER* MAGAZINE

Famous Scientific Illusions
February 1919

*In this original and revolutionizing discussion, Nikola Tesla gives us
something really new to think about. First—Does the moon rotate
on its axis? Second—Is the Franklin pointed lightning rod correct in
theory and operation? Third—Do wireless signals fly thru space by
means of so-called Hertzian waves in the ether, or are they propagated
thru the earth at prodigious velocity by means of earth-bound
oscillations? World-famous conundrums these—questions which have
been answered in many ways by some of the greatest scientists. Dr.
Tesla explains these three predominant scientific fallacies in a masterly
way, so that everyone can understand them.*

*For over a century and a half the whole world, educated and
otherwise, thought that the moon revolved around its axis. Nikola
Tesla in the present highly instructive article disproves that theory
and will convince scientists and all others alike that the moon does
no such thing.*

*For thousands of years it was thought that the sun and stars
revolved around the earth and all kinds of experimental proofs were
furnished to substantiate this theory. The illustrious Galileo thought
different, and everyone today knows that the earth revolves around
the sun.*

*So it is with Tesla's discovery. Tesla also, in the second part of the
present paper, shows us that the ancient and time-worn theory advanced
by Benjamin Franklin as to the lightning conductor is not substantially
correct as viewed by latter day science. It will come as a shock even
to our professors that the lightning rod actually aids the lightning
in hitting the building. The reason is that the lightning rod helps in
ionizing (making conductive) the surrounding air. Mr. Tesla has devised
a lightning conductor with no points, and there is no doubt whatsoever
that his theory is right. Scientists the world over will acknowledge this
very shortly. In a third section of the same paper Tesla explodes still
another popular delusion, viz., that wireless waves follow the curvature
of the earth when messages are transmitted, let us say from a point in*

the United States to a point in Europe. In his revolutionary arguments, supported by facts as well as by logic, Tesla shows why the currents do not travel around the earth but directly thru it. In other words, Tesla maintains that wireless communication is accomplished only thru the medium of the earth itself. His contention seems very sound. If it were not so, let every wireless station, commercial or otherwise, do away with its ground connection. None could then operate as is well known, except perhaps over very limited distances.

Mr. Tesla's present article will arouse world-wide comment due to the revolutionary philosophy contained therein. We are sure our readers will appreciate Mr. Tesla's most timely and illuminating article on this but little understood subject.

The human brain, with all its wonderful capabilities and power, is far from being a faultless apparatus. Most of its parts may be in perfect working order, but some are atrophied, undeveloped or missing altogether. Great men of all classes and professions—scientists, inventors, and hard-headed financiers—have placed themselves on record with impossible theories, inoperative devices, and unrealizable schemes. It is doubtful that there could be found a single work of any one individual free of error. There is no such thing as an infallible brain. Invariably, some cells or fibers are wanting or unresponsive, with the result of impairing judgment, sense of proportion, or some other faculty. A man of genius eminently practical, whose name is a household word, has wasted the best years of his life in a visionary undertaking. A celebrated physicist was incapable of tracing the direction of an electric current according to a childishly simple rule. The writer, who was known to recite entire volumes by heart, has never been able to retain in memory and recapitulate in their proper order the words designating the colors of the rainbow, and can only ascertain them after long and laborious thought, strange as it may seem.

Our organs of reception, too, are deficient and deceptive. As a semblance of life is produced by a rapid succession of inanimate pictures, so many of our perceptions are but trickery of the senses, devoid of reality. The greatest triumphs of man were those in which his mind had to free itself from the influence of delusive appearances. Such was the revelation of Buddha that self is an illusion caused by the persistence and continuity of mental images: the discovery of Copernicus that, contrary to all observation, this planet rotates around the sun; the recognition of Descartes that the human being is an automaton, governed by external influence and the idea that the earth is spherical, which led Columbus to the finding of this continent. And tho the minds of individuals supplement one another and science and experience are continually eliminating fallacies and misconceptions, much of our present knowledge is still incomplete and unreliable. We have sophisms in mathematics which cannot be disproved. Even in pure reasoning, free of the short-comings of symbolic processes, the arc often arrested by doubt which the strongest intelligences have been unable to dispel. Experimental science itself, most positive of all, is not unfailing. In the following I shall consider three exceptionally interesting errors in the interpretation and application of physical phenomena which have for years dominated the minds of experts and men of science.

I. The Illusion of the Axial Rotation of the Moon

It is well known since the discovery of Galileo that the moon, in travelling thru space, always turns the same face towards the earth. This is explained by stating that while passing once around its mother-planet the lunar globe performs just one revolution on its axis. The spinning motion of a heavenly body must necessarily undergo modifications in the course of time, being either retarded by resistances internal or external, or accelerated owing to shrinkage and other causes. An unalterable rotational velocity thru all phases of planetary evolution is manifestly impossible. What wonder, then, that at this very instant of its long existence our satellite should revolve exactly so, and not faster or slower. But many astronomers have accepted as a physical fact that such rotation takes place. It does not, but only appears so; it is an illusion, a most surprising one, too. I will endeavor to make this clear by reference to Fig. 1, in which E represents the earth and M the moon. The movement thru space is such that the arrow, firmly attached to the latter, always occupies the position indicated with reference to the earth. If one imagines himself as looking down on the orbital plane and follows the motion he will become convinced that the moon *does* turn on its axis as it travels around. But in this very act the observer will have deceived himself. To make the delusion complete let him take a washer similarly marked and supporting it rotatably in the center, carry it around a stationary object, constantly keeping the arrow pointing towards the latter. Tho to his bodily vision the disk will revolve on its axis, such movement does not exist. He can dispel the illusion at once by holding the washer fixedly while going around. He will now readily see that the supposed axial rotation is only apparent, the impression being produced by successive changes of position in space.

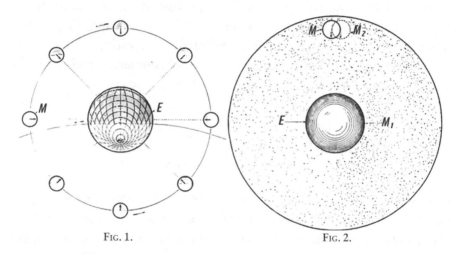

FIG. 1.   FIG. 2.

FIG. 1. It is well known that the moon, M, always turns the same face toward the earth, E, as the black arrows indicate. The parallel rays from the sun illuminate the moon in its successive orbital positions as the unshaded semi-circles indicate. Bearing this in mind, do you believe that the moon rotates on its own axis?

FIG. 2. Tesla's conception of the rotation of the moon, M, around the earth, E; the moon, in this demonstration hypothesis, being considered as embedded in a solid Mass, M_1. If, as commonly believed, the moon rotates, this would be equally true for a portion of the mass M_1, and the part common to both bodies would turn simultaneously in "opposite" directions.

But more convincing proofs can be given that the moon does not, and cannot, revolve on its axis. With this object in view attention is called to Fig. 2, in which both the satellite, M, and earth, E, are shown embedded in a solid mass, M_1 (indicated by stippling) and supposed to rotate so as to impart to the moon its normal translator velocity. Evidently, if the lunar globe could rotate as commonly believed, this would be equally true of any other portion of mass M_1, as the sphere M_2, shown in dotted lines, and then the part common to both bodies would have to turn *simultaneously in opposite directions*. This can be experimentally illustrated in the manner suggested by using instead of one, two overlapping rotatable washers, as may be conveniently represented by circles M and M_2, and carrying them around a center as E, so that the plain and dotted arrows are always pointing towards the same center. No further argument is needed to demonstrate that the two gyrations cannot co-exist or even be pictured in the imagination and reconciled in a purely abstract sense.

The truth is, the so-called "axial rotation" of the moon is a phenomenon deceptive alike to the eye and mind and devoid of physical meaning. It has nothing in common with real mass revolution characterized by effects positive and unmistakable. Volumes have been written on the subject and many erroneous arguments advanced in support of the notion. Thus, it is reasoned, that if the planet did *not* turn on its axis it would expose the whole surface to terrestrial view; as only one-half is visible, it *must* revolve. The first statement is true but the logic of the second is defective, for it admits of only one alternative. The conclusion is not justified as the same appearance can also be produced in another way. The moon does rotate, not on its own, but about an axis passing thru the center of the earth, the true and only one.

The unfailing test of the spinning of a mass is, however, the existence of energy of motion. The moon is not possest of such *vis viva*. If it were the case then a revolving body as M_1 would contain mechanical energy other than that of which we have experimental evidence. Irrespective of this so exact a coincidence between the axial and orbital periods is, in itself, immensely improbable for this is not the permanent condition towards which the system is tending. Any axial rotation of a mass left to itself, retarded by forces external or internal, must cease. Even admitting its perfect control by tides the coincidence would still be miraculous. But when we remember that most of the satellites exhibit this peculiarity, the probability becomes infinitestimal.

Three theories have been advanced for the origin of the moon. According to the oldest suggested by the great German philosopher Kant, and developed by Laplac in his monumental treatise "Mécanique

Céleste," the planets have been thrown off from larger central masses by centrifugal force. Nearly forty years ago Prof. George H. Darwin in a masterful essay on tidal friction furnished mathematical proofs, deemed unrefutable, that the moon had separated from the earth. Recently this established theory has been attacked by Prof. T. J. J. See in a remarkable work on the "Evolution of the Stellar Systems," in which he propounds the view that centrifugal force was altogether inadequate to bring about the separation and that all planets, including the moon, have come from the depths of space and have been captured. Still a third hypothesis of unknown origin exists which has been examined and commented upon by Prof. W. H. Pickering in "Popular Astronomy of 1907," and according to which the moon was torn from the earth when the latter was partially solidified, this accounting for the continents which might not have been formed otherwise. Undoubtedly planets and satellites have originated in both ways and, in my opinion, it is not difficult to ascertain the character of their birth. The following conclusions can be safely drawn:

1. A heavenly body thrown off from a larger one cannot rotate on its axis. The mass, rendered fluid by the combined action of heat and pressure, upon the reduction of the latter immediately stiffens, being at the same time deformed by gravitational pull. The shape becomes permanent upon cooling and solidification and the smaller mass continues to move about the larger one as tho it were rigidly connected to it except for pendular swings or librations due to varying orbital velocity. Such motion precludes the possibility of axial rotation in the strictly physical sense. The moon has never spun around as is well demonstrated by the fact that the most precise measurements have failed to show any measurable flattening in form.

2. If a planetary body in its orbital movement turns the same side towards the central mass this is a positive proof that it has been separated from the latter and is a true satellite.

3. A planet revolving on its axis in its passage around another cannot have been thrown off from the same but must have been captured.

II. THE FALLACY OF FRANKLIN'S POINTED LIGHTNING ROD

The display of atmospheric electricity has since ages been one of the most marvelous spectacles afforded to the sight of man. Its grandeur and power filled him with fear and superstition. For centuries he attributed lightning to agents god-like and supernatural and its purpose in the scheme of this universe remained unknown to him. Now we have learned that the waters of the ocean are raised by the sun and maintained in the atmosphere delicately suspended, that they are wafted to distant regions of the globe where electric forces assert themselves in upsetting the sensitive balance and causing precipitation, thus sustaining all organic life. There is every reason to hope that man will soon be able to control this life-giving flow of water and thereby solve many pressing problems of his existence.

Atmospheric electricity became of special scientific interest in Franklin's time. Faraday had not yet announced his epochal discoveries in magnetic induction but static frictional machines were already generally used in physical laboratories. Franklin's powerful mind at once leaped to the conclusion that frictional and atmospheric electricity were identical. To our present view this inference appears obvious, but in his time the mere thought of it was little short of blasphemy. He investigated the phenomena and argued that if they were of the same nature then the clouds could be drained of their charge exactly as the ball of a static machine, and in 1749 he indicated in a publisht memoir how this could be done by the use of pointed metal rods.

The earliest trials were made by Dalibrand in France, but Franklin himself was the first to obtain a spark by using a kite, in June, 1752. When these atmospheric discharges manifest themselves today in our wireless station we feel annoyed and wish that they would stop, but to the man who discovered them they brought tears of joy.

The lightning conductor in its classical form was invented by Benjamin Franklin in 1755 and immediately upon its adoption proved a success to a degree. As usual, however, its virtues were often exaggerated. So, for instance, it was seriously claimed that in the city of Piatermaritzburg (capital of Natal, South Africa) no lightning strokes occurred after the pointed rods were installed, altho the storms were as frequent as before. Experience has shown that just the opposite is true. A modern city like New York, presenting innumerable sharp points and projections in good contact with the earth, is struck much more often than equivalent area of land. Statistical records, carefully compiled and publisht from time to time, demonstrate that the danger from lightning to property and life has been reduced to a small percentage by Franklin's invention, but the damage by fire amounts, nevertheless, to several million dollars annually. It is astonishing that this device, which has been in universal use for more than one century and a half, should be found to involve a gross fallacy in design and construction which impairs its usefulness and may even render its employment hazardous under certain conditions.

FIG. 3. Diagram used to explain the fallacy of the Franklin pointed lightning rod, and an analogy whereby the author shows in a clear manner how the charged sphere may for illustration be considered as heated to a high degree, and the heat allowed to escape at a known rate.

For explanation of this curious fact I may first refer to Fig. 3, in which s is a metallic sphere of radius r, such as the capacity terminal of a static machine, provided with a sharply pointed pin of length h, as indicated. It is well known that the latter has the property of quickly dissipating the accumulated charge into the air. To examine this action in the light of present knowledge we may liken electric potential to temperature. Imagine that sphere s is heated to T degrees and that the pin or metal bar is a perfect conductor of heat so that its extreme end is at the same temperature T. Then if another sphere of larger radius, v_1, is drawn about the first and the temperature along this boundary is T_1 it is evident that there will be between the end of the bar and its surrounding a difference of temperature $T - T_1$, which will determine the outflow of heat. Obviously, if the adjacent medium was not affected by the hot sphere this temperature difference would be greater and more heat would be given off. Exactly so in the electric system. Let q be the quantity of the charge, then the sphere—and owing to its great conductivity also the pin—will be at the potential q/r. The medium around the point of the pin will be at the potential $q/r_2 = $ q/(r+h) and, consequently, the difference q/r − q/(r+h) = qh/r(r+h). Suppose now that a sphere s of much larger radius R = nr is employed containing a charge Q this difference of potential will be, analogously Qh/R(R+h). According to elementary principles of electro-statics the potentials of the two spheres s and S will be equal if Q = nq in which case Qh/R(R+h) = nqh/nr(nr+h) = qh/r(nr+h).

FIG. 4. Tesla explains the fallacy of the Franklin pointed lightning rod, here illustrated, and shows that usually such a rod could not draw off the electricity in a single cloud in many years. The density of the dots indicates the intensity of the charges.

Thus the difference of potential between the point of the pin and the medium around the same will be smaller in the ratio (r+h)/(nr+h) when the large sphere is used. In many scientific tests and experiments this important observation has been disregarded with the result of causing serious errors. Its significance is that the behavior of the pointed rod entirely depends on the linear dimensions of the electrified body. Its quality to give off the charge may be entirely lost if the latter is very large. For this reason, all points or projections on the surface of a conductor of such vast dimensions as the earth would be quite ineffective were it not for other influences. These will be elucidated with reference to Fig. 4, in which our artist of the Impressionist school has emphasized Franklin's notion that his rod was drawing electricity from the clouds.

If the earth were not surrounded by an atmosphere which is generally oppositely charged it would behave, despite all its irregularities of surface, like a polished sphere. But owing to the electrified masses of air and cloud the distribution is greatly modified. Thus in Fig. 4, the positive charge of the cloud induces in the earth an equivalent opposite charge, the density at the surface of the latter diminishing with the cube of the distance from the static center of the cloud. A brush discharge is then formed at the point of the rod and the action Franklin anticipated takes place. In addition, the surrounding air is ionized and rendered conducting and, eventually, a bolt may hit the building or some other object in the vicinity. The virtue of the pointed end to dissipate the charge, which was uppermost in Franklin's mind is, however, infinitesimal. Careful measurements show that it would take many years before the electricity stored in a single cloud of moderate size would be drawn off or neutralized thru such a lightning conductor. The grounded rod has the quality of rendering harmless most of the strokes it receives, tho occasionally the charge is diverted with damaging results. But, what is very important to note, it invites danger and hazard on account of the fallacy involved in its design. The sharp point which was thought advantageous and indispensable to its operation, is really a defect detracting considerably from the practical value of the device. I have produced a much improved form of lightning protector characterized by the employment of a terminal of considerable area and large radius of curvature which makes impossible undue density of the charge and ionization of the air. These protectors act as quasi-repellents and so far have never been struck tho exposed a long time. Their safety is experimentally demonstrated to greatly exceed that invented by Franklin. By their use property worth millions of dollars which is now annually lost, can be saved.

III. The Singular Misconception of the Wireless

To the popular mind this sensational advance conveys the impression of a single invention but in reality it is an art, the successful practise of which involves the employment of a great many discoveries and improvements. I viewed it as such when I undertook to solve wireless problems and it is due to this fact that my insight into its underlying principles was clear from their very inception.

In the course of development of my induction motors it became desirable to operate them at high speeds and for this purpose I constructed alternators of relatively high frequencies. The striking behavior of the currents soon captivated my attention and in 1889 I started a systematic investigation of their properties and the possibilities of practical application. The first gratifying result of my efforts in this direction was the transmission of electrical energy thru *one wire* without return, of which I gave demonstrations in my lectures and addresses before several scientific bodies here and abroad in 1891 and 1892. During that period, while working with my oscillation transformers and dynamos of frequencies up to 200,000 cycles per second, the idea gradually took hold of me that the earth might be used in place of the wire, thus dispensing with artificial conductors altogether. The immensity of the globe seemed an unsurmountable obstacle but after a prolonged study of the subject I became satisfied that the undertaking was rational, and in my lectures before the Franklin Institute and National Electric Light Association early in 1893 I gave the outline of the system I had conceived. In the latter part of that year, at the Chicago World's Fair, I had the good fortune of meeting Prof. Helmholtz to whom I explained my plan, illustrating it with experiments. On that occasion I asked the celebrated physicist for an expression of opinion on the feasibility of the scheme. He stated unhesitatingly that it was practicable, provided I could perfect apparatus capable of putting it into effect but this, he anticipated, would be extremely difficult to accomplish.

I resumed the work very much encouraged and from that date to 1896 advanced slowly but steadily, making a number of improvements the chief of which was my system of *concatenated tuned circuits* and method of regulation, now universally adopted. In the summer of 1897 Lord Kelvin happened to pass thru New York and honored me by a visit to my laboratory where I entertained him with demonstrations in support of my wireless theory. He was fairly carried away with what he saw but, nevertheless, condemned my project in emphatic terms, qualifying it as something impossible, "an illusion and a snare." I had expected his approval and was pained and surprised. But the next day he returned and gave me a better opportunity for explanation of the advances I had made and of the true principles underlying the system I had evolved. Suddenly he remarked with evident astonishment: "Then you are not making use of Hertz waves?" "Certainly not," I replied, *these are radiations*. No energy could be economically transmitted to a distance by any such agency. In my system the process is one of *true conduction* which, theoretically, can be effected at the greatest distance without appreciable loss." I can never forget the magic change that came over the illustrious philosopher the moment he freed himself from that erroneous impression. The skeptic who would not believe was suddenly transformed into the warmest of supporters. He parted from me not only thoroly convinced of the scientific soundness of the idea but strongly exprest his confidence in its success. In my exposition to him I resorted to the following mechanical analogues of my own and the Hertz wave system.

Imagine the earth to be a bag of rubber filled with water, a small quantity of which is periodically forced in and out of the same by means of a reciprocating pump, as illustrated. If the strokes of the latter are effected in intervals of more than one hour and forty-eight minutes, sufficient for the transmission of the impulse thru the whole mass, the entire bag will expand and contract and corresponding movements will be imparted to pressure gauges or movable pistons with the same intensity, irrespective of distance. By working the pump faster, shorter waves will be produced which, on reaching the opposite end of the bag, may be reflected and give rise to stationary nodes and loops, but in any case, the fluid being incompressible, its inclosure perfectly elastic, and the frequency of oscillations not very high, the energy will be economically transmitted and very little power consumed so long as no work is done in the receivers. This is a crude but correct representation of my wireless system in which, however, I resort to various refinements. Thus, for instance, the pump is made part of a resonant system of great inertia, enormously magnifying the force of the imprest impulses. The receiving devices are similarly conditioned and in this manner the amount of energy collected in them vastly increased.

The Hertz wave system is in many respects the very opposite of this. To explain it by analogy, the piston of the pump is assumed to vibrate to and fro at a terrific rate and the orifice thru which the fluid passes in and out of the cylinder is reduced to a small hole. There is scarcely any movement of the fluid and almost the whole work performed results in the production of radiant heat, of which an infinitesimal part is recovered in a remote locality.

However incredible, it is true that the minds of some of the ablest experts have been from the beginning, and still are, obsest by this monstrous idea, and so it comes that the true wireless art, to which I laid the foundation in 1893, has been retarded in its development for twenty years. This is the reason why the "statics" have proved unconquerable, why the wireless shares are of little value and why the Government has been compelled to interfere.

FIG. 5 & FIG. 6. A section of the earth and its atmospheric envelope drawn to scale. It is obvious that the Hertzian rays cannot traverse so thin a crack between two conducting surfaces for any considerable distance, without being absorbed, says Dr. Tesla, in discussing the ether space wave theory.

We are living on a planet of well-nigh inconceivable dimensions, surrounded by a layer of insulating air above which is a rarefied and conducting atmosphere (Fig. 5). This is providential for if all the air were conducting, the transmission of electrical energy thru the natural media would be impossible. My early experiments have shown that currents of high frequency and great tension readily pass thru an atmosphere but moderately rarefied, so that the insulating stratum is reduced to a small thickness as will be evident by inspection of Fig. 6, in which a part of the earth and its gaseous envelope is shown to scale. If the radius of the sphere

FIG. 7. The theory has been seriously advanced and taught that the radio ether wave oscillations pass around the earth by successive reflections, as here shown. The efficiency of such a reflector cannot be more than 25 percent; the amount of energy recoverable in a 12,000-mile transmission being but one hundred and fifteen billionth part of one watt, with 1,000 kilowatts at the transmitter.

is 12½", then the non-conducting layer is only $1/64$" thick and it will be obvious that the Hertzian rays cannot traverse so thin a crack between two conducting surfaces for any considerable distance, without being absorbed. The theory has been seriously advanced that these radiations pass around the globe by *successive reflections*, but to show the absurdity of this suggestion reference is made to Fig. 7 in which this process is diagrammatically indicated. Assuming that there is no refraction, the rays, as shown on the right, would travel along the sides of a polygon drawn around the solid, and inscribed into the conducting gaseous boundary in which case the length of the side would be about 400 miles. As one-half the circumference of the earth is approximately 12,000 miles long there will be, roughly, thirty deviations. The efficiency of such a reflector cannot be more than 25 percent, so that if none of the energy of the transmitter were lost in other ways, the part recovered would be measured by the fraction $(¼)^{30}$. Let the transmitter radiate Hertz waves at the rate of 1,000 kilowatts. Then about *one hundred and fifteen billionth part of one watt* is all that would be collected in a *perfect* receiver. In truth, the reflections would be much more numerous as shown on the left of the figure, and owing to this and other reasons, on which it is unnecessary to dwell, the amount recovered would be a vanishing quantity.

Consider now the process taking place in the transmission by the instrumentalities and methods of my invention. For this purpose attention is called to Fig. 8, which gives an idea of the mode of propagation of the current waves and is largely self-explanatory. The drawing represents a

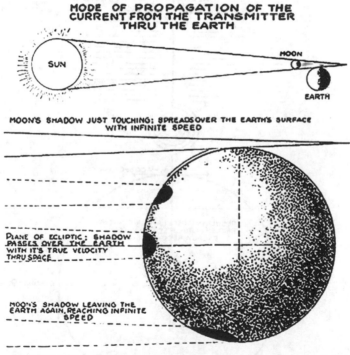

MODE OF PROPAGATION OF THE CURRENT FROM THE TRANSMITTER THRU THE EARTH

SUN

MOON

EARTH

MOON'S SHADOW JUST TOUCHING: SPREADS OVER THE EARTH'S SURFACE WITH INFINITE SPEED

PLANE OF ECLIPTIC: SHADOW PASSES OVER THE EARTH WITH IT'S TRUE VELOCITY THRU SPACE

MOON'S SHADOW LEAVING THE EARTH AGAIN, REACHING INFINITE SPEED

FIG. 8. This diagram illustrates how, during a solar eclipse, the moon's shadow passes over the earth with changing velocity, and should be studied in connection with FIG. 9. The shadow moves downward with infinite velocity at first. Then with its true velocity through space, and finally with infinite velocity again.

solar eclipse with the shadow of the moon just touching the surface of the earth at a point where the transmitter is located. As the shadow moves downward it will spread over the earth's surface, first with infinite and then gradually diminishing velocity until at a distance of about 6,000 miles it will attain its true speed in space. From there on it will proceed with increasing velocity, reaching infinite value at the opposite point of the globe. It hardly need be stated that this is merely an illustration and not an accurate representation in the astronomical sense.

The exact law will be readily understood by reference to Fig. 9, in which a transmitting circuit is shown connected to earth and to an antenna. The transmitter being in action, two effects are

ELECTROMAGNETIC HERTZ WAVES
RADIATED HORIZONTALLY FROM VERTICAL
CONDUCTOR, SLIGHTLY AFFECTED BY
CONDUCTING EARTH SURFACE
ENERGY UNRECOVERABLE

VELOCITY V=
SPEED OF LIGHT

$V_1 = V \operatorname{cosec} 0 = \infty$

RESULTANT OSCILLATORY CURRENT
THROUGH EARTH, SPEED $V = V \operatorname{cosec} \alpha$
ENERGY RECOVERABLE

$V = V \operatorname{cosec} 90° = V$

$V = V \operatorname{cosec} 180° = -\infty$

FIG. 9. Theory.

Tesla's world-wide wireless transmission of electrical signals, as well as light and power, is here illustrated in theory, analogy and realization. Tesla's experiments with 100 foot discharges at potentials of millions of volts have demonstrated that the Hertz waves are infinitesimal in effect and unrecoverable; the recoverable ground waves of Tesla fly "thru the earth". Radio engineers are gradually beginning to see the light and that the laws of propagation laid down by Tesla over a quarter of a century ago form the real and true basis of all wireless transmission today.

produced: Hertz waves pass thru the air, and a current traverses the earth. The former propagate with the speed of light and their energy is *unrecoverable* in the circuit. The latter proceeds with the speed varying as the cosecant of the angle which a radius drawn from any point under consideration forms with the axis of symmetry of the waves. At the origin the speed is infinite but gradually diminishes until a quadrant is traversed, when the velocity is that of light. From there on it again increases, becoming infinite at the antipole. Theoretically the energy of this current is *recoverable* in its entirety, in properly attuned receivers.

Some experts, whom I have credited with better knowledge, have for years contended that my proposals to transmit power without wires are sheer nonsense but I note that they are growing more cautious every day. The latest objection to my system is found in the cheapness of gasoline. These men labor under the impression that the energy flows in all directions and that, therefore, only a minute amount can be recovered in any individual receiver. But this is far from being so. The power is conveyed in only one direction, from the transmitter to the receiver, and none of it is lost elsewhere. It is perfectly practicable to recover at any point of the globe energy enough for driving an airplane, or a pleasure boat or for lighting a dwelling. I am especially sanguine in regard to the lighting of isolated places and believe that a more economical and convenient method can hardly be devised. The future will show whether my foresight is as accurate now as it has proved heretofore.

TESLA'S EGG OF COLUMBUS
MARCH 1919

HOW TESLA PERFORMED THE FEAT
OF COLUMBUS WITHOUT CRACKING THE EGG

Probably one of the most far-reaching and revolutionary discoveries made by Mr. Tesla is the so-called *rotating magnetic field*. This is a new and wonderful manifestation of force—a magnetic cyclone—producing striking phenomena which amazed the world when they were first shown by him. It results from the joint action of two or more alternating-currents definitely related to one another and creating magnetic fluxes, which, by their periodic rise and fall according to a mathematical law, cause a continuous shifting of the lines of force. There is a vast difference between an ordinary electro-magnet and that invented by Tesla. In the former the lines are stationary, in the latter they are made to whirl around at a furious rate. The first attracts a piece of iron and holds it fast; the second causes it to spin in any direction and with any speed desired. Long ago, when Tesla was still a student, he conceived the idea of the rotating magnetic field and this remarkable principle is embodied in his famous *induction motor* and system of transmission of power now in universal use.

In this issue of the *Electrical Experimenter* Mr. Tesla gives a remarkable account of his early efforts and trials as an inventor and of his final success. Unlike other technical advances arrived at thru the usual hit and miss methods and haphazard experimentation, the rotating field was purely the work of scientific imagination. Tesla developed and perfected, entirely in his mind, this great idea in all its details and applications without making one single experiment. Not even the usual first model was used. When the various forms of apparatus he had devised were tried for the first time they worked exactly as he had imagined and he took out some forty fundamental patents covering the whole vast region he had explored. He obtained the first rotations in the summer of 1883 after five years of constant and intense thought on the subject and then undertook the equally difficult task of finding believers in his discovery. The alternating-current was but imperfectly understood and had no standing with engineers or electricians and for a long time Tesla talked to deaf ears. But, ultimately, his pains were rewarded and early in 1887 a company bearing his name was formed for the commercial introduction of the invention.

Dr. Tesla recently told the editors an amusing incident in this connection. He had approached a Wall Street capitalist—a prominent lawyer—with a view of getting financial support and this gentleman called in a friend of his, a well-known engineer at the head of one of the big corporations in New York, to pass upon the merits of the scheme. This man was a practical expert who knew of the failures in the industrial exploitation of alternating-currents and was distinctly prejudiced to a point of not caring even to witness some tests. After several discouraging conferences Mr. Tesla had an inspiration. Everybody has heard of the "Egg of Columbus." The saying goes that at a certain dinner the great explorer asked some scoffers of his project to balance an egg on its end. They tried it in vain. He then took it and cracking the shell slightly by a gentle blow, made it stand upright. This may be a myth but the fact is that he was granted an audience by Isabella, the Queen of Spain, and won her support. There is a suspicion that she was more imprest by his portly bearing than the prospect of his discovery. Whatever it might have been, the Queen pawned her jewels and three ships were equipt for him and so it happened that the Germans got all that was coming to them in this war. But to return to Tesla's reminiscence. He said to these men, "Do you know the story of the Egg of Columbus?" Of course they did. "Well," he continued, "what if I could make an egg stand on the pointed end without cracking the shell?" "If you could do this we would admit that you had gone Columbus one better." "And would you be willing to go out of your way as much as Isabella?" "We have no crown jewels to pawn," said the lawyer, who was a wit, "but there are a few ducats in our buckskins and we might help you to an extent."

Mr. Tesla thus succeeded in capturing the attention and personal interest of these very busy men, extremely conservative and reluctant to go into any new enterprise, and the rest was easy. He arranged for a demonstration the following day. A rotating field-magnet was fastened under the top board of a wooden table and Mr. Tesla provided a copper-plated egg and several brass balls and pivoted iron discs for convincing his prospective associates. He placed the egg on the table and, to their astonishment, it stood on end, but when they found that it was *rapidly spinning* their stupefaction was complete. The brass balls and pivoted iron discs in turn

Fig. 1. This hitherto unpublisht photograph is extremely interesting as it shows not only "Tesla's Electric Egg" apparatus in the center of the background, but also a comprehensive view of a corner at his famous Houston Street laboratory of a decade ago. At the left may be seen a number of Tesla's oscillators or high-frequency generators, while in the rear may be noted a large high-frequency transformer of the spiral type, the diameter of which was little over nine feet. The electric egg apparatus comprising a two-phase A.C. circular core and winding, rests on a table, and this particular model measured about two feet across in making the demonstrations. Tesla applied as much as 200 H.P. from a two-phase alternator to the exciting coils, and so intense was the revolving magnetic field created in the surrounding space, that small delicately pivoted iron discs would revolve in any part of the hall, and a great many other devices could be simultaneously operated from this magnetic field when thus excited. The frequency of the two-phase A.C. energizing the coils was varied from 25 to 300 cycles, the best results being obtained with currents of from 35 to 40 cycles. This laboratory was lighted by Tesla's vacuum tubes, several of which may be seen on the ceiling, and each of which emitted 50 C.P. The coil resting on three legs and observed in the immediate foreground is the primary of a resonant Tesla transformer which collected energy from an oscillatory circuit encircling the laboratory, no matter in what position the transformer was placed. A low tension secondary of one or two turns of heavy cable (not visible) was provided for stepping down the energy collected by "mutual induction," and supplied the current to incandescent lamps, vacuum tubes, motors and other devices. When the circuit around the hall was strongly excited, the secondary furnished energy at the rate of about three-quarters of one horsepower.

were set spinning rapidly by the rotating field, to the amazement of the spectators. No sooner had they regained their composure than Tesla was delighted with the question: "Do you want any money?" "Columbus was never in a worse predicament," said the great inventor, who had parted with his last portrait of George Washington in defraying the expenses of the preparation. Before the meeting adjourned he had a substantial check in his pocket, and it was given with the assurance that there was more to be had in the same bank. That started the ball rolling. Tens of millions of horsepower of Tesla's induction motors are now in use all over the world and their production is rising like a flood.

In 1893 Mr. Albert Schmid, then Superintendent of the Westinghouse Electric and Mfg. Co. constructed a powerful rotating field-ring with an egg made of copper, and larger than that of an ostrich, for Dr. Tesla's personal collection at the Chicago World's Fair. This piece of apparatus was one of the most attractive novelties ever publicly shown and drew enormous crowds every day. Subsequently it was taken to Mr. Tesla's laboratory and served there permanently for demonstrating rotating field phenomena. In his experiments it was practicable to use as much as 200 horsepower for a short time, without overheating the wires and the effects of the magnetic forces were wonderfully fascinating to observe. This is the very rung indicated in the accompanying photograph (Fig. 1), giving a view of Mr. Tesla's former laboratory at 46 E. Houston Street, New York. It is shown in detail in Fig. 2, and the mode of winding is illustrated in diagram (Fig. 3). Originally the two-phase arrangement was provided but Mr. Tesla transformed it to the three- and four-phase when desired. On top of the ring was fastened a thin circular board, slightly hollowed, and provided around its circumference with a guard to prevent the objects from flying off.

Even more interesting than the spinning egg was the exhibition of planetary motion. In this experiment one large, and several small brass balls were usually employed. When the field was energized all the balls would be set spinning, the large one remaining in the center while the small ones revolved around it, like moons about a planet, gradually receding until they reached the outer guard and raced along the same.

But the demonstration which most imprest the audiences was the simultaneous operation of numerous balls, pivoted discs and other devices placed in all sorts of positions and at *considerable distances from the rotating field.* When the currents were turned on and the whole

The smaller balls revolve around the larger centrial egg both moving around the path of the revolving magnetic field

Polished wooden platform placed on 2 phase magnet coils

2 phase magnet winding

Direction of rotating magnetic field

Copper egg

Platform

Iron ring core

Direction of rotating magnetic field

N S N S 2 Balls

To Phase 2

Single ball spinning on its own axis & whirling round the rotating magnetic field

To Phase 1

Fig. 2. Illustrating the polyphase coil and rotating magnetic field which caused copper eggs to spin.
Fig. 3. Insert: detail of coil apparatus showing coil connections to different phases.

animated with motion, it presented an unforgettable spectacle. Mr. Tesla had many vacuum bulbs in which small, light metal discs were pivotally arranged on jewels and these would spin anywhere in the hall when the iron ring was energized.

FIG. 4. This photograph represents a collection of a few of Tesla's wireless lamps, such as he proposes to use In lighting isolated dwellings all over the world from central wireless plants. The two lamps at either corner at the bottom are Illuminated, owing to the fact that a high-frequency oscillator was In operation some distance away when this photograph was being taken. These tubes were filled with various gases for experimental research work in determining which was most efficient.

Rotating fields of 15,000 horsepower are now being turned out by the leading manufacturers and it is very likely that in the near future capacities of 50,000 horsepower will be employed in the steel and other industries and ship propulsion by Tesla's electric drive which, according to Secretary of the Navy Daniels' statement, has proved a great success.

But any student interested in these phenomena can repeat all the classical experiments of Tesla by inexpensive apparatus. For this purpose it is only necessary to make two slip ring connections on an ordinary small direct-current motor or dynamo and to wind an iron ring with four coils as indicated in diagram Fig. 3. No particular rule need be given for the windings but it may be stated that he will get the best results if he will use

FIG. 5. This illustration shows one of Tesla's high-frequency oscillation generators and a bank of his high-frequency lamps lighted by the same. These highly evacuated, gas filled tubes were operated in different ways. In some cases they were connected to one wire only; in other instances to two wires, in the manner of ordinary incandescent lamps. Often, however, they were operated without any connection to wires at all, i.e., by "wireless energy" over quite appreciable distances, which could have been greatly extended with more power. The oscillator comprises a Tesla high-potential transformer which is excited from a condenser and circuit controller, as described in his patents of 1896. The primary exciting element comprised a powerful electro-magnet actuating an armature, and this circuit was connected with 110 volt, 60 cycle A.C. or O.C. When the oscillator was put into operation, the interrupter actuated by the electro-magnet connected to the 110 volt circuit, became simultaneously the spark gap for the high-potential exciting circuit, which included this vibrator, spark gap, a high-tension condenser and the primary of the high frequency Tesla transformer. The lamps were connected to the secondary of the latter the terminals of which are seen in the rear of the machine.

an iron ring of comparatively *small section* and wind it *with as many turns of stout wire as practicable.* He can heavily copper plate an egg but he should bear in mind that Tesla's egg is not as innocent as that of Columbus. The worst that can happen with the latter is that it might be—er—over ripe! But the Tesla egg may explode with disastrous effect because the copper plating is apt to be brought to a high temperature thru the induced currents. The sensible experimenter will, therefore, first suck out the contents of the egg—thus satisfying both his appetite and thirst for knowledge.

Besides the rotating field-apparatus Mr. Tesla had other surprises for his audiences, which were even more wonderful. So, for instance, the coil on three legs, visible in the foreground, was used to operate wireless motors, lamps and other devices, and the spiral coil in the background served to show extraordinary high potential phenomena, as streamers of great length.

THE MOON'S ROTATION
APRIL 1919

We believe the accompanying illustrations and its explanations will dispel all doubts as to whether the moon rotates on its axis or not. Each of the balls, as M, depicts a different position of, and rotates exactly like, the moon keeping always the same face turned towards the center O, representing the earth.

But as you study this diagram, can you conceive that any of the balls turn on their axis? Plainly, this is rendered physically impossible by the spokes. But if you are still unconvinced, Mr. Tesla's experimental proof will surely satisfy you. A body rotating on its axis must contain rotational energy. If the earth's attraction would suddenly erase and cause it to fly off in a tangent, the moon would have no other energy except that of translatory movement, and it would not spin like the ball.

—EDITOR

Since the appearance of my article entitled the "Famous Scientific Illusions" in your February issue, I have received a number of letters criticizing the views I exprest regarding the moon's "axial rotation." These have been partly answered by my statement to the *New York Tribune* of February 23, which allow me to quote:

In your issue of February 2, Mr. Charles E. Manierre, commenting upon my article in the *Electrical Experimenter* for February which appeared in the *Tribune* of January 26, suggests that I give a definition of axial rotation.

I intended to be explicit on this point as may be judged from the following quotation: "The unfailing test of the spinning of a mass is, however, the existence of *energy of motion*. The moon is not possest of such *vis viva*." By this I meant that "axial rotation" is not simply "rotation upon an axis" nonchalantly defined in dictionaries, but is a circular motion in the true physical sense—that is, one in which half the product of the mass with the square of velocity is a definite and positive quantity. The moon is a nearly spherical body, of a radius of about 1,087.5 miles, from which I calculate its volume to be approximately 5,300,216,300 cubic miles. Since its mean density is 3.27, one cubic foot of material composing it weighs close on 205 lbs. Accordingly, the total weight of the satellite is about 79,969,000,000,000,000,000, and its mass 2,483,500,000,000,000,000 terrestrial short tons. Assuming that the moon does physically rotate upon its axis, it performs one revolution in 27 days, 7 hours, 43 minutes and 11 seconds, or 2,360,591 seconds. If, in conformity with mathematical principles, we imagine the entire mass concentrated at a distance from the center equal to two-fifths of the radius, then the calculated rotational velocity is 3.04 feet per second, at which the globe would contain 11,474,000,000,000,000,000 short foot tons of energy sufficient to run 1,000,000,000 horsepower for a period of 1,323 years. Now, I say that there is not enough of that energy in the moon to run a delicate watch.

In astronomical treatises usually the argument is advanced that "if the lunar globe did not turn upon its axis it would expose all parts to terrestrial view. As only a little over one-half is visible it *must* rotate." But this inference is erroneous, for it only admits of one alternative. There are an infinite number of axis besides its own on each of which the moon might turn and still exhibit the same peculiarity.

I have stated in my article that the moon rotates about an axis passing thru the center of the earth, which is not strictly true, but it does not vitiate the conclusions I have drawn. It is well known, of course, that the two bodies revolve around a common center of gravity, which is at a distance of a little over 2,899 miles from the earth's center.

Another mistake in books on astronomy is made in considering this motion equivalent to that of a weight whirled on a string or in a sling. In the first place there is an essential difference between these two devices tho involving the same mechanical principle. If a metal ball, attached to a string, is whirled around and the latter breaks, an axial rotation of the missile results which is definitely related in magnitude and direction to the motion preceding. By way of illustration—if the ball is whirled on the string clockwise ten times per second, then when it flies off, it will rotate on its axis ten times per second, likewise in the direction of a clock. Quite different are the conditions when the ball is thrown from a sling. In this case a *much more rapid* rotation is imparted to it in the *opposite sense*. There is no true analogy to these in the motion of the moon. *If the gravitational string, as it were, would snap, the satellite would go off in a tangent without the slightest swerving or rotation, for there is no moment about the axis and, consequently, no tendency whatever to spinning motion.*

Mr. Manierre is mistaken in his surmise as to what would happen if the earth were suddenly eliminated. Let us suppose that this would occur at the instant when the moon is in *opposition.* Then it would continue on its eliptical path around the sun, presenting to it steadily the face which was always exposed to the earth. If, on the other hand, the latter would disappear at the moment of *conjunction,* the moon would gradually swing around thru 180° and, after a number of oscillations, revolve, again with the same face to the sun. In either case there would be no periodic changes but eternal day and night, respectively, on the sides turned towards, and away from, the luminary.

Some of the arguments advanced by the correspondents are ingenious and not a few comical. None, however, are valid.

One of the writers imagines the earth in the center of a circular orbital plate, having fixedly attached to its periperal portion a disk-shaped moon, in frictional or geared engagement with another disk of the same diameter and freely rotatable on a pivot projecting from an arm entirely independent of the planetary system. The arm being held continuously parallel to itself, the pivoted disk, of course, is made to turn on its axis as the orbital plate is rotated. This is a well-known drive, and the rotation of the pivoted disk is as palpable a fact as that of the orbital plate. But, the moon in this model only revolves about the center of the system *without the slightest angular displacement* on its own axis. The same is true of a cart-wheel to which this writer refers. So long as it advances on the earth's surface it turns on the axle in the true physical sense; when one of its spokes is always kept in a perpendicular position the wheel still *revolves* about the earth's center, *but axial rotation has ceased.* Those who think that it then still exists are laboring under an illusion.

An obvious fallacy is involved in the following abstract reasoning. The orbital plate is assumed to gradually shrink, so that finally the centers of the earth and the satellite coincide when the latter revolves simultaneously about its own and the earth's axis. We may reduce the earth to a mathematical point and the distance between the two planets to the radius of the moon without affecting the system in principle, but a further diminution of the distance is manifestly absurd and of no bearing on the question under consideration.

FIG. 1. If you still think that the moon rotates on its axis, look at the diagram and follow closely the successive positions taken by one of the balls M while it is rotated by a spoke of the wheel. Substitute gravity for the spoke and the analogy solves the moon rotation riddle.

In all the communications I have received, tho different in the manner of presentation, the successive changes of position in space are mistaken for axial rotation. So, for instance, a positive refutation of my arguments is found in the observation that the moon exposes all sides to other planets! It revolves, to be sure, but none of the evidences is a proof that it turns on its axis. Even the well-known experiment with the Foucault pendulum, altho exhibiting similar phenomena as on our globe, would merely demonstrate a motion of the satellite about some axis. The view I have advanced is *not based on a theory* but on facts demonstrable by experiment. It is not a matter of definition as some would have it. *A mass revolving on its axis must be possest of momentum.* If it has none, there is no axial rotation, all appearances to the contrary notwithstanding.

A few simple reflections based on well establisht mechanical principles will make this clear. Consider first the case of two equal weights w and w_1, in Fig. 1, whirled about the center O on a string s as shown. Assuming the latter to break at a both weights will fly off on tangents to their circles of gyration, and, being animated with different velocities, they will rotate around their common center of gravity o. If the weights are whirled n times per second then the speed of the outer and the inner one will be, respectively, $V = 2(R + r) n$ and $V_1 = 2\pi(R - r) n$, and the difference $V - V_1 = 4 \pi r n$, will be the length of the circular path of the outer weight. Inasmuch, however, as there will be equalization of the speeds until the mean value is attained, we shall have $(V - V_1)/2 = 2\pi r n = 2\pi r N$, N being the number of revolutions per second of the weights around their center of gravity. Evidently then, the weights continue to rotate at the original rate and in the same direction. I know this to be a fact from actual experiments. It also follows that a ball, as that shown in the figure, will behave in a similar manner for the two half-spherical masses can be concentrated at their centers of gravity and m and m_1, respectively, which will be at a distance from o equal to $3/8\, r$.

This being understood, imagine a number of balls M carried by as many spokes S radiating from a hub H, as illustrated in Fig. 2, and let this system be rotated n times per second around center O on frictionless bearings. A certain amount of work will be required to bring the structure to this speed, and it will be found that it equals exactly half the product of the masses with the square of the tangential velocity. Now if it be true that the moon rotates in reality on its axis *this must also hold good for* each *of the balls as it performs the same kind of movement.* Therefore, in imparting to the system a given velocity, energy must have been used up in the axial rotation of the balls. Let M be the mass of one of these and R the radius of gyration, then the rotational energy will be $E = \frac{1}{2} M(2 \pi R n)^2$. Since for one complete turn of the wheel every ball makes one revolution on its axis, according to the prevailing theory, the energy of axial rotation of each ball will be $e = \frac{1}{2} M(2 \pi r_1 n)^2$, r_1 being the radius of gyration about the axis and equal to $0.6325\, r$. We can use as large balls as we like, and so make e a considerable percentage of E and yet, it is positively established by experiment that each of the rotating balls contain only the energy E, no power whatever being consumed in the supposed axial rotation, which is, consequently, wholly illusionary. Something even more interesting may, however, be stated. As I have shown before, a ball flying off will rotate at the rate of the wheel and in the same direction. But this whirling motion, unlike that of a projectile, neither adds to, nor detracts from, the energy of the translatory movement which is exactly equal to the work consumed in giving to the mass the observed velocity.

Diagram illustrating the rotation of weights
thrown off by centrifugal force.

From the foregoing it will be seen that in order to make one physical revolution on its axis the moon should have twice its present angular velocity, and then it would contain a quantity of stored energy as given in my above letter to the *New York Tribune* on the assumption that the radius of gyration is $2/5$ that of figure. This, of course, is uncertain, as the distribution of density in the interior is unknown. But from the character of motion of the satellite it may be concluded with certitude *that it is devoid of momentum about its axis*. If it be bisected by a plane tangential to the orbit, the masses of the two halves arc inversely as the distances of their centers of gravity from the earth's center and, therefore, if the latter were to disappear suddenly, no axial rotation, as in the case of a weight thrown off, would ensue.

THE TRUE WIRELESS
MAY, 1919

In this remarkable and complete story of his discovery of the "True Wireless" and the principles upon which transmission and reception, even in the present day systems, are based, Dr. Nikola Tesla shows us that he is indeed the "Father of the Wireless." To him the Hertz wave theory is a delusion; it looks sound from certain angles, but the facts tend to prove that it is hollow and empty. He convinces us that the real Hertz waves are blotted out after they have traveled but a short distance from the sender. It follows, therefore, that the measured antenna current is no indication of the effect, because only a small part of it is effective at a distance. The limited activity of pure Hertz wave transmission and reception is here clearly explained, besides showing definitely that in spite of themselves, the radio engineers of today are employing the original Tesla tuned oscillatory system. He shows by examples with different forms of arials that the signals picked up by the instruments must actually be induced by earth currents—not etheric space waves. Tesla also disproves the "Heaviside layer" theory from his personal observations and tests.

—EDITOR

Ever since the announcement of Maxwell's electro-magnetic theory scientific investigators all the world over had been bent on its experimental verification. They were convinced that it would be done and lived in an atmosphere of eager expectancy, unusually favorable to the reception of any evidence to this end. No wonder then that the publication of Dr. Heinrich Hertz's results caused a thrill as had scarcely ever been experienced before. At that time I was in the midst of pressing work in connection with the commercial introduction of my system of power transmission, but, nevertheless, caught the fire of enthusiasm and fairly burned with desire to behold the miracle with my own eyes. Accordingly, as soon as I had freed myself of these imperative duties and resumed research work in my laboratory on Grand Street, New York, I began, parallel with high-frequency alternators, the construction of several forms of apparatus with the object of exploring the field opened up by Dr. Hertz. Recognizing the limitations of the devices he had employed, I concentrated my attention on the production of a powerful induction coil but made no notable progress until a happy inspiration led me to the invention of the oscillation transformer. In the latter part of 1891 I was already so far advanced in the development of this new principle that I had at my disposal means vastly superior to those of the German physicist. All my previous efforts with Rhumkorf coils had left me unconvinced, and in order to settle my doubts I went over the whole ground once more, very carefully, with these improved appliances. Similar phenomena were noted, greatly magnified in intensity, but they were susceptible of a different and more plausible explanation. I considered this so important that in 1892 I went to Bonn, Germany, to confer with Dr. Hertz in regard to my observations. He seemed disappointed to such a degree that I regretted my trip and parted from him sorrowfully. During the succeeding years I made numerous experiments with the same object, but the results were uniformly negative. In 1900, however, after I had evolved a wireless transmitter which enabled me to obtain electro-magnetic activities of many millions of horsepower, I made a last desperate attempt to prove that the disturbances emanating from the oscillator were ether vibrations akin to those of light, but met again with utter failure. For more than eighteen years I have been reading treatises, reports of scientific transactions, and articles on Hertz-wave telegraphy, to keep myself informed, but they have always imprest me like works of fiction.

The history of science shows that theories are perishable. With every new truth that is revealed we get a better understanding of Nature and our conceptions and views are modified. Dr. Hertz did not discover a new principle. He merely gave material support to a hypothesis which had been long ago formulated. It was a perfectly well-established fact that a circuit, traversed by a periodic current, emitted some kind of space waves, but we were in ignorance as to their character. He apparently gave an experimental proof that they were transversal vibrations in the ether. Most people look upon this as his great accomplishment. To my mind it seems that his immortal merit was not so much in this as in the focusing of the investigators' attention on the processes taking place in the ambient medium. The Hertz-wave theory, by its fascinating hold on the imagination, has stifled creative effort in the wireless art and retarded it for twenty-five years. But, on the other hand, it is impossible to over-estimate the beneficial effects of the powerful stimulus it has given in many directions.

As regards signaling without wires, the application of these radiations for the purpose was quite obvious. When Dr. Hertz was asked whether such a system would be of practical value, he did not think so, and he was correct in his forecast. The best that might have been expected was a method of communication similar to the heliographic and subject to the same or even greater limitations.

In the spring of 1891 I gave my demonstrations with a high-frequency machine before the American Institute of Electrical Engineers at Columbia College, which laid the foundation to a new and far more promising departure. Altho the laws of electrical resonance were well known at that time and my lamented friend, Dr. John Hopkinson, had even indicated their specific application to an alternator in the Proceedings of the Institute of Electrical Engineers, London, Nov. 13, 1889, nothing had been done towards the practical use of this knowledge and it is probable that those experiments of mine were the first public exhibition with resonant circuits, more particularly of high frequency. While the spontaneous success of my lecture was due to spectacular features, its chief import was in showing that all kinds of devices could be operated thru a single wire without return. This was the initial step in the evolution of my wireless system. The idea presented itself to me that it might be possible, under observance of proper conditions of resonance, to transmit electric energy thru the earth, thus dispensing with all artificial conductors. Anyone who might wish to examine impartially the merit of that early suggestion must not view it in the light of present day science. I only need to say that as late as 1893, when I had prepared an elaborate chapter on my wireless system, dwelling on its various instrumentalities and future prospects, Mr. Joseph Wetzler and other friends of mine emphatically protested against its publication on the ground that such idle and far-fetched speculations would injure me in the opinion of conservative business men.

So it came that only a small part of what I had intended to say was embodied in my address of that year before the Franklin Institute and National Electric Light Association under the chapter "On Electrical Resonance." This little salvage from the wreck has earned me the title of "Father of the Wireless" from many well-disposed fellow workers, rather than the invention of scores of appliances which have brought wireless transmission within the reach of every young amateur and which, in a time not distant, will lead to undertakings overshadowing in magnitude and importance all past achievements of the engineer.

FIG. 1. Alternator of 10,000 cycles p.s., capacity 10 k.w., which was employed by Tesla in his first demonstrations of high-frequency phenomena before the American Institute of Electrical Engineers at Columbus College, May 20, 1891.

The popular impression is that my wireless work was begun in 1893, but as a matter of fact I spent the two preceding years in investigations, employing forms of apparatus, some of which were almost like those of today. It was clear to me from the very start that the successful consummation could only be brought about by a number of radical improvements. Suitable high-frequency generators and electrical oscillators had first to be produced. The energy of these had to be transformed in effective transmitters and collected at a distance in proper receivers. Such a system would be manifestly circumscribed to its usefulness if all extraneous interference were not prevented and exclusiveness secured. In time, however, I recognized that devices of this kind, to be most effective and efficient, should be designed with due regard to the physical properties of this planet and the electrical conditions obtaining on the same. I will briefly touch upon the salient advances as they were made in the gradual development of the system.

The high-frequency alternator employed in my first demonstrations is illustrated in Fig. 1. It comprised a field-ring, with 384 pole projections and a disc armature with coils wound in one single layer which were connected in various ways according to requirements. It was an excellent machine for experimental purposes, furnishing sinusoidal currents of from 10,000 to 20,000 cycles per second. The output was comparatively large, due to the fact that as much as 30 amperes per square millimeter could be past thru the coils without injury. The diagram in Fig. 2 shows the circuit arrangements as used in my lecture. Resonant conditions were maintained by means of a condenser subdivided into small sections, the finer adjustments being effected by a movable iron core within an inductance coil. Loosely linked with the latter was a high-tension secondary which was tuned to the primary.

FIG. 2. Diagram illustrating the circuit connections and tuning devices employed by Tesla in his experimental demonstrations before the American Institute of Electrical Engineers with the high-frequency alternator shown in FIG. 1.

The operation of devices thru a single wire without return was puzzling at first because of its novelty, but can be readily explained by suitable analogs. For this purpose reference is made to Figs. 3 and 4.

In the former the low resistance electric conductors are represented by pipes of large section, the alternator by an oscillating piston and the filament of an incandescent lamp by a minute channel connecting the pipes. It will be clear from a glance at the diagram that very slight excursions of the piston would cause the fluid to rush with high velocity thru the small channel and that virtually all the energy of movement would be transformed into heat by friction, similarly to that of the electric current in the lamp filament.

FIG. 3. Electric transmission thru two wires and hydraulic analog.

FIG. 4. Electric transmission thru a single wire hydraulic analog.

The second diagram will now be self-explanatory. Corresponding to the terminal capacity of the electric system an elastic reservoir is employed which dispenses with the necessity of a return pipe. As the piston oscillates the bag expands and contracts, and the fluid is made to surge thru the restricted passage with great speed, this resulting in the generation of heat as in the incandescent lamp. Theoretically considered, the efficiency of conversion of energy should be the same in both cases.

FIG. 5. Illustrating typical arrangements for collecting energy in a system of transmission thru a single wire.

Granted, then, that an economic system of power transmission thru a single wire is practicable, the question arises how to collect the energy in the receivers. With this object attention is called to Fig. 5, in which a conductor is shown excited by an oscillator joined to it at one end. Evidently, as the periodic impulses pass thru the wire, differences of potential will be created along the same as well as at right angles to it in the surrounding medium and either of these may be usefully applied. Thus at a, a circuit comprising an inductance and capacity is resonantly excited in the transverse, and at b, in the longitudinal sense. At c, energy is collected in a circuit parallel to the conductor but not in contact with it, and again at d, in a circuit which is partly sunk into the conductor and may be, or not, electrically connected to the same. It is important to keep these typical dispositions in mind, for however the distant actions of the oscillator might be modified thru the immense extent of the globe the principles involved are the same. Consider now the effect of such a conductor of vast dimensions on a circuit exciting it. The upper diagram of Fig. 6 illustrates a familiar oscillating system comprising a straight rod of self-inductance $2L$ with small terminal capacities c c and a node in the center. In the lower diagram of the figure a large capacity C is attached to the rod at one end with the result of shifting the node to the right, thru a distance corresponding to self-inductance X. As both parts of the system on either side of the node vibrate at the same rate, we have evidently $(L + X) c = (L - X) C$ from which $X = L (C - c)/(C + c)$.

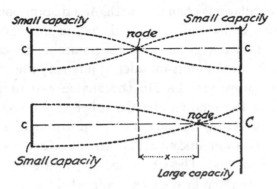

FIG. 6. Diagram elucidating effect of
large capacity on one end.

When the capacity C becomes commensurate to that of the earth, X approximates L, in other words, the node is close to the ground connection. *The exact determination of its position is very important in the calculation of certain terrestrial electrical and geodetic data* and I have devised special means with this purpose in view.

FIG. 7. Transmission of electrical energy thru the earth as illustrated in Tesla's lectures before the Franklin Institute and Electrical Light Association in February and March 1893, and mechanical analog of the same.

My original plan of transmitting energy without wires is shown in the upper diagram of Fig. 7, while the lower one illustrates its mechanical analog, first publisht in my article in the *Century Magazine* of June, 1900. An alternator, preferably of high tension, has one of its terminals connected to the ground and the other to an elevated capacity and impresses its oscillations upon the earth. At a distant point a receiving circuit, likewise connected to ground and to an elevated capacity, collects some of the energy and actuates a suitable device. I suggested a multiplication of such units in order to intensify the effects, an idea which may yet prove valuable. In the analog two tuning forks are provided, one at the sending and the other at the receiving station, each having attached to its lower prong a piston fitting in a cylinder. The two cylinders communicate with a large elastic reservoir filled with an incompressible fluid. The vibrations transmitted to either of the tuning forks excite them by resonance and, thru electrical contacts or otherwise, bring about the desired result. This, I may say, was not a mere mechanical illustration, but a simple representation of my apparatus for submarine signaling, perfected by me in 1892, but not appreciated at that time, altho more efficient than the instruments now in use.

FIG. 8. Tesla's system of wireless transmission thru the earth as actually exposed in his lectures before the Franklin Institute and Electric Light Association in February and March, 1893.

The electric diagram in Fig. 7, which was reproduced from my lecture, was meant only for the exposition of the principle. The arrangement, as I described it in detail, is shown in Fig. 8. In this case an alternator energizes the primary of a transformer, the high-tension secondary of which is connected to the ground and an elevated capacity and tuned to the imprest oscillations. The receiving circuit consists of an inductance connected to the ground and to an elevated terminal without break and is resonantly responsive to the transmitted oscillations. A specific form of receiving device was not mentioned, but I had in mind to transform the received currents and thus make their volume and tension suitable for any purpose. This, in substance, is the system of today and I am not aware of a single authenticated instance of successful transmission at considerable distance by different instrumentalities. It might, perhaps, not be clear to those who have perused my first description of these improvements that, besides making known new and efficient types of apparatus, I gave to the world a wireless system of potentialities far beyond anything before conceived. I made explicit and repeated statements that I contemplated transmission, absolutely unlimited as to terrestrial distance and amount of energy. But, altho I have overcome all obstacles which seemed in the beginning unsurmountable and found elegant solutions of all the problems which confronted me, yet, even at this very day, the majority of experts are still blind to the possibilities which are within easy attainment.

Fig. 9. The forerunner of the audion—the most sensitive wireless detector known, as described by Tesla in his lecture before the Institution of Electrical Engineers, London, February, 1892.

My confidence that a signal could be easily flashed around the globe was strengthened thru the discovery of the "rotating brush," a wonderful phenomenon which I have fully described in my address before the

Institution of Electrical Engineers, London, in 1892, and which is illustrated in Fig. 9. This is undoubtedly the most delicate wireless detector known, but for a long time it was hard to produce and to maintain in the sensitive state. These difficulties do not exist now and I am looking to valuable applications of this device, particularly in connection with the high-speed photographic method, which I suggested, in wireless, as well as in wire, transmission.

Possibly the most important advances during the following three or four years were my system of concatenated tuned circuits and methods of regulation, now universally adopted. The intimate bearing of these inventions on the development of the wireless art will appear from Fig. 10, which illustrates an arrangement described in my U.S. Patent No. 568178 of September 22, 1896, and corresponding dispositions of wireless apparatus. The captions of the individual diagrams are thought sufficiently explicit to dispense with further comment. I will merely remark that in this early record, in addition to indicating how any number of resonant circuits may be linked and regulated, I have shown the advantage of the proper timing of primary impulses and use of harmonics. In a farcical wireless suit in London, some engineers, reckless of their reputation, have claimed that my circuits were not at all attuned; in fact they asserted that I had looked upon resonance as a sort of wild and untamable beast!

It will be of interest to compare my system as first described in a Belgian patent of 1897 with the Hertz-wave system of that period. The significant differences between them will be observed at a glance. The first enables us to transmit economical energy to any distance and is of inestimable value; the latter is capable of a radius of only a few miles and is worthless. In the first there are no spark-gaps and the actions are enormously magnified by resonance. In both transmitter and receiver the currents are transformed and rendered more effective and suitable for the operation of any desired device. Properly constructed, my system is safe against static and other interference and the amount of energy which may be transmitted is *billions of times greater* than with the Hertzian which has none of these virtues, has never been used successfully and of which no trace can be found at present.

A well-advertised expert gave out a statement in 1899 that my apparatus did not work and that it would take 200 years before a message would be flashed across the Atlantic and he even accepted stolidly my

Fig. 10. Tesla's system of concatenated tuned circuits shown and described in U. S. Patent No. 568.178 of September 22, 1896, and corresponding arrangements in wireless transmission.

Fig. 11. Tesla's four circuit tuned system contrasted with the contemporaneous Hertz-wave system.

congratulations on a supposed great feat. But subsequent examination of the records showed that my devices were secretly used all the time and ever since I learned of this I have treated these Borgia-Medici methods with the contempt in which they are held by all fair-minded men. The wholesale appropriation of my inventions was, however, not always without a diverting side. As an example to the point I may mention my oscillation transformer operating with an air gap. This was in turn replaced by a carbon arc, quenched gap, an atmosphere of hydrogen, argon or helium, by a mechanical break with oppositely rotating members, a mercury interrupter or some kind of a vacuum bulb and by such *tours de force* as many new "systems" have been produced. I refer to this of course, without the slightest ill-feeling, let us advance by all means. But I cannot help thinking how much better it would have been if the ingenious men, who have originated these "systems," had invented something of their own instead of depending on me altogether.

Before 1900 two most valuable improvements were made. One of these was my individualized system with transmitters emitting a wave-complex and receivers comprising separate tuned elements cooperatively associated. The underlying principle can be explained in a few words. Suppose that there are n simple vibrations suitable for use in wireless transmission, the probability that any one tune will be struck by an extraneous disturbance is $1/n$. There will then remain $n - 1$ vibrations and the chance that one of these will be excited is $1/(n - 1)$ hence the probability that two tunes would be struck at the same time is $1/n (n - 1)$. Similarly, for a combination of three the chance will be $1/n (n - 1)(n - 2)$ and so on. It will be readily seen that in this manner any desired degree of safety against the statics or other kind of disturbance can be attained provided the receiving apparatus is so designed that its operation is possible only thru the joint action of all the tuned elements. This was a difficult problem which I have successfully solved so that now *any desired number of simultaneous messages is practicable in the transmission thru the earth as well as thru artificial conductors.*

The other invention, of still greater importance, is a peculiar oscillator enabling the transmission of energy without wires in any quantity that may ever be required for industrial use, to any distance, and with very high economy, as was the outcome of years of systematic study and investigation and wonders will be achieved by its means.

FIG. 12. Arrangements of directive circuits described in Tesla's U.S. Patent No. 613,809 of November 8, 1898, on "Method of and Apparatus for Controlling Mechanism of Moving Vessels or Vehicles".

The prevailing misconception of the mechanism involved in the wireless transmission has been responsible for various unwarranted announcements which have misled the public and worked harm. By keeping steadily in mind that the transmission thru the earth is in every respect identical to that thru a straight wire, one will gain a clear understanding of the phenomena and will be able to judge correctly the merits of a new scheme. Without wishing to detract from the value of any plan that has been put forward I may say that they are devoid of novelty. So for instance in Fig. 12 arrangements of transmitting and receiving circuits are illustrated, which I have described in my U.S. Patent No. 613809 of November 8, 1898, on a Method of and Apparatus for Controlling Mechanism of Moving Vessels or Vehicles, and which have been recently dished up as original discoveries. In other patents and technical publications I have suggested conductors in the ground as one of the obvious modifications indicated in Fig. 5.

For the same reason the statics are still the bane of the wireless. There is about as much virtue in the remedies recently proposed as in hair restorers. *A small and compact apparatus has been produced which does away entirely with this trouble*, at least in plants suitably remodelled.

Nothing is more important in the present phase of development of the wireless art than to dispose of the dominating erroneous ideas. With this object I shall advance a few arguments based on my own observations *which prove that Hertz waves have little to do with the results obtained even at small distances.*

Fig. 13. Diagram exposing the fallacy of the gliding wave theory as propounded in wireless textbooks.

In Fig. 13 a transmitter is shown radiating space waves of considerable frequency. It is generally believed that these waves pass along the earth's surface and thus affect the receivers. I can hardly think of anything more improbable than this "gliding wave" theory and the conception of the "guided wireless" which are contrary to all laws of action and reaction. Why should these disturbances cling to a conductor where they are counteracted by induced currents, when they can propagate in all other directions unimpeded? The fact is that the radiations of the transmitter passing along the earth's surface are soon extinguished, the height of the inactive zone indicated in the diagram being some function of the wave length, the bulk of the waves traversing freely the atmosphere. Terrestrial phenomena which I have noted conclusively show that there is no *Heaviside layer*, or if it exists, it is of no effect. It certainly would be unfortunate if the human race were thus imprisoned and forever without power to reach out into the depths of space.

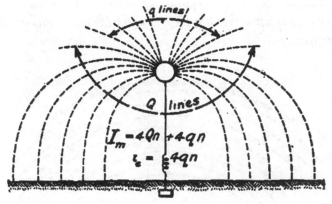

FIG. 14. Diagram explaining the relation between the effective and the measured current in the antenna.

The actions at a distance cannot be proportionate to the height of the antenna and the current in the same. I shall endeavor to make this clear by reference to diagram in Fig. 14. The elevated terminal charged to a high potential induces an equal and opposite charge in the earth and there are thus Q lines giving an average current $I = 4Qn$ which circulates locally and is useless except that it adds to the momentum. A relatively small number of lines Q however, go off to great distance and to these corresponds a mean current of $i_e = 4qn$ to *which is due the action at a distance*. The total average current in the antenna is thus $I_u = 4Qn + 4qn$ and its intensity is no criterion for the performance. The electric efficiency of the antenna is $q/(Q + q)$ and this is often a very small fraction.

Dr. L. W. Austin and Mr. J. L. Hogan have made quantitative measurements which are valuable, but far from supporting the Hertz wave theory they are evidences in disproval of the same, as will be easily perceived by taking the above facts into consideration. Dr.

Austin's researches are especially useful and instructive and I regret that I cannot agree with him on this subject. I do not think that if his receiver was affected by Hertz waves he could ever establish such relations as he has found, but he would be likely to reach these results if the Hertz waves were in a large part eliminated. At great distance the space waves and the current waves are of equal energy, the former being merely an accompanying manifestation of the latter in accordance with the fundamental teachings of Maxwell. It occurs to me here to ask the question—why have the Hertz waves been reduced from the original frequencies to those I have advocated for my system,

FIG. 15. Illustrating one of the general evidences against the space wave transmission.

when in so doing the activity of the transmitting apparatus has been reduced a billion fold? I can invite any expert to perform an experiment such as is illustrated in Fig. 15, which shows the classical Hertz oscillator and my grounded transmitting circuit. It is a fact which I have demonstrated that, altho we may have in the Hertz oscillator an activity thousands of times greater, the effect on the receiver is not to be compared to that of the grounded circuit. This shows that *in the transmission from an airplane we are merely working thru a condenser,* the capacity of which is a function of a logarithmic ratio between the length of the conductor and the distance from the ground. The receiver is affected in exactly the same manner as from an ordinary transmitter, the only difference being that there is a certain modification of the action which can be predetermined from the electrical constants. It is not at all difficult to maintain communication between an airplane and a station on the ground, on the contrary, the feat is very easy.

Fɪɢ. 16. Showing unimportance of relative position of transmitting and receiving antennae in disproval of the Hertz-wave theory.

To mention another experiment in support of my view, I may refer to Fig. 16 in which two grounded circuits are shown excited by oscillations of the Hertzian order. It will be found that the antennas can be out of parallelism without noticeable change in the action on the receiver, this proving that it is due to currents propagated thru the ground and not to space waves.

Fɪɢ. 17. Illustrating influence of obstacle in the path of transmission as evidence against the Hertz-wave theory.

Particularly significant are the results obtained in cases illustrated in Figures 17 and 18. In the former an obstacle is shown in the path of the waves but unless the receiver is within the effective *electrostatic* influence of the mountain range, the signals are not appreciably weakened by the presence of the latter, because the currents pass under it and excite the

FIG. 18. Showing effect of two hills as further proof against the Hertz-wave theory.

circuit in the same way as if it were attached to an energized wire. If, as in Fig. 18, a second range happens to be beyond the receiver, it could only strengthen the Hertz wave effect by reflection, but as a matter of fact it detracts greatly from the intensity of the received impulses because the electric nivean between the mountain is raised, as I have explained in connection with my lightning protector in the *Experimenter* of February.

FIG. 19. Comparing the actions of two forms of transmitter as bearing out the fallacy of the Hertz-wave theory.

Again in Fig. 19 two transmitting circuits, one grounded directly and the other thru an air gap, are shown. It is a common observation that the former is far more effective, which could not be the case in a transmission with Hertz radiations. In like manner if two grounded circuits are observed from day to day the effect is found to increase greatly with the dampness of the ground, and for the same reason also the transmission thru sea-water is more efficient.

FIG. 20. Disproving the Hertz-wave theory by two transmitters, one of great and the other of small energy.

An illuminating experiment is indicated in Fig. 20 in which two grounded transmitters are shown, one with a large and the other with a small terminal capacity. Suppose that the latter be 1/10 of the former but that it is charged to 10 times the potential and let the frequency of the two circuits and therefore the currents in both antennas be exactly the same. The circuit with the smaller capacity will then have 10 times the energy of the other but the effects on the receiver will be in no wise proportionate.

The same conclusions will be reached by transmitting and receiving circuits with wires buried underground. In each case the actions carefully investigated will be found to be due to *earth currents*. Numerous other proofs might be cited which can be easily verified. So for example *oscillations of low frequency* are ever so much more effective in the transmission which is inconsistent with the prevailing idea. My observations in 1900 and the recent transmissions of signals to very great distances are another emphatic disproval.

The Hertz wave theory of wireless transmission may be kept up for a while, but I do not hesitate to say that in a short time it will be recognized as one of the most remarkable and inexplicable aberrations of the scientific mind which has ever been recorded in history.

THE MOON'S ROTATION
JUNE 1919

*In this article Dr. Tesla proves conclusively by theory and experiment
that all the kinetic energy of a rotating mass is purely translational
and that the moon contains absolutely no rotational energy, in other
words, does not rotate on its axis.*

—EDITOR

In revising my article on "The Moon's Rotation," which appeared
in the April issue of the *Electrical Experimenter*, I appended a few
remarks to the original text in further support and elucidation of the
theory advanced. Due to the printer's error these were lost and, in
consequence, I found it necessary to forward another communication
which, unfortunately, was received too late for embodiment in the May
number. Meanwhile many letters have reached me in which certain
phenomena presented by rotating bodies, as the moon's librations of
longitude, are cited as evidences of energy due to spinning motion, i.e.,
proofs of axial rotation of the satellite in the true physical sense. I trust
that the following amplified statement will meet all of the objections
raised and convert to my views those who are still unconvinced.

FIG. 1. In determining the kinetic energy of a rotating mass, this figure shows the selection
of a number of points taken within the straight rod or mass M, at successive distances from
the axis of rotation O. Knowing these values and the speed of rotation the kinetic energy of
the mass is readily computed.

The kinetic energy of a rotating mass can be determined in four ways which are illustrated in diagrams, Figs. 1, 2, 3 and 4 and may be found more or less suitable.

Referring to Fig. 1, the method consists in selecting judiciously a number of points as O_1, O_2, O_3, etc., within the straight rod or mass M, respectively at distances r_1, r_2, r_3, etc., from the axis of rotation O and calculating the square root of the mean square of these distances. Its value being R_g, denoted radius of gyration, the *effective* velocity of the mass at n revolutions per second will be $V_e = 2 \pi R_g n$ and its kinetic energy $E = \frac{1}{2} M V_e^2 = \frac{1}{2} M (2 \pi R_g n)^2$.

Fig. 2. In this case the mass M, rotating n times per second, about an axis O, is divided into numerous elements or small parts at various radii from O. Knowing the kinetic energy of Each part, the whole kinetic energy of the mass is easily determined by taking a summation of the individual quantities.

In Fig. 2 the mass M, rotating n times per second about an axis O at right angles to the plane of the paper, is divided into numerous elements or small parts, most conveniently very thin concentric laminae, as I_1, I_2, I_3, etc., at distances r_1, r_2, r_3, etc., from O. Since the kinetic energy of each part is equal to half the product of its mass and the square of the velocity, the sum of all these elemental energies $E = \frac{1}{2} \Sigma m V^2 = \frac{1}{2} m_1 V_1^2 - \frac{1}{2} m_2 V_2^2 + \frac{1}{2} m_3 V_3^2 + \ldots \ldots = \frac{1}{2} m_1 (2 \pi r_1 n)^2 + \frac{1}{2} m_2 (2 \pi r_2 n)^2 + \frac{1}{2} m_3 (2 \pi r_3 n)^2 + \ldots \ldots$

A different form of expression for the energy of a rotating body may be obtained by determining its moment of inertia. For this purpose the mass M (in Fig. 3), rotating n times per second about an axis O, is separated into minute parts, as m_1, m_2, m_3, etc., respectively at distances r_1, r_2, r_3, etc., from the same. The sum of the products of all these small masses and the squares of their distances is the moment of inertia I, and then $E = \frac{1}{2} I \omega^2$, $\omega = 2 \pi n$ being the angular velocity.

Fig. 3. Another form of expression for the energy of a rotating body may be obtained by determining its moment of inertia. Here the mass M is subdivided into minute parts M_1, M_2, M_3, ... etc. The sum of the products of these masses and the squares of their distances is the moment of inertia, which with the angular speed, gives the kinetic energy E.

It is obvious that in all these instances many points or elements will be required for great accuracy but, as a rule, very few are sufficient in practice.

Still another way to compute the kinetic energy is illustrated in Fig. 4, in which case the quantity I is given in terms of the moment of inertia I_c about another axis parallel to O and passing thru the center of gravity C of mass M. In conformity with this the energy of motion $E = \frac{1}{2} M V^2 + \frac{1}{2} I_c \omega^2$ in which equation V is the velocity of the center of gravity.

The preceding is deemed indispensable as I note that the correspondents, even those who seem thoroly familiar with mechanical principles, fail to make a distinction between theoretical and physical truths which is essential to my argument.

In estimating the kinetic energy of a rotating mass in any of the ways indicated we arrive, thru suitable conceptions and methods of approximation, at expressions which may be made quantitatively precise to any desired degree, but do not truly define the actual condition of the body. To illustrate, when proceeding according to the plan of Fig. 1, we find a certain hypothetical velocity with which the entire mass should move in order to contain the same energy, a state wholly imaginary and irreconcilable with the actual. Only, when *all* particles of the body have the same velocity, does the

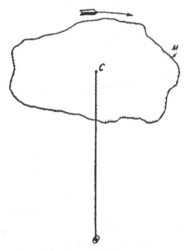

FIG. 4. In this case the motion is resolved into two separate components— one translational about O and the other rotational about C. The total kinetic energy of the mass equals the sum of these two energies.

product $\frac{1}{2} M V^2$ specify a physical fact and is numerically and descriptively accurate. Still more remote from palpable truth is the equation of motion obtained in the manner indicated in Fig. 4, in which the first term represents the kinetic energy of translation of the body as a whole and the second that of its axial rotation. The former would demand a movement of the mass in a definite path and direction, all particles having the same velocity, the latter its simultaneous motion in another path and direction, the particles having different velocities. This abstract idea of angular motion is chiefly responsible for the illusion of the moon's axial rotation, which I shall endeavor to dispel by additional evidences.

With this object attention is called to Fig. 5 showing a system composed of eight balls M, which are carried on spokes S, radiating from a hub H, rotatable around a central axis O in bearings supposed to be frictionless. It is an arrangement similar to that before illustrated with the exception that the balls, instead of forming parts of the spokes, are supported in screw pivots s, which are normally loose but can be tightened so as to permit both free turning and rigid fixing as may be desired. To facilitate observation the spokes are provided with radial marks and the lower sides of the balls are shaded. Assume, first, that the drawing depicts the state of rest, the balls being rotatable without friction, and let an angular velocity $\omega = 2\pi n$ be imparted to the system in the clockwise direction as indicated by the long solid arrow.

FIG. 5. This diagram represents a system composed of 8 balls M, carried on spokes S, and rotating around center O. The balls are freely rotatable on pivots which can be tightened. With this model the fallacy of the moon's rotation on its axis is demonstrable.

Viewing a ball as M, its successive positions 1, 2, 3–8 in space, and also relatively to the spoke, will be just as drawn, and it is evident from an inspection of the diagram that while moving with the angular velocity ω about O, in the clockwise direction, the ball turns, with respect to its axis, at the same angular velocity but in the opposition direction, that of the dotted arrow. The combined result of these two motions is a translatory, movement of the ball such that all particles are animated with the same velocity V, which is that of its center of gravity. In this case, granted that there is absolutely no friction the kinetic energy of each ball will be given by the product of ½ M V² not approximately, but with mathematical rigor. If now the pivots are screwed tight and the balls fixt rigidly to the spokes, this angular motion relatively to their axes becomes physically impossible and then it is found that the kinetic energy of each ball is increased, the increment being exactly the energy of rotation of the ball on its axis. This fact, which is borne out both by theory and experiment, is the foundation of the general notion that a gyrating body—in this instance ball M—presenting always the same face towards the center of motion, *actually* rotates upon its axis in the same sense, as indicated by the short full arrow. *But it does not tho to the eye it seems so.* The fallacy will become manifest on further inquiry.

To begin with, observe that when a mass, say the armature of an electric motor, rotating with the angular velocity ω, is reversed, its speed is $-\omega$ and the difference $\omega - (-\omega) = 2\omega$. Now, in fixing the ball to the spoke, the change of angular velocity is only ω; therefore, an additional velocity ω would have to be imparted to it in order to cause a clockwise rotation of the ball on its axis in the true significance of the word. The kinetic energy would then be equal to the sum of the energies of the translatory and axial motions, not merely in the abstract mathematical meaning, but as a physical fact. I am well aware that, according to the prevailing opinion, when the ball is free of the pivots it does not turn on its axis at all, and only rotates with the angular velocity of the frame when rigidly attached to the same, but the truth will appear upon a closer examination of this kind of movement.

Let the system be rotated as first assumed and illustrated, the balls being perfectly free on the pivots, and imagine the latter to be gradually tightened to cause friction slowly reducing and finally preventing the slip. At the outset all particles of each ball have been moving with the speed of its center of gravity, but as the bearing resistance asserts itself more and more the translatory velocity of the particles nearer to the axis O will be *diminishing*, while that of the diametrically opposite ones will be *increasing*, until the maxima of these changes are attained when the balls are firmly held. In this operation we have thus deprived those parts of the masses which are nearer to the center of motion, of some kinetic energy of *translation* while adding to the energy of those which are farther and, obviously, the gain was greater than the loss so that the *effective* velocity of each ball as a whole was increased. *Only so* have we augmented the kinetic energy of the system, not by causing *axial rotation* of the balls. The energy E of each of these is *solely* that of translatory movement with an effective velocity V_e as above defined such that $E = \frac{1}{2} M V_e^2$. The axial rotations of the ball in either direction are but apparent; *they have no reality whatever* and call for no mechanical effort. It is merely when an extraneous force acts independently to turn the whirling body on its axis that energy comes into play. Incidentally it should be pointed out that in true axial rotation of a rigid and homogenous mass all symmetrically situated particles contribute equally to the momentum which is not the case here. That there exists not even the slightest tendency to such motion can, however, be readily established.

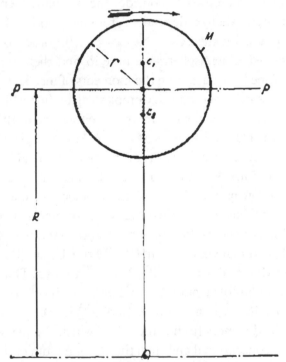

Fig. 6. Diagram showing a ball having mass M, of radius r, rotating about center O, and used in the theoretical analysis of the moon's motion.

For this purpose I would refer to Fig. 6 showing a ball M of radius r, the center C of which is at a distance R from axis O and which is bisected by a tangential plane $p\ p$ as indicated, the lower half sphere being shaded for distinction. The kinetic energy of the ball when whirled n times per second about O is according to the first form of expression $E = \frac{1}{2} M V_e^2 = \frac{1}{2} M (2 \pi R_g n)^2$, M being the mass and R_g the radius of gyration. But, as explained in connection with Fig. 4, we have also $E = \frac{1}{2} M V^2 + \frac{1}{2} I_c \omega^2$, $V = 2 \pi R n$ being the velocity of the center of gravity C and I_c the moment of inertia of the ball, about the parallel axis passing thru C and equal to $\frac{2}{5} M I^2$ so that $E = \frac{1}{2} M (2 \pi R n)^2 + \frac{1}{5} M r_2 (2 \pi n)^2$. Neither of these two expressions for E describes the actual state of the body but the first is certainly preferable conveying, as it does, the idea of a single motion instead of two, one of which moreover is devoid of existence. I shall first undertake to demonstrate that there is no torque or rotary effort about center C and that the kinetic energy of the supposed axial rotation of the ball is mathematically equal to zero. This makes it necessary to consider the two halves separated by the tangential plane $p\,p$ wholly independent from one another. Let c_1 and c_2 be their centers of gravity, then $C\,c_1 = C\,c_2 = \frac{3}{8}\,r$. In order to ascertain the kinetic energy of the hemispheres we have to find their radii of gyration which can be done by determining the moments of inertia I_{c_1} and I_{c_2} about parallel axes passing thru c_1 and c_2. Complex calculation will be avoided by remembering that the moment of inertia of either one of the half spheres about an axis thru C is $I_c = \frac{1}{2} \times \frac{2}{5} M r_2 = \frac{1}{5} M r_2$, and since $M = 2 m$, $I_c = \frac{2}{5} m r_2$. This can be exprest in terms of the moments I_{c_1} and I_{c_2}; namely $I_c = I_{c_1} + m (\frac{3}{8} r)^2 = I_{c_2} + m (\frac{3}{8} r)^2$. Hence $I_{c_1} = I_{c_2} = I_c - m (\frac{3}{8} r)^2 = \frac{2}{5} m r_2 - \frac{9}{64} m r_2 = \frac{83}{320} m r_2$. Following the same rule the moments of inertia of the half spheres about the axis passing thru the center of motion O can be found. Designating the moments for the upper and lower halves of the ball, respectively, I_{o_1} and I_{o_2} we have $I_{o_1} = m (R + \frac{3}{8} r)^2 + I_{c_1} = m (R + \frac{3}{8} r)^2 + \frac{83}{320} m r^2$ and $I_{o_2} = m (R - \frac{3}{8} r)^2 + I_{c_2} = m (R - \frac{3}{8} r)^2 + \frac{83}{320} m r^2$. Thus for the upper half sphere the radius of gyration $R_{g1} = \sqrt{(I_{o_1}/m)} = \sqrt{((R + \frac{3}{8} r)^2 + \frac{83}{320} r_2)}$ and for the lower one $R_{g2} = \sqrt{(I_{o_2}/m)} = \sqrt{((R - \frac{3}{8} r)^2 - \frac{83}{320} r^2)}$.

These are the distances from center O, at which the masses of the half spheres may be concentrated and then the algebraic sum of their energies—which are wholly translatory those of axial rotation being nil—will be exactly equal to the total kinetic energy of the ball as a unit. The significance of this will be understood by reference to Fig. 7 in which the two masses, condensed into points, are represented as attached to independent weightless strings of lengths R_{g1} and R_{g2} which are purposely shown as displaced but should be imagined as coincident. It will be readily seen that if both strings are cut in the same instant the masses will fly off in tangents to their circular orbits, the angular movement becoming rectilinear without any transformation of energy occurring. Let us now inquire what will happen if the two masses are rigidly joined, the connection being assumed imponderable. *Here we come to the real bug in*

Fig. 7. Here two masses *m-m*, are considered as condensed into points, attached to weightless strings of different radii. If both strings are cut, and the masses considered as joined, then there will be no rotation about the common center of gravity.

the question under discussion. Evidently, so long as the whirling motion continues, and both the masses have precisely the same angular velocity, this connecting link will be of no effect whatever, not the slightest turning effort about the common center of gravity of the masses or tendency of equalization of energy between them will exist. The moment the strings are broken and they are thrown off they will begin to rotate but, as pointed out before, this motion neither adds to or detracts from the energy stored. The rotation is, however, not due to an exclusive virtue of angular motion, but to the fact that the tangential velocities of the masses or parts of the body thrown off are different.

To make this clear and to investigate the effects produced, imagine two rifle barrels, as shown in Fig. 8, placed parallel to each other with their axes separated by a distance $R_{g1} - R_{g2}$ and assume that two balls of same diameter, each having mass m, are discharged with muzzle velocities V_1 and V_2 respectively equal to $2 \pi n R_{g1}$ and $2 \pi n R_{g2}$ as in the case just

FIG. 8. To Make the problem shown in Fig. 7 clear, imagine two rifle barrels parallel to each other. If two balls *m-m* are fired simultaneously, joined by a theoretical bond, they will revolve about their common center of gravity, proving that the moon possesses only kinetic energy of translation.

considered. If it be further supposed that at the instant of leaving the barrels the balls are joined by a rigid but weightless link they will rotate about their common center of gravity and in accordance with the statement in my previous article above mentioned, the relation will exist $(V_1 - V_2)/2 = \pi n (R_{g1} - R_{g2})$, n being the number of revolutions per second. The equalization of the speeds and kinetic energies of the balls will be, under these circumstances, very rapid but in two heavenly bodies linked by gravitational attraction, the process might require ages. Now, this whirling movement is real and requires energy which, obviously, must be derived from that originally imparted and, consequently, must reduce the velocity of the balls in the direction of flight by an amount which can be easily calculated. At the moment of discharge the total kinetic energy was $E = \frac{1}{2} m V_1^2 + \frac{1}{2} m V_2^2$ which is evidently equal to $m V_3^2$, V_3 being the effective velocity of the common center of gravity, from which follows that $V_3 = \sqrt{(V_1^2 - V_2^2/2)}$. The speed of revolution of the masses is, of course $(V_1 - V_2/2)$ and the rotational energy of both balls, which must be considered as points, is $e = m (V_1 - V_2)/2$. The kinetic energy of translation in the direction of flight is then $\frac{1}{2} m V_1^2 + \frac{1}{2} m V_2^2 - m ((V_1 - V_2)/2)^2 = m ((V_1 + V_2)/2)^2 = m V_1^2$, $V_1 - (V_1 - V_2)/2$ being the speed of the common center of gravity, so that $V_2 - V_1$ is the loss of velocity in the direction of flight owing to the rotation of the two mass points. If instead of these we would deal with the balls as they are, their rotational energy $e_1 = e + i \omega^2 = m ((V_1 + V_2)/2)^2 + i (2 \pi n)^2$, i being the moment of inertia of each ball about its axis.

As will be seen, we arrive at precisely the same results whether the movement is rectilinear or in a circle. In both cases the total kinetic energy can be divided into two parts, respectively of the same numerical values, but *there is an essential difference*. In angular motion the axial rotation is nothing more than an *abstract conception*; in rectilinear movement it is a *positive event*.

Virtually all satellites rotate in like manner and the probability that the acceleration or retardation of their axial motions—if they ever existed—should come to a stop precisely at a definite angular velocity, is infinitesimal while it is almost absolutely certain that all movement of this kind would ultimately cease. The most plausible view is that no true moon has ever rotated on its axis, for at the time of its birth there must have been some deformation and displacement of its center of gravity thru the attractive force of the mother planet so as to make its peculiar position in space, relative to the latter, in which it persists irrespective of distance, more or less stable. In explanation of this, suppose that one of the balls as M in Fig. 5 is not of homogenous material and that it is similarly supported but on an axis passing thru its center of gravity instead of form. Then, no matter in what position the ball is fixed on the pivots, its kinetic energy and centrifugal pull will be the same. Nevertheless a directive tendency will exist as the two centers do not coincide and there is, consequently, no *dynamic* balance. When permitted to turn freely on the axis of gravity the body, of whatever shape it may be, will tend to place itself so that the line joining the two centers points to O and there may be two positions of stability but, generally, if the center of gravity is not greatly displaced, the heavier side will swing outwardly. Such condition may obtain in the moon if it had solidified before receding from the earth to great distance, when the arrangement of the masses in its interior became subject to gravitational forces of its own, vastly greater than the terrestrial. It has been suggested that the planet is egg-shaped or ellipsoidal but the departure from spherical form must be inconsiderable. It may even be a perfect sphere with the centers of gravity and symmetry coinciding and still rotate as it does. Whatever be its origin and past history, the fact is, that at present all its parts have the same angular velocity as though it were rigidly connected with the earth. This state must endure forever unless forces from without the luna-terrestrial system bring about different conditions and thus the hope of the star-gazers that its other side may become visible some day must be indefinitely deferred.

A motion of this character, as I have shown, precludes the possibility of axial rotation. The easiest way to free ourselves of this illusion is to conceive the satellite subdivided into minute and entirely independent parts, as dust particles, which have different orbital, but rigorously the same angular, velocities. One must at once recognize that the kinetic energy of such an agglomeration is solely translational, there being absolutely no tendency to axial rotation. This makes it also perfectly clear why the moon, provided its distance does not greatly increase, *must* always turn the same face to us even *without any inherent directive tendency* nor so much as *the slightest effort from the earth.*

Referring to the librations of longitude, I do not see that they have any bearing on this question. In astronomical treatises the axial rotation of the moon is accepted as a material fact and it is thought that its angular velocity is constant while that of the orbital movement is not, this resulting in an apparent oscillation revealing more of its surface to our view. To a degree this may be true, but I hold that the mere change of orbital velocity, as will be evident from what has been stated before could not produce these phenomena, for no matter how fast or slow the gyration, the position of the body relative to the center of attraction remains the same. The real cause of these axial displacements is the changing distance of the moon from the earth owing to which the tangential components of velocity of its parts are varied. In *apogee*, when the planet recedes, the radial component of velocity decreases while the tangential increases but, as the decrement of the former is the same for all parts, this is more pronounced in the regions facing the earth than in those turned away from it, the consequence being an axial displacement exposing more of the eastern side. In *perigee*, on the contrary, the radial component increases and the effect is just the opposite with the result that more of the western side is seen. The moon actually swings on the axis passing thru its center of gravity on which it is supported like a ball on a string. The forces involved in these pendular movements are incomparably smaller than those required to effect changes in orbital velocity. If we estimate the radius of gyration of the satellite at 600 miles and its mean distance from the earth at 240,000 miles, then the energy necessary to rotate it once in a month would be only $(600 / 240,000)^2 = 1/160,000$ of the kinetic energy of the orbital movement.

ELECTRICAL OSCILLATORS
JULY 1919

Mr. Tesla makes a very important contribution to the electrical arts with this article. The pioneer of all high-frequency apparatus divulges much that is new and startling in these pages. Few people realize the enormous value of Mr. Tesla's machines and the many different important uses to which they can be applied in our everyday lives. New and startling uses are being found every year for these machines.

It is characteristic of Mr. Tesla that he has developed and actually built an astounding variation of these machines, and we regret that we can publish only a very few of the more important models. Most of the Tesla coils shown have never been publisht before.

—EDITOR

Few fields have been opened up the exploration of which has proved as fruitful as that of high-frequency currents. Their singular properties and the spectacular character of the phenomena they presented immediately commanded universal attention. Scientific men became interested in their investigation, engineers were attracted by their commercial possibilities, and physicians recognized in them a long-sought means for effective treatment of bodily ills. Since the publication of my first researches in 1891, hundreds of volumes have been written on the subject and many invaluable results obtained thru the medium of this new agency. Yet, the art is only in its infancy and the future has incomparably bigger things in store.

From the very beginning I felt the necessity of producing efficient apparatus to meet a rapidly growing demand and during the eight years succeeding my original announcements I developed not less than fifty types of these transformers or electrical oscillators, each complete in every detail and refined to such a degree that I could not materially improve any one of them today. Had I been guided by practical considerations I might have built up an immense and profitable business, incidentally rendering important services to the world. But the force of circumstances and the ever enlarging vista of

Electrical Oscillators

Fig. 2. Small Tesla Coil for gas engine ignition and similar uses

Fig. 3. Tesla Transformer, 12 inch spark, chiefly for wireless

Fig. 5. Later type of Tesla Transformer

Fig. 1. Oscillator with detachable transformer for experimental purposes

Fig. 4. Tesla Oscillator in action, generating undamped waves

Fig. 6. Small oscillator for production of ozone

Fig. 7. Large Tesla Transformer for various purposes

Fig. 8. Tesla Transformer with rotary break for wireless

Fig. 9. Tesla Transformer with mercury interrupter

Fig. 10. Large Tesla Transformer with hermetically sealed mercury interrupter

Fig. 12. Another type of Tesla Transformer with sealed mercury interrupter

Fig. 11. Tesla Transformer with sealed mercury interrupter for low tension work greater achievements turned my efforts in other directions. And so it comes that instruments will shortly be placed on the market which, oddly enough, were perfected twenty years ago!

These oscillators are expressly intended to operate on direct and alternating lighting circuits and to generate damped and undamped oscillations or currents of any frequency, volume and tension within the widest limits. They are compact, self-contained, require no care for long periods of time and will be found very convenient and useful for various purposes as, wireless telegraphy and telephony; conversion of electrical energy; formation of chemical compounds thru fusion and combination; synthesis of gases; manufacture of ozone; lighting; welding; municipal, hospital, and domestic sanitation and sterilization, and numerous other applications in scientific laboratories and industrial institutions.

While these transformers have never been described before, the general principles underlying them were fully set forth in my publisht articles and patents, more particularly those of September 22, 1896, and it is thought, therefore, that the appended photographs of a few types, together with a short explanation, will convey all the information that may be desired.

The essential parts of such an oscillator are: condenser, a self-induction coil for charging the same to a high potential, a circuit controller, and a transformer which is energized by the oscillatory discharges of the condenser. There are at least three, but usually four, five or six, circuits in tune and the regulation is effected in several ways, most frequently merely by means of an adjusting screw. Under favorable conditions an efficiency as high as 85% is attainable, that is to say, that percentage of energy supplied can be recovered in the secondary of the transformer. While the chief virtue of this kind of apparatus is obviously due to the wonderful powers of the condenser, special qualities result from concatenation of circuits under observance of accurate harmonic relations, and minimization of frictional and other losses which has been one of the principal objects of the design.

Broadly, the instruments can be divided into two classes: one in which the circuit controller comprises solid contacts, and the other in which the make and break is effected by mercury. Figures 1 to 8, inclusive, belong to the first, and the remaining ones to the second class. The former are capable of an appreciably higher efficiency on account of the fact that the losses involved in the make and break are reduced to the minimum and the resistance component of the damping factor is very small. The latter are preferable for purposes requiring larger output and a great number of breaks per second. The operation of the motor and circuit controller of course consumes a certain amount of energy which, however, is the less significant the larger the capacity of the machine.

In Fig. 1 is shown one of the earliest forms of oscillator constructed for experimental purposes. The condenser is contained in a square box of mahogany upon which is mounted the self-induction or charging coil wound, as will be noted, in two sections connected in multiple or series according to whether the tension of the supply circuit is 110 or 220 volts. From the box protrude four brass columns carrying a plate with the spring contacts and adjusting screws as well as two massive terminals for the reception of the primary of the transformer. Two of the columns serve as condenser connections while the other pair is employed to join the binding posts of the switch in front to the self-inductance and condenser. The primary coil consists of a few turns of copper ribbon to the ends of which are soldered short rods fitting into the terminals referred to. The secondary is made in two parts, wound in a manner to reduce as much as possible the distributed capacity and at the same time enable the coil to withstand a very high-pressure between its terminals at the center, which are connected to binding posts on two rubber columns projecting from the primary. The circuit connections may be slightly varied but ordinarily they are as diagrammatically illustrated in the *Electrical Experimenter* for May on page 89, relating to my oscillation transformer photograph of which appeared on page 16 of the same number. The operation is as follows: When the switch is thrown on, the current from the supply circuit rushes thru the self-induction coil, magnetizing the iron core within and separating the contacts of the controller. The high tension induced current then charges the condenser and upon closure of the contacts the accumulated energy is released thru the primary, giving rise to a long series of oscillations which excite the tuned secondary circuit.

This device has proved highly serviceable in carrying on laboratory experiments of all kinds. For instance, in studying phenomena of impedance, the transformer was removed and a bent copper bar inserted in the terminals. The latter was often replaced by a large circular loop to exhibit inductive effects at a distance or to excite resonant circuits used in various investigations and measurements. A transformer suitable for any desired performance could be readily improvised and attached to the terminals and in this way much time and labor was saved. Contrary to what might be naturally expected, little trouble was experienced with the contacts, altho the currents thru them were heavy, namely, proper conditions of resonance existing, the great flow occurs only when the circuit is closed and no destructive arcs can develop. Originally I employed platinum and iridium tips but later replaced them by some of meteorite and finally of tungsten. The last have given the best satisfaction, permitting working for hours and days without interruption.

Fig. 2 illustrates a small oscillator designed for certain specific uses. The underlying idea was to attain great activities during minute intervals of time each succeeded by a comparatively long period of inaction. With this object a large self-induction and a quick-acting break were employed owing to which arrangement the condenser was charged to a very high potential. Sudden secondary currents and sparks of great volume were thus obtained, eminently suitable for welding thin wires, flashing lamp filaments, igniting explosive mixtures and kindred applications. The instrument was also adapted for battery use and in this form was a very effective igniter for gas engines on which a patent bearing number 609,250 was granted to me August 16, 1898.

Fig. 3 represents a large oscillator of the first class intended for wireless experiments, production of Röntgen rays and scientific research in general. It comprises a box containing two condensers of the same capacity on which are supported the charging coil and transformer. The automatic circuit controller, hand switch and connecting posts are mounted on the front plate of the inductance spool as is also one of the contact springs. The condenser box is equipt with three terminals, the two external ones serving merely for connection while the middle one carries a contact bar with a screw for regulating the interval during which the circuit is closed. The vibrating spring itself, the sole function of which is to cause periodic interruptions, can be adjusted in its strength as well as distance from the iron core in the center of the charging coil by four screws visible on the top plate so that any desired conditions of mechanical control might be secured. The primary coil of the transformer is of copper sheet and taps are made at suitable points for the purpose of varying, at will, the number of turns. As in Fig. 1 the inductance coil is wound in two sections to adapt the instrument both to 110 and 220 volt circuits and several secondaries were provided to suit the various wave lengths of the primary. The output was approximately 500 watt with damped waves of about 50,000 cycles per second. For short periods of time undamped oscillations were produced in screwing the vibrating spring tight against the iron core and separating the contacts by the adjusting screw *which also performed the function of a key.* With this oscillator I made a number of important observations and it was one of the machines exhibited at a lecture before the New York Academy of Sciences in 1897.

Fig. 4 is a photograph of a type of transformer in every respect similar to the one illustrated in the May, 1919, issue of the *Electrical Experimenter* to which reference has already been made. It contains the identical essential parts, disposed in like manner, but was specially designed for use on supply circuits of higher tension, from 220 to 500 volts or more. The usual adjustments are made in setting the contact spring and shifting the iron core within the inductance coil up and down by means of two screws. In order to prevent injury thru a short-circuit, fuses are inserted in the lines. The instrument was photographed in action, generating undamped oscillations from a 220 volt lighting circuit.

Fig. 5 shows a later form of transformer principally intended to replace Rhumkorf coils. In this instance a primary is employed, having a much greater number of turns and the secondary is closely linked with the same. The currents developed in the latter, having a tension of from 10,000 to 30,000 volts, are used to charge condensers and operate an independent high-frequency coil as customary. The controlling mechanism is of somewhat different construction but the core and contact spring are both adjustable as before.

Fig. 6 is a small instrument of this type, particularly intended for ozone production or sterilization. It is remarkably efficient for its size and can be connected either to a 110 or 220 volt circuit, direct or alternating, preferably the former.

In Fig. 7 is shown a photograph of a larger transformer of this kind. The construction and disposition of the parts is as before but there are two condensers in the box, one of which is connected in the circuit as in the previous cases, while the other is in shunt to the primary coil. In this manner currents of great volume are produced in the latter and the secondary effects are accordingly magnified. The introduction of an additional tuned circuit secures also other advantages but the adjustments are rendered more difficult and for this reason it is desirable to use such an instrument in the production of currents of a definite and unchanging frequency.

Fig. 8 illustrates a transformer with rotary break. There are two condensers of the same capacity in the box which can be connected in series or multiple. The charging inductances are in the form of two long spools upon which are supported the secondary terminals. A small direct-current motor, the speed of which can be varied within wide limits, is employed to drive a specially constructed make and break. In other features the oscillator is like the one illustrated in Fig. 3 and its operation will be readily understood from the foregoing. This transformer was used in my wireless experiments and frequently also for lighting the laboratory by my vacuum tubes and was likewise exhibited at my lecture before the New York Academy of Sciences above mentioned.

Coming now to machines of the second class, Fig. 9 shows an oscillatory transformer comprising a condenser and charging inductance enclosed in a box, a transformer and a mercury circuit controller, the latter being of a construction described for the first time in my patent No. 609,251 of August 16, 1898. It consists of a motor driven hollow pulley containing a small quantity of mercury which is thrown outwardly against the walls of the vessel by centrifugal force and entrains a contact wheel which periodically closes and opens the condenser circuit. By means of adjusting screws above the pulley, the depth of immersion of the vanes and consequently, also, the duration of each contact can be varied at desire and thus the intensity of the effects and their character controlled. This form of break has given thoro satisfaction, working continuously with currents of from 20 to 25 amperes. The number of interruptions is usually from 500 to 1,000 per second but higher frequencies are practicable. The space occupied is about 10" x 8" x 10" and the output approximately ½ K.W.

In the transformer just described the break is exposed to the atmosphere and a slow oxidation of the mercury takes place. This disadvantage is overcome in the instrument shown in Fig. 10, which consists of a perforated metal box containing the condenser and charging inductance and carrying on the top a motor driving the break, and a transformer. The mercury break is of a kind to be described and operates on the principle of a jet which establishes, intermittently, contact with a rotating wheel in the interior of the pulley. The stationary parts are supported in the vessel on a bar passing thru the long hollow shaft of the motor and a mercury seal is employed to effect hermetic closure of the chamber enclosing the circuit controller. The current is led into the interior of the pulley thru two sliding rings on the top which are in series with the condenser and primary. The exclusion of the oxygen is a decided improvement, the deterioration of the metal and attendant trouble being eliminated and perfect working conditions continuously maintained.

Fig. 11 is a photograph of a similar oscillator with hermetically inclosed mercury break. In this machine the stationary parts of the interrupter in the interior of the pulley were supported on a tube thru which was led an insulated wire connecting to one terminal of the break while the other was in contact with the vessel. The sliding rings were, in this manner, avoided and the construction simplified. The instrument was designed for oscillations of lower tension and frequency requiring primary currents of comparatively smaller amperage and was used to excite other resonant circuits.

Fig. 12 shows an improved form of oscillator of the kind described in Fig. 10, in which the supporting bar thru the hollow motor shaft was done away with, the device pumping the mercury being kept in position by gravity, as will be more fully explained with reference to another figure. Both the capacity of the condenser and primary turns were made variable with the view of producing oscillations of several frequencies.

Fig. 13 is a photographic view of another form of oscillatory transformer with hermetically sealed mercury interrupter, and Fig. 14

diagrams showing the circuit connections and arrangement of parts reproduced from my patent, No. 609,245, of August 16, 1898, describing this particular device. The condenser, inductance, transformer and circuit controller are disposed as before, but the latter is of different construction, which will be clear from an inspection of Fig. 14. The hollow pulley *a* is secured to a shaft *c* which is mounted in a vertical bearing passing thru the stationary field-magnet *d* of the motor. In the interior of the vessel is supported, on frictionless bearings, a body *h* of

Fig. 13. Tesla oscillator with magnetically controlled, sealed mercury interrupter.

magnetic material which is surrounded by a dome *b* in the center of a laminated iron ring, with pole-pieces *o o* wound with energizing coils *p*. The ring is supported on four columns and, when magnetized, keeps the body *h* in position while the pulley is rotated. The latter is of steel, but the dome is preferably made of German silver burnt black by acid or nickeled. The body *h* carries a short tube *k* bent, as indicated, to

FIG. 14. Electrical oscillator, illustrated in Fig. 13, showing details and circuit connections.

catch the fluid as it is whirled around, and project it against the teeth of a wheel fastened to the pulley. This wheel is insulated and contact from it to the external circuit is established thru a mercury cup. As the pulley is rapidly rotated a jet of the fluid is thrown against the wheel, *thus making and breaking contact about 1,000 times per second.* The instrument works silently and, owing to the absence of all deteriorating agents, keeps continually clean and in perfect condition. The number of interruptions per second may be much greater, however, so as to make the currents suitable for wireless telephony and like purposes.

FIG. 15. Tesla transformer with gravity controlled, sealed mercury interrupter.

A modified form of oscillator is represented in Figs. 15 and 16, the former being a photographic view and the latter a diagrammatic illustration showing the arrangement of the interior parts of the controller. In this instance the shaft *b* carrying the vessel *a* is hollow and supports, in frictionless bearings, a spindle *j* to which is fastened a weight *k*. Insulated from the latter, but mechanically fixt to it, is a curved arm ʟ upon which is supported, freely rotatable, a break-wheel with projections Q Q. The wheel is in electrical connection with the

Fig. 16. Electrical oscillator, illustrated in Fig. 15, showing details of motor and break mechanism.

external circuit thru a mercury cup and an insulated plug supported from the top of the pulley. Owing to the inclined position of the motor the weight k keeps the break-wheel in place by the force of gravity and as the pulley is rotated the circuit, including the condenser and primary coil of the transformer, is rapidly made and broken.

Fig. 17 shows a similar instrument in which, however, the make and break device is a jet of mercury impinging against an insulated toothed wheel carried on an insulated stud in the center of the cover of the pulley as shown. Connection to the condenser circuit is made by brushes bearing on this plug.

Fig. 17. Tesla transformer with adjustable mercury contoller.

Fig. 18. Tesla transformer with mercury jet interrupter.

Fig. 18 is a photograph of another transformer with a mercury circuit controller of the wheel type, modified in some features on which it is unnecessary to dwell. These are but a few of the oscillatory transformers I have perfected and constitute only a small part of my high-frequency apparatus of which I hope to give a full description, when I shall have freed myself of pressing duties, at some future date.

The Problem of Increasing Human Energy

WITH SPECIAL REFERENCE TO THE HARNESSING OF THE SUN'S ENERGY

The Onward Movement of Man— the Energy of the Movement— the Three Ways of Increasing Human Energy

As Originally Published in *Century Magazine*, June 1900

Of all the endless variety of phenomena which nature presents to our senses, there is none that fills our minds with greater wonder than that inconceivably complex movement which, in its entirety, we designate as human life. Its mysterious origin is veiled in the forever impenetrable mist of the past, its character is rendered incomprehensible by its infinite intricacy, and its destination is hidden in the unfathomable depths of the future. Whence does it come? What is it? Whither does it tend? are the great questions which the sages of all times have endeavored to answer.

Modern science says: The sun is the past, the earth is the present, the moon is the future. From an incandescent mass we have originated, and into a frozen mass we shall turn. Merciless is the law of nature, and rapidly and irresistibly we are drawn to our doom. Lord Kelvin, in his profound meditations, allows us only a short span of life, something like six million years, after which time the sun's bright light will have ceased to shine, and its life-giving heat will have ebbed away, and our own earth will be a lump of ice, hurrying on through the eternal night. But do not let us despair. There will still be left upon it a glimmering spark of life, and there will be a chance to kindle a new fire on some distant star. This wonderful possibility seems, indeed, to exist, judging from Professor Dewar's beautiful experiments with liquid air, which show that germs of organic life are not destroyed by cold, no matter how intense; consequently they may be transmitted through the interstellar space. Meanwhile the cheering lights of science and art, ever increasing in intensity, illuminate our path, and marvels they disclose, and the enjoyments they offer, make us measurably forgetful of the gloomy future.

Though we may never be able to comprehend human life, we know certainly that it is a movement, of whatever nature it be. The existence of movement unavoidably implies a body which is being moved and a force which is moving it. Hence, wherever there is life, there is a mass moved by a force. All mass possesses inertia, all force tends to persist.

Owing to this universal property and condition, a body, be it at rest or in motion, tends to remain in the same state, and a force, manifesting itself anywhere and through whatever cause, produces an equivalent opposing force, and as an absolute necessity of this it follows that every movement in nature must be rhythmical. Long ago this simple truth was clearly pointed out by Herbert Spencer, who arrived at it through a somewhat different process of reasoning. It is borne out in everything we perceive— in the movement of a planet, in the surging and ebbing of the tide, in the reverberations of the air, the swinging of a pendulum, the oscillations of an electric current, and in the infinitely varied phenomena of organic life. Does not the whole of human life attest to it? Birth, growth, old age, and death of an individual, family, race, or nation, what is it all but a rhythm? All life-manifestation, then, even in its most intricate form, as exemplified in man, however involved and inscrutable, is only a movement, to which the same general laws of movement which govern throughout the physical universe must be applicable.

FIG. 1. Burning the nitrogen of the atmosphere. Note to Fig. 1.—This result is produced by the discharge of an electrical oscillator giving twelve million volts. The electrical pressure, alternating one hundred thousand times per second, excites the normally inert nitrogen, causing it to combine with the oxygen. The flame-like discharge shown in the photograph measures sixty-five feet across.

When we speak of man, we have a conception of humanity as a whole, and before applying scientific methods to the investigation of his movement we must accept this as a physical fact. But can anyone doubt today that all the millions of individuals and all the innumerable types and characters constitute an entity, a unit? Though free to think and act, we are held together, like the stars in the firmament, with ties inseparable. These ties cannot be seen, but we can feel them. I cut myself in the finger, and it pains me: this finger is part of me. I see a friend hurt, and it hurts me, too: my friend and I are one. And now I see stricken down an enemy, a lump of matter which, of all the lumps of matter in the universe, I care least for, and it still grieves me. Does this not prove that each of us is only part of a whole?

For ages this idea has been proclaimed in the consummately wise teachings of religion, probably not alone as a means of insuring peace and harmony among men, but as a deeply founded truth. The Buddhist expresses it in one way, the Christian in another, but both say the same: We are all one. Metaphysical proofs are, however, not the only ones which we are able to bring forth in support of this idea. Science, too, recognizes this connectedness of separate individuals, though not quite in the same sense as it admits that the suns, planets, and moons of a constellation are one body, and there can be no doubt that it will be experimentally confirmed in times to come, when our means and methods for investigating psychical and other states and phenomena shall have been brought to great perfection. Still more: this one human being lives on and on. The individual is ephemeral, races and nations come and pass away, but man remains. Therein lies the profound difference between the individual and the whole. Therein, too, is to be found the partial explanation of many of those marvelous phenomena of heredity which are the result of countless centuries of feeble but persistent influence.

Conceive, then, man as a mass urged on by a force. Though this movement is not of a translatory character, implying change of place, yet the general laws of mechanical movement are applicable to it, and the energy associated with this mass can be measured, in accordance with well-known principles, by half the product of the mass with the square of a certain velocity. So, for instance, a cannonball which is at rest possesses a certain amount of energy in the form of heat, which we measure in a similar way. We imagine the ball to consist of innumerable minute particles, called atoms or molecules, which vibrate or whirl around one another. We determine their masses and velocities, and from them the energy of each of these minute systems, and adding them all together, we get an idea of the total heat-energy contained in the ball, which is only seemingly at rest. In this purely theoretical estimate this energy may then be calculated by multiplying half of the total mass—that is half of the sum of all the small masses—with the square of a velocity which is determined from the velocities of the separate particles. In like manner we may conceive of human energy being measured by half the human mass multiplied with the square of the velocity which we are not yet able to compute. But our deficiency in this knowledge will not vitiate the truth of the deductions I shall draw, which rest on the firm basis that the same laws of mass and force govern throughout nature.

Man, however, is not an ordinary mass, consisting of spinning atoms and molecules, and containing merely heat-energy. He is a mass possessed of certain higher qualities by reason of the creative principle of life with which he is endowed. His mass, as the water in an ocean wave, is being continuously exchanged, new taking the place of the old. Not only this, but he grows, propagates, and dies, thus altering his mass independently, both in bulk and density. What is most wonderful of all, he is capable of increasing or diminishing his velocity of movement by the mysterious power he possesses by appropriating more or less energy from other substance, and turning it into motive energy. But in any given moment we may ignore these slow changes and assume that human energy is measured by half the product of man's mass with the square of a certain hypothetical velocity. However we may compute this velocity, and whatever we may take as the standard of its measure, we must, in harmony with this conception, come to the conclusion that the great problem of science is, and always will be, to increase the energy thus defined. Many years ago, stimulated by the perusal of that deeply interesting work, Draper's "History of the Intellectual Development of Europe," depicting so vividly human movement, I recognized that to solve this eternal problem must ever be the chief task of the man of science.

Some results of my own efforts to this end I shall endeavor briefly to describe here.

DIAGRAM *a*. The three ways of increasing human energy.

Let, then, in diagram *a*, *M* represent the mass of man. This mass is impelled in one direction by a force *f*, which is resisted by another partly frictional and partly negative force *R*, acting in a direction exactly opposite, and retarding the movement of the mass. Such an antagonistic force is present in every movement and must be taken into consideration. The difference between these two forces is the effective force which imparts a velocity *V* to the mass *M* in the direction of the arrow on the line representing the force *f*. In accordance with the preceding, the human energy will then be given by the product $\frac{1}{2} MV2 = \frac{1}{2} MV \times V$, in which *M* is the total mass of man in the ordinary interpretation of the term "mass," and *V* is a certain hypothetical velocity, which, in the present state of science, we are unable exactly to define and determine. To increase the human energy is, therefore, equivalent to increasing this product, and there are, as will readily be seen, only three ways possible to attain this result, which are illustrated in the above diagram. The first way shown in the top figure, is to increase the mass (as indicated by the dotted circle), leaving the two opposing forces the same. The second way is to reduce the retarding force R to a smaller value R, leaving the mass and the impelling force the same, as diagrammatically shown in the middle figure. The third way, which is illustrated in the last figure, is to increase the impelling force F to a higher value F, while the mass and the retarding force R remain unaltered. Evidently fixed limits exist as regards increase of mass and reduction of retarding force, but the impelling force can be increased indefinitely. Each of these three possible solutions presents a different aspect of the main problem of increasing human energy, which is thus divided into three distinct problems, to be successively considered.

The First Problem:
How to Increase the Human Mass—
the Burning of Atmospheric Nitrogen

Viewed generally, there are obviously two ways of increasing the mass of mankind: first, by aiding and maintaining those forces and conditions which tend to increase it; and, second, by opposing and reducing those which tend to diminish it. The mass will be increased by careful attention to health, by substantial food, by moderation, by regularity of habits, by promotion of marriage, by conscientious attention to children, and, generally stated, by the observance of all the many precepts and laws of religion and hygiene. But in adding new mass to the old, three cases again present themselves. Either the mass added is of the same velocity as the old, or it is of a smaller or of a higher velocity. To gain an idea of the relative importance of these cases, imagine a train composed of, say, one hundred locomotives running on a track, and suppose that, to increase the energy of the moving mass, four more locomotives are added to the train. If these four move at the same velocity at which the train is going, the total energy will be increased four percent; if they are moving at only one half of that velocity, the increase will amount to only one percent; if they are moving at twice that velocity, the increase of energy will be sixteen percent. This simple illustration shows that it is of greatest importance to add mass of a higher velocity. Stated more to the point, if, for example, the children be of the same degree of enlightenment as the parents—that is, mass of the "same velocity"— the energy will simply increase proportionately to the number added. If they are less intelligent or advanced, or mass of "smaller velocity" there will be a very slight gain in the energy; but if they are further advanced, or mass of "higher velocity," then the new generation will add very considerably to the sum total of human energy. Any addition of mass of "smaller velocity," beyond that indispensable amount required by the law expressed in the proverb, *Mens sana in corpore sano*," should be strenuously opposed. For instance, the mere development of muscle, as aimed at in some of our colleges, I consider equivalent to adding mass of "smaller velocity," and I would not commend it, although my views were different when I was a student myself. Moderate exercise, insuring the right balance between mind and body, and the highest efficiency of performance, is, of course, a prime requirement. The above example shows that the most important result to be attained is the education, or the increase of the "velocity," of the mass newly added.

Conversely, it scarcely need be stated that everything that is against the teachings of religion and the laws of hygiene is tending to decrease the mass. Whisky, wine, tea, coffee, tobacco, and other such stimulants are responsible for the shortening of the lives of many, and ought to be used with moderation. But I do not think that rigorous measures of suppression of habits followed through many generations are commendable. It is wiser to preach moderation than abstinence. We have become accustomed to these stimulants, and if such reforms are to be effected, they must be slow and gradual. Those who are devoting their energies to such ends could make themselves far more useful by turning their efforts in other directions, as, for instance, toward providing pure water.

For every person who perishes from the effects of a stimulant, at least a thousand die from the consequences of drinking impure water. This precious fluid, which daily infuses new life into us, is likewise the chief vehicle through which disease and death enter our bodies. The germs of destruction it conveys are enemies all the more terrible as they perform their fatal work unperceived. They seal our doom while we live and enjoy. The majority of people are so ignorant or careless in drinking water, and the consequences of this are so disastrous, that a philanthropist can scarcely use his efforts better than by endeavoring to enlighten those who are thus injuring themselves. By systematic purification and sterilization of the drinking water the human mass would be very considerably increased. It should be made a rigid rule— which might be enforced by law—to boil or to sterilize otherwise the drinking water in every household and public place. The mere filtering does not afford sufficient security against infection. All ice for internal uses should be artificially prepared from water thoroughly sterilized. The importance of eliminating germs of disease from the city water is generally recognized, but little is being done to improve the existing conditions, as no satisfactory method of sterilizing great quantities of water has yet been brought forward. By improved electrical appliances we are now enabled to produce ozone cheaply and in large amounts, and this ideal disinfectant seems to offer a happy solution of the important question.

Gambling, business rush, and excitement, particularly on the exchanges, are causes of much mass reduction, all the more so because the individuals concerned represent units of higher value. Incapacity of observing the first symptoms of an illness, and careless neglect of the same, are important factors of mortality. In noting carefully every new sign of approaching danger, and making conscientiously every possible effort to avert it, we are not only following wise laws of hygiene in the interest of our well-being and the success of our labors, but we are also complying with a higher moral duty. Everyone should consider his body as a priceless gift from one whom he loves above all, as a marvelous work of art, of indescribable beauty and mastery beyond human conception, and so delicate and frail that a word, a breath, a look, nay, a thought, may injure it. Uncleanness, which breeds disease and death, is not only a self destructive but highly immoral habit. In keeping our bodies free from infection, healthful, and pure, we are expressing our reverence for the high principle with which they are endowed. He who follows the precepts of hygiene in this spirit is proving himself, so far, truly religious. Laxity of morals is a terrible evil, which poisons both mind and body, and which is responsible for a great reduction of the human mass in some countries. Many of the present customs and tendencies are productive of similar hurtful results. For example, the society life, modern education and pursuits of women, tending to draw them away from their household duties and make men out of them, must needs detract from the elevating ideal they represent, diminish the artistic creative power, and cause sterility and a general weakening of the race. A thousand other evils might be mentioned, but all put together, in their bearing upon the problem under discussion, they could not equal a single one, the want of food, brought on by poverty, destitution, and famine. Millions of individuals die yearly for want of food, thus keeping down the mass. Even in our enlightened communities, and notwithstanding the many charitable efforts, this is still, in all probability, the chief evil. I do not mean here absolute want of food, but want of healthful nutriment.

How to provide good and plentiful food is, therefore, a most important question of the day. On the general principles the raising of cattle as a means of providing food is objectionable, because, in the sense interpreted above, it must undoubtedly tend to the addition of mass of a "smaller velocity." It is certainly preferable to raise vegetables, and I think, therefore, that vegetarianism is a commendable departure from the established barbarous habit. That we can subsist on plant food and perform our work even to advantage is not a theory, but a well-demonstrated fact. Many races living almost exclusively on vegetables are of superior physique and strength. There is no doubt that some plant food, such as oatmeal, is more economical than meat, and superior to it in regard to both mechanical and mental performance. Such food, moreover, taxes our digestive organs decidedly less, and, in making us more contented and sociable, produces an amount of good difficult to estimate. In view of these facts every effort should be made to stop the wanton and cruel slaughter of animals, which must be destructive to our morals. To free ourselves from animal instincts and appetites, which keep us down, we should begin at the very root from which we spring: we should effect a radical reform in the character of the food.

There seems to be no *philosophical* necessity for food. We can conceive of organized beings living without nourishment, and deriving all the energy they need for the performance of their life functions from the ambient medium. In a crystal we have the clear evidence of the existence of a formative life-principle, and though we cannot understand the life of a crystal, it is none the less a living being. There may be, besides crystals, other such individualized, material systems of beings, perhaps of gaseous constitution, or composed of substance still more tenuous. In view of this possibility—nay, probability—we cannot apodictically deny the existence of organized beings on a planet merely because the conditions on the same are unsuitable for the existence of life as we conceive it. We cannot even, with positive assurance, assert that some of them might not be present here, in this our world, in the very midst of us, for their constitution and life-manifestation may be such that we are unable to perceive them.

The production of artificial food as a means for causing an increase of the human mass naturally suggests itself, but a direct attempt of this kind to provide nourishment does not appear to me rational, at least not for the present. Whether we could thrive on such food is very doubtful. We are the result of ages of continuous adaptation, and we cannot radically change without unforeseen and, in all probability, disastrous consequences. So uncertain an experiment should not be tried. By far the best way, it seems to me, to meet the ravages of the evil, would be to find ways of increasing the productivity of the soil. With this object the preservation of forests is of an importance which cannot be overestimated, and in this connection, also, the utilization of water-power for purposes of electrical transmission, dispensing in many ways with the necessity of burning wood, and tending thereby to forest preservation, is to be strongly advocated. But there are limits in the improvement to be effected in this and similar ways.

To increase materially the productivity of the soil, it must be more effectively fertilized by artificial means. The question of food-production resolves itself, then, into the question how best to fertilize the soil. What it is that made the soil is still a mystery. To explain its origin is probably equivalent to explaining the origin of life itself. The rocks, disintegrated by moisture and heat and wind and weather, were in themselves not capable of maintaining life. Some unexplained condition arose, and some new principle came into effect, and the first layer capable of sustaining low organisms, like mosses, was formed. These, by their life and death, added more of the life-sustaining quality to the soil, and higher organisms could then subsist, and so on and on, until at last highly developed plant and animal life could flourish.

But though the theories are, even now, not in agreement as to how fertilization is effected, it is a fact, only too well ascertained, that the soil cannot indefinitely sustain life, and some way must be found to supply it with the substances which have been abstracted from it by the plants. The chief and most valuable among these substances are compounds of nitrogen, and the cheap production of these is, therefore, the key for the solution of the all-important food problem. Our atmosphere contains an inexhaustible amount of nitrogen, and could we but oxidize it and produce these compounds, an incalculable benefit for mankind would follow.

Long ago this idea took a powerful hold on the imagination of scientific men, but an efficient means for accomplishing this result could not be devised. The problem was rendered extremely difficult by the extraordinary inertness of the nitrogen, which refuses to combine even with oxygen. But here electricity comes to our aid: the dormant affinities of the element are awakened by an electric current of the proper quality. As a lump of coal which has been in contact with oxygen for centuries without burning will combine with it when once ignited, so nitrogen, excited by electricity, will burn. I did not succeed, however, in producing electrical discharges exciting very effectively the atmospheric nitrogen until a comparatively recent date, although I showed, in May, 1891, in a scientific lecture, a novel form of discharge or electrical flame named "St. Elmo's hotfire," which, besides being capable of generating ozone in abundance, also possessed, as I pointed out on that occasion, distinctly the quality of exciting chemical affinities. This discharge or flame was then only three or four inches long, its chemical action was likewise very feeble, and consequently the process of oxidation of nitrogen was wasteful. How to intensify this action was the question. Evidently electric currents of a peculiar kind had to be produced in order to render the process of nitrogen combustion more efficient.

The first advance was made in ascertaining that the chemical activity of the discharge was very considerably increased by using currents of extremely high frequency or rate of vibration. This was an important improvement, but practical considerations soon set a definite limit to the progress in this direction. Next, the effects of the electrical pressure of the current impulses, of their wave-form and other characteristic features, were investigated. Then the influence of the atmospheric pressure and temperature and of the presence of water and other bodies was studied, and thus the best conditions for causing the most intense chemical action of the discharge and securing the highest efficiency of the process were gradually ascertained. Naturally, the improvements were not quick in coming; still, little by little, I advanced. The flame grew larger and larger, and its oxidizing action grew more intense. From an insignificant brush-discharge a few inches long it developed into a marvelous electrical phenomenon, a roaring blaze, devouring the nitrogen of the atmosphere and measuring sixty or seventy feet across. Thus slowly, almost imperceptibly, possibility became accomplishment. All is not yet done, by any means, but to what a degree my efforts have been rewarded an idea may be gained from an inspection of Fig. 1 (p. 162), which, with its title, is self explanatory. The flame-like discharge visible is produced by the intense electrical oscillations which pass through the coil shown, and violently agitate the electrified molecules of the air. By this means a strong affinity is created between the two normally indifferent constituents of the atmosphere, and they combine readily, even if no further provision is made for intensifying the chemical action of the discharge. In the manufacture of nitrogen compounds by this method, of course, every possible means bearing upon the intensity of this action and the efficiency of the process will be taken advantage of, and, besides, special arrangements will be provided for the fixation of the compounds formed, as they are generally unstable, the nitrogen becoming again inert after a little lapse of time. Steam is a simple and effective means for fixing permanently the compounds. The result illustrated makes it practicable to oxidize the atmospheric nitrogen in unlimited quantities, merely by the use of cheap mechanical power and simple electrical apparatus. In this manner many compounds of nitrogen may be manufactured all over the world, at a small cost, and in any desired amount, and by means of these compounds the soil can be fertilized and its productiveness indefinitely increased. An abundance of cheap and healthful food, not artificial, but such as we are accustomed to, may thus be obtained. This new and inexhaustible source of food-supply will be of incalculable benefit to mankind, for it will enormously contribute to the increase of the human mass, and thus add immensely to human energy. Soon, I hope, the world will see the beginning of an industry which, in time to come, will, I believe, be in importance next to that of iron.

THE SECOND PROBLEM:
HOW TO REDUCE THE FORCE RETARDING
THE HUMAN MASS—THE ART OF TELAUTOMATICS

As before stated, the force which retards the onward movement of man is partly frictional and partly negative. To illustrate this distinction I may name, for example, ignorance, stupidity, and imbecility as some of the purely frictional forces, or resistances devoid of any directive tendency. On the other hand, visionariness, insanity, self-destructive tendency, religious fanaticism, and the like, are all forces of a negative character, acting in definite directions. To reduce or entirely overcome these dissimilar retarding forces, radically different methods must be employed. One knows, for instance, what a fanatic may do, and one can take preventive measures, can enlighten, convince, and possibly direct him, turn his vice into virtue; but one does not know, and never can know, what a brute or an imbecile may do, and one must deal with him as with a mass, inert, without mind, let loose by the mad elements. A negative force always implies some quality, not infrequently a high one, though badly directed, which it is possible to turn to good advantage; but a directionless, frictional force involves unavoidable loss. Evidently, then, the first and general answer to the above question is: turn all negative force in the right direction and reduce all frictional force.

There can be no doubt that, of all the frictional resistances, the one that most retards human movement is ignorance. Not without reason said that man of wisdom, Buddha: "Ignorance is the greatest evil in the world." The friction which results from ignorance, and which is greatly increased owing to the numerous languages and nationalities, can be reduced only by the spread of knowledge and the unification of the heterogeneous elements of humanity. No effort could be better spent. But however ignorance may have retarded the onward movement of man in times past, it is certain that, nowadays, negative forces have become of greater importance. Among these there is one of far greater moment than any other. It is called organized warfare. When we consider the millions of individuals, often the ablest in mind and body, the flower of humanity, who are compelled to a life of inactivity and unproductiveness, the immense sums of money daily required for the maintenance of armies and war apparatus, representing ever so much of human energy, all the effort uselessly spent in the production of arms and implements of destruction, the loss of life and the fostering of a barbarous spirit, we are appalled at the inestimable loss to mankind which the existence of these deplorable conditions must involve. What can we do to combat best this great evil?

Law and order absolutely require the maintenance of organized force. No community can exist and prosper without rigid discipline. Every country must be able to defend itself, should the necessity arise. The conditions of today are not the result of yesterday, and a radical change cannot be effected tomorrow. If the nations would at once disarm, it is more than likely that a state of things worse than war itself would follow. Universal peace is a beautiful dream, but not at once realizable. We have seen recently that even the noble effort of the man invested with the greatest worldly power has been virtually without effect. And no wonder, for the establishment of universal peace is, for the time being, a physical impossibility. War is a negative force, and cannot be turned in a positive direction without passing through the intermediate phases. It is a problem of making a wheel, rotating one way, turn in the opposite direction without slowing it down, stopping it, and speeding it up again the other way.

It has been argued that the perfection of guns of great destructive power will stop warfare. So I myself thought for a long time, but now I believe this to be a profound mistake. Such developments will greatly modify, but not arrest it. On the contrary, I think that every new arm that is invented, every new departure that is made in this direction, merely invites new talent and skill, engages new effort, offers new incentive, and so only gives a fresh impetus to further development. Think of the discovery of gunpowder. Can we conceive of any more radical departure than was effected by this innovation? Let us imagine ourselves living in that period: would we not have thought then that warfare was at an end, when the armor of the knight became an object of ridicule, when bodily strength and skill, meaning so much before, became of comparatively little value? Yet gunpowder did not stop warfare: quite the opposite—it acted as a most powerful incentive. Nor do I believe that warfare can ever be arrested by any scientific or ideal development, so long as similar conditions to those prevailing now exist, because war has itself become a science, and because war involves some of the most sacred sentiments of which man is capable. In fact, it is doubtful whether men who would not be ready to fight for a high principle would be good for anything at all. It is not the mind which makes man, nor is it the body; it is mind and body. Our virtues and our failings are inseparable, like force and matter. When they separate, man is no more.

Another argument, which carries considerable force, is frequently made, namely, that war must soon become impossible because the means of defense are outstripping the means of attack. This is only in accordance with a fundamental law which may be expressed by the statement that it is easier to destroy than to build. This law defines human capacities and human conditions. Were these such that it would be easier to build than to destroy, man would go on unresisted, creating and accumulating without limit. Such conditions are not of this earth. A being which could do this would not be a man: it might be a god. Defense will always have the advantage over attack, but this alone, it seems to me, can never stop war. By the use of new principles of defense we can render harbors impregnable against attack, but we cannot by such means prevent two warships meeting in battle on the high sea. And then, if we follow this idea to its ultimate development, we are led to the conclusion that it would be better for mankind if attack and defense were just oppositely related; for if every country, even the smallest, could surround itself with a wall absolutely impenetrable, and could defy the rest of the world, a state of things would surely be brought on which would be extremely unfavorable to human progress. It is by abolishing all the barriers which separate nations and countries that civilization is best furthered.

Again, it is contended by some that the advent of the flying-machine must bring on universal peace. This, too, I believe to be an entirely erroneous view. The flying-machine is certainly coming, and very soon, but the conditions will remain the same as before. In fact, I see no reason why a ruling power, like Great Britain, might not govern the air as well as the sea. Without wishing to put myself on record as a prophet, I do not hesitate to say that the next years will see the establishment of an "air-power," and its center may be not far from New York. But, for all that, men will fight on merrily.

The ideal development of the war principle would ultimately lead to the transformation of the whole energy of war into purely potential, explosive energy, like that of an electrical condenser. In this form the war-energy could be maintained without effort; it would need to be much smaller in amount, while incomparably more effective.

As regards the security of a country against foreign invasion, it is interesting to note that it depends only on the relative, and not the absolute, number of the individuals or magnitude of the forces, and that, if every country should reduce the war-force in the same ratio, the security would remain unaltered. An international agreement with the object of reducing to a minimum the war-force which, in view of the present still imperfect education of the masses, is absolutely indispensable, would, therefore, seem to be the first rational step to take toward diminishing the force retarding human movement.

Fortunately, the existing conditions cannot continue indefinitely, for a new element is beginning to assert itself. A change for the better is eminent, and I shall now endeavor to show what, according to my ideas, will be the first advance toward the establishment of peaceful relations between nations, and by what means it will eventually be accomplished.

Let us go back to the early beginning, when the law of the stronger was the only law. The light of reason was not yet kindled, and the weak was entirely at the mercy of the strong. The weak individual then began to learn how to defend himself. He made use of a club, stone, spear, sling, or bow and arrow, and in the course of time, instead of physical strength, intelligence became the chief deciding factor in the battle. The wild character was gradually softened by the awakening of noble sentiments, and so, imperceptibly, after ages of continued progress, we have come from the brutal fight of the unreasoning animal to what we call the "civilized warfare" of today, in which the combatants shake hands, talk in a friendly way, and smoke cigars in the entr'actes, ready to engage again in deadly conflict at a signal. Let pessimists say what they like, here is an absolute evidence of great and gratifying advance.

But now, what is the next phase in this evolution? Not peace as yet, by any means. The next change which should naturally follow from modern developments should be the continuous diminution of the number of individuals engaged in battle. The apparatus will be one of specifically great power, but only a few individuals will be required to operate it. This evolution will bring more and more into prominence a machine or mechanism with the fewest individuals as an element of warfare, and the absolutely unavoidable consequence of this will be the abandonment of large, clumsy, slowly moving, and unmanageable units. Greatest possible speed and maximum rate of energy-delivery by the war apparatus will be the main object. The loss of life will become smaller and smaller, and finally, the number of the individuals continuously diminishing, merely machines will meet in a contest without bloodshed, the nations being simply interested, ambitious spectators. When this happy condition is realized, peace will be assured. But, no matter to what degree of perfection rapid-fire guns, high-power cannon, explosive projectiles, torpedo boats, or other implements of war may be brought, no matter how destructive they may be made, that condition can never be reached through any such development. All such implements require men for their operation; men are indispensable parts of the machinery. Their object is to kill and to destroy. Their power resides in their capacity for doing evil. So long as men meet in battle, there will be bloodshed. Bloodshed will ever keep up barbarous passion. To break this fierce spirit, a radical departure must be made, an entirely new principle must be introduced, something that never existed before in warfare—a principle which will forcibly, unavoidably, turn the battle into a mere spectacle, a play, a contest without loss of blood. To bring on this result, men must be dispensed with: machine must fight machine. But how to accomplish that which seems impossible? The answer is simple enough: produce a machine capable of acting as though it were part of a human being—no mere mechanical contrivance, comprising levers, screws, wheels, clutches, and nothing more, but a machine embodying a higher principle, which will enable it to perform its duties as though it had intelligence, experience, judgment, a mind!

This conclusion is the result of my thoughts and observations which have extended through virtually my whole life, and I shall now briefly describe how I came to accomplish that which at first seemed an unrealizable dream.

A long time ago, when I was a boy, I was afflicted with a singular trouble, which seems to have been due to an extraordinary excitability of the retina. It was the appearance of images which, by their persistence, marred the vision of real objects and interfered with thought. When a word was said to me, the image of the object which it designated would appear vividly before my eyes, and many times it was impossible for me to tell whether the object I saw was real or not. This caused me great discomfort and anxiety, and I tried hard to free myself of the spell. But for a long time I tried in vain, and it was not, as I clearly recollect, until I was about twelve years old that I succeeded for the first time, by an effort of the will, in banishing an image which presented itself. My happiness will never be as complete as it was then, but, unfortunately (as I thought at that time), the old trouble returned, and with it my anxiety. Here it was that the observations to which I refer began. I noted, namely, that whenever the image of an object appeared before my eyes I had seen something that reminded me of it. In the first instances I thought this to be purely accidental, but soon I convinced myself that it was not so. A visual impression, consciously or unconsciously received, invariably preceded the appearance of the image. Gradually the desire arose in me to find out, every time, what caused the images to appear, and the satisfaction of this desire soon became a necessity. The next observation I made was that, just as these images followed as a result of something I had seen, so also the thoughts which I conceived were suggested in like manner. Again, I experienced the same desire to locate the image which caused the thought, and this search for the original visual impression soon grew to be a second nature. My mind became automatic, as it were, and in the course of years of continued, almost unconscious performance, I acquired the ability of locating every time and, as a rule, instantly the visual impression which started the thought. Nor is this all. It was not long before I was aware that also all my movements were prompted in the same way, and so, searching, observing, and verifying continuously, year by year, I have, by every thought and every act of mine, demonstrated, and do so daily, to my absolute satisfaction, that I am an automaton endowed with power of movement, which merely responds to external stimuli beating upon my sense organs, and thinks and acts and moves accordingly. I remember only one or two cases in all my life in which I was unable to locate the first impression which prompted a movement or a thought, or even a dream.

Fig. 2. The first practical Teleautomaton. A machine having all its bodily or translatory movements and the operations of the interior mechanism controlled from a distance without wires. The crewless boat shown in the photograph contains its own motive power, propelling and steering machinery, and numerous other accessories, all of which are controlled by transmitting from a distance, without wires, electrical oscillations to a circuit carried by the boat and adjusted to respond only to these oscillations.

With these experiences it was only natural that, long ago, I conceived the idea of constructing an automaton which would mechanically represent me, and which would respond, as I do myself, but, of course, in a much more primitive manner, to external influences. Such an automaton evidently had to have motive power, organs for locomotion, directive organs, and one or more sensitive organs so adapted as to be excited by external stimuli. This machine would, I reasoned, perform its movements in the manner of a living being, for it would have all the chief mechanical characteristics or elements of the same. There was still the capacity for growth, propagation, and, above all, the mind which would be wanting to make the model complete. But growth was not necessary in this case, since a machine could be manufactured full-grown, so to speak. As to the capacity for propagation, it could likewise be left out of consideration, for in the mechanical model it merely signified a process of manufacture. Whether the automation be of flesh and bone, or of wood and steel, it mattered little, provided it could perform all the duties required of it like an intelligent being. To do so, it had to have an element corresponding to the mind, which would effect the control of all its movements and operations, and cause it to act, in any unforeseen case that might present itself, with knowledge, reason, judgment, and experience. But this element I could easily embody in it by conveying to it my own intelligence, my own understanding. So this invention was evolved, and so a new art came into existence, for which the name "teleautomatics" has been suggested, which means the art of controlling the movements and operations of distant automatons.

This principle evidently was applicable to any kind of machine that moves on land or in the water or in the air. In applying it practically for the first time, I selected a boat (see Fig. 2). A storage battery placed within it furnished the motive power. The propeller, driven by a motor, represented the locomotive organs. The rudder, controlled by another motor likewise driven by the battery, took the place of the directive organs. As to the sensitive organ, obviously the first thought was to utilize a device responsive to rays of light, like a selenium cell, to represent the human eye. But upon closer inquiry I found that, owing to experimental and other difficulties, no thoroughly satisfactory control of the automaton could be effected by light, radiant heat, Hertzian radiations, or by rays in general, that is, disturbances which pass in straight lines through space. One of the reasons was that any obstacle coming between the operator and the distant automaton would place it beyond his control. Another reason was that the sensitive device representing the eye would have to be in a definite position with respect to the distant controlling apparatus, and this necessity would impose great limitations in the control. Still another and very important reason was that, in using rays, it would be difficult, if not impossible, to give to the automaton individual features or characteristics distinguishing it from other machines of this kind. Evidently the automaton should respond only to an individual call, as a person responds to a name. Such considerations led me to conclude that the sensitive device of the machine should correspond to the ear rather than the eye of a human being, for in this case its actions could be controlled irrespective of intervening obstacles, regardless of its position relative to the distant controlling apparatus, and, last, but not least, it would remain deaf and unresponsive, like a faithful servant, to all calls but that of its master. These requirements made it imperative to use, in the control of the automaton, instead of light or other rays, waves or disturbances which propagate in all directions through space, like sound, or which follow a path of least resistance, however curved. I attained the result aimed at by means of an electric circuit placed within the boat, and adjusted, or "tuned," exactly to electrical vibrations of the proper kind transmitted to it from a distant "electrical oscillator." This circuit, in responding, however feebly, to the transmitted vibrations, affected magnets and other contrivances, through the medium of which were controlled the movements of the propeller and rudder, and also the operations of numerous other appliances.

By the simple means described the knowledge, experience, judgment—the mind, so to speak—of the distant operator were embodied in that machine, which was thus enabled to move and to perform all its operations with reason and intelligence. It behaved just like a blindfolded person obeying directions received through the ear.

The automatons so far constructed had "borrowed minds," so to speak, as each merely formed part of the distant operator who conveyed to it his intelligent orders; but this art is only in the beginning. I purpose to show that, however impossible it may now seem, an automaton may be contrived which will have its "own mind," and by this I mean that it will be able, independent of any operator, left entirely to itself, to perform, in response to external influences affecting its sensitive organs, a great variety of acts and operations as if it had intelligence. It will be able to follow a course laid out or to obey orders given far in advance; it will be capable of distinguishing between what it ought and what it ought not to do, and of making experiences or, otherwise stated, of recording impressions which will definitely affect its subsequent actions. In fact, I have already conceived such a plan.

Although I evolved this invention many years ago and explained it to my visitors very frequently in my laboratory demonstrations, it was not until much later, long after I had perfected it, that it became known, when, naturally enough, it gave rise to much discussion and to sensational reports. But the true significance of this new art was not grasped by the majority, nor was the great force of the underlying principle recognized. As nearly as I could judge from the numerous comments which appeared, the results I had obtained were considered as entirely impossible. Even the few who were disposed to admit the practicability of the invention saw in it merely an automobile torpedo, which was to be used for the purpose of blowing up battleships, with doubtful success. The general impression was that I contemplated simply the steering of such a vessel by means of Hertzian or other rays. There are torpedoes steered electrically by wires, and there are means of communicating without wires, and the above was, of course, an obvious inference. Had I accomplished nothing more than this, I should have made a small advance indeed. But the art I have evolved does not contemplate merely the change of direction of a moving vessel; it affords means of absolutely controlling, in every respect, all the innumerable translatory movements, as well as the operations of all the internal organs, no matter how many, of an individualized automaton. Criticisms to the effect that the control of the automaton could be interfered with were made by people who do not even dream of the wonderful results which can be accomplished by use of electrical vibrations. The world moves slowly, and new truths are difficult to see. Certainly, by the use of this principle, an arm for attack as well as defense may be provided, of a destructiveness all the greater as the principle is applicable to submarine and aërial vessels. There is virtually no restriction as to the amount of explosive it can carry, or as to the distance at which it can strike, and failure is almost impossible. But the force of this new principle does not wholly reside in its destructiveness. Its advent introduces into warfare an element which never existed before—a fighting-machine without men as a means of attack and defense. The continuous development in this direction must ultimately make war a mere contest of machines without men and without loss of life—a condition which would have been impossible without this new departure, and which, in my opinion, must be reached as preliminary to permanent peace. The future will either bear out or disprove these views. My ideas on this subject have been put forth with deep conviction, but in a humble spirit.

The establishment of permanent peaceful relations between nations would most effectively reduce the force retarding the human mass, and would be the best solution of this great human problem. But will the dream of universal peace ever be realized? Let us hope that it will. When all darkness shall be dissipated by the light of science, when all nations shall be merged into one, and patriotism shall be identical with religion, when there shall be one language, one country, one end, then the dream will have become reality.

THE THIRD PROBLEM:
HOW TO INCREASE THE FORCE
ACCELERATING THE HUMAN MASS—
THE HARNESSING OF THE SUN'S ENERGY

Of the three possible solutions of the main problem of increasing human energy, this is by far the most important to consider, not only because of its intrinsic significance, but also because of its intimate bearing on all the many elements and conditions which determine the movement of humanity. In order to proceed systematically, it would be necessary for me to dwell on all those considerations which have guided me from the outset in my efforts to arrive at a solution, and which have led me, step by step, to the results I shall now describe.

As a preliminary study of the problem an analytical investigation, such as I have made, of the chief forces which determine the onward movement, would be of advantage, particularly in conveying an idea of that hypothetical "velocity" which, as explained in the beginning, is a measure of human energy; but to deal with this specifically here, as I would desire, would lead me far beyond the scope of the present subject. Suffice it to state that the resultant of all these forces is always in the direction of reason, which therefore, determines, at any time, the direction of human movement. This is to say that every effort which is scientifically applied, rational, useful, or practical, must be in the direction in which the mass is moving. The practical, rational man, the observer, the man of business, he who reasons, calculates, or determines in advance, carefully applies his effort so that when coming into effect it will be in the direction of the movement, making it thus most efficient, and in this knowledge and ability lies the secret of his success. Every new fact discovered, every new experience or new element added to our knowledge and entering into the domain of reason, affects the same and, therefore, changes the direction of movement, which, however, must always take place along the resultant of all those efforts which, at that time, we designate as reasonable, that is, self-preserving, useful, profitable, or practical. These efforts concern our daily life, our necessities and comforts, our work and business, and it is these which drive man onward.

But looking at all this busy world about us, on all this complex mass as it daily throbs and moves, what is it but an immense clock-work driven by a spring? In the morning, when we rise, we cannot fail to note that all the objects about us are manufactured by machinery: the water we use is lifted by steam-power; the trains bring our breakfast from distant localities; the elevators in our dwelling and our office building, the cars that carry us there, are all driven by power; in all our daily errands, and in our very life-pursuit, we depend upon it; all the objects we see tell us of it; and when we return to our machine-made dwelling at night, lest we should forget it, all the material comforts of our home, our cheering stove and lamp, remind us of how much we depend on power. And when there is an accidental stoppage of the machinery, when the city is snowbound, or the life-sustaining movement otherwise temporarily arrested, we are affrighted to realize how impossible it would be for us to live the life we live without motive power. Motive power means work. To increase the force accelerating human movement means, therefore, to perform more work.

So we find that the three possible solutions of the great problem of increasing human energy are answered by the three words: *food, peace, work.* Many a year I have thought and pondered, lost myself in speculations and theories, considering man as a mass moved by a force, viewing his inexplicable movement in the light of a mechanical one, and applying the simple principles of mechanics to the analysis of the same until I arrived at these solutions, only to realize that they were taught to me in my early childhood. These three words sound the key-notes of the Christian religion. Their scientific meaning and purpose are now clear to me: food to increase the mass, peace to diminish the retarding force, and work to increase the force accelerating human movement. These are the only three solutions which are possible of that great problem, and all of them have one object, one end, namely, to increase human energy. When we recognize this, we cannot help wondering how profoundly wise and scientific and how immensely practical the Christian religion is, and in what a marked contrast it stands in this respect to other religions. It is unmistakably the result of practical experiment and scientific observation which have extended through the ages, while other religions seem to be the outcome of merely abstract reasoning. Work, untiring effort, useful and accumulative, with periods of rest and recuperation aiming at higher efficiency, is its chief and ever-recurring command. Thus we are inspired both by Christianity and Science to do our utmost toward increasing the performance of mankind. This most important of human problems I shall now specifically consider.

The Source of Human Energy— the Three Ways of Drawing Energy from the Sun

First let us ask: Whence comes all the motive power? What is the spring that drives all? We see the ocean rise and fall, the rivers flow, the wind, rain, hail, and snow beat on our windows, the trains and steamers come and go; we hear the rattling noise of carriages, the voices from the street; we feel, smell, and taste; and we think of all this. And all this movement, from the surging of the mighty ocean to that subtle movement concerned in our thought, has but one common cause. All this energy emanates from one single center, one single source—the sun. The sun is the spring that drives all. The sun maintains all human life and supplies all human energy. Another answer we have now found to the above great question: To increase the force accelerating human movement means to turn to the uses of man more of the sun's energy. We honor and revere those great men of bygone times whose names are linked with immortal achievements, who have proved themselves benefactors of humanity—the religious reformer with his wise maxims of life, the philosopher with his deep truths, the mathematician with his formulæ, the physicist with his laws, the discoverer with his principles and secrets wrested from nature, the artist with his forms of the beautiful; but who honors him, the greatest of all—who can tell the name of him—who first turned to use the sun's energy to save the effort of a weak fellow-creature? That was man's first act of scientific philanthropy, and its consequences have been incalculable.

From the very beginning three ways of drawing energy from the sun were open to man. The savage, when he warmed his frozen limbs at a fire kindled in some way, availed himself of the energy of the sun stored in the burning material. When he carried a bundle of branches to his cave and burned them there, he made use of the sun's stored energy transported from one to another locality. When he set sail to his canoe, he utilized the energy of the sun applied to the atmosphere or the ambient medium. There can be no doubt that the first is the oldest way. A fire, found accidentally, taught the savage to appreciate its beneficial heat. He then very likely conceived of the idea of carrying the glowing members to his abode. Finally he learned to use the force of a swift current of water or air. It is characteristic of modern development that progress has been effected in the same order. The utilization of the energy stored in wood or coal, or, generally speaking, fuel, led to the steam-engine. Next a great stride in advance was made in energy transportation by the use of electricity, which permitted the transfer of energy from one locality to another without transporting the material. But as to the utilization of the energy of the ambient medium, no radical step forward has as yet been made known.

The ultimate results of development in these three directions are: first, the burning of coal by a cold process in a battery; second, the efficient utilization of the energy of the ambient medium; and, third the transmission without wires of electrical energy to any distance. In whatever way these results may be arrived at, their practical application will necessarily involve an extensive use of iron, and this invaluable metal will undoubtedly be an essential element in the further development along these three lines. If we succeed in burning coal by a cold process and thus obtain electrical energy in an efficient and inexpensive manner, we shall require in many practical uses of this energy electric motors—that is, iron. If we are successful in deriving energy from the ambient medium, we shall need, both in the obtainment and utilization of the energy, machinery—again, iron. If we realize the transmission of electrical energy without wires on an industrial scale, we shall be compelled to use extensively electric generators—once more, iron. Whatever we may do, iron will probably be the chief means of accomplishment in the near future, possibly more so than in the past. How long its reign will last is difficult to tell, for even now aluminum is looming up as a threatening competitor. But for the time being, next to providing new resources of energy, it is of the greatest importance to making improvements in the manufacture and utilization of iron. Great advances are possible in these latter directions, which, if brought about, would enormously increase the useful performance of mankind.

GREAT POSSIBILITIES OFFERED BY IRON FOR INCREASING HUMAN PERFORMANCE— ENORMOUS WASTE IN IRON MANUFACTURE

Iron is by far the most important factor in modern progress. It contributes more than any other industrial product to the force accelerating human movement. So general is the use of this metal, and so intimately is it connected with all that concerns our life, that it has become as indispensable to us as the very air we breathe. Its name is synonymous with usefulness. But, however great the influence of iron may be on the present human development, it does not add to the force urging man onward nearly as much as it might. First of all, its manufacture as now carried on is connected with an appalling waste of fuel—that is, waste of energy. Then, again, only a part of all the iron produced is applied for useful purposes. A good part of it goes to create frictional resistances, while still another large part is the means of developing negative forces greatly retarding human movement. Thus the negative force of war is almost wholly represented in iron. It is impossible to estimate with any degree of accuracy the magnitude of this greatest of all retarding forces, but it is certainly very considerable. If the present positive impelling force due to all useful applications of iron be represented by ten, for instance, I should not think it exaggeration to estimate the negative force of war, with due consideration of all its retarding influences and results, at, say, six. On the basis of this estimate the effective impelling force of iron in the positive direction would be measured by the difference of these two numbers, which is four. But if, through the establishment of universal peace, the manufacture of war machinery should cease, and all struggle for supremacy between nations should be turned into healthful, ever active and productive commercial competition, then the positive impelling force due to iron would be measured by the sum of those two numbers, which is sixteen—that is, this force would have four times its present value. This example is, of course, merely intended to give an idea of the immense increase in the useful performance of mankind which would result from a radical reform of the iron industries supplying the implements of warfare.

A similar inestimable advantage in the saving of energy available to man would be secured by obviating the great waste of coal which is inseparably connected with the present methods of manufacturing iron. In some countries, such as Great Britain, the hurtful effects of this squandering of fuel are beginning to be felt. The price of coal is constantly rising, and the poor are made to suffer more and more. Though we are still far from the dreaded "exhaustion of the coal-fields," philanthropy commands us to invent novel methods of manufacturing iron, which will not involve such barbarous waste of this valuable material from which we derive at present most of our energy. It is our duty to coming generations to leave this store of energy intact for them, or at least not to touch it until we shall have perfected processes for burning coal more efficiently. Those who are coming after us will need fuel more than we do. We should be able to manufacture the iron we require by using the sun's energy, without wasting any coal at all. As an effort to this end the idea of smelting iron ores by electric currents obtained from the energy of falling water has naturally suggested itself to many. I have myself spent much time in endeavoring to evolve such a practical process, which would enable iron to be manufactured at small cost. After a prolonged investigation of the subject, finding that it was unprofitable to use the currents generated directly for smelting the ore, I devised a method which is far more economical.

Economical Production of Iron
by a New Process

The industrial project, as I worked it out six years ago, contemplated the employment of the electric currents derived from the energy of a waterfall, not directly for smelting the ore, but for decomposing water for a preliminary step. To lessen the cost of the plant, I proposed to generate the currents in exceptionally cheap and simple dynamos, which I designed for this sole purpose. The hydrogen liberated in the electrolytic decomposition was to be burned or recombined with oxygen, not with that from which it was separated, but with that of the atmosphere. Thus very nearly the total electrical energy used up in the decomposition of the water would be recovered in the form of heat resulting from the recombination of the hydrogen. This heat was to be applied to the smelting of ore. The oxygen gained as a byproduct of the decomposition of the water I intended to use for certain other industrial purposes, which would probably yield good financial returns, inasmuch as this is the cheapest way of obtaining this gas in large quantities. In any event, it could be employed to burn all kinds of refuse, cheap hydrocarbon, or coal of the most inferior quality which could not be burned in air or be otherwise utilized to advantage, and thus again a considerable amount of heat would be made available for the smelting of the ore. To increase the economy of the process I contemplated, furthermore, using an arrangement such that the hot metal and the products of combustion, coming out of the furnace, would give up their heat upon the cold ore going into the furnace, so that comparatively little of the heat energy would be lost in the smelting. I calculated that probably forty thousand pounds of iron could be produced per horse-power per annum by this method. Liberal allowances were made for those losses which are unavoidable, the above quantity being about half of that theoretically obtainable. Relying on this estimate and on practical data with reference to a certain kind of sand ore existing in abundance in the region of the Great Lakes, including cost of transportation and labor, I found that in some localities iron could be manufactured in this manner cheaper than by any of the adopted methods. This result would be obtained all the more surely if the oxygen obtained from the water, instead of being used for smelting of ore, as assumed, should be more profitably employed. Any new demand for this gas would secure a higher revenue from the plant, thus cheapening the iron. This project was advanced merely in the interest of industry. Some day, I hope, a beautiful industrial butterfly will come out of the dusty and shriveled chrysalis.

The production of iron from sand ores by a process of magnetic separation is highly commendable in principle, since it involves no waste of coal; but the usefulness of this method is largely reduced by the necessity of melting the iron afterward. As to the crushing of iron ore, I would consider it rational only if done by water-power, or by energy otherwise obtained without consumption of fuel. An electrolytic cold process, which would make it possible to extract iron cheaply, and also to mold it into the required forms without any fuel consumption, would, in my opinion, be a very great advance in iron manufacture. In common with some other metals, iron has so far resisted electrolytic treatment, but there can be no doubt that such a cold process will ultimately replace in metallurgy the present crude method of casting, and thus obviating the enormous waste of fuel necessitated by the repeated heating of metal in the foundries.

Up to a few decades ago the usefulness of iron was based almost wholly on its remarkable mechanical properties, but since the advent of the commercial dynamo and electric motor its value to mankind has been greatly increased by its unique magnetic qualities. As regards the latter, iron has been greatly improved of late. The signal progress began about thirteen years ago, when I discovered that in using soft Bessemer steel instead of wrought iron, as then customary, in an alternating motor, the performance of the machine was doubled. I brought this fact to the attention of Mr. Albert Schmid, to whose untiring efforts and ability is largely due the supremacy of American electrical machinery, and who was then superintendent of an industrial corporation engaged in this field. Following my suggestion, he constructed transformers of steel, and they showed the same marked improvement. The investigation was then systematically continued under Mr. Schmid's guidance, the impurities being gradually eliminated from the "steel" (which was only such in name, for in reality it was pure soft iron), and soon a product resulted which admitted of little further improvement.

The Coming Age of Aluminium—
Doom of the Copper Industry—
the Great Civilizing Potency of the New Metal

With the advances made in iron of late years we have arrived virtually at the limits of improvement. We cannot hope to increase very materially its tensile strength, elasticity, hardness, or malleability, nor can we expect to make it much better as regards its magnetic qualities. More recently a notable gain was secured by the mixture of a small percentage of nickel with the iron, but there is not much room for further advance in this direction. New discoveries may be expected, but they cannot greatly add to the valuable properties of the metal, though they may considerably reduce the cost of manufacture. The immediate future of iron is assured by its cheapness and its unrivaled mechanical and magnetic qualities. These are such that no other product can compete with it now. But there can be no doubt that, at a time not very distant, iron, in many of its now uncontested domains, will have to pass the scepter to another: the coming age will be the age of aluminum. It is only seventy years since this wonderful metal was discovered by Woehler, and the aluminum industry, scarcely forty years old, commands already the attention of the entire world. Such rapid growth has not been recorded in the history of civilization before. Not long ago aluminum was sold at the fanciful price of thirty or forty dollars per pound; today it can be had in any desired amount for as many cents. What is more, the time is not far off when this price, too, will be considered fanciful, for great improvements are possible in the methods of its manufacture. Most of the metal is now produced in the electric furnace by a process combining fusion and electrolysis, which offers a number of advantageous features, but involves naturally a great waste of the electrical energy of the current. My estimates show that the price of aluminum could be considerably reduced by adopting in its manufacture a method similar to that proposed by me for the production of iron. A pound of aluminum requires for fusion only about seventy percent of the heat needed for melting a pound of iron, and inasmuch as its weight is only about one third of that of the latter, a volume of aluminum four times that of iron could be obtained from a given amount of heat-energy. But a cold electrolytic process of manufacture is the ideal solution, and on this I have placed my hope.

The absolutely unavoidable consequence of the advancement of the aluminum industry will be the annihilation of the copper industry. They cannot exist and prosper together, and the latter is doomed beyond any hope of recovery. Even now it is cheaper to convey an electric current through aluminum wires than through copper wires; aluminum castings cost less, and in many domestic and other uses copper has no chance of successfully competing. A further material reduction of the price of aluminum cannot but be fatal to copper. But the progress of the former will not go on unchecked, for, as it ever happens in such cases, the larger industry will absorb the smaller one: the giant copper interests will control the pygmy aluminum interests, and the slow-pacing copper will reduce the lively gait of aluminum. This will only delay, not avoid the impending catastrophe.

Aluminum, however, will not stop at downing copper. Before many years have passed it will be engaged in a fierce struggle with iron, and in the latter it will find an adversary not easy to conquer. The issue of the contest will largely depend on whether iron shall be indispensable in electric machinery. This the future alone can decide.

The magnetism as exhibited in iron is an isolated phenomenon in nature. What it is that makes this metal behave so radically different from all other materials in this respect has not yet been ascertained, though many theories have been suggested. As regards magnetism, the molecules of the various bodies behave like hollow beams partly filled with a heavy fluid and balanced in the middle in the manner of a see-saw. Evidently some disturbing influence exists in nature which causes each molecule, like such a beam, to tilt either one or the other way. If the molecules are tilted one way, the body is magnetic; if they are tilted the other way, the body is non-magnetic; but both positions are stable, as they would be in the case of the hollow beam, owing to the rush of the fluid to the lower end. Now, the wonderful thing is that the molecules of all known bodies went one way, while those of iron went the other way. This metal, it would seem, has an origin entirely different from that of the rest of the globe. It is highly improbable that we shall discover some other and cheaper material which will equal or surpass iron in magnetic qualities.

Unless we should make a radical departure in the character of the electric currents employed, iron will be indispensable. Yet the advantages it offers are only apparent. So long as we use feeble magnetic forces it is by far superior to any other material; but if we find ways of producing great magnetic forces, then better results will be obtainable without it. In fact, I have already produced electric transformers in which no iron is employed, and which are capable of performing ten times as much work per pound of weight as those of iron. This result is attained by using electric currents of a very high rate of vibration, produced in novel ways, instead of the ordinary currents now employed in the industries. I have also succeeded in operating electric motors without iron by such rapidly vibrating currents, but the results, so far, have been inferior to those obtained with ordinary motors constructed of iron, although theoretically the former should be capable of performing incomparably more work per unit of weight than the latter. But the seemingly insuperable difficulties which are now in the way may be overcome in the end, and then iron will be done away with, and all electric machinery will be manufactured of aluminum, in all probability, at prices ridiculously low. This would be a severe, if not fatal, blow to iron. In many other branches of industry, as ship-building, or wherever lightness of structure is required, the progress of the new metal will be much quicker. For such uses it is eminently suitable, and is sure to supersede iron sooner or later. It is highly probable that in the course of time we shall be able to give it many of those qualities which make iron so valuable.

While it is impossible to tell when this industrial revolution will be consummated, there can be no doubt that the future belongs to aluminum, and that in times to come it will be the chief means of increasing human performance. It has in this respect capacities greater by far than those of any other metal. I should estimate its civilizing potency at fully one hundred times that of iron. This estimate, though it may astonish, is not at all exaggerated. First of all, we must remember that there is thirty times as much aluminium as iron in bulk, available for the uses of man. This in itself offers great possibilities. Then, again, the new metal is much more easily workable, which adds to its value. In many of its properties it partakes of the character of a precious metal, which gives it additional worth. Its electric conductivity, which, for a given weight, is greater than that of any other metal, would be alone sufficient to make it one of the most important factors in future human progress. Its extreme lightness makes it far more easy to transport the objects manufactured. By virtue of this property it will revolutionize naval construction, and in facilitating transport and travel it will add enormously to the useful performance of mankind. But its greatest civilizing property will be, I believe, in aërial travel, which is sure to be brought about by means of it. Telegraphic instruments will slowly enlighten the barbarian. Electric motors and lamps will do it more quickly, but quicker than anything else the flying-machine will do it. By rendering travel ideally easy it will be the best means for unifying the heterogeneous elements of humanity. As the first step toward this realization we should produce a lighter storage-battery or get more energy from coal.

Efforts Toward Obtaining More Energy from Coal— the Electric Transmission—the Gas-engine— the Cold-coal Battery

I remember that at one time I considered the production of electricity by burning coal in a battery as the greatest achievement toward advancing civilization, and I am surprised to find how much the continuous study of these subjects has modified my views. It now seems to me that to burn coal, however efficiently, in a battery would be a mere makeshift, a phase in the evolution toward something much more perfect. After all, in generating electricity in this manner, we should be destroying material, and this would be a barbarous process. We ought to be able to obtain the energy we need without consumption of material. But I am far from underrating the value of such an efficient method of burning fuel. At the present time most motive power comes from coal, and, either directly or by its products, it adds vastly to human energy. Unfortunately, in all the processes now adopted, the larger portion of the energy of the coal is uselessly dissipated. The best steam-engines utilize only a small part of the total energy. Even in gas-engines, in which, particularly of late, better results are obtainable, there is still a barbarous waste going on. In our electric-lighting systems we scarcely utilize one third of one percent, and in lighting by gas a much smaller fraction of the total energy of the coal. Considering the various uses of coal throughout the world, we certainly do not utilize more than two percent of its energy theoretically available. The man who should stop this senseless waste would be a great benefactor of humanity, though the solution he would offer could not be a permanent one, since it would ultimately lead to the exhaustion of the store of material. Efforts toward obtaining more energy from coal are now being made chiefly in two directions—by generating electricity and by producing gas for motive-power purposes. In both of these lines notable success has already been achieved. The advent of the alternating-current system of electric power-transmission marks an epoch in the economy of energy available to man from coal. Evidently all electrical energy obtained from a waterfall, saving so much fuel, is a net gain to mankind, which is all the more effective as it is secured with little expenditure of human effort, and as this most perfect of all known methods of deriving energy from the sun contributes in many ways to the advancement of civilization. But electricity enables us also to get from coal much more energy than was practicable in the old ways. Instead of transporting the coal to distant places of consumption, we burn it near the mine, develop electricity in the dynamos, and transmit the current to remote localities, thus effecting a considerable saving. Instead of driving the machinery in a factory in the old wasteful way of belts and shafting, we generate electricity by steam-power and operate electric motors. In this manner it is not uncommon to obtain two or three times as much effective motive power from the fuel, besides securing many other important advantages. It is in this field as much as in the transmission of energy to great distance that the alternating system, with its ideally simple machinery, is bringing about an industrial revolution. But in many lines this progress has not been yet fully felt. For example, steamers and trains are still being propelled by the direct application of steam-power to shafts or axles. A much greater percentage of the heatenergy of the fuel could be transformed into motive energy by using, in place of the adopted marine engines and locomotives, dynamos driven by specially designed high-pressure steam- or gas-engines and by utilizing the electricity generated for the propulsion. A gain of fifty to one hundred percent in the effective energy derived from the coal could be secured in this manner. It is difficulty to understand why a fact so plain and obvious is not receiving more attention from engineers. In ocean steamers such an improvement would be particularly desirable, as it would do away with noise and increase materially the speed and the carrying capacity of the liners.

Still more energy is now being obtained from coal by the latest improved gas-engine, the economy of which is, on the average, probably twice that of the best steam-engine. The introduction of the gas-engine is very much facilitated by the importance of the gas industry. With the increasing use of the electric light more and more of the gas is utilized for heating and motive-power purposes. In many instances gas is manufactured close to the coal-mine and conveyed to distant places of consumption, a considerable saving both in cost of transportation and in utilization of the energy of the fuel being thus effected. In the present state of the mechanical and electrical arts the most rational way of deriving energy from coal is evidently to manufacture gas close to the coal store, and to utilize it, either on the spot or elsewhere, to generate electricity for industrial uses in dynamos driven by gas engines. The commercial success of such a plant is largely dependent upon the production of gas-engines of great nominal horse-power, which, judging from the keen activity in this field will soon be forthcoming. Instead of consuming coal directly, as usual, gas should be manufactured from it and burned to economize energy.

But all such improvements cannot be more than passing phases in the evolution toward something far more perfect, for ultimately we must succeed in obtaining electricity from coal in a more direct way, involving no great loss of heat-energy. Whether coal can be oxidized by a cold process is still a question. Its combination with oxygen always involves heat, and whether the energy of the combination of the carbon with another element can be turned directly into electrical energy has not yet been determined. Under certain conditions nitric acid will burn the carbon, generating an electric current, but the solution does not remain cold. Other means of oxidizing coal have been proposed, but they have offered no promise of leading to an efficient process. My own lack of success has been complete, though perhaps not quite so complete as that of some who have "perfected" the cold-coal battery. This problem is essentially one for the chemist to solve. It is not for the physicist, who determines all his results in advance, so that, when the experiment is tried, it cannot fail. Chemistry, though a positive science, does not yet admit of a solution by such positive methods as those which are available in the treatment of many physical problems. The result, if possible, will be arrived at through patient trying rather than through deduction or calculation. The time will soon come, however, when the chemist will be able to follow a course clearly mapped out beforehand, and when the process of his arriving at a desired result will be purely constructive. The cold-coal battery would give a great impetus to electrical development; it would lead very shortly to a practical flying-machine, and would enormously enhance the introduction of the automobile. But these and many other problems will be better solved, and in a more scientific manner, by a light storage battery.

Energy from the Medium— the Windmill and the Solar Engine— Motive Power from Terrestrial Heat— Electricity from Natural Sources

Besides fuel, there is abundant material from which we might eventually derive power. An immense amount of energy is locked up in limestone, for instance, and machines can be driven by liberating the carbonic acid through sulphuric acid or otherwise. I once constructed such an engine, and it operated satisfactorily.

But, whatever our resources of primary energy may be in the future, we must, to be rational, obtain it without consumption of any material. Long ago I came to this conclusion, and to arrive at this result only two ways, as before indicated, appeared possible—either to turn to use the energy of the sun stored in the ambient medium, or to transmit, through the medium, the sun's energy to distant places from some locality where it was obtainable without consumption of material. At that time I at once rejected the latter method as entirely impracticable, and turned to examine the possibilities of the former.

It is difficult to believe, but it is, nevertheless, a fact, that since time immemorial man has had at his disposal a fairly good machine which has enabled him to utilize the energy of the ambient medium. This machine is the windmill. Contrary to popular belief, the power obtainable from wind is very considerable. Many a deluded inventor has spent years of his life in endeavoring to "harness the tides," and some have even proposed to compress air by tide- or wave-power for supplying energy, never understanding the signs of the old windmill on the hill, as it sorrowfully waved its arms about and bade them stop. The fact is that a wave- or tide-motor would have, as a rule, but a small chance of competing commercially with the windmill, which is by far the better machine, allowing a much greater amount of energy to be obtained in a simpler way. Wind-power has been, in old times, of inestimable value to man, if for nothing else but for enabling him to cross the seas, and it is even now a very important factor in travel and transportation. But there are great limitations in this ideally simple method of utilizing the sun's energy. The machines are large for a given output, and the power is intermittent, thus necessitating the storage of energy and increasing the cost of the plant.

A far better way, however, to obtain power would be to avail ourselves of the sun's rays, which beat the earth incessantly and supply energy at a maximum rate of over four million horsepower per square mile. Although the average energy received per square mile in any locality during the year is only a small fraction of that amount, yet an inexhaustible source of power would be opened up by the discovery of some efficient method of utilizing the energy of the rays. The only rational way known to me at the time when I began the study of this subject was to employ some kind of heat- or thermodynamic-engine, driven by a volatile fluid evaporated in a boiler by the heat of the rays. But closer investigation of this method, and calculation, showed that, notwithstanding the apparently vast amount of energy received from the sun's rays, only a small fraction of that energy could be actually utilized in this manner. Furthermore, the energy supplied through the sun's radiations is periodical, and the same limitations as in the use of the windmill I found to exist here also. After a long study of this mode of obtaining motive power from the sun, taking into account the necessarily large bulk of the boiler, the low efficiency of the heat-engine, the additional cost of storing the energy and other drawbacks, I came to the conclusion that the "solar engine," a few instances excepted, could not be industrially exploited with success.

Another way of getting motive power from the medium without consuming any material would be to utilize the heat contained in the earth, the water, or the air for driving an engine. It is a well-known fact that the interior portions of the globe are very hot, the temperature rising, as observations show, with the approach to the center at the rate of approximately 1° C for every hundred feet of depth. The difficulties of sinking shafts and placing boilers at depths of, say, twelve thousand feet, corresponding to an increase in temperature of about 120° C, are not insuperable, and we could certainly avail ourselves in this way of the internal heat of the globe. In fact, it would not be necessary to go to any depth at all in order to derive energy from the stored terrestrial heat. The superficial layers of the earth and the air strata close to the same are at a temperature sufficiently high to evaporate some extremely volatile substances, which we might use in our boilers instead of water. There is no doubt that a vessel might be propelled on the ocean by an engine driven by such a volatile fluid, no other energy being used but the heat abstracted from the water. But the amount of power which could be obtained in this manner would be, without further provision, very small.

Electricity produced by natural causes is another source of energy which might be rendered available. Lightning discharges involve great amounts of electrical energy, which we could utilize by transforming and storing it. Some years ago I made known a method of electrical transformation which renders the first part of this task easy, but the storing of the energy of lightning discharges will be difficult to accomplish. It is well known, furthermore, that electric currents circulate constantly through the earth, and that there exists between the earth and any air stratum a difference of electrical pressure, which varies in proportion to the height. In recent experiments I have discovered two novel facts of importance in this connection. One of these facts is that an electric current is generated in a wire extending from the ground to a great height by the axial, and probably also by the translatory, movement of the earth. No appreciable current, however, will flow continuously in the wire unless the electricity is allowed to leak out into the air. Its escape is greatly facilitated by providing at the elevated end of the wire a conducting terminal of great surface, with many sharp edges or points. We are thus enabled to get a continuous supply of electrical energy by merely supporting a wire at a height, but, unfortunately, the amount of electricity which can be so obtained is small.

The second fact which I have ascertained is that the upper air strata are permanently charged with electricity opposite to that of the earth. So, at least, I have interpreted my observations, from which it appears that the earth, with its adjacent insulating and outer conducting envelope, constitutes a highly charged electrical condenser containing, in all probability, a great amount of electrical energy which might be turned to the uses of man, if it were possible to reach with a wire to great altitudes.

It is possible, and even probable, that there will be, in time, other resources of energy opened up, of which we have no knowledge now. We may even find ways of applying forces such as magnetism or gravity for driving machinery without using any other means. Such realizations, though highly improbable, are not impossible. An example will best convey an idea of what we can hope to attain and what we can never attain. Imagine a disk of some homogeneous material turned perfectly true and arranged to turn in frictionless bearings on a horizontal shaft above the ground. This disk, being under the above conditions perfectly balanced, would rest in any position. Now, it is possible that we may learn how to make such a disk rotate continuously and perform work by the force of gravity without any further effort on our part; but it is perfectly impossible for the disk to turn and to do work without any force from the outside. If it could do so, it would be what is designated scientifically as a "perpetuum mobile," a machine creating its own motive power. To make the disk rotate by the force of gravity we have only to invent a screen against this force. By such a screen we could prevent this force from acting on one half of the disk, and the rotation of the latter would follow. At least, we cannot deny such a possibility until we know exactly the nature of the force of gravity. Suppose that this force were due to a movement comparable to that of a stream of air passing from above toward the center of the earth. The effect of such a stream upon both halves of the disk would be equal, and the latter would not rotate ordinarily; but if one half should be guarded by a plate arresting the movement, then it would turn.

A Departure from Known Methods— Possibility of a "Self-acting" Engine or Machine, Inanimate, Yet Capable, Like a Living Being, or Deriving Energy from the Medium— the Ideal way of Obtaining Motive Power

When I began the investigation of the subject under consideration, and when the preceding or similar ideas presented themselves to me for the first time, though I was then unacquainted with a number of the facts mentioned, a survey of the various ways of utilizing the energy of the medium convinced me, nevertheless, that to arrive at a thoroughly satisfactory practical solution a radical departure from the methods then known had to be made. The windmill, the solar engine, the engine driven by terrestrial heat, had their limitations in the amount of power obtainable. Some new way had to be discovered which would enable us to get more energy. There was enough heat-energy in the medium, but only a small part of it was available for the operation of an engine in the ways then known. Besides, the energy was obtainable only at a very slow rate. Clearly, then, the problem was to discover some new method which would make it possible both to utilize more of the heat-energy of the medium and also to draw it away from the same at a more rapid rate.

I was vainly endeavoring to form an idea of how this might be accomplished, when I read some statements from Carnot and Lord Kelvin (then Sir William Thomson) which meant virtually that it is impossible for an inanimate mechanism or self-acting machine to cool a portion of the medium below the temperature of the surrounding, and operate by the heat abstracted. These statements interested me intensely. Evidently a living being could do this very thing, and since the experiences of my early life which I have related had convinced me that a living being is only an automaton, or, otherwise stated, a "self-acting-engine," I came to the conclusion that it was possible to construct a machine which would do the same. As the first step toward this realization I conceived the following mechanism. Imagine a thermopile consisting of a number of bars of metal extending from the earth to the outer space beyond the atmosphere. The heat from below, conducted upward along these metal bars, would cool the earth or the sea or the air, according to the location of the lower parts of the bars, and the result, as is well known, would be an electric current circulating in these bars. The two terminals of the thermopile could now be joined through an electric motor, and, theoretically, this motor would run on and on, until the media below would be cooled down to the temperature of the outer space. This would be an inanimate engine which, to all evidence, would be cooling a portion of the medium below the temperature of the surrounding, and operating by the heat abstracted.

But was it not possible to realize a similar condition without necessarily going to a height? Conceive, for the sake of illustration, an enclosure T, as illustrated in diagram b, such that energy could not be transferred across it except through a channel or path O, and that, by some means or other, in this enclosure a medium were maintained which would have little energy, and that on the outer side of the same there would be the ordinary ambient medium with much energy. Under these assumptions the energy would flow through the

DIAGRAM *b*. Obtaining energy from the ambient medium. *A*, medium with little energy; *B*, *B*, ambient medium with much energy; *O*, path of the energy.

path *O*, as indicated by the arrow, and might then be converted on its passage into some other form of energy. The question was: Could such a condition be attained? Could we produce artificially such a "sink" for the energy of the ambient medium to flow in? Suppose that an extremely low temperature could be maintained by some process in a given space; the surrounding medium would then be compelled to give off heat, which could be converted into mechanical or other form of energy, and utilized. By realizing such a plan, we should be enabled to get at any point of the globe a continuous supply of energy, day and night. More than this, reasoning in the abstract, it would seem possible to cause a quick circulation of the medium, and thus draw the energy at a very rapid rate.

Here, then, was an idea which, if realizable, afforded a happy solution of the problem of getting energy from the medium. But was it realizable? I convinced myself that it was so in a number of ways, of which one is the following. As regards heat, we are at a high level, which may be represented by the surface of a mountain lake considerably above the sea, the level of which may mark the absolute zero of temperature existing in the interstellar space. Heat, like water, flows from high to low level, and, consequently, just as we can let the water of the lake run down to the sea, so we are able to let heat from the earth's surface travel up into the cold region above. Heat, like water, can perform work in flowing down, and if we had any doubt as to whether we could derive energy from the medium by means of a thermopile, as before described, it would be dispelled by this analogue. But can we produce cold in a given portion of the space and cause the heat to flow in continually? To create such a "sink," or "cold hole," as we might say, in the medium, would be equivalent to producing in the lake a space either empty or filled with something much lighter than water. This we could do by placing in the lake a tank, and pumping all the water out of the latter. We know, then, that the water, if allowed to flow back into the tank, would, theoretically, be able to perform exactly the same amount of work which was used in pumping it out, but not a bit more. Consequently nothing could be gained in this double operation of first raising the water and then letting it fall down. This would mean that it is impossible to create such a sink in the medium. But let us reflect a moment. Heat, though following certain general laws of mechanics, like a fluid, is not such; it is energy which may be converted into other forms of energy as it passes from a high to a low level. To make our mechanical analogy complete and true, we must, therefore, assume that the water, in its passage into the tank, is converted into something else, which may be taken out of it without using any, or by using very little, power. For example, if heat be represented in this analogue by the water of the lake, the oxygen and hydrogen composing the water may illustrate other forms of energy into which the heat is transformed in passing from hot to cold. If the process of heat transformation were absolutely perfect, no heat at all would arrive at the low level, since all of it would be converted into other forms of energy. Corresponding to this ideal case, all the water flowing into the tank would be decomposed into oxygen and hydrogen before reaching the bottom, and the result would be that water would continually flow in, and yet the tank would remain entirely empty, the gases formed escaping. We would thus produce, by expending initially a certain amount of work to create a sink for the heat or, respectively, the water to flow in, a condition enabling us to get any amount of energy without further effort. This would be an ideal way of obtaining motive power. We do not know of any such absolutely perfect process of heat-conversion, and consequently some heat will generally reach the low level, which means to say, in our mechanical analogue, that some water will arrive at the bottom of the tank, and a gradual and slow filling of the latter will take place, necessitating continuous pumping out. But evidently there will be less to pump out than flows in, or, in other words, less energy will be needed to maintain the initial condition than is developed by the fall, and this is to say that some energy will be gained from the medium. What is not converted in flowing down can just be raised up with its own energy, and what is converted is clear gain. Thus the virtue of the principle I have discovered resides wholly in the conversion of the energy on the downward flow.

First Efforts to Produce the Self-acting Engine— the Mechanical Oscillator— Work of Dewar and Linde—Liquid Air

Having recognized this truth, I began to devise means for carrying out my idea, and, after long thought, I finally conceived a combination of apparatus which should make possible the obtaining of power from the medium by a process of continuous cooling of atmospheric air. This apparatus, by continually transforming heat into mechanical work, tended to become colder and colder, and if it only were practicable to reach a very low temperature in this manner, then a sink for the heat could be produced, and energy could be derived from the medium. This seemed to be contrary to the statements of Carnot and Lord Kelvin before referred to, but I concluded from the theory of the process that such a result could be attained. This conclusion I reached, I think, in the latter part of 1883, when I was in Paris, and it was at a time when my mind was being more and more dominated by an invention which I had evolved during the preceding year, and which has since become known under the name of the "rotating magnetic field." During the few years which followed I elaborated further the plan I had imagined, and studied the working conditions, but made little headway. The commercial introduction in this country of the invention before referred to required most of my energies until 1889, when I again took up the idea of the self-acting machine. A closer investigation of the principles involved, and calculation, now showed that the result I aimed at could not be reached in a practical manner by ordinary machinery, as I had in the beginning expected. This led me, as a next step, to the study of a type of engine generally designated as "turbine," which at first seemed to offer better chances for a realization of the idea. Soon I found, however, that the turbine, too, was unsuitable. But my conclusions showed that if an engine of a peculiar kind could be brought to a high degree of perfection, the plan I had conceived was realizable, and I resolved to proceed with the development of such an engine, the primary object of which was to secure the greatest economy of transformation of heat into mechanical energy. A characteristic feature of the engine was that the work-performing piston was not connected with anything else, but was perfectly free to vibrate at an enormous rate. The mechanical difficulties encountered in the construction of this engine were greater than I had anticipated, and I made slow progress. This work was continued until early in 1892, when I went to London, where I saw Professor Dewar's admirable experiments with liquefied gases. Others had liquefied gases before, and notably Ozlewski and Pictet had performed creditable early experiments in this line, but there was such a vigor about the work of Dewar that even the old appeared new. His experiments showed, though in a way different from that I had imagined, that it was possible to reach a very low temperature by transforming heat into mechanical work, and I returned, deeply impressed with what I had seen, and more than ever convinced that my plan was practicable. The work temporarily interrupted was taken up anew, and soon I had in a fair state of perfection the engine which I have named "the mechanical oscillator." In this machine I succeeded in doing away with all packings, valves, and lubrication, and in producing so rapid a vibration of the piston that shafts of tough steel, fastened to the same and vibrated longitudinally, were torn asunder. By combining this engine with a dynamo of special design I produced a highly efficient electrical generator, invaluable in measurements and determinations of physical quantities on account of the unvarying rate of oscillation obtainable by its means. I exhibited several types of this machine, named "mechanical and electrical oscillator," before the Electrical Congress at the World's Fair in Chicago during the summer of 1893, in a lecture which, on account of other pressing work, I was unable to prepare for publication. On that occasion I exposed the principles of the mechanical oscillator, but the original purpose of this machine is explained here for the first time.

In the process, as I had primarily conceived it, for the utilization of the energy of the ambient medium, there were five essential elements in combination, and each of these had to be newly designed and perfected, as no such machines existed. The mechanical oscillator was the first element of this combination, and having perfected this, I turned to the next, which was an air-compressor of a design in certain respects resembling that of the mechanical oscillator. Similar difficulties in the construction were again encountered, but the work was pushed vigorously, and at the close of 1894 I had completed these two elements of the combination, and thus produced an apparatus for compressing air, virtually to any desired pressure, incomparably simpler, smaller, and more efficient than the ordinary. I was just beginning work on the third element, which together with the first two would give a refrigerating machine of exceptional efficiency and simplicity, when a misfortune befell me in the burning of my laboratory, which crippled my labors and delayed me. Shortly afterward Dr. Carl Linde announced the liquefaction of air by a self-cooling process, demonstrating that it was practicable to proceed with the cooling until liquefaction of the air took place. This was the only experimental proof which I was still wanting that energy was obtainable from the medium in the manner contemplated by me.

The liquefaction of air by a self-cooling process was not, as popularly believed, an accidental discovery, but a scientific result which could not have been delayed much longer, and which, in all probability, could not have escaped Dewar. This fascinating advance, I believe, is largely due to the powerful work of this great Scotchman. Nevertheless, Linde's is an immortal achievement. The manufacture of liquid air has been carried on for four years in Germany, on a scale much larger than in any other country, and this strange product has been applied for a variety of purposes. Much was expected of it in the beginning, but so far it has been an industrial *ignis fatuus*. By the use of such machinery as I am perfecting, its cost will probably be greatly lessened, but even then its commercial success will be questionable. When used as a refrigerant it is uneconomical, as its temperature is unnecessarily low. It is as expensive to maintain a body at a very low temperature as it is to keep it very hot; it takes coal to keep air cold. In oxygen manufacture it cannot yet compete with the electrolytic method. For use as an explosive it is unsuitable, because its low temperature again condemns it to a small efficiency, and for motive-power purposes its cost is still by far too high. It is of interest to note, however, that in driving an engine by liquid air a certain amount of energy may be gained from the engine, or, stated otherwise, from the ambient medium which keeps the engine warm, each two hundred pounds of iron-casting of the latter contributing energy at the rate of about one effective horsepower during one hour. But this gain of the consumer is offset by an equal loss of the producer.

Much of this task on which I have labored so long remains to be done. A number of mechanical details are still to be perfected and some difficulties of a different nature to be mastered, and I cannot hope to produce a self-acting machine deriving energy from the ambient medium for a long time yet, even if all my expectations should materialize. Many circumstances have occurred which have retarded my work of late, but for several reasons the delay was beneficial.

One of these reasons was that I had ample time to consider what the ultimate possibilities of this development might be. I worked for a long time fully convinced that the practical realization of this method of obtaining energy from the sun would be of incalculable industrial value, but the continued study of the subject revealed the fact that while it will be commercially profitable if my expectations are well founded, it will not be so to an extraordinary degree.

Discovery of Unexpected Properties of the Atmosphere— Strange Experiments—Transmission of Electrical Energy Through One Wire Without Return—Transmissions Through the Earth Without any Wire

Another of these reasons was that I was led to recognize the transmission of electrical energy to any distance through the media as by far the best solution of the great problem of harnessing the sun's energy for the uses of man. For a long time I was convinced that such a transmission on an industrial scale could never be realized, but a discovery which I made changed my view. I observed that under certain conditions the atmosphere, which is normally a high insulator, assumes conducting properties, and so becomes capable of conveying any amount of electrical energy. But the difficulties in the way of a practical utilization of this discovery for the purpose of transmitting electrical energy without wires were seemingly insuperable. Electrical pressures of many millions of volts had to be produced and handled; generating apparatus of a novel kind, capable of withstanding the immense electrical stresses, had to be invented and perfected, and a complete safety against the dangers of the high-tension currents had to be attained in the system before its practical introduction could be even thought of. All this could not be done in a few weeks or months, or even years. The work required patience and constant application, but the improvements came, though slowly. Other valuable results were, however, arrived at in the course of this long-continued work, of which I shall endeavor to give a brief account, enumerating the chief advances as they were successively effected.

The discovery of the conducting properties of the air, though unexpected, was only a natural result of experiments in a special field which I had carried on for some years before. It was, I believe, during 1889 that certain possibilities offered by extremely rapid electrical oscillations determined me to design a number of special machines adapted for their investigation. Owing to the peculiar requirements, the construction of these machines was very difficult, and consumed much time and effort; but my work on them was generously rewarded, for I reached by their means several novel and important results. One of the earliest observations I made with these new machines was that electrical oscillations of an extremely high rate act in an extraordinary manner upon the human organism. Thus, for instance, I demonstrated that powerful electrical discharges of several hundred thousand volts, which at that time were considered absolutely deadly, could be passed through the body without inconvenience or hurtful consequences. These oscillations produced other specific physiological effects, which, upon my announcement, were eagerly taken up by skilled physicians and further investigated. This new field has proved itself fruitful beyond expectation, and in the few years which have passed since, it has been developed to such an extent that it now forms a legitimate and important department of medical science. Many results, thought impossible at that time, are now readily obtainable with these oscillations, and many experiments undreamed of then can now be readily performed by their means. I still remember with pleasure how, nine years ago, I passed the discharge of a powerful induction-coil through my body to demonstrate before a scientific society the comparative harmlessness of very rapidly vibrating electric currents, and I can still recall the astonishment of my audience. I would now undertake, with much less apprehension that I had in that experiment, to transmit through my body with such currents the entire electrical energy of the dynamos now working at Niagara—forty or fifty thousand horsepower. I have produced electrical oscillations which were of such intensity that when circulating through my arms and chest they have melted wires which joined my hands, and still I felt no inconvenience. I have energized with such oscillations a loop of heavy copper wire so powerfully that masses of metal, and even objects of an electrical resistance specifically greater than that of human tissue, brought close to or placed within the loop, were heated to a high temperature and melted, often with the violence of an explosion, and yet into this very space in which this terribly-destructive turmoil was going on I have repeatedly thrust my head without feeling anything or experiencing injurious after-effects.

Another observation was that by means of such oscillations light could be produced in a novel and more economical manner, which promised to lead to an ideal system of electric illumination by vacuum-tubes, dispensing with the necessity of renewal of lamps or incandescent filaments, and possibly also with the use of wires in the interior of buildings. The efficiency of this light increases in

FIG. 3. Experiment to illustrate the supplying of electrical energy through a single wire without return. An ordinary incandescent lamp, connected with one or both of its terminals to the wire forming the upper free end of the coil shown in the photograph, is lighted by electrical vibrations conveyed to it through the coil from an electrical oscillator, which is worked only to one fifth of one per cent. of its full capacity.

proportion to the rate of the oscillations, and its commercial success is, therefore, dependent on the economical production of electrical vibrations of transcending rates. In this direction I have met with gratifying success of late, and the practical introduction of this new system of illumination is not far off.

The investigations led to many other valuable observations and results, one of the more important of which was the demonstration of the

Fig. 4. Experiment to illustrate the transmission of electrical energy through the earth without wire. The coil shown in the photograph has its lower end or terminal connected to the ground, and is exactly attuned to the vibrations of a distant electrical oscillator. The lamp lighted is in an independent wire loop, energized by induction from the coil excited by the electrical vibrations transmitted to it through the ground from the oscillator, which is worked only to five percent of its full capacity.

practicability of supplying electrical energy through one wire without return. At first I was able to transmit in this novel manner only very small amounts of electrical energy, but in this line also my efforts have been rewarded with similar success.

The photograph shown in Fig. 3 illustrates, as its title explains, an actual transmission of this kind effected with apparatus used in other experiments here described. To what a degree the appliances have been perfected since my first demonstrations early in 1891 before a scientific society, when my apparatus was barely capable of lighting one lamp (which result was considered wonderful), will appear when I state that I have now no difficulty in lighting in this manner four or five hundred lamps, and could light many more. In fact, there is no limit to the amount of energy which may in this way be supplied to operate any kind of electrical device.

After demonstrating the practicability of this method of transmission, the thought naturally occurred to me to use the earth as a conductor, thus dispensing with all wires. Whatever electricity may be, it is a fact that it behaves like an incompressible fluid, and the earth may be looked upon as an immense reservoir of electricity, which, I thought, could be disturbed effectively by a properly designed electrical machine. Accordingly, my next efforts were directed toward perfecting a special apparatus which would be highly effective in creating a disturbance of electricity in the earth. The progress in this new direction was necessarily very slow and the work discouraging, until I finally succeeded in perfecting a novel kind of transformer or induction-coil, particularly suited for this special purpose. That it is practicable, in this manner, not only to transmit minute amounts of electrical energy for operating delicate electrical devices, as I contemplated at first, but also electrical energy in appreciable quantities, will appear from an inspection of Fig. 4, which illustrates an actual experiment of this kind performed with the same apparatus. The result obtained was all the more remarkable as the top end of the coil was not connected to a wire or plate for magnifying the effect.

"Wireless" Telegraphy—The Secret of Tuning— Errors in the Hertzian Investigations— A Receiver of Wonderful Sensitiveness

As the first valuable result of my experiments in this latter line a system of telegraphy without wires resulted, which I described in two scientific lectures in February and March, 1893. It is mechanically illustrated in diagram *c*, the upper part of which shows the electrical arrangement as I described it then, while the lower part illustrates its mechanical analogue. The system is extremely simple in principle. Imagine two tuning-forks F, F_1, one at the sending- and the other at the receiving-station respectively, each having attached to its lower prong a minute piston *p*, fitting in a cylinder. Both the cylinders communicate with a large reservoir *R*, with elastic walls, which is supposed to be closed and filled with a light and incompressible fluid. By striking repeatedly one of the prongs of the tuning-fork *F*, the small piston *p* below would be vibrated, and its vibrations, transmitted through the fluid, would reach

DIAGRAM *c*. "Wireless" telegraphy mechanically illustrated.

the distant fork F_1, which is "tuned" to the fork F, or, stated otherwise, of exactly the same note as the latter. The fork F_1 would now be set vibrating, and its vibration would be intensified by the continued action of the distant fork F until its upper prong, swinging far out, would make an electrical connection with a stationary contact c'', starting in this manner some electrical or other appliances which may be used for recording the signals. In this simple way messages could be exchanged between the two stations, a similar contact c' being provided for this purpose, close to the upper prong of the fork F, so that the apparatus at each station could be employed in turn as receiver and transmitter.

The electrical system illustrated in the upper figure of diagram c is exactly the same in principle, the two wires or circuits ESP and $E_1S_1P_1$, which extend vertically to a height, representing the two tuning-forks with the pistons attached to them. These circuits are connected with the ground by plates E, E_1, and to two elevated metal sheets P, P_1, which store electricity and thus magnify considerably the effect. The closed reservoir R, with elastic walls, is in this case replaced by the earth, and the fluid by electricity. Both of these circuits are "tuned" and operate just like the two tuning-forks. Instead of striking the fork F at the sending station, electrical oscillations are produced in the vertical sending- or transmitting-wire ESP, as by the action of a source S, included in this wire, which spread through the ground and

FIG. 5. Photographic view of the coils responding to electrical oscillations.
The picture shows a number of coils, differently attuned and responding to the
vibrations transmitted to them through the earth from an electrical oscillator.
The large coil on the right, discharging strongly, is tuned to the fundamental
vibration, which is fifty thousand per second; the two larger vertical coils to twice
that number; the smaller white wire coil to four times that number, and the
remaining small coils to higher tones. The vibrations produced by the oscillator
were so intense that they affected perceptibly a small coil tuned to the twenty-
sixth higher tone.

reach the distant vertical receiving-wire $E_1S_1P_1$, exciting corresponding
electrical oscillations in the same. In the latter wire or circuit is
included a sensitive device or receiver S_1, which is thus set in action and
made to operate a relay or other appliance. Each station is, of course,
provided both with a source of electrical oscillations S and a sensitive
receiver S_1, and a simple provision is made for using each of the two
wires alternately to send and to receive the messages.

The exact attunement of the two circuits secures great advantages,
and, in fact, it is essential in the practical use of the system. In this
respect many popular errors exist, and, as a rule, in the technical
reports on this subject circuits and appliances are described as
affording these advantages when from their very nature it is evident

that this is impossible. In order to attain the best results it is essential that the length of each wire or circuit, from the ground connection to the top, should be equal to one quarter of the wavelength of the electrical vibration in the wire, or else equal to that length multiplied by an odd number. Without the observation of this rule it is virtually impossible to prevent the interference and insure the privacy of messages. Therein lies the secret of tuning. To obtain the most satisfactory results it is, however, necessary to resort to electrical vibrations of low pitch. The Hertzian spark apparatus, used generally by experimenters, which produces oscillations of a very high rate, permits no effective tuning, and slight disturbances are sufficient to render an exchange of messages impracticable. But scientifically designed, efficient appliances allow nearly perfect adjustment. An experiment performed with the improved apparatus repeatedly referred to, and intended to convey an idea of this feature, is illustrated in Fig. 5, which is sufficiently explained by its note.

Since I described these simple principles of telegraphy without wires I have had frequent occasion to note that the identical features and elements have been used, in the evident belief that the signals are being transmitted to considerable distance by "Hertzian" radiations. This is only one of many misapprehensions to which the investigations of the lamented physicist have given rise. About thirty-three years ago Maxwell, following up a suggestive experiment made by Faraday in 1845, evolved an ideally simple theory which intimately connected light, radiant heat, and electrical phenomena, interpreting them as being all due to vibrations of a hypothetical fluid of inconceivable tenuity, called the ether. No experimental verification was arrived at until Hertz, at the suggestion of Helmholtz, undertook a series of experiments to this effect. Hertz proceeded with extraordinary ingenuity and insight, but devoted little energy to the perfection of his old-fashioned apparatus. The consequence was that he failed to observe the important function which the air played in his experiments, and which I subsequently discovered. Repeating his experiments and reaching different results, I ventured to point out this oversight. The strength of the proofs brought forward by Hertz in support of Maxwell's theory resided in the correct estimate of the rates of vibration of the circuits he used. But I ascertained that he could not have obtained the rates he thought he was getting. The vibrations with identical apparatus he employed are, as a rule, much slower, this being due to the presence of air, which produces a dampening effect upon a rapidly vibrating electric circuit of high pressure, as a fluid does upon a vibrating tuning-fork. I have, however, discovered since that time other causes of error, and I have long ago ceased to look upon his results as being an experimental verification of the poetical conceptions of Maxwell. The work of the great German physicist has acted as an immense stimulus to contemporary electrical research, but it has likewise, in a measure, by its fascination, paralyzed the scientific mind, and thus hampered independent inquiry. Every new phenomenon which was discovered was made to fit the theory, and so very often the truth has been unconsciously distorted.

When I advanced this system of telegraphy, my mind was dominated by the idea of effecting communication to any distance through the earth or environing medium, the practical consummation of which I considered of transcendent importance, chiefly on account of the moral effect which it could not fail to produce universally. As the first effort to this end I proposed, at that time, to employ relay-stations with tuned circuits, in the hope of making thus practicable signaling over vast distances, even with apparatus of very moderate power then at my command. I was confident, however, that with properly designed machinery signals could be transmitted to any point of the globe, no matter what the distance, without the necessity of using such intermediate stations. I gained this conviction through the discovery of a singular electrical phenomenon, which I described early in 1892, in lectures I delivered before some scientific societies abroad, and which I have called a "rotating brush." This is a bundle of light which is formed, under certain conditions, in a vacuum-bulb, and which is of a sensitiveness to magnetic and electric influences bordering, so to speak, on the supernatural. This light-bundle is rapidly rotated by the earth's magnetism as many as twenty thousand times per second, the rotation in these parts being opposite to what it would be in the southern hemisphere, while in the region of the magnetic equator it should not rotate at all. In its most sensitive state, which is difficult to obtain, it is responsive to electric or magnetic influences to an incredible degree. The mere stiffening of the muscles of the arm and consequent slight electrical change in the body of an observer standing at some distance from it, will perceptibly affect it. When in this highly sensitive state it is capable of indicating the slightest magnetic and electric changes taking place in the earth. The observation of this wonderful phenomenon impressed me strongly that communication at any distance could be easily effected by its means, provided that apparatus could be perfected capable of producing an electric or magnetic change of state, however small, in the terrestrial globe or environing medium.

DEVELOPMENT OF A NEW PRINCIPLE— THE ELECTRICAL OSCILLATOR—PRODUCTION OF IMMENSE ELECTRICAL MOVEMENTS—THE EARTH RESPONDS TO MAN— INTERPLANETARY COMMUNICATION NOW PROBABLE

I resolved to concentrate my efforts upon this venturesome task, though it involved great sacrifice, for the difficulties to be mastered were such that I could hope to consummate it only after years of labor. It meant delay of other work to which I would have preferred to devote myself, but I gained the conviction that my energies could not be more usefully employed for I recognized that an efficient apparatus for, the production of powerful electrical oscillations, as was needed for that specific purpose, was the key to the solution of other most important electrical and, in fact, human problems. Not only was communication, to any distance, without wires possible by its means, but, likewise, the transmission of energy in great amounts, the burning of the atmospheric nitrogen, the production of an efficient illuminant, and many other results of inestimable scientific and industrial value. Finally, however, I had the satisfaction of accomplishing the task undertaken by the use of a new principle, the virtue of which is based on the marvelous properties of the electrical condenser. One of these is that it can discharge or explode its stored energy in an inconceivably short time. Owing to this it is unequaled in explosive violence. The explosion of dynamite is only the breath of a consumptive compared with its discharge. It is the means of producing the strongest current, the highest electrical pressure, the greatest commotion in the medium. Another of its properties, equally valuable, is that its discharge may vibrate at any rate desired up to many millions per second.

FIG. 6. Photographic view of the essential parts of the electrical oscillator used in the experiments described.

I had arrived at the limit of rates obtainable in other ways when the happy idea presented itself to me to resort to the condenser. I arranged such an instrument so as to be charged and discharged alternately in rapid succession through a coil with a few turns of stout wire, forming the primary of a transformer or induction-coil. Each time the condenser was discharged the current would quiver in the primary wire and induce corresponding oscillations in the secondary. Thus a transformer or induction-coil on new principles was evolved, which I have called "the electrical oscillator," partaking of those unique qualities which characterize the condenser, and enabling results to be attained impossible by other means. Electrical effects of any desired character and of intensities undreamed of before are now easily producible by perfected apparatus of this kind, to which frequent reference has been made, and the essential parts of which are shown in Fig. 6. For certain purposes a strong inductive effect is required; for others the greatest possible suddenness; for others again, an exceptionally high rate of vibration or extreme pressure; while for

FIG. 7. Experiment to illustrate an inductive effect of an electrical oscillator of great power. The photograph shows three ordinary incandescent lamps lighted to full candle-power by currents induced in a local loop consisting of a single wire forming a square of fifty feet each side, which includes the lamps, and which is at a distance of one hundred feet from the primary circuit energized by the oscillator. The loop likewise includes an electrical condenser, and is exactly attuned to the vibrations of the oscillator, which is worked at less than five percent of its total capacity.

certain other objects immense electrical movements are necessary. The photographs in Figs. 7, 8, 9, and 10, of experiments performed with such an oscillator, may serve to illustrate some of these features and convey an idea of the magnitude of the effects actually produced. The completeness of the titles of the figures referred to makes a further description of them unnecessary. However extraordinary the results shown may appear, they are but trifling compared with those which are attainable by apparatus designed on these same principles. I have produced electrical discharges the actual path of which, from end to end, was probably more than one hundred feet long; but it would not be difficult to reach lengths one hundred times as great. I have produced electrical movements occurring at the rate of approximately one hundred thousand horse-power, but rates of one, five, or ten million

FIG. 8. **Experiment to illustrate the capacity of the oscillator for producing electrical explosions of great power.** Note to Fig. 8.—The coil, partly shown in the photograph, creates an alternative movement of electricity from the earth into a large reservoir and back at a rate of one hundred thousand alternations per second. The adjustments are such that the reservoir is filled full and bursts at each alternation just at the moment when the electrical pressure reaches the maximum. The discharge escapes with a deafening noise, striking an unconnected coil twenty-two feet away, and creating such a commotion of electricity in the earth that sparks an inch long can be drawn from a water main at a distance of three hundred feet from the laboratory.

horse-power are easily practicable. In these experiments effects were developed incomparably greater than any ever produced by human agencies, and yet these results are but an embryo of what is to be.

That communication without wires to any point of the globe is practicable with such apparatus would need no demonstration, but through a discovery which I made I obtained absolute certitude. Popularly explained, it is exactly this: When we raise the voice and hear an echo in reply, we know that the sound of the voice must have reached a distant wall, or boundary, and must have been reflected from the same. Exactly as the sound, so an electrical wave is reflected, and the same evidence which is afforded by an echo is offered by an electrical phenomenon known as a "stationary" wave—that is, a

FIG. 9. Experiment to illustrate the capacity of the oscillator for creating a great electrical movement. The ball shown in the photograph, covered with a polished metallic coating of twenty square feet of surface, represents a large reservoir of electricity, and the inverted tin pan underneath, with a sharp rim, a big opening through which the electricity can escape before filling the reservoir. The quantity of electricity set in movement is so great that, although most of it escapes through the rim of the pan or opening provided, the ball or reservoir is nevertheless alternately emptied and filled to over-flowing (as is evident from the discharge escaping on the top of the ball) one hundred and fifty thousand times per second.

wave with fixed nodal and ventral regions. Instead of sending sound-vibrations toward a distant wall, I have sent electrical vibrations toward the remote boundaries of the earth, and instead of the wall the earth has replied. In place of an echo I have obtained a stationary electrical wave, a wave reflected from afar.

FIG. 10. Photographic view of an experiment to illustrate an effect of an electrical oscillator delivering energy at a rate of seventy-five thousand horse-power. The discharge, creating a strong draft owing to the heating of the air, is carried upward through the open roof of the building. The greatest width across is nearly seventy feet. The pressure is over twelve million volts, and the current alternates one hundred and thirty thousand times per second.

Stationary waves in the earth mean something more than mere telegraphy without wires to any distance. They will enable us to attain many important specific results impossible otherwise. For instance, by their use we may produce at will, from a sending-station, an electrical effect in any particular region of the globe; we may determine the relative position or course of a moving object, such as a vessel at sea, the distance traversed by the same, or its speed; or we may send over the earth a wave of electricity traveling at any rate we desire, from the pace of a turtle up to lightning speed.

With these developments we have every reason to anticipate that in a time not very distant most telegraphic messages across the oceans will be transmitted without cables. For short distances we need a "wireless" telephone, which requires no expert operators. The greater the spaces to be bridged, the more rational becomes communication without wires. The cable is not only an easily damaged and costly instrument, but it limits us in the speed of transmission by reason of a certain electrical property inseparable from its construction. A properly designed plant for effecting communication without wires ought to have many times the working capacity of a cable, while it will involve incomparably less expense. Not a long time will pass, I believe, before communication by cable will become obsolete, for not only will signaling by this new method be quicker and cheaper, but also much safer. By using some new means for isolating the messages which I have contrived, an almost perfect privacy can be secured.

I have observed the above effects so far only up to a limited distance of about six hundred miles, but inasmuch as there is virtually no limit to the power of the vibrations producible with such an oscillator, I feel quite confident of the success of such a plant for effecting transoceanic communication. Nor is this all. My measurements and calculations have shown that it is perfectly practicable to produce on our globe, by the use of these principles, an electrical movement of such magnitude that, without the slightest doubt, its effect will be perceptible on some of our nearer planets, as Venus and Mars. Thus from mere possibility interplanetary communication has entered the stage of probability. In fact, that we can produce a distinct effect on one of these planets in this novel manner, namely, by disturbing the electrical condition of the earth, is beyond any doubt. This way of effecting such communication is, however, essentially different from all others which have so far been proposed by scientific men. In all the previous instances only a minute fraction of the total energy reaching the planet—as much as it would be possible to concentrate in a reflector—could be utilized by the supposed observer in his instrument. But by the means I have developed he would be enabled to concentrate the larger portion of the entire energy transmitted to the planet in his instrument, and the chances of affecting the latter are thereby increased many millionfold.

Besides machinery for producing vibrations of the required power, we must have delicate means capable of revealing the effects of feeble influences exerted upon the earth. For such purposes, too, I have perfected new methods. By their use we shall likewise be able, among other things, to detect at considerable distance the presence of an iceberg or other object at sea. By their use, also, I have discovered some terrestrial phenomena still unexplained. That we can send a message to a planet is certain, that we can get an answer is probable: man is not the only being in the Infinite gifted with a mind.

TRANSMISSION OF ELECTRICAL ENERGY TO ANY DISTANCE WITHOUT WIRES—NOW PRACTICABLE— THE BEST MEANS OF INCREASING THE FORCE ACCELERATING THE HUMAN MASS

The most valuable observation made in the course of these investigations was the extraordinary behavior of the atmosphere toward electric impulses of excessive electromotive force. The experiments showed that the air at the ordinary pressure became distinctly conducting, and this opened up the wonderful prospect of transmitting large amounts of electrical energy for industrial purposes to great distances without wires, a possibility which, up to that time, was thought of only as a scientific dream. Further investigation revealed the important fact that the conductivity imparted to the air by these electrical impulses of many millions of volts increased very rapidly with the degree of rarefaction, so that air strata at very moderate altitudes, which are easily accessible, offer, to all experimental evidence, a perfect conducting path, better than a copper wire, for currents of this character.

Thus the discovery of these new properties of the atmosphere not only opened up the possibility of transmitting, without wires, energy in large amounts, but, what was still more significant, it afforded the certitude that energy could be transmitted in this manner economically. In this new system it matters little—in fact, almost nothing—whether the transmission is effected at a distance of a few miles or of a few thousand miles.

While I have not, as yet, actually effected a transmission of a considerable amount of energy, such as would be of industrial importance, to a great distance by this new method, I have operated several model plants under exactly the same conditions which will exist in a large plant of this kind, and the practicability of the system is thoroughly demonstrated. The experiments have shown conclusively that, with two terminals maintained at an elevation of not more than thirty thousand to thirty-five thousand feet above sea-level, and with an electrical pressure of fifteen to twenty million volts, the energy of thousands of horse-power can be transmitted over distances which may be hundreds and, if necessary, thousands of miles. I am hopeful, however, that I may be able to reduce very considerably the elevation of the terminals now required, and with this object I am following up an idea which promises such a realization. There is, of course, a popular prejudice against using an electrical pressure of millions of volts, which may cause sparks to fly at distances of hundreds of feet, but, paradoxical as it may seem, the system, as I have described it in a technical publication, offers greater personal safety than most of the ordinary distribution circuits now used in the cities. This is, in a measure, borne out by the fact that, although I have carried on such experiments for a number of years, no injury has been sustained either by me or any of my assistants.

But to enable a practical introduction of the system, a number of essential requirements are still to be fulfilled. It is not enough to develop appliances by means of which such a transmission can be effected. The machinery must be such as to allow the transformation and transmission of electrical energy under highly economic and practical conditions. Furthermore, an inducement must be offered to those who are engaged in the industrial exploitation of natural sources of power, as waterfalls, by guaranteeing greater returns on the capital invested than they can secure by local development of the property.

From that moment when it was observed that, contrary to the established opinion, low and easily accessible strata of the atmosphere are capable of conducting electricity, the transmission of electrical energy without wires has become a rational task of the engineer, and one surpassing all others in importance. Its practical consummation would mean that energy would be available for the uses of man at any point of the globe, not in small amounts such as might be derived from the ambient medium by suitable machinery, but in quantities virtually unlimited, from waterfalls. Export of power would then become the chief source of income for many happily situated countries, as the United States, Canada, Central and South America, Switzerland, and Sweden. Men could settle down everywhere, fertilize and irrigate the soil with little effort, and convert barren deserts into gardens, and thus the entire globe could be transformed and made a fitter abode for mankind. It is highly probable that if there are intelligent beings on Mars they have long ago realized this very idea, which would explain the changes on its surface noted by astronomers. The atmosphere on that planet, being of considerably smaller density than that of the earth, would make the task much more easy.

It is probable that we shall soon have a self-acting heat-engine capable of deriving moderate amounts of energy from the ambient medium. There is also a possibility—though a small one—that we may obtain electrical energy direct from the sun. This might be the case if the Maxwellian theory is true, according to which electrical vibrations of all rates should emanate from the sun. I am still investigating this subject. Sir William Crookes has shown in his beautiful invention known as the "radiometer" that rays may produce by impact a mechanical effect, and this may lead to some important revelation as to the utilization of the sun's rays in novel ways. Other sources of energy may be opened up, and new methods of deriving energy from the sun discovered, but none of these or similar achievements would equal in importance the transmission of power to any distance through the medium. I can conceive of no technical advance which would tend to unite the various elements of humanity more effectively than this one, or of one which would more add to and more economize human energy. It would be the best means of increasing the force accelerating the human mass. The mere moral influence of such a radical departure would be incalculable. On the other hand if at any point of the globe energy can be obtained in limited quantities from the ambient medium by means of a self-acting heat-engine or otherwise, the conditions will remain the same as before. Human performance will be increased, but men will remain strangers as they were.

I anticipate that many, unprepared for these results, which, through long familiarity, appear to me simple and obvious, will consider them still far from practical application. Such reserve, and even opposition, of some is as useful a quality and as necessary an element in human progress as the quick receptivity and enthusiasm of others. Thus, a mass which resists the force at first, once set in movement, adds to the energy. The scientific man does not aim at an immediate result. He does not expect that his advanced ideas will be readily taken up. His work is like that of the planter—for the future. His duty is to lay the foundation for those who are to come, and point the way. He lives and labors and hopes with the poet who says:

> Schaff' das Tagwerk meiner Hände,
> Hohes Glück, dass ich's vollende!
> Lass, o lass mich nicht ermatten!
> Nein, es sind nicht leere Träume:
> Jetzt nur Stangen, diese Bäume
> Geben einst noch Frucht und Schatten.*

---

* "Daily work my hands' employment, /To complete is pure enjoyment! /Let, oh, let me never falter! /No! there is no empty dreaming: /Lo! these trees, but bare poles seeming, /Yet will yield both fruit and shelter!" Goethe's "Hope." Translated by William Gibson, Com. U. S. N.

The Inventions, Researches, and Writings of Nikola Tesla

By Thomas Commerford Martin

PREFACE

The electrical problems of the present day lie largely in the economical transmission of power and in the radical improvement of the means and methods of illumination. To many workers and thinkers in the domain of electrical invention, the apparatus and devices that are familiar, appear cumbrous and wasteful, and subject to severe limitations. They believe that the principles of current generation must be changed, the area of current supply be enlarged, and the appliances used by the consumer be at once cheapened and simplified. The brilliant successes of the past justify them in every expectancy of still more generous fruition.

The present volume is a simple record of the pioneer work done in such departments up to date, by Mr. Nikola Tesla, in whom the world has already recognized one of the foremost of modern electrical investigators and inventors. No attempt whatever has been made here to emphasize the importance of his researches and discoveries. Great ideas and real inventions win their own way, determining their own place by intrinsic merit. But with the conviction that Mr. Tesla is blazing a path that electrical development must follow for many years to come, the compiler has endeavored to bring together all that bears the impress of Mr. Tesla's genius, and is worthy of preservation. Aside from its value as showing the scope of his inventions, this volume may be of service as indicating the range of his thought. There is intellectual profit in studying the push and play of a vigorous and original mind.

Although the lively interest of the public in Mr. Tesla's work is perhaps of recent growth, this volume covers the results of full ten years. It includes his lectures, miscellaneous articles[Pg vi] and discussions, and makes note of all his inventions thus far known, particularly those bearing on polyphase motors and the effects obtained with currents of high potential and high frequency. It will be seen that Mr. Tesla has ever pressed forward, barely pausing for an instant to work out in detail the utilizations that have at once been obvious to him of the new principles he has elucidated. Wherever possible his own language has been employed.

It may be added that this volume is issued with Mr. Tesla's sanction and approval, and that permission has been obtained for the re-publication in it of such papers as have been read before various technical societies of this country and Europe. Mr. Tesla has kindly favored the author by looking over the proof sheets of the sections embodying his latest researches. The work has also enjoyed the careful revision of the author's friend and editorial associate, Mr. Joseph Wetzler, through whose hands all the proofs have passed.

December, 1893.

Part I

Polyphase Currents

CHAPTER I

BIOGRAPHICAL AND INTRODUCTORY

As an introduction to the record contained in this volume of Mr. Tesla's investigations and discoveries, a few words of a biographical nature will, it is deemed, not be out of place, nor other than welcome.

Nikola Tesla was born in 1857 at Smiljan, Lika, a borderland region of Austro-Hungary, of the Serbian race, which has maintained against Turkey and all comers so unceasing a struggle for freedom. His family is an old and representative one among these Switzers of Eastern Europe, and his father was an eloquent clergyman in the Greek Church. An uncle is today Metropolitan in Bosnia. His mother was a woman of inherited ingenuity, and delighted not only in skillful work of the ordinary household character, but in the construction of such mechanical appliances as looms and churns and other machinery required in a rural community. Nikola was educated at Gospich in the public school for four years, and then spent three years in the Real Schule. He was then sent to Carstatt, Croatia, where he continued his studies for three years in the Higher Real Schule. There for the first time he saw a steam locomotive. He graduated in 1873, and, surviving an attack of cholera, devoted himself to experimentation, especially in electricity and magnetism. His father would have had him maintain the family tradition by entering the Church, but native genius was too strong, and he was allowed to enter the Polytechnic School at Gratz, to finish his studies, and with the object of becoming a professor of mathematics and physics. One of the machines there experimented with was a Gramme dynamo, used as a motor. Despite his instructor's perfect demonstration of the fact that it was impossible to operate a dynamo without commutator or brushes, Mr. Tesla could not be convinced that such accessories were necessary or desirable. He had already seen with quick intuition that a way could be found to dispense with them; and from that time he may be said to have begun work on the ideas that fructified ultimately in his rotating field motors.

In the second year of his Gratz course, Mr. Tesla gave up the notion of becoming a teacher, and took up the engineering curriculum. His studies ended, he returned home in time to see his father die, and then went to Prague and Buda-Pesth to study languages, with the object of qualifying himself broadly for the practice of the engineering profession. For a short time he served as an assistant in the Government Telegraph Engineering Department, and then became associated with M. Puskas, a personal and family friend, and other exploiters of the telephone in Hungary. He made a number of telephonic inventions, but found his opportunities of benefiting by them limited in various ways. To gain a wider field of action, he pushed on to Paris and there secured employment as an electrical engineer with one of the large companies in the new industry of electric lighting.

It was during this period, and as early as 1882, that he began serious and continued efforts to embody the rotating field principle in operative apparatus. He was enthusiastic about it; believed it to mark a new departure in the electrical arts, and could think of nothing else. In fact, but for the solicitations of a few friends in commercial circles who urged him to form a company to exploit the invention, Mr. Tesla, then a youth of little worldly experience, would have sought an immediate opportunity to publish his ideas, believing them to be worthy of note as a novel and radical advance in electrical theory as well as destined to have a profound influence on all dynamo electric machinery.

At last he determined that it would be best to try his fortunes in America. In France he had met many Americans, and in contact with them learned the desirability of turning every new idea in electricity to practical use. He learned also of the ready encouragement given in the United States to any inventor who could attain some new and valuable result. The resolution was formed with characteristic quickness, and abandoning all his prospects in Europe, he at once set his face westward.

Arrived in the United States, Mr. Tesla took off his coat the day he arrived, in the Edison Works. That place had been a goal of his ambition, and one can readily imagine the benefit and stimulus derived from association with Mr. Edison, for whom Mr. Tesla has always had the strongest admiration. It was impossible, however, that, with his own ideas to carry out, and his own inventions to develop, Mr. Tesla could long remain in even the most delightful employ; and, his work now attracting attention, he left the Edison ranks to join a company intended to make and sell an arc lighting system based on some of his inventions in that branch of the art. With unceasing diligence he brought the system to perfection, and saw it placed on the market. But the thing which most occupied his time and thoughts, however, all through this period, was his old discovery of the rotating field principle for alternating current work, and the application of it in motors that have now become known the world over.

Strong as his convictions on the subject then were, it is a fact that he stood very much alone, for the alternating current had no well recognized place. Few electrical engineers had ever used it, and the majority were entirely unfamiliar with its value, or even its essential features. Even Mr. Tesla himself did not, until after protracted effort and experimentation, learn how to construct alternating current apparatus of fair efficiency. But that he had accomplished his purpose was shown by the tests of Prof. Anthony, made in the winter of 1887–8, when Tesla motors in the hands of that distinguished expert gave an efficiency equal to that of direct current motors. Nothing now stood in the way of the commercial development and introduction of such motors, except that they had to be constructed with a view to operating on the circuits then existing, which in this country were all of high frequency.

The first full publication of his work in this direction—outside his patents—was a paper read before the American Institute of Electrical Engineers in New York, in May, 1888 (read at the suggestion of Prof. Anthony and the present writer), when he exhibited motors that had been in operation long previous, and with which his belief that brushes and commutators could be dispensed with, was triumphantly proved to be correct. The section of this volume devoted to Mr. Tesla's inventions in the utilization of polyphase currents will show how thoroughly from the outset he had mastered the fundamental idea and applied it in the greatest variety of ways.

Having noted for years the many advantages obtainable with alternating currents, Mr. Tesla was naturally led on to experiment with them at higher potentials and higher frequencies than were common or approved of. Ever pressing forward to determine in even the slightest degree the outlines of the unknown, he was rewarded very quickly in this field with results of the most surprising nature. A slight acquaintance with some of these experiments led the compiler of this volume to urge Mr. Tesla to repeat them before the American Institute of Electrical

Engineers. This was done in May, 1891, in a lecture that marked, beyond question, a distinct departure in electrical theory and practice, and all the results of which have not yet made themselves fully apparent. The New York lecture, and its successors, two in number, are also included in this volume, with a few supplementary notes.

Mr. Tesla's work ranges far beyond the vast departments of polyphase currents and high potential lighting. The "Miscellaneous" section of this volume includes a great many other inventions in arc lighting, transformers, pyro-magnetic generators, thermo-magnetic motors, third-brush regulation, improvements in dynamos, new forms of incandescent lamps, electrical meters, condensers, unipolar dynamos, the conversion of alternating into direct currents, etc. It is needless to say that at this moment Mr. Tesla is engaged on a number of interesting ideas and inventions, to be made public in due course. The present volume deals simply with his work accomplished to date.

CHAPTER II

A NEW SYSTEM OF ALTERNATING CURRENT MOTORS AND TRANSFORMERS

The present section of this volume deals with polyphase currents, and the inventions by Mr. Tesla, made known thus far, in which he has embodied one feature or another of the broad principle of rotating field poles or *resultant attraction* exerted on the armature. It is needless to remind electricians of the great interest aroused by the first enunciation of the rotating field principle, or to dwell upon the importance of the advance from a single alternating current, to methods and apparatus which deal with more than one. Simply prefacing the consideration here attempted of the subject, with the remark that in nowise is the object of this volume of a polemic or controversial nature, it may be pointed out that Mr. Tesla's work has not at all been fully understood or realized up to date. To many readers, it is believed, the analysis of what he has done in this department will be a revelation, while it will at the same time illustrate the beautiful flexibility and range of the principles involved. It will be seen that, as just suggested, Mr. Tesla did not stop short at a mere rotating field, but dealt broadly with the shifting of the resultant attraction of the magnets. It will be seen that he went on to evolve the "multiphase" system with many ramifications and turns; that he showed the broad idea of motors employing currents of differing phase in the armature with direct currents in the field; that he first described and worked out the idea of an armature with a body of iron and coils closed upon themselves; that he worked out both synchronizing and torque motors; that he explained and illustrated how machines of ordinary construction might be adapted to his system; that he employed condensers in field and armature circuits, and went to the bottom of the fundamental principles, testing, approving or rejecting, it would appear, every detail that inventive ingenuity could hit upon.

Now that opinion is turning so emphatically in favor of lower frequencies, it deserves special note that Mr. Tesla early recognized the importance of the low frequency feature in motor work. In fact his first motors exhibited publicly—and which, as Prof. Anthony showed in his tests in the winter of 1887–8, were the equal of direct current motors in efficiency, output and starting torque—were of the low frequency type. The necessity arising, however, to utilize these motors in connection with the existing high frequency circuits, our survey reveals in an interesting manner Mr. Tesla's fertility of resource in this direction. But that, after exhausting all the possibilities of this field, Mr. Tesla returns to low frequencies, and insists on the superiority of his polyphase system in alternating current distribution, need not at all surprise us, in view of the strength of his convictions, so often expressed, on this subject. This is, indeed, significant, and may be regarded as indicative of the probable development next to be witnessed.

Incidental reference has been made to the efficiency of rotating field motors, a matter of much importance, though it is not the intention to dwell upon it here. Prof. Anthony in his remarks before the American Institute of Electrical Engineers, in May, 1888, on the two small Tesla motors then shown, which he had tested, stated that one gave an efficiency of about 50 percent and the other a little over sixty percent. In 1889, some tests were reported from Pittsburgh, made by Mr. Tesla and Mr. Albert Schmid, on motors up to 10 H. P. and weighing about 850 pounds. These machines showed an efficiency of nearly 90 percent. With some larger motors it was then found practicable to obtain an efficiency, with the three wire system, up to as high as 94 and 95 percent. These interesting figures, which, of course, might be supplemented by others more elaborate and of later date, are cited to show that the efficiency of the system has not had to wait until the present late day for any demonstration of its commercial usefulness. An invention is none the less beautiful because it may lack utility, but it must be a pleasure to any inventor to know that the ideas he is advancing are fraught with substantial benefits to the public.

CHAPTER III

THE TESLA ROTATING MAGNETIC FIELD—
MOTORS WITH CLOSED CONDUCTORS—
SYNCHRONIZING MOTORS—
ROTATING FIELD TRANSFORMERS

The best description that can be given of what he attempted, and succeeded in doing, with the rotating magnetic field, is to be found in Mr. Tesla's brief paper explanatory of his rotary current, polyphase system, read before the American Institute of Electrical Engineers, in New York, in May, 1888, under the title "A New System of Alternate Current Motors and Transformers." As a matter of fact, which a perusal of the paper will establish, Mr. Tesla made no attempt in that paper to describe all his work. It dealt in reality with the few topics enumerated in the caption of this chapter. Mr. Tesla's reticence was no doubt due largely to the fact that his action was governed by the wishes of others with whom he was associated, but it may be worth mention that the compiler of this volume—who had seen the motors running, and who was then chairman of the Institute Committee on Papers and Meetings—had great difficulty in inducing Mr. Tesla to give the Institute any paper at all. Mr. Tesla was overworked and ill, and manifested the greatest reluctance to an exhibition of his motors, but his objections were at last overcome. The paper was written the night previous to the meeting, in pencil, very hastily, and under the pressure just mentioned.

In this paper casual reference was made to two special forms of motors not within the group to be considered. These two forms were: 1. A motor with one of its circuits in series with a transformer, and the other in the secondary of the transformer. 2. A motor having its armature circuit connected to the generator, and the field coils closed upon themselves. The paper in its essence is as follows, dealing with a few leading features of the Tesla system, namely, the rotating magnetic field, motors with closed conductors, synchronizing motors, and rotating field transformers:

The subject which I now have the pleasure of bringing to your notice is a novel system of electric distribution and transmission of power by means of alternate currents, affording peculiar advantages, particularly in the way of motors, which I am confident will at once establish the superior adaptability of these currents to the transmission of power and will show that many results heretofore unattainable can be reached by their use; results which are very much desired in the practical operation of such systems, and which cannot be accomplished by means of continuous currents.

Before going into a detailed description of this system, I think it necessary to make a few remarks with reference to certain conditions existing in continuous current generators and motors, which, although generally known, are frequently disregarded.

In our dynamo machines, it is well known, we generate alternate currents which we direct by means of a commutator, a complicated device and, it may be justly said, the source of most of the troubles experienced in the operation of the machines. Now, the currents so directed cannot be utilized in the motor, but they must—again by means of a similar unreliable device—be reconverted into their original state of alternate currents. The function of the commutator is entirely external, and in no way does it affect the internal working of the machines. In reality, therefore, all machines are alternate current machines, the currents appearing as continuous only in the external circuit during their transit from generator to motor. In view simply of this fact, alternate currents would commend themselves as a more direct application of electrical energy, and the employment of continuous currents would only be justified if we had dynamos which would primarily generate, and motors which would be directly actuated by, such currents.

But the operation of the commutator on a motor is twofold; first, it reverses the currents through the motor, and secondly, it effects automatically, a progressive shifting of the poles of one of its magnetic constituents. Assuming, therefore, that both of the useless operations in the systems, that is to say, the directing of the alternate currents on the generator and reversing the direct currents on the motor, be eliminated, it would still be necessary, in order to cause a rotation of the motor, to produce a progressive shifting of the poles of one of its elements, and the question presented itself—How to perform this operation by the direct action of alternate currents? I will now proceed to show how this result was accomplished.

Fig. 1.

Fig. 1a.

Fig. 2.

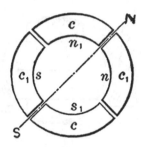

Fig. 2a.

In the first experiment a drum-armature was provided with two coils at right angles to each other, and the ends of these coils were connected to two pairs of insulated contact-rings as usual. A ring was then made of thin insulated plates of sheet-iron and wound with four coils, each two opposite coils being connected together so as to produce free poles on diametrically opposite sides of the ring. The remaining free ends of the coils were then connected to the contact-rings of the generator armature so as to form two independent circuits, as indicated in Fig. 9. It may now be seen what results were secured in this combination, and with this view I would refer to the diagrams, Figs. 1 to 8a. The field of the generator being independently excited, the rotation of the armature sets up currents in the coils $c \, c_1$, varying in strength and direction in the well-known manner. In the position shown in Fig. 1, the current in coil c is nil, while coil c_1 is traversed by its maximum current, and the connections may be such that the ring is magnetized by the coils c_1 c_1, as indicated by the letters N S in Fig. 1a, the magnetizing effect of the coils c c being nil, since these coils are included in the circuit of coil C.

In Fig. 2, the armature coils are shown in a more advanced position, one-eighth of one revolution being completed. Fig. 2a illustrates the corresponding magnetic condition of the ring. At this moment the coil c_1 generates a current of the same direction as previously, but weaker, producing the poles n_1 s_1 upon the ring; the coil C also generates a current of the same direction, and the connections may be such that the coils c c produce the poles n s, as shown in Fig. 2a. The resulting polarity is indicated by the letters N S, and it will be observed that the poles of the ring have been shifted one-eighth of the periphery of the same.

FIG. 3.

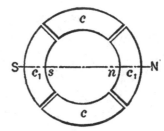

FIG. 3a.

In Fig. 3 the armature has completed one quarter of one revolution. In this phase the current in coil C is a maximum, and of such direction as to produce the poles N S in Fig. 3a, whereas the current in coil c_1 is nil, this coil being at its neutral position. The poles N S in Fig. 3a are thus shifted one quarter of the circumference of the ring.

FIG. 4.

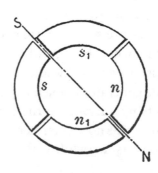

FIG. 4a.

Fig. 4 shows the coils c c in a still more advanced position, the armature having completed three-eighths of one revolution. At that moment the coil c still generates a current of the same direction as before, but of less strength, producing the comparatively weaker poles n s in Fig. 4a. The current in the coil c_1 is of the same strength, but opposite direction. Its effect is, therefore, to produce upon the ring the poles n_1 s_1, as indicated, and a polarity, N S, results, the poles now being shifted three-eighths of the periphery of the ring.

FIG. 5.

FIG. 5a.

In Fig. 5 one half of one revolution of the armature is completed, and the resulting magnetic condition of the ring is indicated in Fig. 5a. Now the current in coil c is nil, while the coil c_1 yields its maximum current, which is of the same direction as previously; the magnetizing effect is, therefore, due to the coils, c_1 c_1 alone, and, referring to Fig. 5a, it will be observed that the poles N S are shifted one half of the circumference of the ring. During the next half revolution the operations are repeated, as represented in the Figs. 6 to 8a.

FIG. 6.

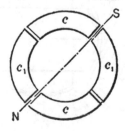

FIG. 6a.

A reference to the diagrams will make it clear that during one revolution of the armature the poles of the ring are shifted once around its periphery, and, each revolution producing like effects, a rapid whirling of the poles in harmony with the rotation of the armature is the result. If the connections of either one of the circuits in the ring are reversed, the shifting of the poles is made to progress in the opposite direction, but the operation is identically the same. Instead of using four wires, with like result, three wires may be used, one forming a common return for both circuits.

This rotation or whirling of the poles manifests itself in a series of curious phenomena. If a delicately pivoted disc of steel or other magnetic metal is approached to the ring it is set in rapid rotation, the direction of rotation varying with the position of the disc. For instance, noting the direction outside of the ring it will be found that inside the ring it turns in an opposite direction, while it is unaffected if placed in a position symmetrical to the ring. This is easily explained. Each time that a pole approaches, it induces an opposite pole in the nearest point on the disc, and an attraction is produced upon that point; owing to this, as the pole is shifted further away from the disc a tangential pull is exerted upon the same, and the action being constantly repeated, a more or less rapid rotation of the disc is the result. As the pull is exerted mainly upon that part which is nearest to the ring, the rotation outside and inside, or right and left, respectively, is in opposite directions, Fig. 9. When placed symmetrically to the ring, the pull on the opposite sides of the disc being equal, no rotation results. The action is based on the magnetic inertia of iron; for this reason a disc of hard steel is much more affected

FIG. 7.

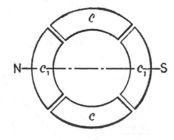

FIG. 7a.

than a disc of soft iron, the latter being capable of very rapid variations of magnetism. Such a disc has proved to be a very useful instrument in all these investigations, as it has enabled me to detect any irregularity in the action. A curious effect is also produced upon iron filings. By placing some upon a paper and holding them externally quite close to the ring, they are set in a vibrating motion, remaining in the same place, although the paper may be moved back and forth; but in lifting the paper to a certain height which seems to be dependent on the intensity of the poles and the speed of rotation, they are thrown away in a direction always opposite to the supposed movement of the poles. If a paper with filings is put flat upon the ring and the current turned on suddenly, the existence of a magnetic whirl may easily be observed.

To demonstrate the complete analogy between the ring and a revolving magnet, a strongly energized electro-magnet was rotated by mechanical power, and phenomena identical in every particular to those mentioned above were observed.

Obviously, the rotation of the poles produces corresponding inductive effects and may be utilized to generate currents in a closed conductor placed within the influence of the poles. For this purpose it is convenient to wind a ring with two sets of superimposed coils forming respectively the primary and secondary circuits, as shown in Fig. 10. In order to secure the most economical results the magnetic circuit should be completely closed, and with this object in view the construction may be modified at will.

FIG. 8.

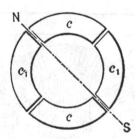

FIG. 8a.

The inductive effect exerted upon the secondary coils will be mainly due to the shifting or movement of the magnetic action; but there may also be currents set up in the circuits in consequence of the variations in the intensity of the poles. However, by properly designing the generator and determining the magnetizing effect of the primary coils, the latter element may be made to disappear. The intensity of the poles being maintained constant, the action of the apparatus will be perfect, and the same result will be secured as though the shifting were effected by means of a commutator with an infinite number of bars. In such case the theoretical relation between the energizing effect of each set of primary coils and their resultant magnetizing effect may be expressed by the equation of a circle having its center coinciding with that of an orthogonal system of axes, and in which the radius represents the resultant and the co-ordinates both of its components. These are then respectively the sine and cosine of the angle a between the radius and one of the axes $(O\,X)$. Referring to Fig. 11, we have $r^2 = x^2 + y^2$; where $x = r \cos a$, and $y = r \sin a$.

Assuming the magnetizing effect of each set of coils in the transformer to be proportional to the current—which may be admitted for weak degrees of magnetization—then $x = K\,c$ and $y = K\,c^i$, where K is a constant and c and c^i the current in both sets of coils respectively. Supposing, further, the field of the generator to be uniform, we have for constant speed $c^i = K^i$ $\sin a$ and $c = K^i \sin (90° + a) = K^i \cos a$, where K^i is a constant. See Fig. 12.

Therefore, $x = K\,c = K\,K^i \cos a;$
          $y = K\,c^i = K\,K^i \sin a;$ and
          $K\,K^i = r.$

FIG. 9.

That is, for a uniform field the disposition of the two coils at right angles will secure the theoretical result, and the intensity of the shifting poles will be constant. But from $r^2 = x^2 + y^2$ it follows that for $y = 0$, $r = x$; it follows that the joint magnetizing effect of both sets of coils should be equal to the effect of one set when at its maximum action. In transformers and in a certain class of motors the fluctuation of the poles is not of great importance, but in another class of these motors it is desirable to obtain the theoretical result.

In applying this principle to the construction of motors, two typical forms of motor have been developed. First, a form having a comparatively small rotary effort at the start but maintaining a perfectly uniform speed at all loads, which motor has been termed synchronous. Second, a form possessing a great rotary effort at the start, the speed being dependent on the load.

FIG. 10.

These motors may be operated in three different ways: 1. By the alternate currents of the source only. 2. By a combined action of these and of induced currents. 3. By the joint action of alternate and continuous currents.

The simplest form of a synchronous motor is obtained by winding a laminated ring provided with pole projections with four coils, and connecting the same in the manner before indicated. An iron disc having a segment cut away on each side may be used as an armature. Such a motor is shown in Fig. 9. The disc being arranged to rotate freely within the ring in close proximity to the projections, it is evident that as the poles are shifted it will, owing to its tendency to place itself in such a position as to embrace the greatest number of the lines of force, closely follow the movement of the poles, and its motion will be synchronous with that of the armature of the generator; that is, in the peculiar disposition shown in Fig. 9, in which the armature produces by one revolution two current impulses in each of the circuits. It is evident that if, by one revolution of the armature, a greater number of impulses is produced, the speed of the motor will be correspondingly increased. Considering that the attraction exerted upon the disc is greatest when the same is in close proximity to the poles, it follows that such a motor will maintain exactly the same speed at all loads within the limits of its capacity.

To facilitate the starting, the disc may be provided with a coil closed upon itself. The advantage secured by such a coil is evident. On the start the currents set up in the coil strongly energize the disc and increase the attraction exerted upon the same by the ring, and currents being generated in the coil as long as the speed of the armature is inferior to that of the poles, considerable work may be performed by such a motor even if the speed be below normal. The intensity of the poles being constant, no currents will be generated in the coil when the motor is turning at its normal speed.

FIG. 11.

FIG. 12.

Instead of closing the coil upon itself, its ends may be connected to two insulated sliding rings, and a continuous current supplied to these from a suitable generator. The proper way to start such a motor is to close the coil upon itself until the normal speed is reached, or nearly so, and then turn on the continuous current. If the disc be very strongly energized by a continuous current the motor may not be able to start, but if it be weakly energized, or generally so that the magnetizing effect of the ring is preponderating, it will start and reach the normal speed. Such a motor will maintain absolutely the same speed at all loads. It has also been found that if the motive power of the generator is not excessive, by checking the motor the speed of the generator is diminished in synchronism with that of the motor. It is characteristic of this form of motor that it cannot be reversed by reversing the continuous current through the coil.

The synchronism of these motors may be demonstrated experimentally in a variety of ways. For this purpose it is best to employ a motor consisting of a stationary field magnet and an armature arranged to

FIG. 13.

rotate within the same, as indicated in Fig. 13. In this case the shifting of the poles of the armature produces a rotation of the latter in the opposite direction. It results therefrom that when the normal speed is reached, the poles of the armature assume fixed positions relatively to the field magnet, and the same is magnetized by induction, exhibiting a distinct pole on each of the pole-pieces. If a piece of soft iron is approached to the field magnet, it will at the start be attracted with a rapid vibrating motion produced by the reversals of polarity of the magnet, but as the speed of the armature increases, the vibrations become less and less frequent and finally entirely cease. Then the iron is weakly but permanently attracted, showing that synchronism is reached and the field magnet energized by induction.

The disc may also be used for the experiment. If held quite close to the armature it will turn as long as the speed of rotation of the poles exceeds that of the armature; but when the normal speed is reached, or very nearly so, it ceases to rotate and is permanently attracted.

A crude but illustrative experiment is made with an incandescent lamp. Placing the lamp in circuit with the continuous current generator and in series with the magnet coil, rapid fluctuations are observed in the light in consequence of the induced currents set up in the coil at the start; the speed increasing, the fluctuations occur at longer intervals, until they entirely disappear, showing that the motor has attained its normal speed. A telephone receiver affords a most sensitive instrument; when connected to any circuit in the motor the synchronism may be easily detected on the disappearance of the induced currents.

In motors of the synchronous type it is desirable to maintain the quantity of the shifting magnetism constant, especially if the magnets are not properly subdivided.

To obtain a rotary effort in these motors was the subject of long thought. In order to secure this result it was necessary to make such a disposition that while the poles of one element of the motor are shifted by the alternate currents of the source, the poles produced upon the other elements should always be maintained in the proper relation to the former, irrespective of the speed of the motor. Such a condition exists in a continuous current motor; but in a synchronous motor, such as described, this condition is fulfilled only when the speed is normal.

The object has been attained by placing within the ring a properly subdivided cylindrical iron core wound with several independent coils closed upon themselves. Two coils at right angles as in Fig. 14, are sufficient, but a greater number may be advantageously employed. It results from this disposition that when the poles of the ring are shifted, currents are generated in the closed armature coils. These currents are the most intense at or near the points of the greatest density of the lines of force, and their effect is to produce poles upon the armature at right angles to those of the ring, at least theoretically so; and since this action is entirely independent of the speed—that is, as far as the location of the poles is concerned—a continuous pull is exerted upon the periphery of the armature. In many respects these motors are similar to the continuous current motors. If load is put on, the speed, and also the resistance of the motor, is diminished and more current is made to pass through the energizing coils, thus increasing the effort. Upon the load being taken off, the counter-electromotive force increases and less current passes through the primary or energizing coils. Without any load the speed is very nearly equal to that of the shifting poles of the field magnet.

FIG. 14.

It will be found that the rotary effort in these motors fully equals that of the continuous current motors. The effort seems to be greatest when both armature and field magnet are without any projections; but as in such dispositions the field cannot be concentrated, probably the best results will be obtained by leaving pole projections on one of the elements only. Generally, it may be stated the projections diminish the torque and produce a tendency to synchronism.

A characteristic feature of motors of this kind is their property of being very rapidly reversed. This follows from the peculiar action of the motor. Suppose the armature to be rotating and the direction of rotation of the poles to be reversed. The apparatus then represents a dynamo machine, the power to drive this machine being the momentum stored up in the armature and its speed being the sum of the speeds of the armature and the poles.

If we now consider that the power to drive such a dynamo would be very nearly proportional to the third power of the speed, for that reason alone the armature should be quickly reversed. But simultaneously with the reversal another element is brought into action, namely, as the movement of the poles with respect to the armature is reversed, the motor acts like a transformer in which the resistance of the secondary circuit would be abnormally diminished by producing in this circuit an additional electromotive force. Owing to these causes the reversal is instantaneous.

If it is desirable to secure a constant speed, and at the same time a certain effort at the start, this result may be easily attained in a variety of ways. For instance, two armatures, one for torque and the other for synchronism, may be fastened on the same shaft and any desired preponderance may be given to either one, or an armature may be wound for rotary effort, but a more or less pronounced tendency to synchronism may be given to it by properly constructing the iron core; and in many other ways.

As a means of obtaining the required phase of the currents in both the circuits, the disposition of the two coils at right angles is the simplest, securing the most uniform action; but the phase may be obtained in many other ways, varying with the machine employed. Any of the dynamos at present in use may be easily adapted for this purpose by making connections to proper points of the generating coils. In closed circuit armatures, such as used in the continuous current systems, it is best to make four derivations from equi-distant points or bars of the commutator, and to connect the same to four insulated sliding rings on the shaft. In this case each of the motor circuits is connected to two diametrically opposite bars of the commutator. In such a disposition the motor may also be operated at half the potential and on the three-wire plan, by connecting the motor circuits in the proper order to three of the contact rings.

In multipolar dynamo machines, such as used in the converter systems, the phase is conveniently obtained by winding upon the armature two series of coils in such a manner that while the coils of one set or series are at their maximum production of current, the coils of the other will be at their neutral position, or nearly so, whereby both sets of coils may be subjected simultaneously or successively to the inducing action of the field magnets.

Generally the circuits in the motor will be similarly disposed, and various arrangements may be made to fulfill the requirements; but the simplest and most practicable is to arrange primary circuits on stationary parts of the motor, thereby obviating, at least in certain forms, the employment of sliding contacts. In such a case the magnet coils are connected alternately in both the circuits; that is, 1, 3, 5 . . . in one, and 2, 4, 6 . . . in the other, and the coils of each set of series may be connected all in the same manner, or alternately in opposition; in the latter case a motor with half the number of poles will result, and its action will be correspondingly modified. The Figs. 15, 16, and 17 show three different phases, the magnet coils in each circuit being connected alternately in opposition. In this case there will be always four poles, as in Figs. 15 and 17; four pole projections will be neutral; and in Fig. 16 two adjacent pole projections will have the same polarity. If the coils are connected in the same manner there will be eight alternating poles, as indicated by the letters n' s' in Fig. 15.

Fig. 15. Fig. 16. Fig. 17.

The employment of multipolar motors secures in this system an advantage much desired and unattainable in the continuous current system, and that is, that a motor may be made to run exactly at a predetermined speed irrespective of imperfections in construction, of the load, and, within certain limits, of electromotive force and current strength.

In a general distribution system of this kind the following plan should be adopted. At the central station of supply a generator should be provided having a considerable number of poles. The motors operated from this generator should be of the synchronous type, but possessing sufficient rotary effort to insure their starting. With the observance of proper rules of construction it may be admitted that the speed of each motor will be in some inverse proportion to its size, and the number of poles should be chosen accordingly. Still, exceptional demands may modify this rule. In view of this, it will be advantageous to provide each motor with a greater number of pole projections or coils, the number being preferably a multiple of two and three. By this means, by simply changing the connections of the coils, the motor may be adapted to any probable demands.

If the number of the poles in the motor is even, the action will be harmonious and the proper result will be obtained; if this is not the case, the best plan to be followed is to make a motor with a double number of poles and connect the same in the manner before indicated, so that half the number of poles result. Suppose, for instance, that the generator has twelve poles, and it would be desired to obtain a speed equal to 12/7 of the speed of the generator. This would require a motor with seven pole projections or magnets, and such a motor could not be properly connected in the circuits unless fourteen armature coils would be provided, which would necessitate the employment of sliding contacts. To avoid this, the motor should be provided with fourteen magnets and seven connected in each circuit, the magnets in each circuit alternating among themselves. The armature should have fourteen closed coils. The action of the motor will not be quite as perfect as in the case of an even number of poles, but the drawback will not be of a serious nature.

However, the disadvantages resulting from this unsymmetrical form will be reduced in the same proportion as the number of the poles is augmented.

If the generator has, say, n, and the motor n_1 poles, the speed of the motor will be equal to that of the generator multiplied by n/n_1.

The speed of the motor will generally be dependent on the number of the poles, but there may be exceptions to this rule. The speed may be modified by the phase of the currents in the circuit or by the character of the current impulses or by intervals between each or between groups of impulses. Some of the possible cases are indicated in the diagrams, Figs. 18, 19, 20 and 21, which are self-explanatory. Fig. 18 represents the condition generally existing, and which secures the best result. In such a case, if the typical form of motor illustrated in Fig. 9 is employed, one complete wave in each circuit will produce one revolution of the motor. In Fig. 19 the same result will be effected by one wave in each circuit, the impulses being successive; in Fig. 20 by four, and in Fig. 21 by eight waves.

FIG. 18.   FIG. 19.   FIG. 20.   FIG. 21.

By such means any desired speed may be attained, that is, at least within the limits of practical demands. This system possesses this advantage, besides others, resulting from simplicity. At full loads the motors show an efficiency fully equal to that of the continuous current motors. The transformers present an additional advantage in their capability of operating motors. They are capable of similar modifications in construction, and will facilitate the introduction of motors and their adaptation to practical demands. Their efficiency should be higher than that of the present transformers, and I base my assertion on the following:

In a transformer, as constructed at present, we produce the currents in the secondary circuit by varying the strength of the primary or exciting currents. If we admit proportionality with respect to the iron core the inductive effect exerted upon the secondary coil will be proportional to the numerical sum of the variations in the strength of the exciting current per unit of time; whence it follows that for a given variation any prolongation of the primary current will result in a proportional loss. In order to obtain rapid variations in the strength of the current, essential to efficient induction, a great number of undulations are employed; from this practice various disadvantages result. These are: Increased cost and diminished efficiency of the generator; more waste of energy in heating the cores, and also diminished output of the transformer, since the core is not properly utilized, the reversals being too rapid. The inductive effect is also very small in certain phases, as will be apparent from a graphic representation, and there may be periods of inaction, if there are intervals between the succeeding current impulses or waves. In producing a shifting of the poles in a transformer, and thereby inducing currents, the induction is of the ideal character, being always maintained at its maximum action. It is also reasonable to assume that by a shifting of the poles less energy will be wasted than by reversals.

CHAPTER IV

MODIFICATIONS AND EXPANSIONS OF
THE TESLA POLYPHASE SYSTEMS

In his earlier papers and patents relative to polyphase currents, Mr. Tesla devoted himself chiefly to an enunciation of the broad lines and ideas lying at the basis of this new work; but he supplemented this immediately by a series of other striking inventions which may be regarded as modifications and expansions of certain features of the Tesla systems. These we shall now proceed to deal with.

FIG. 22.

FIG. 23.

In the preceding chapters we have thus shown and described the Tesla electrical systems for the transmission of power and the conversion and distribution of electrical energy, in which the motors and the transformers contain two or more coils or sets of coils, which were connected up in independent circuits with corresponding coils of an alternating current generator, the operation of the system being brought about by the co-operation of the alternating currents in the independent circuits in progressively moving or shifting the poles or points of maximum magnetic effect of the motors or converters. In these systems two independent conductors are employed for each of the independent circuits connecting the generator with the devices for converting the transmitted currents into mechanical energy or into electric currents of another character. This, however, is not always necessary. The two or more circuits may have a single return path or wire in common, with a loss, if any, which is so extremely slight that it may be disregarded entirely. For the sake of illustration, if the generator have two independent coils and the motor two coils or two sets of coils in corresponding relations to its operative elements one terminal of each generator coil is connected to the corresponding terminals of the motor coils through two independent conductors, while the opposite terminals of the respective coils are both connected to one return wire. The following description deals with the modification. Fig. 22 is a diagrammatic illustration of a generator and single motor constructed and electrically connected in accordance with the invention. Fig. 23 is a diagram of the system as it is used in operating motors or converters, or both, in parallel, while Fig. 24 illustrates diagrammatically the manner of operating two or more motors or converters, or both, in series. Referring to Fig. 22, A A designate the poles of the field magnets of an alternating-current generator, the armature of which, being in this case cylindrical in form and mounted on a shaft, c, is wound longitudinally with coils B B'. The shaft c carries three insulated contact-rings, a b c, to two of which, as b c, one terminal of each coil, as e d, is connected. The remaining terminals, f g, are both connected to the third ring, a.

A motor in this case is shown as composed of a ring, H, wound with four coils, I I J J, electrically connected, so as to co-operate in pairs, with a tendency to fix the poles of the ring at four points ninety degrees apart. Within the magnetic ring H is a disc or cylindrical core wound with two coils, G G', which may be connected to form two closed circuits. The terminals j k of the two sets or pairs of coils are connected, respectively, to the binding-posts E' F', and the other terminals, h i, are connected to a single binding-post, D'. To operate the motor, three line-wires are used to connect the terminals of the generator with those of the motor.

So far as the apparent action or mode of operation of this arrangement is concerned, the single wire D, which is, so to speak, a common return-wire for both circuits, may be regarded as two

Fig. 24.

independent wires. In the illustration, with the order of connection shown, coil B′ of the generator is producing its maximum current and coil B its minimum; hence the current which passes through wire e, ring b, brush b′, line-wire E, terminal E′, wire j, coils I I, wire or terminal D′, line-wire D, brush a′, ring a, and wire f, fixes the polar line of the motor midway between the two coils I I; but as the coil B′ moves from the position indicated it generates less current, while coil B, moving into the field, generates more. The current from coil B passes through the devices and wires designated by the letters d, c, c′ F, F′ k, J J, i, D′, D, a′, a, and g, and the position of the poles of the motor will be due to the resultant effect of the currents in the two sets of coils—that is, it will be advanced in proportion to the advance or forward movement of the armature coils. The movement of the generator-armature through one-quarter of a revolution will obviously bring coil B′ into its neutral position and coil B into its position of maximum effect, and this shifts the poles ninety degrees, as they are fixed solely by coils B. This action is repeated for each quarter of a complete revolution.

When more than one motor or other device is employed, they may be run either in parallel or series. In Fig. 23 the former arrangement is shown. The electrical device is shown as a converter, L, of which the two sets of primary coils p r are connected, respectively, to the mains F E, which are electrically connected with the two coils of the generator. The cross-circuit wires l m, making these connections, are then connected to the common return-wire D. The secondary coils p′ p″ are in circuits n o, including, for example, incandescent lamps. Only one converter is shown entire in this figure, the others being illustrated diagrammatically.

When motors or converters are to be run in series, the two wires E F are led from the generator to the coils of the first motor or converter, then continued on to the next, and so on through the whole series, and are then joined to the single wire D, which completes both circuits through the generator. This is shown in Fig. 24, in which J I represent the two coils or sets of coils of the motors.

There are, of course, other conditions under which the same idea may be carried out. For example, in case the motor and generator each has three independent circuits, one terminal of each circuit is connected to a line-wire, and the other three terminals to a common return-conductor. This arrangement will secure similar results to those attained with a generator and motor having but two independent circuits, as above described.

When applied to such machines and motors as have three or more induced circuits with a common electrical joint, the three or more terminals of the generator would be simply connected to those of the motor. Mr. Tesla states, however, that the results obtained in this manner show a lower efficiency than do the forms dwelt upon more fully above.

CHAPTER V

UTILIZING FAMILIAR TYPES OF GENERATOR OF THE CONTINUOUS CURRENT TYPE

The preceding descriptions have assumed the use of alternating current generators in which, in order to produce the progressive movement of the magnetic poles, or of the resultant attraction of independent field magnets, the current generating coils are independent or separate. The ordinary forms of continuous current dynamos may, however, be employed for the same work, in accordance with a method of adaptation devised by Mr. Tesla. As will be seen, the modification involves but slight changes in their construction, and presents other elements of economy.

On the shaft of a given generator, either in place of or in addition to the regular commutator, are secured as many pairs of insulated collecting-rings as there are circuits to be operated. Now, it will be understood that in the operation of any dynamo electric generator the currents in the coils in their movement through the field of force undergo different phases—that is to say, at different positions of the coils the currents have certain directions and certain strengths—and that in the Tesla motors or transformers it is necessary that the currents in the energizing coils should undergo a certain order of variations in strength and direction. Hence, the further step—viz., the connection between the induced or generating coils of the machine and the contact-rings from which the currents are to be taken off—will be determined solely by what order of variations of strength and direction in the currents is desired for producing a given result in the electrical translating device. This may be accomplished in various ways; but in the drawings we give typical instances only of the best and most practicable ways of applying the invention to three of the leading types of machines in widespread use, in order to illustrate the principle.

Fig. 25 is a diagram illustrative of the mode of applying the invention to the well-known type of "closed" or continuous circuit machines. Fig. 26 is a similar diagram embodying an armature with separate coils connected diametrically, or what is generally called an "open-circuit"

FIG. 25.

machine. Fig. 27 is a diagram showing the application of the invention to a machine the armature-coils of which have a common joint.

Referring to Fig. 25, let A represent a Tesla motor or transformer which, for convenience, we will designate as a "converter." It consists of an annular core, B, wound with four independent coils, C and D, those diametrically opposite being connected together so as to co-operate in pairs in establishing free poles in the ring, the tendency of each pair being to fix the poles at ninety degrees from the other. There may be an armature, E, within the ring, which is wound with coils closed upon themselves. The object is to pass through coils C D currents of such relative strength and direction as to produce a progressive shifting or movement of the points of maximum magnetic effect around the ring, and to thereby maintain a rotary movement of the armature. There are therefore secured to the shaft F of the generator, four insulated contact-rings, *a b c d*, upon which bear the collecting-brushes *a' b' c' d'*, connected by wires G G H H, respectively, with the terminals of coils C and D.

Assume, for sake of illustration, that the coils D D are to receive the maximum and coils C C at the same instant the minimum current, so that the polar line may be midway between the coils D D. The rings *a b* would therefore be connected to the continuous armature-coil at its neutral points with respect to the field, or the point corresponding with that of the ordinary commutator brushes, and between which exists the greatest difference of potential; while rings *c d* would be connected to two points in the coil, between which exists no difference of potential. The best results will be obtained by making these connections at points equidistant from one another, as shown. These connections are easiest made by using wires L between the rings and the loops or wires J, connecting the coil I to the segments of the commutator K. When the converters are made in this manner, it is evident that the phases of the currents in the sections of the generator coil will be reproduced in the converter coils. For example, after turning through an arc of ninety degrees the conductors L L, which before conveyed the maximum current, will receive the minimum current by reason of the change in the position of their coils, and it is evident that for the same reason the current in these coils has gradually fallen from the maximum to the minimum in passing through the arc of ninety degrees. In this special plan of connections, the rotation of the magnetic poles of the converter

FIG. 26.

FIG. 27.

will be synchronous with that of the armature coils of the generator, and the result will be the same, whether the energizing circuits are derivations from a continuous armature coil or from independent coils, as in Mr. Tesla's other devices.

In Fig. 25, the brushes M M are shown in dotted lines in their proper normal position. In practice these brushes may be removed from the commutator and the field of the generator excited by an external source of current; or the brushes may be allowed to remain on the commutator and to take off a converted current to excite the field, or to be used for other purposes.

In a certain well-known class of machines known as the "open circuit," the armature contains a number of coils the terminals of which connect to commutator segments, the coils being connected across the armature in pairs. This type of machine is represented in Fig. 26. In this machine each pair of coils goes through the same phases as the coils in some of the generators already shown, and it is obviously only necessary to utilize them in pairs or sets to operate a Tesla converter by extending the segments of the commutators belonging to each pair of coils and causing a collecting brush to bear on the continuous portion of each segment. In this way two or more circuits may be taken off from the generator, each including one or more pairs or sets of coils as may be desired.

In Fig. 26 I I represent the armature coils, T T the poles of the field magnet, and F the shaft carrying the commutators, which are extended to form continuous portions $a\,b\,c\,d$. The brushes bearing on the continuous portions for taking off the alternating currents are represented by $a'\,b'\,c'\,d'$. The collecting brushes, or those which may be used to take off the direct current, are designated by M M. Two pairs of the armature coils and their commutators are shown in the figure as being utilized; but all may be utilized in a similar manner.

There is another well-known type of machine in which three or more coils, A' B' C', on the armature have a common joint, the free ends being connected to the segments of a commutator. This form of generator is illustrated in Fig. 27. In this case each terminal of the generator is connected directly or in derivation to a continuous ring, $a\,b\,c$, and collecting brushes, $a'\,b'\,c'$, bearing thereon, take off the alternating currents that operate the motor. It is preferable in this case to employ a motor or transformer with three energizing coils, A" B" C", placed symmetrically with those of the generator, and the circuits from the latter are connected to the terminals of such coils either directly—as when they are stationary—or by means of brushes e' and contact rings e. In this, as in the other cases, the ordinary commutator may be used on the generator, and the current taken from it utilized for exciting the generator field-magnets or for other purposes.

CHAPTER VI

METHOD OF OBTAINING DESIRED SPEED OF MOTOR OR GENERATOR

With the object of obtaining the desired speed in motors operated by means of alternating currents of differing phase, Mr. Tesla has devised various plans intended to meet the practical requirements of the case, in adapting his system to types of multipolar alternating current machines yielding a large number of current reversals for each revolution.

For example, Mr. Tesla has pointed out that to adapt a given type of alternating current generator, you may couple rigidly two complete machines, securing them together in such a way that the requisite difference in phase will be produced; or you may fasten two armatures to the same shaft within the influence of the same field and with the requisite angular displacement to yield the proper difference in phase between the two currents; or two armatures may be attached to the same shaft with their coils symmetrically disposed, but subject to the influence of two sets of field magnets duly displaced; or the two sets of coils may be wound on the same armature alternately or in such manner that they will develop currents the phases of which differ in time sufficiently to produce the rotation of the motor.

Another method included in the scope of the same idea, whereby a single generator may run a number of motors either at its own rate of speed or all at different speeds, is to construct the motors with fewer poles than the generator, in which case their speed will be greater than that of the generator, the rate of speed being higher as the number of their poles is relatively less. This may be understood from an example, taking a generator that has two independent generating coils which revolve between two pole pieces oppositely magnetized; and a motor with energizing coils that produce at any given time two magnetic poles in one element that tend to set up a rotation of the motor. A generator thus constructed yields four reversals, or impulses, in each revolution, two in each of its independent circuits; and the effect upon the motor is to shift the magnetic poles through three hundred and sixty degrees. It is obvious that if the four reversals in the same order could be produced by each half-revolution of the generator the motor would make two revolutions to the generator's one. This would be readily accomplished by adding two intermediate poles to the generator or altering it in any of the other equivalent ways above indicated. The same rule applies to generators and motors with multiple poles. For instance, if a generator be constructed with two circuits, each of which produces twelve reversals of current to a revolution, and these currents be directed through the independent energizing-coils of a motor, the coils of which are so applied as to produce twelve magnetic poles at all times, the rotation of the two will be synchronous; but if the motor-coils produce but six poles, the movable element will be rotated twice while the generator rotates once; or if the motor have four poles, its rotation will be three times as fast as that of the generator.

Fig. 28.   Fig. 29.

These features, so far as necessary to an understanding of the principle, are here illustrated. Fig. 28 is a diagrammatic illustration of a generator constructed in accordance with the invention. Fig. 29 is a similar view of a correspondingly constructed motor. Fig. 30 is a diagram of a generator of modified construction. Fig. 31 is a diagram of a motor of corresponding character. Fig. 32 is a diagram of a system containing a generator and several motors adapted to run at various speeds.

FIG. 30.

In Fig. 28, let *c* represent a cylindrical armature core wound longitudinally with insulated coils A A, which are connected up in series, the terminals of the series being connected to collecting-rings *a a* on the shaft G. By means of this shaft the armature is mounted to rotate between the poles of an annular field-magnet D, formed with polar projections wound with coils E, that magnetize the said projections. The coils E are included in the circuit of a generator F, by means of which the field-magnet is energized. If thus constructed, the machine is a well-known form of alternating-current generator. To adapt it to his system, however, Mr. Tesla winds on armature C a second set of coils B B intermediate to the first, or, in other words, in such positions that while the coils of one set are in the relative positions to the poles of the field-magnet to produce the maximum current, those of the other set will be in the position in which they produce the minimum current. The coils B are connected, also, in series and to two connecting-rings, secured generally to the shaft at the opposite end of the armature.

The motor shown in Fig. 29 has an annular field-magnet H, with four pole-pieces wound with coils I. The armature is constructed similarly to the generator, but with two sets of two coils in closed circuits to correspond with the reduced number of magnetic poles in the field. From the foregoing it is evident that one revolution of the armature of the generator producing eight current impulses in each circuit will produce two revolutions of the motor-armature.

The application of the principle of this invention is not, however, confined to any particular form of machine. In Figs. 30 and 31 a generator and motor of another well-known type are shown. In Fig. 30, J J are magnets disposed in a circle and wound with coils K, which are in circuit with a generator which supplies the current that maintains the field of force. In the usual construction of these machines the armature-conductor L is carried by a suitable frame, so as to be rotated in face of the magnets J J, or between these magnets and another similar set in front of them. The magnets are energized so as to be of alternately opposite polarity throughout the series, so that as the conductor C is rotated the current impulses combine or are added to one another, those produced by the conductor in any given position being all in the same direction. To adapt such a machine to his system, Mr. Tesla adds a second set of induced conductors M, in all respects similar to the first, but so placed in reference to it that the currents produced in each will differ by a quarter-phase. With such relations it is evident that as the current decreases in conductor L it increases in conductor M, and conversely, and that any of the forms of Tesla motor invented for use in this system may be operated by such a generator.

FIG. 31.

Fig. 31 is intended to show a motor corresponding to the machine in Fig. 30. The construction of the motor is identical with that of the generator, and if coupled thereto it will run synchronously therewith. J' J' are the field-magnets, and K' the coils thereon. L' is one of the armature-conductors and M' the other.

Fig. 32 shows in diagram other forms of machine. The generator N in this case is shown as consisting of a stationary ring O, wound with twenty-four coils P P', alternate coils being connected in series in two circuits. Within this ring is a disc or drum Q, with projections Q' wound with energizing-coils included in circuit with a generator R. By driving this disc or cylinder alternating currents are produced in the coils P and P', which are carried off to run the several motors.

FIG. 32.

The motors are composed of a ring or annular field-magnet s, wound with two sets of energizing-coils T T', and armatures U, having projections U' wound with coils V, all connected in series in a closed circuit or each closed independently on itself.

Suppose the twelve generator-coils P are wound alternately in opposite directions, so that any two adjacent coils of the same set tend to produce a free pole in the ring O between them and the twelve coils P' to be similarly wound. A single revolution of the disc or cylinder Q, the twelve polar projections of which are of opposite polarity, will therefore produce twelve current impulses in each of the circuits w w'. Hence the motor X, which has sixteen coils or eight free poles, will make one and a half turns to the generator's one. The motor Y, with twelve coils or six poles, will rotate with twice the speed of the generator, and the motor Z, with eight coils or four poles, will revolve three times as fast as the generator. These multipolar motors have a peculiarity which may be often utilized to great advantage. For example, in the motor X, Fig. 32, the eight poles may be either alternately opposite or there may be at any given time alternately two like and two opposite poles. This is readily attained by making the proper electrical connections. The effect of such a change, however, would be the same as reducing the number of poles one-half, and thereby doubling the speed of any given motor.

It is obvious that the Tesla electrical transformers which have independent primary currents may be used with the generators described. It may also be stated with respect to the devices we now describe that the most perfect and harmonious action of the generators and motors is obtained when the numbers of the poles of each are even and not odd. If this is not the case, there will be a certain unevenness of action which is the less appreciable as the number of poles is greater; although this may be in a measure corrected by special provisions which it is not here necessary to explain. It also follows, as a matter of course, that if the number of the poles of the motor be greater than that of the generator the motor will revolve at a slower speed than the generator.

In this chapter, we may include a method devised by Mr. Tesla for avoiding the very high speeds which would be necessary with large generators. In lieu of revolving the generator armature at a high rate of speed, he secures the desired result by a rotation of the magnetic poles of one element of the generator, while driving the other at a different speed. The effect is the same as that yielded by a very high rate of rotation.

In this instance, the generator which supplies the current for operating the motors or transformers consists of a subdivided ring or annular core wound with four diametrically-opposite coils, E E′, Fig. 33. Within the ring is mounted a cylindrical armature-core wound longitudinally with

FIG. 33.

two independent coils, F F', the ends of which lead, respectively, to two pairs of insulated contact or collecting rings, D D' G G', on the armature shaft. Collecting brushes *d d' g g'* bear upon these rings, respectively, and convey the currents through the two independent line-circuits M M'. In the main line there may be included one or more motors or transformers, or both. If motors be used, they are of the usual form of Tesla construction with independent coils or sets of coils J J', included, respectively, in the circuits M M'. These energizing-coils are wound on a ring or annular field or on pole pieces thereon, and produce by the action of the alternating currents passing through them a progressive shifting of the magnetism from pole to pole. The cylindrical armature H of the motor is wound with two coils at right angles, which form independent closed circuits.

If transformers be employed, one set of the primary coils, as N N, wound on a ring or annular core is connected to one circuit, as M', and the other primary coils, N N', to the circuit M. The secondary coils K K' may then be utilized for running groups of incandescent lamps P P'.

With this generator an exciter is employed. This consists of two poles, A A, of steel permanently magnetized, or of iron excited by a battery or other generator of continuous currents, and a cylindrical armature core mounted on a shaft, B, and wound with two longitudinal coils, C C'. One end of each of these coils is connected to the collecting-rings *b c*, respectively, while the other ends are both connected to a ring, *a*. Collecting-brushes *b' c'* bear on the rings *b c*, respectively, and conductors L L convey the currents therefrom through the coils E and E of the generator. L' is a common return-wire to brush *a'*. Two independent circuits are thus formed, one including coils C of the exciter and E E of the generator, the other coils C' of the exciter and E' E' of the generator. It results from this that the operation of the exciter produces a progressive movement of the magnetic poles of the annular field-core of the generator, the shifting or rotary movement of the poles being synchronous with the rotation of the exciter armature. Considering the operative conditions of a system thus established, it will be found that when the exciter is driven so as to energize the field of the generator, the armature of the latter, if left free to turn, would rotate at a speed practically the same as that of the exciter. If under such conditions the coils F F' of the generator armature be closed upon themselves or short-circuited, no currents, at least theoretically, will be generated in these armature coils. In practice the presence of slight currents is observed, the existence of which is attributable to more or less pronounced fluctuations in the intensity of the magnetic poles of the generator ring. So, if the armature-coils F F′ be closed through the motor, the latter will not be turned as long as the movement of the generator armature is synchronous with that of the exciter or of the magnetic poles of its field. If, on the contrary, the speed of the generator armature be in any way checked, so that the shifting or rotation of the poles of the field becomes relatively more rapid, currents will be induced in the armature coils. This obviously follows from the passing of the lines of force across the armature conductors. The greater the speed of rotation of the magnetic poles relatively to that of the armature the more rapidly the currents developed in the coils of the latter will follow one another, and the more rapidly the motor will revolve in response thereto, and this continues until the armature generator is stopped entirely, as by a brake, when the motor, if properly constructed, runs at the speed with which the magnetic poles of the generator rotate.

The effective strength of the currents developed in the armature coils of the generator is dependent upon the strength of the currents energizing the generator and upon the number of rotations per unit of time of the magnetic poles of the generator; hence the speed of the motor armature will depend in all cases upon the relative speeds of the armature of the generator and of its magnetic poles. For example, if the poles are turned two thousand times per unit of time and the armature is turned eight hundred, the motor will turn twelve hundred times, or nearly so. Very slight differences of speed may be indicated by a delicately balanced motor.

Let it now be assumed that power is applied to the generator armature to turn it in a direction opposite to that in which its magnetic poles rotate. In such case the result would be similar to that produced by a generator the armature and field magnets of which are rotated in opposite directions, and by reason of these conditions the motor armature will turn at a rate of speed equal to the sum of the speeds of the armature and magnetic poles of the generator, so that a comparatively low speed of the generator armature will produce a high speed in the motor.

It will be observed in connection with this system that on diminishing the resistance of the external circuit of the generator armature by checking the speed of the motor or by adding translating devices in multiple arc in the secondary circuit or circuits of the transformer the strength of the current in the armature circuit is greatly increased. This is due to two causes: first, to the great differences in the speeds of the motor and generator, and, secondly, to the fact that the apparatus follows the analogy of a transformer, for, in proportion as the resistance of the armature or secondary circuits is reduced, the strength of the currents in the field or primary circuits of the generator is increased and the currents in the armature are augmented correspondingly. For similar reasons the currents in the armature-coils of the generator increase very rapidly when the speed of the armature is reduced when running in the same direction as the magnetic poles or conversely.

It will be understood from the above description that the generator armature may be run in the direction of the shifting of the magnetic poles, but more rapidly, and that in such case the speed of the motor will be equal to the difference between the two rates.

Chapter VII

Regulator for Rotary Current Motors

An interesting device for regulating and reversing has been devised by Mr. Tesla for the purpose of varying the speed of polyphase motors. It consists of a form of converter or transformer with one element capable of movement with respect to the other, whereby the inductive relations may be altered, either manually or automatically, for the purpose of varying the strength of the induced current. Mr. Tesla prefers to construct this device in such manner that the induced or secondary element may be movable with respect to the other; and the invention, so far as relates merely to the construction of the device itself, consists, essentially, in the combination, with two opposite magnetic poles, of an armature wound with an insulated coil and mounted on a shaft, whereby it may be turned to the desired extent within the field produced by the poles. The normal position of the core of the secondary element is that in which it most completely closes the magnetic circuit between the poles of the primary element, and in this position its coil is in its most effective position for the inductive action upon it of the primary coils; but by turning the movable core to either side, the induced currents delivered by its coil become weaker until, by a movement of the said core and coil through 90°, there will be no current delivered.

Fig. 34 is a view in side elevation of the regulator. Fig. 35 is a broken section on line $x\,x$ of Fig. 34. Fig. 36 is a diagram illustrating the most convenient manner of applying the regulator to ordinary forms of motors, and Fig. 37 is a similar diagram illustrating the application of the device to the Tesla alternating-current motors. The regulator may be constructed in many ways to secure the desired result; but that which is, perhaps, its best form is shown in Figs. 34 and 35.

A represents a frame of iron. B B are the cores of the inducing or primary coils C C. D is a shaft mounted on the side bars, D′, and on which is secured a sectional iron core, E, wound with an induced or secondary coil, F, the convolutions of which are parallel with the axis of the shaft.

FIG. 34.

FIG. 35.

The ends of the core are rounded off so as to fit closely in the space between the two poles and permit the core E to be turned to and held at any desired point. A handle, G, secured to the projecting end of the shaft D, is provided for this purpose.

FIG. 36.

In Fig. 36 let H represent an ordinary alternating current generator, the field-magnets of which are excited by a suitable source of current, I. Let J designate an ordinary form of electro-magnetic motor provided with an armature, K, commutator L, and field-magnets M. It is well known that such a motor, if its field-magnet cores be divided up into insulated sections, may be practically operated by an alternating current; but in using this regulator with such a motor, Mr. Tesla includes one element of the motor only—say the armature-coils—in the main circuit of the generator, making the connections through the brushes and the commutator in the usual way. He also includes one of the elements of the regulator—say the stationary coils—in the same circuit, and in the circuit with the secondary or movable coil of the regulator he connects up the field-coils of the motor. He also prefers to use flexible conductors to make the connections from the secondary coil of the regulator, as he thereby avoids the use of sliding contacts or rings without interfering with the requisite movement of the core E.

If the regulator be in its normal position, or that in which its magnetic circuit is most nearly closed, it delivers its maximum induced current, the phases of which so correspond with those of the primary current that the motor will run as though both field and armature were excited by the main current.

To vary the speed of the motor to any rate between the minimum and maximum rates, the core E and coils F are turned in either direction to an extent which produces the desired result, for in its normal position the convolutions of coil F embrace the maximum number of lines of force, all of which act with the same effect upon the coil; hence it will deliver its maximum current; but by turning the coil F out of its position of maximum effect the number of lines of force embraced by it is diminished. The inductive effect is therefore impaired, and the current delivered by coil F will continue to diminish in proportion to the angle at which the coil F is turned until, after passing through an angle of ninety degrees, the convolutions of the coil will be at right angles to those of coils C C, and the inductive effect reduced to a minimum.

Incidentally to certain constructions, other causes may influence the variation in the strength of the induced currents. For example, in the present case it will be observed that by the first movement of coil F a certain portion of its convolutions are carried beyond the line of the direct influence of the lines of force, and that the magnetic path or circuit for the lines is impaired; hence the inductive effect would be reduced. Next, that after moving through a certain angle, which is obviously determined by the relative dimensions of the bobbin or coil F, diagonally opposite portions of the coil will be simultaneously included in the field, but in such positions that the lines which produce a current-impulse in one portion of the coil in a certain direction will produce in the diagonally opposite portion a corresponding impulse in the opposite direction; hence portions of the current will neutralize one another.

As before stated, the mechanical construction of the device may be greatly varied; but the essential conditions of the principle will be fulfilled in any apparatus in which the movement of the elements with respect to one another effects the same results by varying the inductive relations of the two elements in a manner similar to that described.

It may also be stated that the core E is not indispensable to the operation of the regulator; but its presence is obviously beneficial. This regulator, however, has another valuable property in its capability of reversing the motor, for if the coil F be turned through a half-revolution, the position of its convolutions relatively to the two coils c c and to the lines of force is reversed, and consequently the phases of the current will be reversed. This will produce a rotation of the motor in an opposite direction. This form of regulator is also applied with great advantage to Mr. Tesla's system of utilizing alternating currents, in which the magnetic poles of the field of a motor are progressively shifted by means of the combined effects upon the field of magnetizing coils included in independent circuits, through which pass alternating currents in proper order and relations to each other.

In Fig. 37, let P represent a Tesla generator having two independent coils, P′ and P″, on the armature, and T a diagram of a motor having two independent energizing coils or sets of coils, R R′. One of the circuits from the generator, as S′ S′, includes one set, R′ R′, of the energizing coils of the motor, while the other circuit, as S S, includes the primary coils of the regulator. The secondary coil of the regulator includes the other coils, R R, of the motor.

While the secondary coil of the regulator is in its normal position, it produces its maximum current, and the maximum rotary effect is imparted to the motor; but this effect will be diminished in proportion to the angle at which the coil F of the regulator is turned. The motor will also be reversed by reversing the position of the coil with reference to the coils c c, and thereby reversing the phases of the current produced by the generator. This changes the direction of the movement of the shifting poles which the armature follows.

One of the main advantages of this plan of regulation is its economy of power. When the induced coil is generating its maximum current,

FIG. 37.

the maximum amount of energy in the primary coils is absorbed; but as the induced coil is turned from its normal position the self-induction of the primary coils reduces the expenditure of energy and saves power.

It is obvious that in practice either coils c c or coil f may be used as primary or secondary, and it is well understood that their relative proportions may be varied to produce any desired difference or similarity in the inducing and induced currents.

Chapter VIII

Single Circuit, Self-Starting Synchronizing Motors

In the first chapters of this section we have, bearing in mind the broad underlying principle, considered a distinct class of motors, namely, such as require for their operation a special generator capable of yielding currents of differing phase. As a matter of course, Mr. Tesla recognizing the desirability of utilizing his motors in connection with ordinary systems of distribution, addressed himself to the task of inventing various methods and ways of achieving this object. In the succeeding chapters, therefore, we witness the evolution of a number of ideas bearing upon this important branch of work. It must be obvious to a careful reader, from a number of hints encountered here and there, that even the inventions described in these chapters to follow do not represent the full scope of the work done in these lines. They might, indeed, be regarded as exemplifications.

We will present these various inventions in the order which to us appears the most helpful to an understanding of the subject by the majority of readers. It will be naturally perceived that in offering a series of ideas of this nature, wherein some of the steps or links are missing, the descriptions are not altogether sequential; but anyone who follows carefully the main drift of the thoughts now brought together will find that a satisfactory comprehension of the principles can be gained.

As is well known, certain forms of alternating-current machines have the property, when connected in circuit with an alternating current generator, of running as a motor in synchronism therewith; but, while the alternating current will run the motor after it has attained a rate of speed synchronous with that of the generator, it will not start it. Hence, in all instances heretofore where these "synchronizing motors," as they are termed, have been run, some means have been adopted to bring the motors up to synchronism with the generator, or approximately so, before the alternating current of the generator is applied to drive them. In some instances mechanical appliances have been utilized for this purpose. In others special and complicated forms of motor have been constructed. Mr. Tesla has discovered a much more simple method or plan of operating synchronizing motors, which requires practically no other apparatus than the motor itself. In other words, by a certain change in the circuit connections of the motor he converts it at will from a double circuit motor, or such as have been already described, and which will start under the action of an alternating current, into a synchronizing motor, or one which will be run by the generator only when it has reached a certain speed of rotation synchronous with that of the generator. In this manner he is enabled to extend very greatly the applications of his system and to secure all the advantages of both forms of alternating current motor.

The expression "synchronous with that of the generator," is used here in its ordinary acceptation—that is to say, a motor is said to synchronize with the generator when it preserves a certain relative speed determined by its number of poles and the number of alternations produced per revolution of the generator. Its actual speed, therefore, may be faster or slower than that of the generator; but it is said to be synchronous so long as it preserves the same relative speed.

In carrying out this invention Mr. Tesla constructs a motor which has a strong tendency to synchronism with the generator. The construction preferred is that in which the armature is provided with polar projections. The field-magnets are wound with two sets of coils, the terminals of which are connected to a switch mechanism, by means of which the line-current may be carried directly through these coils or indirectly through paths by which its phases are modified. To start such a motor, the switch is turned on to a set of contacts which includes in one motor circuit a dead resistance, in the other an inductive resistance, and, the two circuits being in derivation, it is obvious that the difference in phase of the current in such circuits will set up a rotation of the motor. When the speed of the motor has thus been brought to the desired rate the switch is shifted to throw the main current directly through the motor-circuits, and although the currents in both circuits will now be of the same phase the motor will continue to revolve, becoming a true synchronous motor. To secure greater efficiency, the armature or its polar projections are wound with coils closed on themselves.

In the accompanying diagrams, Fig. 38 illustrates the details of the plan above set forth, and Figs. 39 and 40 modifications of the same.

Referring to Fig. 38, let A designate the field-magnets of a motor, the polar projections of which are wound with coils B C included in independent circuits, and D the armature with polar projections wound with coils E closed upon themselves, the motor in these respects being similar in construction to those described already, but having on account of the polar projections on the armature core, or other similar and well-known features, the properties of a synchronizing-motor. L L' represents the conductors of a line from an alternating current generator G.

Near the motor is placed a switch the action of which is that of the one shown in the diagrams, which is constructed as follows: F F' are two conducting plates or arms, pivoted at their ends and connected by an insulating cross-bar, H, so as to be shifted in parallelism. In the path of the bars F F' is the contact 2, which forms one terminal of the circuit through coils C, and the contact 4, which is one terminal of the circuit through coils B. The opposite end of the wire of coils C is connected to the wire L or bar F', and the corresponding end of coils B is connected to wire L' and bar F; hence if the bars be shifted so as to bear on contacts 2 and 4 both sets of coils B C will be included in the circuit L L' in multiple arc or derivation. In the path of the levers F F' are two other contact terminals, 1 and 3. The contact 1 is connected to contact 2 through an artificial resistance, I, and contact 3 with contact 4 through a self-induction coil, J, so that when the switch levers are shifted upon the points 1 and 3 the circuits of coils B and C will be connected in multiple arc or derivation to the circuit L L', and will include the resistance and self-induction coil respectively. A third position of the switch is that in which the levers F and F' are shifted out of contact with both sets of points. In this case the motor is entirely out of circuit.

The purpose and manner of operating the motor by these devices are as follows: The normal position of the switch, the motor being out of circuit, is off the contact points. Assuming the generator to be running, and that it is desired to start the motor, the switch is shifted until its levers rest upon points 1 and 3. The two motor-circuits are thus connected with the generator circuit; but by reason

Figs. 38, 39, and 40.

of the presence of the resistance I in one and the self-induction coil J in the other the coincidence of the phases of the current is disturbed sufficiently to produce a progression of the poles, which starts the motor in rotation. When the speed of the motor has run up to synchronism with the generator, or approximately so, the switch is shifted over upon the points 2 and 4, thus cutting out the coils I and J, so that the currents in both circuits have the same phase; but the motor now runs as a synchronous motor.

It will be understood that when brought up to speed the motor will run with only one of the circuits B or C connected with the main or generator circuit, or the two circuits may be connected in series. This latter plan is preferable when a current having a high number of alternations per unit of time is employed to drive the motor. In such case the starting of the motor is more difficult, and the dead and inductive resistances must take up a considerable proportion of the electromotive force of the circuits. Generally the conditions are so adjusted that the electromotive force used in each of the motor circuits is that which is required to operate the motor when its circuits are in series. The plan followed in this case is illustrated in Fig. 39. In this instance the motor has twelve poles and the armature has polar projections D wound with closed coils E. The switch used is of substantially the same construction as that shown in the previous figure. There are, however, five contacts, designated as 5, 6, 7, 8, and 9. The motor-circuits B C, which include alternate field-coils, are connected to the terminals in the following order: One end of circuit C is connected to contact 9 and to contact 5 through a dead resistance, I. One terminal of circuit B is connected to contact 7 and to contact 6 through a self-induction coil, J. The opposite terminals of both circuits are connected to contact 8.

One of the levers, as F, of the switch is made with an extension, f, or otherwise, so as to cover both contacts 5 and 6 when shifted into the position to start the motor. It will be observed that when in this position and with lever F' on contact 8 the current divides between the two circuits B C, which from their difference in electrical character produce a progression of the poles that starts the motor in rotation. When the motor has attained the proper speed, the switch is shifted so that the levers cover the contacts 7 and 9, thereby connecting circuits B and C in series. It is found that by this disposition the motor is maintained in rotation in synchronism with the generator. This principle of operation, which consists in converting by a change of connections or otherwise a double-circuit motor, or one operating by a progressive shifting of the poles, into an ordinary synchronizing motor may be carried out in many other ways. For instance, instead of using the switch shown in the previous figures, we may use a temporary ground circuit between the generator and motor, in order to start the motor, in substantially the manner indicated in Fig. 40. Let G in this figure represent an ordinary alternating-current generator with, say, two poles, M M', and an armature wound with two coils, N N', at right angles and connected in series. The motor has, for example, four poles wound with coils B C, which are connected in series, and an armature with polar projections D wound with closed coils E E. From the common joint or union between the two circuits of both the generator and the motor an earth connection is established, while the terminals or ends of these circuits are connected to the line. Assuming that the motor is a synchronizing motor or one that has the capability of running in synchronism with the generator, but not of starting, it may be started by the above-described apparatus by closing the ground connection from both generator and motor. The system thus becomes one with a two-circuit generator and motor, the ground forming a common return for the currents in the two circuits L and L'. When by this arrangement of circuits the motor is brought to speed, the ground connection is broken between the motor or generator, or both, ground-switches P P' being employed for this purpose. The motor then runs as a synchronizing motor.

In describing the main features which constitute this invention illustrations have necessarily been omitted of the appliances used in conjunction with the electrical devices of similar systems—such, for instance, as driving-belts, fixed and loose pulleys for the motor, and the like; but these are matters well understood.

Mr. Tesla believes he is the first to operate electro-magnetic motors by alternating currents in any of the ways herein described—that is to say, by producing a progressive movement or rotation of their poles or points of greatest magnetic attraction by the alternating currents until they have reached a given speed, and then by the same currents producing a simple alternation of their poles, or, in other words, by a change in the order or character of the circuit connections to convert a motor operating on one principle to one operating on another.

CHAPTER IX

CHANGE FROM DOUBLE CURRENT TO
SINGLE CURRENT MOTOR

A description is given elsewhere of a method of operating alternating current motors by first rotating their magnetic poles until they have attained synchronous speed, and then alternating the poles. The motor is thus transformed, by a simple change of circuit connections from one operated by the action of two or more independent energizing currents to one operated either by a single current or by several currents acting as one. Another way of doing this will now be described.

At the start the magnetic poles of one element or field of the motor are progressively shifted by alternating currents differing in phase and passed through independent energizing circuits, and short circuit the coils of the other element. When the motor thus started reaches or passes the limit of speed synchronous with the generator, Mr. Tesla connects up the coils previously short-circuited with a source of direct current and by a change of the circuit connections produces a simple alternation of the poles. The motor then continues to run in synchronism with the generator. The motor here shown in Fig. 41 is one of the ordinary forms, with field-cores either laminated or solid and with a cylindrical laminated armature wound, for example, with the coils A B at right angles. The shaft of the armature carries three collecting or contact rings C D E. (Shown, for better illustration, as of different diameters.)

One end of coil A connects to one ring, as C, and one end of coil B connects with ring D. The remaining ends are connected to ring E. Collecting springs or brushes F G H bear upon the rings and lead to the contacts of a switch, to be presently described. The field-coils have their terminals in binding-posts K K, and may be either closed upon themselves or connected with a source of direct current L, by means of a switch M. The main or controlling switch has five contacts a b c d e and two levers f g, pivoted and connected by an insulating cross-bar h, so as to move in parallelism. These levers are connected to the line wires from a source of alternating currents N. Contact a is connected to brush G and coil B through a dead resistance R and wire P. Contact *b* is connected with brush F and coil A through a self-induction coil S and wire O. Contacts *c* and *e* are connected to brushes G F, respectively, through the wires P O, and contact *d* is directly connected with brush H. The lever *f* has a widened end, which may span the contacts *a b*. When in such position and with lever *g* on contact *d*, the alternating currents divide between the two motor-coils, and by reason of their different self-induction a difference of current-phase is obtained that starts the motor in rotation. In starting, the field-coils are short circuited.

FIG. 41.

When the motor has attained the desired speed, the switch is shifted to the position shown in dotted lines—that is to say, with the levers $f g$ resting on points $c e$. This connects up the two armature coils in series, and the motor will then run as a synchronous motor. The field-coils are thrown into circuit with the direct current source when the main switch is shifted.

CHAPTER X

MOTOR WITH "CURRENT LAG" ARTIFICIALLY SECURED

One of the general ways followed by Mr. Tesla in developing his rotary phase motors is to produce practically independent currents differing primarily in phase and to pass these through the motor-circuits. Another way is to produce a single alternating current, to divide it between the motor-circuits, and to effect artificially a lag in one of these circuits or branches, as by giving to the circuits different self-inductive capacity, and in other ways. In the former case, in which the necessary difference of phase is primarily effected in the generation of currents, in some instances, the currents are passed through the energizing coils of both elements of the motor—the field and armature; but a further result or modification may be obtained by doing this under the conditions hereinafter specified in the case of motors in which the lag, as above stated, is artificially secured.

Figs. 42 to 47, inclusive, are diagrams of different ways in which the invention is carried out; and Fig. 48, a side view of a form of motor used by Mr. Tesla for this purpose.

A B in Fig. 42 indicate the two energizing circuits of a motor, and c D two circuits on the armature. Circuit or coil A is connected in series with circuit or coil C, and the two circuits B D are similarly connected. Between coils A and c is a contact-ring e, forming one terminal of the latter, and a brush a, forming one terminal of the former. A ring d and brush c similarly connect coils B and D. The opposite terminals of the field-coils connect to one binding post h of the motor, and those of the armature coils are similarly connected to the opposite binding post i through a contact-ring f and brush g. Thus each motor-circuit while in derivation to the other includes one armature and one field coil. These circuits are of different self-induction, and may be made so in various ways. For the sake of clearness, an artificial resistance R is shown in one of these circuits, and in the other a self-induction coil s. When an alternating current is passed through this motor it divides between its two energizing-circuits. The higher self-induction of one circuit produces a greater retardation or lag in the current therein than in the other. The difference of phase between the two currents effects the rotation or shifting of the points of maximum magnetic effect that secures the rotation of the armature. In certain respects this plan of including both armature and field coils in

Fig. 42.

circuit is a marked improvement. Such a motor has a good torque at starting; yet it has also considerable tendency to synchronism, owing to the fact that when properly constructed the maximum magnetic effects in both armature and field coincide—a condition which in the usual construction of these motors with closed armature coils is not readily attained. The motor thus constructed exhibits too, a better regulation of current from no load to load, and there is less difference between the apparent and real energy expended in running it. The true synchronous speed of this form of motor is that of the generator when both are alike—that is to say, if the number of the coils on the armature and on the field is x, the motor will run normally at the same speed as a generator driving it if the number of field magnets or poles of the same be also x.

Fig. 43 shows a somewhat modified arrangement of circuits. There is in this case but one armature coil E, the winding of which maintains effects corresponding to the resultant poles produced by the two field-circuits.

FIG. 43.

Fig. 44 represents a disposition in which both armature and field are wound with two sets of coils, all in multiple arc to the line or main circuit. The armature coils are wound to correspond with the

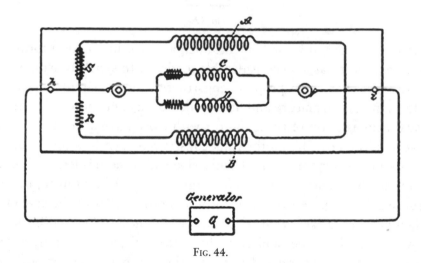

FIG. 44.

field-coils with respect to their self-induction. A modification of this plan is shown in Fig. 45—that is to say, the two field coils and two armature coils are in derivation to themselves and in series with one another. The armature coils in this case, as in the previous figure, are wound for different self-induction to correspond with the field coils.

FIG. 45.

Another modification is shown in Fig. 46. In this case only one armature-coil, as D, is included in the line-circuit, while the other, as C, is short-circuited.

In such a disposition as that shown in Fig. 43, or where only one armature-coil is employed, the torque on the start is somewhat reduced, while the tendency to synchronism is somewhat increased.

FIG. 46.

In such a disposition as shown in Fig. 46, the opposite conditions would exist. In both instances, however, there is the advantage of dispensing with one contact-ring.

In Fig. 46 the two field-coils and the armature-coil D are in multiple arc. In Fig. 47 this disposition is modified, coil D being shown in series with the two field-coils.

FIG. 47.

FIG. 48.

Fig. 48 is an outline of the general form of motor in which this invention is embodied. The circuit connections between the armature and field coils are made, as indicated in the previous figures, through brushes and rings, which are not shown.

CHAPTER XI

ANOTHER METHOD OF TRANSFORMATION FROM A TORQUE TO A SYNCHRONIZING MOTOR

In a preceding chapter we have described a method by which Mr. Tesla accomplishes the change in his type of rotating field motor from a torque to a synchronizing motor. As will be observed, the desired end is there reached by a change in the circuit connections at the proper moment. We will now proceed to describe another way of bringing about the same result. The principle involved in this method is as follows:

If an alternating current be passed through the field coils only of a motor having two energizing circuits of different self-induction and the armature coils be short-circuited, the motor will have a strong torque, but little or no tendency to synchronism with the generator; but if the same current which energizes the field be passed also through the armature coils the tendency to remain in synchronism is very considerably increased. This is due to the fact that the maximum magnetic effects produced in the field and armature more nearly coincide. On this principle Mr. Tesla constructs a motor having independent field circuits of different self-induction, which are joined in derivation to a source of alternating currents. The armature is wound with one or more coils, which are connected with the field coils through contact rings and brushes, and around the armature coils a shunt is arranged with means for opening or closing the same. In starting this motor the shunt is closed around the armature coils, which will therefore be in closed circuit. When the current is directed through the motor, it divides between the two circuits (it is not necessary to consider any case where there are more than two circuits used), which, by reason of their different self-induction, secure a difference of phase between the two currents in the two branches, that produces a shifting or rotation of the poles. By the alternations of current, other currents are induced in the closed— or short-circuited—armature coils and the motor has a strong torque. When the desired speed is reached, the shunt around the armature-coils is opened and the current directed through both armature and field coils. Under these conditions the motor has a strong tendency to synchronism.

In Fig. 49, A and B designate the field coils of the motor. As the circuits including these coils are of different self-induction, this is represented by a resistance coil R in circuit with A, and a self-induction coil S in circuit with B. The same result may of course be secured by the winding of the coils. C is the armature circuit, the terminals of which are rings a b. Brushes c d bear on these rings and connect with the line and field circuits. D is the shunt or short circuit around the armature. E is the switch in the shunt.

FIGS. 49, 50, and 51.

It will be observed that in such a disposition as is illustrated in Fig. 49, the field circuits A and B being of different self-induction, there will always be a greater lag of the current in one than the other, and that, generally, the armature phases will not correspond with either, but with the resultant of both. It is therefore important to observe the proper rule in winding the armature. For instance, if the motor have eight poles—four in each circuit—there will be four resultant poles, and hence the armature winding should be such as to produce four poles, in order to constitute a true synchronizing motor.

The diagram, Fig. 50, differs from the previous one only in respect to the order of connections. In the present case the armature-coil, instead of being in series with the field-coils, is in multiple arc therewith. The armature-winding may be similar to that of the field—that is to say, the armature may have two or more coils wound or adapted for different self-induction and adapted, preferably, to produce the same difference of phase as the field-coils. On starting the motor the shunt is closed around both coils. This is shown in Fig. 51, in which the armature coils are F G. To indicate their different electrical character, there are shown in circuit with them, respectively, the resistance R′ and the self-induction coil S′. The two armature coils are in series with the field-coils and the same disposition of the shunt or short-circuit D is used. It is of advantage in the operation of motors of this kind to construct or wind the armature in such manner that when short-circuited on the start it will have a tendency to reach a higher speed than that which synchronizes with the generator. For example, a given motor having, say, eight poles should run, with the armature coil short-circuited, at two thousand revolutions per minute to bring it up to synchronism. It will generally happen, however, that this speed is not reached, owing to the fact that the armature and field currents do not properly correspond, so that when the current is passed through the armature (the motor not being quite up to synchronism) there is a liability that it will not "hold on," as it is termed. It is preferable, therefore, to so wind or construct the motor that on the start, when the armature coils are short-circuited, the motor will tend to reach a speed higher than the synchronous—as for instance, double the latter. In such case the difficulty above alluded to is not felt, for the motor will always hold up to synchronism if the synchronous speed—in the case supposed of two thousand revolutions—is reached or passed. This may be accomplished in various ways; but for all practical purposes the following will suffice: On the armature are wound two sets of coils. At the start only one of these is short-circuited, thereby producing a number of poles on the armature, which will tend to run the speed up above the synchronous limit. When such limit is reached or passed, the current is directed through the other coil, which, by increasing the number of armature poles, tends to maintain synchronism.

Fig. 52.

In Fig. 52, such a disposition is shown. The motor having, say, eight poles contains two field-circuits A and B, of different self-induction. The armature has two coils F and G. The former is closed upon itself, the latter connected with the field and line through contact-rings *a b*, brushes *c d*, and a switch E. On the start the coil F alone is active and the motor tends to run at a speed above the synchronous; but when the coil G is connected to the circuit the number of armature poles is increased, while the motor is made a true synchronous motor. This disposition has the advantage that the closed armature-circuit imparts to the motor torque when the speed falls off, but at the same time the conditions are such that the motor comes out of synchronism more readily. To increase the tendency to synchronism, two circuits may be used on the armature, one of which is short-circuited on the start and both connected with the external circuit after the synchronous speed is reached or passed. This disposition is shown in Fig. 53. There are three contact-rings *a b e* and three brushes *c d f*, which connect the armature circuits with the external

Fig. 53.

circuit. On starting, the switch H is turned to complete the connection between one binding-post P and the field-coils. This short-circuits one of the armature-coils, as G. The other coil F is out of circuit and open. When the motor is up to speed, the switch H is turned back, so that the connection from binding-post P to the field coils is through the coil G, and switch K is closed, thereby including coil F in multiple arc with the field coils. Both armature coils are thus active.

From the above-described instances it is evident that many other dispositions for carrying out the invention are possible.

CHAPTER XII

"MAGNETIC LAG" MOTOR

The following description deals with another form of motor, namely, depending on "magnetic lag" or hysteresis, its peculiarity being that in it the attractive effects or phases while lagging behind the phases of current which produce them, are manifested simultaneously and not successively. The phenomenon utilized thus at an early stage by Mr. Tesla, was not generally believed in by scientific men, and Prof. Ayrton was probably first to advocate it or to elucidate the reason of its supposed existence.

Fig. 54 is a side view of the motor, in elevation. Fig. 55 is a part-sectional view at right angles to Fig. 54. Fig. 56 is an end view in elevation and part section of a modification, and Fig. 57 is a similar view of another modification.

FIG. 54.   FIG. 55.

In Figs. 54 and 55, A designates a base or stand, and B B the supporting-frame of the motor. Bolted to the supporting-frame are two magnetic cores or pole-pieces c c', of iron or soft steel. These may be subdivided or laminated, in which case hard iron or steel plates or bars should be used, or they should be wound with closed coils. D is a circular disc armature, built up of sections or plates of iron and mounted in the frame between the pole-pieces $c\,c'$, curved to conform to the circular shape thereof. This disc may be wound with a number of closed coils E. F F are the main energizing coils, supported by the supporting-frame, so as to include within their magnetizing influence both the pole-pieces $c\,c'$ and the armature D. The pole-pieces $c\,c'$ project out beyond the coils F F on opposite sides, as indicated in the drawings. If an alternating current be passed through the coils F F, rotation of the armature will be produced, and this rotation is explained by the following apparent action, or mode of operation: An impulse of current in the coils F F establishes two polarities in the motor. The protruding end of pole-piece C, for instance, will be of one sign, and the corresponding end of pole-piece c' will be of the opposite sign. The armature also exhibits two poles at right angles to the coils F F, like poles to those in the pole-pieces being on the same side of the coils. While the current is flowing there is no appreciable tendency to rotation developed; but after each current impulse ceases or begins to fall, the magnetism in the armature and in the ends of the pole-pieces $c\,c'$ lags or continues to manifest itself, which produces a rotation of the armature by the repellent force between the more closely approximating points of maximum magnetic effect. This effect is continued by the reversal of current, the polarities of field and armature being simply reversed. One or both of the elements—the armature or field—may be wound with closed induced coils to intensify this effect. Although in the illustrations but one of the fields is shown, each element of the motor really constitutes a field, wound with the closed coils, the currents being induced mainly in those convolutions or coils which are parallel to the coils F F.

A modified form of this motor is shown in Fig. 56. In this form G is one of two standards that support the bearings for the armature-shaft. H H are uprights or sides of a frame, preferably magnetic, the ends $c\,c'$ of which are bent in the manner indicated, to conform to the shape of the armature D and form field-magnet poles. The construction of the armature may be the same as in the previous figure, or it may be simply a magnetic disc or cylinder, as shown, and a coil or coils F F are secured

FIG. 56.

in position to surround both the armature and the poles *c c'*. The armature is detachable from its shaft, the latter being passed through the armature after it has been inserted in position. The operation of this form of motor is the same in principle as that previously described and needs no further explanation.

One of the most important features in alternating current motors is, however, that they should be adapted to and capable of running efficiently on the alternating circuits in present use, in which almost without exception the generators yield a very high number of alternations. Such a motor, of the type under consideration, Mr. Tesla has designed by a development of the principle of the motor shown in Fig. 56, making a multipolar motor, which is illustrated in Fig. 57. In the construction of this motor he employs an annular magnetic frame J, with inwardly-extending ribs or projections K, the ends of which all bend or turn in one direction and are generally shaped to conform to the curved surface of the armature. Coils F F are wound from one part K to the one next adjacent, the ends or loops of each coil or group of wires being carried over toward the shaft, so as to form U-shaped groups of convolutions at each end of the armature. The pole-pieces *c c'*, being substantially concentric with the armature, form ledges, along which the coils are laid and should project to some extent beyond the coils, as shown. The cylindrical or drum armature D is of the same construction as in the other motors described, and is mounted to

Fig. 57.

rotate within the annular frame J and between the U-shaped ends or bends of the coils F. The coils F are connected in multiple or in series with a source of alternating currents, and are so wound that with a current or current impulse of given direction they will make the alternate pole-pieces c of one polarity and the other pole-pieces c′ of the opposite polarity. The principle of the operation of this motor is the same as the other above described, for, considering any two pole-pieces c c′, a current impulse passing in the coil which bridges them or is wound over both tends to establish polarities in their ends of opposite sign and to set up in the armature core between them a polarity of the same sign as that of the nearest pole-piece c. Upon the fall or cessation of the current impulse that established these polarities the magnetism which lags behind the current phase, and which continues to manifest itself in the polar projections c c′ and the armature, produces by repulsion a rotation of the armature. The effect is continued by each reversal of the current. What occurs in the case of one pair of pole-pieces occurs simultaneously in all, so that the tendency to rotation of the armature is measured by the sum of all the forces exerted by the pole-pieces, as above described. In this motor also the magnetic lag or effect is intensified by winding one or both cores with closed induced coils. The armature core is shown as thus wound. When closed coils are used, the cores should be laminated.

It is evident that a pulsatory as well as an alternating current might be used to drive or operate the motors above described.

It will be understood that the degree of subdivision, the mass of the iron in the cores, their size and the number of alternations in the current employed to run the motor, must be taken into consideration in order to properly construct this motor. In other words, in all such motors the proper relations between the number of alternations and the mass, size, or quality of the iron must be preserved in order to secure the best results.

CHAPTER XIII

METHOD OF OBTAINING DIFFERENCE OF PHASE BY MAGNETIC SHIELDING

In that class of motors in which two or more sets of energizing magnets are employed, and in which by artificial means a certain interval of time is made to elapse between the respective maximum or minimum periods or phases of their magnetic attraction or effect, the interval or difference in phase between the two sets of magnets is limited in extent. It is desirable, however, for the economical working of such motors that the strength or attraction of one set of magnets should be maximum, at the time when that of the other set is minimum, and conversely; but these conditions have not heretofore been realized except in cases where the two currents have been obtained from independent sources in the same or different machines. Mr. Tesla has therefore devised a motor embodying conditions that approach more nearly the theoretical requirements of perfect working, or in other words, he produces artificially a difference of magnetic phase by means of a current from a single primary source sufficient in extent to meet the requirements of practical and economical working. He employs a motor with two sets of energizing or field magnets, each wound with coils connected with a source of alternating or rapidly-varying currents, but forming two separate paths or circuits. The magnets of one set are protected to a certain extent from the energizing action of the current by means of a magnetic shield or screen interposed between the magnet and its energizing coil. This shield is properly adapted to the conditions of particular cases, so as to shield or protect the main core from magnetization until it has become itself saturated and no longer capable of containing all the lines of force produced by the current. It will be seen that by this means the energizing action begins in the protected set of magnets a certain arbitrarily-determined period of time later than in the other, and that by this means alone or in conjunction with other means or devices heretofore employed a practical difference of magnetic phase may readily be secured.

Fig. 58 is a view of a motor, partly in section, with a diagram illustrating the invention. Fig. 59 is a similar view of a modification of the same.

FIG. 58.

In Fig. 58, which exhibits the simplest form of the invention, A A is the field-magnet of a motor, having, say, eight poles or inwardly-projecting cores B and C. The cores B form one set of magnets and are energized by coils D. The cores C, forming the other set are energized by coils E, and the coils are connected, preferably, in series with one another, in two derived or branched circuits, F G, respectively, from a suitable source of current. Each coil E is surrounded by a magnetic shield H, which is preferably composed of an annealed, insulated, or oxidized iron wire wrapped or wound on the coils in the manner indicated so as to form a closed magnetic circuit around the coils and between the same and the magnetic cores C. Between the pole-pieces or cores B C is mounted the armature K, which, as is usual in this type of machines, is wound with coils L closed upon themselves. The operation resulting from this disposition is as follows: If a current impulse be directed through the two circuits of the motor, it will quickly energize the cores B, but not so the cores C, for the reason that in passing through the coils E there is encountered the influence of the closed magnetic circuits formed by the shields H. The first effect is to retard effectively the current impulse in circuit G, while at the same time the proportion of current which does pass does not magnetize the cores C, which are shielded or screened by the shields H. As the increasing electromotive force then urges more current through the coils E, the iron wire H becomes magnetically saturated and incapable of carrying all the lines of force, and hence ceases to protect the cores C, which becomes magnetized, developing their maximum effect after an interval of time subsequent to the similar manifestation of strength in the other set of magnets, the extent of which is arbitrarily determined by the thickness of the shield H, and other well-understood conditions.

From the above it will be seen that the apparatus or device acts in two ways. First, by retarding the current, and, second, by retarding the magnetization of one set of the cores, from which its effectiveness will readily appear.

FIG. 59.

Many modifications of the principle of this invention are possible. One useful and efficient application of the invention is shown in Fig. 59. In this figure a motor is shown similar in all respects to that above described, except that the iron wire H, which is wrapped around the coils E, is in this case connected in series with the coils D. The iron-wire coils H, are connected and wound, so as to have little or no self-induction, and being added to the resistance of the circuit F, the action of the current in that circuit will be accelerated, while in the other circuit G it will be retarded. The shield H may be made in many forms, as will be understood, and used in different ways, as appears from the foregoing description.

As a modification of his type of motor with "shielded" fields, Mr. Tesla has constructed a motor with a field-magnet having two sets of poles or inwardly-projecting cores and placed side by side, so as practically to form two fields of force and alternately disposed—that is to say, with the poles of one set or field opposite the spaces between the other. He then connects the free ends of one set of poles by means of laminated iron bands or bridge-pieces of considerably smaller cross-section than the cores themselves, whereby the cores will all form parts of complete magnetic circuits. When the coils on each set of magnets are connected in multiple circuits or branches from a source of alternating currents, electromotive forces are set up in or impressed upon each circuit simultaneously; but the coils on the magnetically bridged or shunted cores will have, by reason of the closed magnetic circuits, a high self-induction, which retards the current, permitting at the beginning of each impulse but little current to pass. On the other hand, no such opposition being encountered in the other set of coils, the current passes freely through them, magnetizing the poles on which they are wound. As soon, however, as the laminated bridges become saturated and incapable of carrying all the lines of force which the rising electromotive force, and consequently increased current, produce, free poles are developed at the ends of the cores, which, acting in conjunction with the others, produce rotation of the armature.

The construction in detail by which this invention is illustrated is shown in the accompanying drawings.

Fig. 60 is a view in side elevation of a motor embodying the principle. Fig. 61 is a vertical cross-section of the motor. A is the frame of the motor, which should be built up of sheets of iron punched out to the desired shape and bolted together with insulation between the sheets. When complete, the frame makes a field-magnet with inwardly projecting pole-pieces B and C. To adapt them to the requirements of this particular case these pole-pieces are out of line with one another, those marked B surrounding one end of the armature and the others,

FIG. 60.   FIG. 61.

as c, the opposite end, and they are disposed alternately—that is to say, the pole-pieces of one set occur in line with the spaces between those of the other sets.

The armature D is of cylindrical form, and is also laminated in the usual way and is wound longitudinally with coils closed upon themselves. The pole-pieces c are connected or shunted by bridge-pieces E. These may be made independently and attached to the pole-pieces, or they may be parts of the forms or blanks stamped or punched out of sheet-iron. Their size or mass is determined by various conditions, such as the strength of the current to be employed, the mass or size of the cores to which they are applied, and other familiar conditions.

Coils F surround the pole-pieces B, and other coils G are wound on the pole-pieces c. These coils are connected in series in two circuits, which are branches of a circuit from a generator of alternating currents, and they may be so wound, or the respective circuits in which they are included may be so arranged, that the circuit of coils G will have, independently of the particular construction described, a higher self-induction than the other circuit or branch.

The function of the shunts or bridges E is that they shall form with the cores c a closed magnetic circuit for a current up to a predetermined strength, so that when saturated by such current and unable to carry more lines of force than such a current produces they will to no further appreciable extent interfere with the development, by a stronger current, of free magnetic poles at the ends of the cores c.

In such a motor the current is so retarded in the coils G, and the manifestation of the free magnetism in the poles c is so delayed beyond the period of maximum magnetic effect in poles B, that a strong torque is produced and the motor operates with approximately the power developed in a motor of this kind energized by independently generated currents differing by a full quarter phase.

CHAPTER XIV

TYPE OF TESLA SINGLE-PHASE MOTOR

Up to this point, two principal types of Tesla motors have been described: First, those containing two or more energizing circuits through which are caused to pass alternating currents differing from one another in phase to an extent sufficient to produce a continuous progression or shifting of the poles or points of greatest magnetic effect, in obedience to which the movable element of the motor is maintained in rotation; second, those containing poles, or parts of different magnetic susceptibility, which under the energizing influence of the same current or two currents coinciding in phase will exhibit differences in their magnetic periods or phases. In the first class of motors the torque is due to the magnetism established in different portions of the motor by currents from the same or from independent sources, and exhibiting time differences in phase. In the second class the torque results from the energizing effects of a current upon different parts of the motor which differ in magnetic susceptibility—in other words, parts which respond in the same relative degree to the action of a current, not simultaneously, but after different intervals of time.

FIG. 62.   FIG. 63.   FIG. 64.

In another Tesla motor, however, the torque, instead of being solely the result of a time difference in the magnetic periods or phases of the poles or attractive parts to whatever cause due, is produced by an angular displacement of the parts which, though movable with respect to one another, are magnetized simultaneously, or approximately so, by the same currents. This principle of operation has been embodied practically in a motor in which the necessary angular displacement between the points of greatest magnetic attraction in the two elements of the motor—the armature and field—is obtained by the direction of the lamination of the magnetic cores of the elements.

Fig. 62 is a side view of such a motor with a portion of its armature core exposed. Fig. 63 is an end or edge view of the same. Fig. 64 is a central cross-section of the same, the armature being shown mainly in elevation.

Let A A designate two plates built up of thin sections or laminae of soft iron insulated more or less from one another and held together by bolts a and secured to a base B. The inner faces of these plates contain recesses or grooves in which a coil or coils D are secured obliquely to the direction of the laminations. Within the coils D is a disc E, preferably composed of a spirally-wound iron wire or ribbon or a series of concentric rings and mounted on a shaft F, having bearings in the plates A A. Such a device when acted upon by an alternating current is capable of rotation and constitutes a motor, the operation of which may be explained in the following manner: A current or current-impulse traversing the coils D tends to magnetize the cores A A and E, all of which are within the influence of the field of the coils. The poles thus established would naturally lie in the same line at right angles to the coils D, but in the plates A they are deflected by reason of the direction of the laminations, and appear at or near the extremities of these plates. In the disc, however, where these conditions are not present, the poles or points of greatest attraction are on a line at right angles to the plane of the coils; hence there will be a torque established by this angular displacement of the poles or magnetic lines, which starts the disc in rotation, the magnetic lines of the armature and field tending toward a position of parallelism. This rotation is continued and maintained by the reversals of the current in coils D D, which change alternately the polarity of the field-cores A A. This rotary tendency or effect will be greatly increased by winding the disc with conductors G, closed upon themselves and having a radial direction, whereby the magnetic intensity of the poles of the disc will be greatly increased by the energizing effect of the currents induced in the coils G by the alternating currents in coils D.

The cores of the disc and field may or may not be of different magnetic susceptibility—that is to say, they may both be of the same kind of iron, so as to be magnetized at approximately the same instant by the coils D; or one may be of soft iron and the other of hard, in order that a certain time may elapse between the periods of their magnetization. In either case rotation will be produced; but unless the disc is provided with the closed energizing coils it is desirable that the above-described difference of magnetic susceptibility be utilized to assist in its rotation.

The cores of the field and armature may be made in various ways, as will be well understood, it being only requisite that the laminations in each be in such direction as to secure the necessary angular displacement of the points of greatest attraction. Moreover, since the disc may be considered as made up of an infinite number of radial arms, it is obvious that what is true of a disc holds for many other forms of armature.

CHAPTER XV

MOTORS WITH CIRCUITS OF
DIFFERENT RESISTANCE

As has been pointed out elsewhere, the lag or retardation of the phases of an alternating current is directly proportional to the self-induction and inversely proportional to the resistance of the circuit through which the current flows. Hence, in order to secure the proper differences of phase between the two motor-circuits, it is desirable to make the self-induction in one much higher and the resistance much lower than the self-induction and resistance, respectively, in the other. At the same time the magnetic quantities of the two poles or sets of poles which the two circuits produce should be approximately equal. These requirements have led Mr. Tesla to the invention of a motor having the following general characteristics: The coils which are included in that energizing circuit which is to have the higher self-induction are made of coarse wire, or a conductor of relatively low resistance, and with the greatest possible length or number of turns. In the other set of coils a comparatively few turns of finer wire are used, or a wire of higher resistance. Furthermore, in order to approximate the magnetic quantities of the poles excited by these coils, Mr. Tesla employs in the self-induction circuit cores much longer than those in the other or resistance circuit.

Fig. 65 is a part sectional view of the motor at right angles to the shaft. Fig. 66 is a diagram of the field circuits.

In Fig. 66, let A represent the coils in one motor circuit, and B those in the other. The circuit A is to have the higher self-induction. There are, therefore, used a long length or a large number of turns of coarse wire in forming the coils of this circuit. For the circuit B, a smaller conductor is employed, or a conductor of a higher resistance than copper, such as German silver or iron, and the coils are wound with fewer turns. In applying these coils to a motor, Mr. Tesla builds up a field-magnet of plates C, of iron and steel, secured together in the usual manner by bolts D. Each plate is formed with four (more or less) long cores E, around which is a space to receive the coil and an equal number of short projections F to receive the coils of the resistance-circuit. The plates are generally annular in shape, having an open space in the center for receiving the armature G, which Mr. Tesla prefers to wind with closed coils. An alternating current divided between the two circuits is retarded as to its phases in the circuit A to a much greater extent than in the

FIG. 65.   FIG. 66.

circuit B. By reason of the relative sizes and disposition of the cores and coils the magnetic effect of the poles E and F upon the armature closely approximate.

An important result secured by the construction shown here is that these coils which are designed to have the higher self-induction are almost completely surrounded by iron, and that the retardation is thus very materially increased.

CHAPTER XVI

MOTOR WITH EQUAL MAGNETIC ENERGIES IN FIELD AND ARMATURE

L et it be assumed that the energy as represented in the magnetism in the field of a given rotating field motor is ninety and that of the armature ten. The sum of these quantities, which represents the total energy expended in driving the motor, is one hundred; but, assuming that the motor be so constructed that the energy in the field is represented by fifty, and that in the armature by fifty, the sum is still one hundred; but while in the first instance the product is nine hundred, in the second it is two thousand five hundred, and as the energy developed is in proportion to these products it is clear that those motors are the most efficient—other things being equal—in which the magnetic energies developed in the armature and field are equal. These results Mr. Tesla obtains by using the same amount of copper or ampere turns in both elements when the cores of both are equal, or approximately so, and the same current energizes both; or in cases where the currents in one element are induced to those of the other he uses in the induced coils an excess of copper over that in the primary element or conductor.

FIG. 67.

The conventional figure of a motor here introduced, Fig. 67, will give an idea of the solution furnished by Mr. Tesla for the specific problem. Referring to the drawing, A is the field-magnet, B the armature, c the field coils, and D the armature-coils of the motor.

Generally speaking, if the mass of the cores of armature and field be equal, the amount of copper or ampere turns of the energizing coils on both should also be equal; but these conditions will be modified in different forms of machine. It will be understood that these results are most advantageous when existing under the conditions presented where the motor is running with its normal load, a point to be well borne in mind.

CHAPTER XVII

MOTORS WITH COINCIDING
MAXIMA OF MAGNETIC EFFECT
IN ARMATURE AND FIELD

In this form of motor, Mr. Tesla's object is to design and build machines wherein the maxima of the magnetic effects of the armature and field will more nearly coincide than in some of the types previously under consideration. These types are: First, motors having two or more energizing circuits of the same electrical character, and in the operation of which the currents used differ primarily in phase; second, motors with a plurality of energizing circuits of different electrical character, in or by means of which the difference of phase is produced artificially, and, third, motors with a plurality of energizing circuits, the currents in one being induced from currents in another. Considering the structural and operative conditions of any one of them—as, for example, that first named—the armature which is mounted to rotate in obedience to the co-operative influence or action of the energizing circuits has coils wound upon it which are closed upon themselves and in which currents are induced by the energizing-currents with the object and result of energizing the armature-core; but under any such conditions as must exist in these motors, it is obvious that a certain time must elapse between the manifestations of an energizing current impulse in the field coils, and the corresponding magnetic state or phase in the armature established by the current induced thereby; consequently a given magnetic influence or effect in the field which is the direct result of a primary current impulse will have become more or less weakened or lost before the corresponding effect in the armature indirectly produced has reached its maximum. This is a condition unfavorable to efficient working in certain cases— as, for instance, when the progress of the resultant poles or points of maximum attraction is very great, or when a very high number of alternations is employed—for it is apparent that a stronger tendency to rotation will be maintained if the maximum magnetic attractions or conditions in both armature and field coincide, the energy developed by a motor being measured by the product of the magnetic quantities of the armature and field.

FIG. 68.

To secure this coincidence of maximum magnetic effects, Mr. Tesla has devised various means, as explained below. Fig. 68 is a diagrammatic illustration of a Tesla motor system in which the alternating currents proceed from independent sources and differ primarily in phase.

A designates the field-magnet or magnetic frame of the motor; B B, oppositely located pole-pieces adapted to receive the coils of one energizing circuit; and c c, similar pole-pieces for the coils of the other energizing circuit. These circuits are designated, respectively, by D E, the conductor D″ forming a common return to the generator G. Between these poles is mounted an armature—for example, a ring or annular armature, wound with a series of coils F, forming a closed circuit or circuits. The action or operation of a motor thus constructed is now well understood. It will be observed, however, that the magnetism of poles B, for example, established by a current impulse in the coils thereon, precedes the magnetic effect set up in the armature by the induced current in coils F. Consequently the mutual attraction between the armature and field-poles is considerably reduced. The same conditions will be found to exist if, instead of assuming the poles B or c as acting independently, we regard the ideal resultant of both acting together, which is the real condition. To remedy this, the motor field is constructed with secondary poles B′ c′, which are situated between the others. These pole-pieces are wound with coils D′ E′, the former in derivation to the coils D, the latter to coils E. The main or primary coils D and E are wound for a different self-induction from that of the coils D′ and E′, the relations being so fixed that if the currents in D and E differ, for example, by a quarter-phase, the currents in each secondary coil, as D′ E′, will differ from those in its appropriate primary D or E by, say, forty-five degrees, or one-eighth of a period.

Now, assuming that an impulse or alternation in circuit or branch E is just beginning, while in the branch D it is just falling from maximum, the conditions are those of a quarter-phase difference. The ideal resultant of the attractive forces of the two sets of poles B c therefore may be considered as progressing from poles B to poles c, while the impulse in E is rising to maximum, and that in D is falling to zero or minimum. The polarity set up in the armature, however, lags behind the manifestations of field magnetism, and hence the maximum points of attraction in armature and field, instead of coinciding, are angularly displaced. This effect is counteracted by the supplemental poles B′ c′. The magnetic phases of these poles succeed those of poles B c by the same, or nearly the same, period of time as elapses between the effect of the poles B c and the corresponding induced effect in the armature; hence the magnetic conditions of poles B′ c′ and of the armature more nearly coincide and a better result is obtained. As poles B′ c′ act in conjunction with the poles in the armature established by poles B c, so in turn poles c B act similarly with the poles set up by B′ c′, respectively. Under such conditions the retardation of the magnetic effect of the armature and that of the secondary poles will bring the maximum of the two more nearly into coincidence and a correspondingly stronger torque or magnetic attraction secured.

In such a disposition as is shown in Fig. 68 it will be observed that as the adjacent pole-pieces of either circuit are of like polarity they will have a certain weakening effect upon one another. Mr. Tesla therefore prefers to remove the secondary poles from the direct influence of the others. This may be done by constructing a motor with two independent sets of fields, and with either one or two armatures electrically connected, or by using two armatures and one field. These modifications are illustrated further on.

FIG. 69.

Fig. 69 is a diagrammatic illustration of a motor and system in which the difference of phase is artificially produced. There are two coils D D

Fig. 70.

in one branch and two coils E E in another branch of the main circuit from the generator G. These two circuits or branches are of different self-induction, one, as D, being higher than the other. This is graphically indicated by making coils D much larger than coils E. By reason of the difference in the electrical character of the two circuits, the phases of current in one are retarded to a greater extent than the other. Let this difference be thirty degrees. A motor thus constructed will rotate under the action of an alternating current; but as happens in the case previously described the corresponding magnetic effects of the armature and field do not coincide owing to the time that elapses between a given magnetic effect in the armature and the condition of the field that produces it. The secondary or supplemental poles B′ C′ are therefore availed of. There being thirty degrees difference of phase between the currents in coils D E, the magnetic effect of poles B′ C′ should correspond to that produced by a current differing from the current in coils D or E by fifteen degrees. This we can attain by winding each supplemental pole B′ C′ with two coils H H′. The coils H are included in a derived circuit having the same self-induction as circuit D, and coils H′ in a circuit having the same self-induction as circuit E, so that if these circuits differ by thirty degrees the magnetism of poles B′ C′ will correspond to that produced by a current differing from that in either D or E by fifteen degrees. This is true in all other cases. For example, if in Fig. 68 the coils D′ E′ be replaced by the coils H H′ included in the derived circuits, the magnetism of the poles B′ C′ will correspond in effect or phase, if it may be so termed, to that produced by a current differing from that in either circuit D or E by forty-five degrees, or one-eighth of a period.

FIG. 71.

This invention as applied to a derived circuit motor is illustrated in Figs. 70 and 71. The former is an end view of the motor with the armature in section and a diagram of connections, and Fig. 71 a vertical section through the field. These figures are also drawn to show one of the dispositions of two fields that may be adopted in carrying out the principle. The poles B B C C are in one field, the remaining poles in the other. The former are wound with primary coils I J and secondary coils I′ J′, the latter with coils K L. The primary coils I J are in derived circuits, between which, by reason of their different self-induction, there is a difference of phase, say, of thirty degrees. The coils ɪ′ ᴋ are in circuit with one another, as also are coils ᴊ′ ʟ, and there should be a difference of phase between the currents in coils ᴋ and ʟ and their corresponding primaries of, say, fifteen degrees. If the poles ʙ ᴄ are at right angles, the armature-coils should be connected directly across, or a single armature core wound from end to end may be used; but if the poles ʙ ᴄ be in line there should be an angular displacement of the armature coils, as will be well understood.

The operation will be understood from the foregoing. The maximum magnetic condition of a pair of poles, as ʙ′ ʙ′, coincides closely with the maximum effect in the armature, which lags behind the corresponding condition in poles ʙ ʙ.

CHAPTER XVIII

MOTOR BASED ON THE DIFFERENCE OF PHASE IN THE MAGNETIZATION OF THE INNER AND OUTER PARTS OF AN IRON CORE

It is well known that if a magnetic core, even if laminated or subdivided, be wound with an insulated coil and a current of electricity be directed through the coil, the magnetization of the entire core does not immediately ensue, the magnetizing effect not being exhibited in all parts simultaneously. This may be attributed to the fact that the action of the current is to energize first those laminae or parts of the core nearest the surface and adjacent to the exciting-coil, and from thence the action progresses toward the interior. A certain interval of time therefore elapses between the manifestation of magnetism in the external and the internal sections or layers of the core. If the core be thin or of small mass, this effect may be inappreciable; but in the case of a thick core, or even of a comparatively thin one, if the number of alternations or rate of change of the current strength be very great, the time interval occurring between the manifestations of magnetism in the interior of the core and in those parts adjacent to the coil is more marked. In the construction of such apparatus as motors which are designed to be run by alternating or equivalent currents—such as pulsating or undulating currents generally—Mr. Tesla found it desirable and even necessary to give due consideration to this phenomenon and to make special provisions in order to obviate its consequences. With the specific object of taking advantage of this action or effect, and to render it more pronounced, he constructs a field magnet in which the parts of the core or cores that exhibit at different intervals of time the magnetic effect imparted to them by alternating or equivalent currents in an energizing coil or coils, are so placed with relation to a rotating armature as to exert thereon their attractive effect successively in the order of their magnetization. By this means he secures a result similar to that which he had previously attained in other forms or types of motor in which by means of one or more alternating currents he has produced the rotation or progression of the magnetic poles.

FIGS. 72 and 73.

This new mode of operation will now be described. Fig. 72 is a side elevation of such motor. Fig. 73 is a side elevation of a more practicable and efficient embodiment of the invention. Fig. 74 is a central vertical section of the same in the plane of the axis of rotation.

FIG. 74.

Referring to Fig. 72, let x represent a large iron core, which may be composed of a number of sheets or laminae of soft iron or steel. Surrounding this core is a coil y, which is connected with a source e of rapidly varying currents. Let us consider now the magnetic conditions existing in this core at any point, as *b*, at or near the center, and any other point, as *a*, nearer the surface. When a current impulse is started in the magnetizing coil y, the section or part at *a*, being close to the coil, is immediately energized, while the section or part at *b*, which, to use a convenient expression, is "protected" by the intervening sections or layers between *a* and *b*, does not at once exhibit its magnetism. However, as the magnetization of a increases, *b* becomes also affected, reaching finally its maximum strength some time later than *a*. Upon the weakening of the current the magnetization of a first diminishes, while *b* still exhibits its maximum strength; but the continued weakening of a is attended by a subsequent weakening of *b*. Assuming the current to be an alternating one, a will now be reversed, while *b* still continues of the first imparted polarity. This action continues the magnetic condition of *b*, following that of a in the manner above described. If an armature—for instance, a simple disc F, mounted to rotate freely on an axis—be brought into proximity to the core, a movement of rotation will be imparted to the disc, the direction depending upon its position relatively to the core, the tendency being to turn the portion of the disc nearest to the core from a to *b*, as indicated in Fig. 72.

This action or principle of operation has been embodied in a practicable form of motor, which is illustrated in Fig. 73. Let a in that figure represent a circular frame of iron, from diametrically opposite points of the interior of which the cores project. Each core is composed of three main parts B, B and c, and they are similarly formed with a straight portion or body *e*, around which the energizing coil is wound, a curved arm or extension *c*, and an inwardly projecting pole or end d. Each core is made up of two parts B B, with their polar extensions reaching in one direction, and a part *c* between the other two, and with its polar extension reaching in the opposite direction. In order to lessen in the cores the circulation of currents induced therein, the several sections are insulated from one another in the manner usually followed in such cases. These cores are wound with coils D, which are connected in the same circuit, either in parallel or series, and supplied with an alternating or a pulsating current, preferably the former, by a generator E, represented diagrammatically. Between the cores or their polar extensions is mounted a cylindrical or similar armature F, wound with magnetizing coils G, closed upon themselves.

The operation of this motor is as follows: When a current impulse or alternation is directed through the coils D, the sections B B of the cores, being on the surface and in close proximity to the coils, are immediately energized. The sections C, on the other hand, are protected from the magnetizing influence of the coil by the interposed layers of iron B B. As the magnetism of B B increases, however, the sections c are also energized; but they do not attain their maximum strength until a certain time subsequent to the exhibition by the sections B B of their maximum. Upon the weakening of the current the magnetic strength of B B first diminishes, while the sections c have still their maximum strength; but as B B continue to weaken the interior sections are similarly weakened. B B may then begin to exhibit an opposite polarity, which is followed later by a similar change on C, and this action continues. B B and c may therefore be considered as separate field-magnets, being extended so as to act on the armature in the most efficient positions, and the effect is similar to that in the other forms of Tesla motor—viz., a rotation or progression of the maximum points of the field of force. Any armature—such, for instance, as a disc—mounted in this field would rotate from the pole first to exhibit its magnetism to that which exhibits it later.

It is evident that the principle here described may be carried out in conjunction with other means for securing a more favorable or efficient action of the motor. For example, the polar extensions of the sections c may be wound or surrounded by closed coils. The effect of these coils will be to still more effectively retard the magnetization of the polar extensions of C.

Chapter XIX

Another Type of Tesla Induction Motor

It will have been gathered by all who are interested in the advance of the electrical arts, and who follow carefully, step by step, the work of pioneers, that Mr. Tesla has been foremost to utilize inductive effects in permanently closed circuits, in the operation of alternating motors. In this chapter one simple type of such a motor is described and illustrated, which will serve as an exemplification of the principle.

Let it be assumed that an ordinary alternating current generator is connected up in a circuit of practically no self-induction, such, for example, as a circuit containing incandescent lamps only. On the operation of the machine, alternating currents will be developed in the circuit, and the phases of these currents will theoretically coincide with the phases of the impressed electromotive force. Such currents may be regarded and designated as the "unretarded currents."

It will be understood, of course, that in practice there is always more or less self-induction in the circuit, which modifies to a corresponding extent these conditions; but for convenience this may be disregarded in the consideration of the principle of operation, since the same laws apply. Assume next that a path of currents be formed across any two points of the above circuit, consisting, for example, of the primary of an induction device. The phases of the currents passing through the primary, owing to the self-induction of the same, will not coincide with the phases of the impressed electromotive force, but will lag behind, such lag being directly proportional to the self-induction and inversely proportional to the resistance of the said coil. The insertion of this coil will also cause a lagging or retardation of the currents traversing and delivered by the generator behind the impressed electromotive force, such lag being the mean or resultant of the lag of the current through the primary alone and of the "unretarded current" in the entire working circuit. Next consider the conditions imposed by the association in inductive relation with the primary coil, of a secondary coil. The current generated in the secondary coil will react upon the primary current, modifying the retardation of the same, according to the amount of self-induction and resistance in the secondary circuit. If the secondary circuit has but little self-induction—as, for instance, when it contains incandescent lamps only—it will increase the actual difference of phase between its own and the primary current, first, by diminishing the lag between the primary current and the impressed electromotive force, and, second, by its own lag or retardation behind the impressed electromotive force. On the other hand, if the secondary circuit have a high self-induction, its lag behind the current in the primary is directly increased, while it will be still further increased if the primary have a very low self-induction. The better results are obtained when the primary has a low self-induction.

FIG. 75.   FIG. 76.

Fig. 75 is a diagram of a Tesla motor embodying this principle. Fig. 76 is a similar diagram of a modification of the same. In Fig. 75 let A designate the field-magnet of a motor which, as in all these motors, is built up of sections or plates. B C are polar projections upon which the coils are wound. Upon one pair of these poles, as C, are wound primary coils D, which are directly connected to the circuit of an alternating current generator G. On the same poles are also wound secondary coils F, either side by side or over or under the primary coils, and these are connected with other coils E, which surround the poles B B. The currents in both primary and secondary coils in such a motor will be retarded or will lag behind the impressed electromotive force; but to secure a proper difference in phase between the primary and secondary currents themselves, Mr. Tesla increases the resistance of the circuit of the secondary and reduces as much as practicable its self-induction. This is done by using for the secondary circuit, particularly in the coils E, wire of comparatively small diameter and having but few turns around the cores; or by using some conductor of higher specific resistance, such as German silver; or by introducing at some point in the secondary circuit an artificial resistance R. Thus the self-induction of the secondary is kept down and its resistance increased, with the result of decreasing the lag between the impressed electro-motive force and the current in the primary coils and increasing the difference of phase between the primary and secondary currents.

In the disposition shown in Fig. 76, the lag in the secondary is increased by increasing the self-induction of that circuit, while the increasing tendency of the primary to lag is counteracted by inserting therein a dead resistance. The primary coils D in this case have a low self-induction and high resistance, while the coils E F, included in the secondary circuit, have a high self-induction and low resistance. This may be done by the proper winding of the coils; or in the circuit including the secondary coils E F, we may introduce a self-induction coil s, while in the primary circuit from the generator G and including coils D, there may be inserted a dead resistance R. By this means the difference of phase between the primary and secondary is increased. It is evident that both means of increasing the difference of phase—namely, by the special winding as well as by the supplemental or external inductive and dead resistance—may be employed conjointly.

In the operation of this motor the current impulses in the primary coils induce currents in the secondary coils, and by the conjoint action of the two the points of greatest magnetic attraction are shifted or rotated.

In practice it is found desirable to wind the armature with closed coils in which currents are induced by the action thereon of the primaries.

CHAPTER XX

COMBINATIONS OF SYNCHRONIZING MOTOR AND TORQUE MOTOR

In the preceding descriptions relative to synchronizing motors and methods of operating them, reference has been made to the plan adopted by Mr. Tesla, which consists broadly in winding or arranging the motor in such manner that by means of suitable switches it could be started as a multiple-circuit motor, or one operating by a progression of its magnetic poles, and then, when up to speed, or nearly so, converted into an ordinary synchronizing motor, or one in which the magnetic poles were simply alternated. In some cases, as when a large motor is used and when the number of alternations is very high, there is more or less difficulty in bringing the motor to speed as a double or multiple-circuit motor, for the plan of construction which renders the motor best adapted to run as a synchronizing motor impairs its efficiency as a torque or double-circuit motor under the assumed conditions on the start. This will be readily understood, for in a large synchronizing motor the length of the magnetic circuit of the polar projections, and their mass, are so great that apparently considerable time is required for magnetization and demagnetization. Hence with a current of a very high number of alternations the motor may not respond properly. To avoid this objection and to start up a synchronizing motor in which these conditions obtain, Mr. Tesla has combined two motors, one a synchronizing motor, the other a multiple-circuit or torque motor, and by the latter he brings the first-named up to speed, and then either throws the whole current into the synchronizing motor or operates jointly both of the motors.

This invention involves several novel and useful features. It will be observed, in the first place, that both motors are run, without commutators of any kind, and, secondly, that the speed of the torque motor may be higher than that of the synchronizing motor, as will be the case when it contains a fewer number of poles or sets of poles, so that the motor will be more readily and easily brought up to speed. Thirdly, the synchronizing motor may be constructed so as to have a much more pronounced tendency to synchronism without lessening the facility with which it is started.

FIG. 77.

FIG. 78.   FIG. 79.

Fig. 77 is a part sectional view of the two motors; Fig. 78 an end view of the synchronizing motor; Fig. 79 an end view and part section of the torque or double-circuit motor; Fig. 80 a diagram of the circuit connections employed; and Figs. 81, 82, 83, 84 and 85 are diagrams of modified dispositions of the two motors.

Inasmuch as neither motor is doing any work while the current is acting upon the other, the two armatures are rigidly connected, both being mounted upon the same shaft A, the field-magnets B of the synchronizing and c of the torque motor being secured to the same base D. The preferably larger synchronizing motor has polar projections on its armature, which rotate in very close proximity to the poles of the field, and in other respects it conforms to the conditions that are necessary to secure synchronous action. The pole-pieces of the armature are, however, wound with closed coils E, as this obviates the employment of sliding contacts. The smaller or torque motor, on the other hand, has, preferably, a cylindrical armature F, without polar projections and wound with closed coils G. The field-coils of the torque motor are connected up in two series H and I, and the alternating current from the generator is directed through or divided between these two circuits in any manner to produce a progression of the poles or points of maximum magnetic effect. This result is secured by connecting the two motor-circuits in derivation with the circuit from the generator, inserting in one motor circuit a dead resistance and in the other a self-induction coil, by which means a difference in phase between the two divisions of the current is secured. If both motors have the same number of field poles, the torque motor for a given number of alternations will tend to run at double the speed of the other, for, assuming the connections to be such as to give the best results, its poles are divided into two series and the number of poles is virtually reduced one-half, which being acted upon by the same number of alternations tend to rotate the armature at twice the speed. By this means the main armature is more easily brought to or above the required speed. When the speed necessary for synchronism is imparted to the main motor, the current is shifted from the torque motor into the other.

FIG. 80.

A convenient arrangement for carrying out this invention is shown in Fig. 80, in which J J are the field coils of the synchronizing, and H I the field coils of the torque motor. L L' are the conductors of the main line. One end of, say, coils H is connected to wire L through a self-induction coil M. One end of the other set of coils I is connected to the same wire through a dead resistance N. The opposite ends of these two circuits are connected to the contact m of a switch, the handle or lever of which is in connection with the line-wire L'. One end of the field circuit of the synchronizing motor is connected to the wire L. The other terminates in the switch-contact n. From the diagram it will be readily seen that if the lever P be turned upon contact m, the torque motor will start by reason of the difference of phase between the currents in its two energizing circuits. Then when the desired speed is attained, if the lever P be shifted upon contact n the entire current will pass through the field coils of the synchronizing motor and the other will be doing no work.

The torque motor may be constructed and operated in various ways, many of which have already been touched upon. It is not necessary that one motor be cut out of circuit while the other is in, for both may be acted upon by current at the same time, and Mr. Tesla has devised various dispositions or arrangements of the two motors for accomplishing this. Some of these arrangements are illustrated in Figs. 81 to 85.

Referring to Fig. 81, let T designate the torque or multiple circuit motor and S the synchronizing motor, L L' being the line-wires from a source of alternating current. The two circuits of the torque motor of different degrees of self-induction, and designated by N M, are connected in derivation to the wire L. They are then joined and connected to the energizing circuit of the synchronizing motor, the opposite terminal of which is connected to wire L'. The two motors are thus in series. To start them Mr. Tesla short-circuits the synchronizing motor by a switch P', throwing the whole current through the torque motor. Then when the desired speed is reached the switch P' is opened, so that the current passes through both motors. In such an arrangement as this it is obviously desirable for economical and other reasons that a proper relation between the speeds of the two motors should be observed.

In Fig. 82 another disposition is illustrated. S is the synchronizing motor and T the torque motor, the circuits of both being in parallel. W is a circuit also in derivation to the motor circuits and containing a switch P". S' is a switch in the synchronizing motor circuit. On the start the switch

FIG. 81, 82, 83, 84, AND 85.

s′ is opened, cutting out the motor s. Then p″ is opened, throwing the entire current through the motor т, giving it a very strong torque. When the desired speed is reached, switch s′ is closed and the current divides between both motors. By means of switch p″ both motors may be cut out.

In Fig. 83 the arrangement is substantially the same, except that a switch т′ is placed in the circuit which includes the two circuits of the torque motor. Fig. 84 shows the two motors in series, with a shunt around both containing a switch s т. There is also a shunt around the synchronizing motor s, with a switch p′. In Fig. 85 the same disposition is shown; but each motor is provided with a shunt, in which are switches p′ and т″, as shown.

CHAPTER XXI

MOTOR WITH A CONDENSER
IN THE ARMATURE CIRCUIT

We now come to a new class of motors in which resort is had to condensers for the purpose of developing the required difference of phase and neutralizing the effects of self-induction. Mr. Tesla early began to apply the condenser to alternating apparatus, in just how many ways can only be learned from a perusal of other portions of this volume, especially those dealing with his high frequency work.

Certain laws govern the action or effects produced by a condenser when connected to an electric circuit through which an alternating or in general an undulating current is made to pass. Some of the most important of such effects are as follows: First, if the terminals or plates of a condenser be connected with two points of a circuit, the potentials of which are made to rise and fall in rapid succession, the condenser allows the passage, or more strictly speaking, the transference of a current, although its plates or armatures may be so carefully insulated as to prevent almost completely the passage of a current of unvarying strength or direction and of moderate electromotive force. Second, if a circuit, the terminals of which are connected with the plates of the condenser, possess a certain self-induction, the condenser will overcome or counteract to a greater or less degree, dependent upon well-understood conditions, the effects of such self-induction. Third, if two points of a closed or complete circuit through which a rapidly rising and falling current flows be shunted or bridged by a condenser, a variation in the strength of the currents in the branches and also a difference of phase of the currents therein is produced. These effects Mr. Tesla has utilized and applied in a variety of ways in the construction and operation of his motors, such as by producing a difference in phase in the two energizing circuits of an alternating current motor by connecting the two circuits in derivation and connecting up a condenser in series in one of the circuits. A further development, however, possesses certain novel features of practical value and involves a knowledge of facts less generally understood. It comprises the use of a condenser or condensers in connection with the induced or armature circuit of a motor and certain details of the construction of such motors. In an alternating current motor of the type particularly referred to above, or in any other which has an armature coil or circuit closed upon itself, the latter represents not only an inductive resistance, but one which is periodically varying in value, both of which facts complicate and render difficult the attainment of the conditions best suited to the most efficient working conditions; in other words, they require, first, that for a given inductive effect upon the armature there should be the greatest possible current through the armature or induced coils, and, second, that there should always exist between the currents in the energizing and the induced circuits a given relation of phase. Hence whatever tends to decrease the self-induction and increase the current in the induced circuits will, other things being equal, increase the output and efficiency of the motor, and the same will be true of causes that operate to maintain the mutual attractive effect between the field magnets and armature at its maximum. Mr. Tesla secures these results by connecting with the induced circuit or circuits a condenser, in the manner described below, and he also, with this purpose in view, constructs the motor in a special manner.

FIG. 86.

Referring to the drawings, Fig. 86, is a view, mainly diagrammatic, of an alternating current motor, in which the present principle is applied. Fig. 87 is a central section, in line with the shaft, of a special form of armature core. Fig. 88 is a similar section of a modification of the same. Fig. 89 is one of the sections of the core detached. Fig. 90 is a diagram showing a modified disposition of the armature or induced circuits.

Fig. 87.

The general plan of the invention is illustrated in Fig. 86. A A in this figure represent the frame and field magnets of an alternating current motor, the poles or projections of which are wound with coils B and C, forming independent energizing circuits connected either to the same or to independent sources of alternating currents, so that the currents flowing through the circuits, respectively, will have a difference of phase. Within the influence of this field is an armature core D, wound with coils E. In motors of this description heretofore these coils have been closed upon themselves, or connected in a closed series; but in the present case each coil or the connected series of coils terminates in the opposite plates of a condenser F. For this purpose the ends of the series of coils are brought out through the shaft to collecting rings G, which are connected to the condenser by contact brushes H and suitable conductors, the condenser being independent of the machine. The armature coils are wound or connected in such manner that adjacent coils produce opposite poles.

The action of this motor and the effect of the plan followed in its construction are as follows: The motor being started in operation and the coils of the field magnets being traversed by alternating currents, currents are induced in the armature coils by one set of field coils, as B, and the poles thus established are acted upon by the other set, as C. The armature coils, however, have necessarily a high self-induction, which opposes the flow of the currents thus set up. The condenser F not only permits the passage or transference of these currents, but also counteracts the effects of self-induction, and by a proper adjustment of the capacity of the condenser, the self-induction of the coils, and the periods of the currents, the condenser may be made to overcome entirely the effect of self-induction.

It is preferable on account of the undesirability of using sliding contacts of any kind, to associate the condenser with the armature directly, or make it a part of the armature. In some cases Mr. Tesla builds up the armature of annular plates K K, held by bolts L between heads M, which are secured to the driving shaft, and in the hollow space thus formed he places a condenser F, generally by winding the two insulated plates spirally around the shaft. In other cases he utilizes the plates of the core itself as

FIG. 88.   FIG. 89.

the plates of the condenser. For example, in Figs. 88 and 89, N is the driving shaft, M M are the heads of the armature-core, and K K' the iron plates of which the core is built up. These plates are insulated from the shaft and from one another, and are held together by rods or bolts L. The bolts pass through a large hole in one plate and a small hole in the one next adjacent, and so on, connecting electrically all of plates K, as one armature of a condenser, and all of plates K' as the other.

To either of the condensers above described the armature coils may be connected, as explained by reference to Fig. 86.

In motors in which the armature coils are closed upon themselves—as, for example, in any form of alternating current motor in which one armature coil or set of coils is in the position of maximum induction with respect to the field coils or poles, while the other is in the position of minimum induction—the coils are best connected in one series, and two points of the circuit thus formed are bridged by a

Fig. 90.

condenser. This is illustrated in Fig. 90, in which E represents one set of armature coils and E' the other. Their points of union are joined through a condenser F. It will be observed that in this disposition the self-induction of the two branches E and E' varies with their position relatively to the field magnet, and that each branch is alternately the predominating source of the induced current. Hence the effect of the condenser F is twofold. First, it increases the current in each of the branches alternately, and, secondly, it alters the phase of the currents in the branches, this being the well-known effect which results from such a disposition of a condenser with a circuit, as above described. This effect is favorable to the proper working of the motor, because it increases the flow of current in the armature circuits due to a given inductive effect, and also because it brings more nearly into coincidence the maximum magnetic effects of the coacting field and armature poles.

It will be understood, of course, that the causes that contribute to the efficiency of condensers when applied to such uses as the above must be given due consideration in determining the practicability and efficiency of the motors. Chief among these is, as is well known, the periodicity of the current, and hence the improvements described are more particularly adapted to systems in which a very high rate of alternation or change is maintained.

Although this invention has been illustrated in connection with a special form of motor, it will be understood that it is equally applicable to any other alternating current motor in which there is a closed armature coil wherein the currents are induced by the action of the field, and the feature of utilizing the plates or sections of a magnetic core for forming the condenser is applicable, generally, to other kinds of alternating current apparatus.

CHAPTER XXII

MOTOR WITH CONDENSER
IN ONE OF THE FIELD CIRCUITS

If the field or energizing circuits of a rotary phase motor be both derived from the same source of alternating currents and a condenser of proper capacity be included in one of the same, approximately, the desired difference of phase may be obtained between the currents flowing directly from the source and those flowing through the condenser; but the great size and expense of condensers for this purpose that would meet the requirements of the ordinary systems of comparatively low potential are particularly prohibitory to their employment.

Another, now well-known, method or plan of securing a difference of phase between the energizing currents of motors of this kind is to induce by the currents in one circuit those in the other circuit or circuits; but as no means had been proposed that would secure in this way between the phases of the primary or inducing and the secondary or induced currents that difference—theoretically ninety degrees—that is best adapted for practical and economical working, Mr. Tesla devised a means which renders practicable both the above described plans or methods, and by which he is enabled to obtain an economical and efficient alternating current motor. His invention consists in placing a condenser in the secondary or induced circuit of the motor above described and raising the potential of the secondary currents to such a degree that the capacity of the condenser, which is in part dependent on the potential, need be quite small. The value of this condenser is determined in a well-understood manner with reference to the self-induction and other conditions of the circuit, so as to cause the currents which pass through it to differ from the primary currents by a quarter phase.

Fig. 91 illustrates the invention as embodied in a motor in which the inductive relation of the primary and secondary circuits is secured by winding them inside the motor partly upon the same cores; but the

Fig. 91.

invention applies, generally, to other forms of motor in which one of the energizing currents is induced in any way from the other.

Let A B represent the poles of an alternating current motor, of which c is the armature wound with coils D, closed upon themselves, as is now the general practice in motors of this kind. The poles A, which alternate with poles B, are wound with coils of ordinary or coarse wire E in such direction as to make them of alternate north and south polarity, as indicated in the diagram by the characters N S. Over these coils, or in other inductive relation to the same, are wound long fine-wire coils F F, and in the same direction throughout as the coils E. These coils are secondaries, in which currents of very high potential are induced. All the coils E in one series are connected, and all the secondaries F in another.

On the intermediate poles B are wound fine-wire energizing coils G, which are connected in series with one another, and also with the series of secondary coils F, the direction of winding being such that a current-impulse induced from the primary coils E imparts the same magnetism to the poles B as that produced in poles A by the primary impulse. This condition is indicated by the characters N' S'.

In the circuit formed by the two sets of coils F and G is introduced a condenser H; otherwise this circuit is closed upon itself, while the free ends of the circuit of coils E are connected to a source of alternating currents. As the condenser capacity which is needed in any particular motor of this kind is dependent upon the rate of alternation or the potential, or both, its size or cost, as before explained, may be brought within economical limits for use with the ordinary circuits if the potential of the secondary circuit in the motor be sufficiently high. By giving to the condenser proper values, any desired difference of phase between the primary and secondary energizing circuits may be obtained.

Chapter XXIII

Tesla Polyphase Transformer

Applying the polyphase principle to the construction of transformers as well to the motors already noticed, Mr. Tesla has invented some very interesting forms, which he considers free from the defects of earlier and, at present, more familiar forms. In these transformers he provides a series of inducing coils and corresponding induced coils, which are generally wound upon a core closed upon itself, usually a ring of laminated iron.

The two sets of coils are wound side by side or superposed or otherwise placed in well-known ways to bring them into the most effective relations to one another and to the core. The inducing or primary coils wound on the core are divided into pairs or sets by the proper electrical connections, so that while the coils of one pair or set co-operate in fixing the magnetic poles of the core at two given diametrically opposite points, the coils of the other pair or set—assuming, for sake of illustration, that there are but two—tend to fix the poles ninety degrees from such points. With this induction device is used an alternating current generator with coils or sets of coils to correspond with those of the converter, and the corresponding coils of the generator and converter are then connected up in independent circuits. It results from this that the different electrical phases in the generator are attended by corresponding magnetic changes in the converter; or, in other words, that as the generator coils revolve, the points of greatest magnetic intensity in the converter will be progressively shifted or whirled around.

Fig. 92 is a diagrammatic illustration of the converter and the electrical connections of the same. Fig. 93 is a horizontal central cross-section of Fig. 92. Fig. 94 is a diagram of the circuits of the entire system, the generator being shown in section.

Mr. Tesla uses a core, A, which is closed upon itself—that is to say, of an annular cylindrical or equivalent form—and as the efficiency of the apparatus is largely increased by the subdivision of this core, he makes

FIGS. 92 and 93.

it of thin strips, plates, or wires of soft iron electrically insulated as far as practicable. Upon this core are wound, say, four coils, B B B′ B′, used as primary coils, and for which long lengths of comparatively fine wire are employed. Over these coils are then wound shorter coils of coarser wire, c c c′ c′, to constitute the induced or secondary coils. The construction of this or any equivalent form of converter may be carried further, as above pointed out, by inclosing these coils with iron—as, for example, by winding over the coils layers of insulated iron wire.

The device is provided with suitable binding posts, to which the ends of the coils are led. The diametrically opposite coils B B and B′ B′ are connected, respectively, in series, and the four terminals are connected to the binding posts. The induced coils are connected together in any desired manner. For example, as shown in Fig. 94, *c c* may be connected in multiple arc when a quantity current is desired—as for running a group of incandescent lamps—while c' c' may be independently

Fig. 94.

connected in series in a circuit including arc lamps or the like. The generator in this system will be adapted to the converter in the manner illustrated. For example, in the present case there are employed a pair of ordinary permanent or electro-magnets, E E, between which is mounted a cylindrical armature on a shaft, F, and wound with two coils, G G'. The terminals of these coils are connected, respectively, to four insulated contact or collecting rings, H H H' H', and the four line circuit wires L connect the brushes K, bearing on these rings, to the converter in the order shown. Noting the results of this combination, it will be observed that at a given point of time the coil G is in its neutral position and is generating little or no current, while the other coil, G', is in a position where it exerts its maximum effect. Assuming coil G to be connected in circuit with coils B B of the converter, and coil G' with coils B' B', it is evident that the poles of the ring A will be determined by coils B' B' alone; but as the armature of the generator revolves, coil G develops more current and coil G' less, until G reaches its maximum and G' its neutral position. The obvious result will be to shift the poles of the ring A through one-quarter of its periphery. The movement of the coils through the next quarter of a turn—during which coil G′ enters a field of opposite polarity and generates a current of opposite direction and increasing strength, while coil G, in passing from its maximum to its neutral position generates a current of decreasing strength and same direction as before—causes a further shifting of the poles through the second quarter of the ring. The second half-revolution will obviously be a repetition of the same action. By the shifting of the poles of the ring A, a powerful dynamic inductive effect on the coils c c′ is produced. Besides the currents generated in the secondary coils by dynamo-magnetic induction, other currents will be set up in the same coils in consequence of many variations in the intensity of the poles in the ring A. This should be avoided by maintaining the intensity of the poles constant, to accomplish which care should be taken in designing and proportioning the generator and in distributing the coils in the ring A, and balancing their effect. When this is done, the currents are produced by dynamo-magnetic induction only, the same result being obtained as though the poles were shifted by a commutator with an infinite number of segments.

The modifications which are applicable to other forms of converter are in many respects applicable to this, such as those pertaining more particularly to the form of the core, the relative lengths and resistances of the primary and secondary coils, and the arrangements for running or operating the same.

Chapter XXIV

A Constant Current Transformer
with Magnetic Shield Between Coils
of Primary and Secondary

Mr. Tesla has applied his principle of magnetic shielding of parts to the construction also of transformers, the shield being interposed between the primary and secondary coils. In transformers of the ordinary type it will be found that the wave of electromotive force of the secondary very nearly coincides with that of the primary, being, however, in opposite sign. At the same time the currents, both primary and secondary, lag behind their respective electromotive forces; but as this lag is practically or nearly the same in the case of each it follows that the maximum and minimum of the primary and secondary currents will nearly coincide, but differ in sign or direction, provided the secondary be not loaded or if it contain devices having the property of self-induction. On the other hand, the lag of the primary behind the impressed electromotive force may be diminished by loading the secondary with a non-inductive or dead resistance—such as incandescent lamps— whereby the time interval between the maximum or minimum periods of the primary and secondary currents is increased. This time interval, however, is limited, and the results obtained by phase difference in the operation of such devices as the Tesla alternating current motors can only be approximately realized by such means of producing or securing this difference, as above indicated, for it is desirable in such cases that there should exist between the primary and secondary currents, or those which, however produced, pass through the two circuits of the motor, a difference of phase of ninety degrees; or, in other words, the current in one circuit should be a maximum when that in the other circuit is a minimum. To attain to this condition more perfectly, an increased retardation of the secondary current is secured in the following manner: Instead of bringing the primary and secondary coils or circuits of a transformer into the closest possible relations, as has hitherto been done, Mr. Tesla protects in a measure the secondary from the inductive action or effect of the primary by surrounding either the primary or the secondary with a comparatively thin magnetic shield or screen. Under these modified conditions, as long as the primary current has a small value, the shield protects the secondary; but as soon as the primary current has reached a certain strength, which is arbitrarily determined, the protecting magnetic shield becomes saturated and the inductive action upon the secondary begins. It results, therefore, that the secondary current begins to flow at a certain fraction of a period later than it would without the interposed shield, and since this retardation may be obtained without necessarily retarding the primary current also, an additional lag is secured, and the time interval between the maximum or minimum periods of the primary and secondary currents is increased. Such a transformer may, by properly proportioning its several elements and determining the proper relations between the primary and secondary windings, the thickness of the magnetic shield, and other conditions, be constructed to yield a constant current at all loads.

FIG. 95.

Fig. 95 is a cross-section of a transformer embodying this improvement. Fig. 96 is a similar view of a modified form of transformer, showing diagrammatically the manner of using the same.

A A is the main core of the transformer, composed of a ring of soft annealed and insulated or oxidized iron wire. Upon this core is wound the secondary circuit or coil B B. This latter is then covered with a layer or layers of annealed and insulated iron wires c c, wound in a direction at right angles to the secondary coil. Over the whole is then wound the primary coil or wire D D. From the nature of this construction it will be obvious that as long as the shield formed by the wires *c* is below magnetic saturation the secondary coil or circuit is effectually protected or shielded from the inductive influence of the primary, although on open circuit it may exhibit some electromotive force. When the strength of the primary reaches a certain value, the shield C, becoming saturated, ceases to protect the secondary from inductive action, and current is in consequence developed therein. For similar reasons, when the primary current weakens, the weakening of the secondary is retarded to the same or approximately the same extent.

Fig. 96.

The specific construction of the transformer is largely immaterial. In Fig. 96, for example, the core A is built up of thin insulated iron plates or discs. The primary circuit D is wound next the core A. Over this is applied the shield C, which in this case is made up of thin strips or plates of iron properly insulated and surrounding the primary, forming a closed magnetic circuit. The secondary B is wound over the shield *c*. In Fig. 96, also, E is a source of alternating or rapidly changing currents. The primary of the transformer is connected with the circuit of the generator. F is a two-circuit alternating current motor, one of the circuits being connected with the main circuit from the source E, and the other being supplied with currents from the secondary of the transformer.

Part II

The Tesla Effects with High Frequency and High Potential Currents

Chapter XXV

Introduction—
The Scope of the Tesla Lectures

Before proceeding to study the three Tesla lectures here presented, the reader may find it of some assistance to have his attention directed to the main points of interest and significance therein. The first of these lectures was delivered in New York, at Columbia College, before the American Institute of Electrical Engineers, May 20, 1891. The urgent desire expressed immediately from all parts of Europe for an opportunity to witness the brilliant and unusual experiments with which the lecture was accompanied, induced Mr. Tesla to go to England early in 1892, when he appeared before the Institution of Electrical Engineers, and a day later, by special request, before the Royal Institution. His reception was of the most enthusiastic and flattering nature on both occasions. He then went, by invitation, to France, and repeated his novel demonstrations before the Société Internationale des Electriciens, and the Société Française de Physique. Mr. Tesla returned to America in the fall of 1892, and in February, 1893, delivered his third lecture before the Franklin Institute of Philadelphia, in fulfilment of a long standing promise to Prof. Houston. The following week, at the request of President James I. Ayer, of the National Electric Light Association, the same lecture was re-delivered in St. Louis. It had been intended to limit the invitations to members, but the appeals from residents in the city were so numerous and pressing that it became necessary to secure a very large hall. Hence it came about that the lecture was listened to by an audience of over 5,000 people, and was in some parts of a more popular nature than either of its predecessors. Despite this concession to the need of the hour and occasion, Mr. Tesla did not hesitate to show many new and brilliant experiments, and to advance the frontier of discovery far beyond any point he had theretofore marked publicly.

We may now proceed to a running review of the lectures themselves. The ground covered by them is so vast that only the leading ideas and experiments can here be touched upon; besides, it is preferable that the lectures should be carefully gone over for their own sake, it being more than likely that each student will discover a new beauty or stimulus in them. Taking up the course of reasoning followed by Mr. Tesla in his first lecture, it will be noted that he started out with the recognition of the fact, which he has now experimentally demonstrated, that for the production of light waves, primarily, electrostatic effects must be brought into play, and continued study has led him to the opinion that all electrical and magnetic effects may be referred to electrostatic molecular forces. This opinion finds a singular confirmation in one of the most striking experiments which he describes, namely, the production of a veritable flame by the agitation of electrostatically charged molecules. It is of the highest interest to observe that this result points out a way of obtaining a flame which consumes no material and in which no chemical action whatever takes place. It also throws a light on the nature of the ordinary flame, which Mr. Tesla believes to be due to electrostatic molecular actions, which, if true, would lead directly to the idea that even chemical affinities might be electrostatic in their nature and that, as has already been suggested, molecular forces in general may be referable to one and the same cause. This singular phenomenon accounts in a plausible manner for the unexplained fact that buildings are frequently set on fire during thunder storms without having been at all struck by lightning. It may also explain the total disappearance of ships at sea.

One of the striking proofs of the correctness of the ideas advanced by Mr. Tesla is the fact that, notwithstanding the employment of the most powerful electromagnetic inductive effects, but feeble luminosity is obtainable, and this only in close proximity to the source of disturbance; whereas, when the electrostatic effects are intensified, the same initial energy suffices to excite luminosity at considerable distances from the source. That there are only electrostatic effects active seems to be clearly proved by Mr. Tesla's experiments with an induction coil operated with alternating currents of very high frequency. He shows how tubes may be made to glow brilliantly at considerable distances from any object when placed in a powerful, rapidly alternating, electrostatic field, and he describes many interesting phenomena observed in such a field. His experiments open up the possibility of lighting an apartment by simply creating in it such an electrostatic field, and this, in a certain way, would appear to be the ideal method of lighting a room, as it would allow the illuminating device to be freely moved about. The power with which these exhausted tubes, devoid of any electrodes, light up is certainly remarkable.

That the principle propounded by Mr. Tesla is a broad one is evident from the many ways in which it may be practically applied. We need only refer to the variety of the devices shown or described, all of which are novel in character and will, without doubt, lead to further important results at the hands of Mr. Tesla and other investigators. The experiment, for instance, of lighting up a single filament or block of refractory material with a single wire, is in itself sufficient to give Mr. Tesla's work the stamp of originality, and the numerous other experiments and effects which may be varied at will, are equally new and interesting. Thus, the incandescent filament spinning in an unexhausted globe, the well-known Crookes experiment on open circuit, and the many others suggested, will not fail to interest the reader. Mr. Tesla has made an exhaustive study of the various forms of the discharge presented by an induction coil when operated with these rapidly alternating currents, starting from the thread-like discharge and passing through various stages to the true electric flame.

A point of great importance in the introduction of high tension alternating current which Mr. Tesla brings out is the necessity of carefully avoiding all gaseous matter in the high tension apparatus. He shows that, at least with very rapidly alternating currents of high potential, the discharge may work through almost any practicable thickness of the best insulators, if air is present. In such cases the air included within the apparatus is violently agitated and by molecular bombardment the parts may be so greatly heated as to cause a rupture of the insulation. The practical outcome of this is, that, whereas with steady currents, any kind of insulation may be used, with rapidly alternating currents oils will probably be the best to employ, a fact which has been observed, but not until now satisfactorily explained. The recognition of the above fact is of special importance in the construction of the costly commercial induction coils which are often rendered useless in an unaccountable manner. The truth of these views of Mr. Tesla is made evident by the interesting experiments illustrative of the behavior of the air between charged surfaces, the luminous streams formed by the charged molecules appearing even when great thicknesses of the best insulators are interposed between the charged surfaces. These luminous streams afford in themselves a very interesting study for the experimenter. With these rapidly alternating currents they become far more powerful and produce beautiful light effects when they issue from a wire, pinwheel or other object attached to a terminal of the coil; and it is interesting to note that they issue from a ball almost as freely as from a point, when the frequency is very high.

From these experiments we also obtain a better idea of the importance of taking into account the capacity and self-induction in the apparatus employed and the possibilities offered by the use of condensers in conjunction with alternate currents, the employment of currents of high frequency, among other things, making it possible to reduce the condenser to practicable dimensions. Another point of interest and practical bearing is the fact, proved by Mr. Tesla, that for alternate currents, especially those of high frequency, insulators are required possessing a small specific inductive capacity, which at the same time have a high insulating power.

Mr. Tesla also makes interesting and valuable suggestion in regard to the economical utilization of iron in machines and transformers. He shows how, by maintaining by continuous magnetization a flow of lines through the iron, the latter may be kept near its maximum permeability and a higher output and economy may be secured in such apparatus. This principle may prove of considerable commercial importance in the development of alternating systems. Mr. Tesla's suggestion that the same result can be secured by heating the iron by hysteresis and eddy currents, and increasing the permeability in this manner, while it may appear less practical, nevertheless opens another direction for investigation and improvement.

The demonstration of the fact that with alternating currents of high frequency, sufficient energy may be transmitted under practicable conditions through the glass of an incandescent lamp by electrostatic or electromagnetic induction may lead to a departure in the construction of such devices. Another important experimental result achieved is the operation of lamps, and even motors, with the discharges of condensers, this method affording a means of converting direct or alternating currents. In this connection Mr. Tesla advocates the perfecting of apparatus capable of generating electricity of high tension from heat energy, believing this to be a better way of obtaining electrical energy for practical purposes, particularly for the production of light.

While many were probably prepared to encounter curious phenomena of impedance in the use of a condenser discharged disruptively, the experiments shown were extremely interesting on account of their paradoxical character. The burning of an incandescent lamp at any candle power when connected across a heavy metal bar, the existence of nodes on the bar and the possibility of exploring the bar by means of an ordinary Cardew voltmeter, are all peculiar developments, but perhaps the most interesting observation is the phenomenon of impedance observed in the lamp with a straight filament, which remains dark while the bulb glows.

Mr. Tesla's manner of operating an induction coil by means of the disruptive discharge, and thus obtaining enormous differences of potential from comparatively small and inexpensive coils, will be appreciated by experimenters and will find valuable application in laboratories. Indeed, his many suggestions and hints in regard to the construction and use of apparatus in these investigations will be highly valued and will aid materially in future research.

The London lecture was delivered twice. In its first form, before the Institution of Electrical Engineers, it was in some respects an amplification of several points not specially enlarged upon in the New York lecture, but brought forward many additional discoveries and new investigations. Its repetition, in another form, at the Royal Institution, was due to Prof. Dewar, who with Lord Rayleigh, manifested a most lively interest in Mr. Tesla's work, and whose kindness illustrated once more the strong English love of scientific truth and appreciation of its votaries. As an indefatigable experimenter, Mr. Tesla was certainly nowhere more at home than in the haunts of Faraday, and as the guest of Faraday's successor. This Royal Institution lecture summed up the leading points of Mr. Tesla's work, in the high potential, high frequency field, and we may here avail ourselves of so valuable a summarization, in a simple form, of a subject by no means easy of comprehension until it has been thoroughly studied.

In these London lectures, among the many notable points made was first, the difficulty of constructing the alternators to obtain the very high frequencies needed. To obtain the high frequencies it was necessary to provide several hundred polar projections, which were necessarily small and offered many drawbacks, and this the more as exceedingly high peripheral speeds had to be resorted to. In some of the first machines both armature and field had polar projections. These machines produced a curious noise, especially when the armature was started from the state of rest, the field being charged. The most efficient machine was found to be one with a drum armature, the iron body of which consisted of very thin wire annealed with special care. It was, of course, desirable to avoid the employment of iron in the armature, and several machines of this kind, with moving or stationary conductors were constructed, but the results obtained were not quite satisfactory, on account of the great mechanical and other difficulties encountered.

The study of the properties of the high frequency currents obtained from these machines is very interesting, as nearly every experiment discloses something new. Two coils traversed by such a current attract or repel each other with a force which, owing to the imperfection of our sense of touch, seems continuous. An interesting observation, already noted under another form, is that a piece of iron, surrounded by a coil through which the current is passing appears to be continuously magnetized. This apparent continuity might be ascribed to the deficiency of the sense of touch, but there is evidence that in currents of such high frequencies one of the impulses preponderates over the other.

As might be expected, conductors traversed by such currents are rapidly heated, owing to the increase of the resistance, and the heating effects are relatively much greater in the iron. The hysteresis losses in iron are so great that an iron core, even if finely subdivided, is heated in an incredibly short time. To give an idea of this, an ordinary iron wire $1/16$ inch in diameter inserted within a coil having 250 turns, with a current estimated to be five amperes passing through the coil, becomes within two seconds' time so hot as to scorch wood. Beyond a certain frequency, an iron core, no matter how finely subdivided, exercises a dampening effect, and it was easy to find a point at which the impedance of a coil was not affected by the presence of a core consisting of a bundle of very thin well annealed and varnished iron wires.

Experiments with a telephone, a conductor in a strong magnetic field, or with a condenser or arc, seem to afford certain proof that sounds far above the usually accepted limit of hearing would be perceived if produced with sufficient power. The arc produced by these currents possesses several interesting features. Usually it emits a note the pitch of which corresponds to twice the frequency of the current, but if the frequency be sufficiently high it becomes noiseless, the limit of audition being determined principally by the linear dimensions of the arc. A curious feature of the arc is its persistency, which is due partly to the inability of the gaseous column to cool and increase considerably in resistance, as is the case with low frequencies, and partly to the tendency of such a high frequency machine to maintain a constant current.

In connection with these machines the condenser affords a particularly interesting study. Striking effects are produced by proper adjustments of capacity and self-induction. It is easy to raise the electromotive force of the machine to many times the original value by simply adjusting the capacity of a condenser connected in the induced circuit. If the condenser be at some distance from the machine, the difference of potential on the terminals of the latter may be only a small fraction of that on the condenser.

But the most interesting experiences are gained when the tension of the currents from the machine is raised by means of an induction coil. In consequence of the enormous rate of change obtainable in the primary current, much higher potential differences are obtained than with coils operated in the usual ways, and, owing to the high frequency, the secondary discharge possesses many striking peculiarities. Both the electrodes behave generally alike, though it appears from some observations that one current impulse preponderates over the other, as before mentioned.

The physiological effects of the high tension discharge are found to be so small that the shock of the coil can be supported without any inconvenience, except perhaps a small burn produced by the discharge upon approaching the hand to one of the terminals. The decidedly smaller physiological effects of these currents are thought to be due either to a different distribution through the body or to the tissues acting as condensers. But in the case of an induction coil with a great many turns the harmlessness is principally due to the fact that but little energy is available in the external circuit when the same is closed through the experimenter's body, on account of the great impedance of the coil.

In varying the frequency and strength of the currents through the primary of the coil, the character of the secondary discharge is greatly varied, and no less than five distinct forms are observed: A weak, sensitive thread discharge, a powerful flaming discharge, and three forms of brush or streaming discharges. Each of these possesses certain noteworthy features, but the most interesting to study are the latter.

Under certain conditions the streams, which are presumably due to the violent agitation of the air molecules, issue freely from all points of the coil, even through a thick insulation. If there is the smallest air space between the primary and secondary, they will form there and surely injure the coil by slowly warming the insulation. As they form even with ordinary frequencies when the potential is excessive, the air-space must be most carefully avoided. These high frequency streamers differ in aspect and properties from those produced by a static machine. The wind produced by them is small and should altogether cease if still considerably higher frequencies could be obtained. A peculiarity is that they issue as freely from surfaces as from points. Owing to this, a metallic vane, mounted in one of the terminals of the coil so as to rotate freely, and having one of its sides covered with insulation, is spun rapidly around. Such a vane would not rotate with a steady potential, but with a high frequency coil it will spin, even if it be entirely covered with insulation, provided the insulation on one side be either thicker or of a higher specific inductive capacity. A Crookes electric radiometer is also spun around when connected to one of the terminals of the coil, but only at very high exhaustion or at ordinary pressures.

There is still another and more striking peculiarity of such a high frequency streamer, namely, it is hot. The heat is easily perceptible with frequencies of about 10,000, even if the potential is not excessively high. The heating effect is, of course, due to the molecular impacts and collisions. Could the frequency and potential be pushed far enough, then a brush could be produced resembling in every particular a flame and giving light and heat, yet without a chemical process taking place.

The hot brush, when properly produced, resembles a jet of burning gas escaping under great pressure, and it emits an extraordinary strong smell of ozone. The great ozonizing action is ascribed to the fact that the agitation of the molecules of the air is more violent in such a brush than in the ordinary streamer of a static machine. But the most powerful brush discharges were produced by employing currents of much higher frequencies than it was possible to obtain by means of the alternators. These currents were obtained by disruptively discharging a condenser and setting up oscillations. In this manner currents of a frequency of several hundred thousand were obtained.

Currents of this kind, Mr. Tesla pointed out, produce striking effects. At these frequencies, the impedance of a copper bar is so great that a potential difference of several hundred volts can be maintained between two points of a short and thick bar, and it is possible to keep an ordinary incandescent lamp burning at full candle power by attaching the terminals of the lamp to two points of the bar no more than a few inches apart. When the frequency is extremely high, nodes are found to exist on such a bar, and it is easy to locate them by means of a lamp.

By converting the high tension discharges of a low frequency coil in this manner, it was found practicable to keep a few lamps burning on the ordinary circuit in the laboratory, and by bringing the undulation to a low pitch, it was possible to operate small motors.

This plan likewise allows of converting high tension discharges of one direction into low tension unidirectional currents, by adjusting the circuit so that there are no oscillations. In passing the oscillating discharges through the primary of a specially constructed coil, it is easy to obtain enormous potential differences with only few turns of the secondary.

Great difficulties were at first experienced in producing a successful coil on this plan. It was found necessary to keep all air, or gaseous matter in general, away from the charged surfaces, and oil immersion was resorted to. The wires used were heavily covered with gutta-percha and wound in oil, or the air was pumped out by means of a Sprengel pump. The general arrangement was the following: An ordinary induction coil, operated from a low frequency alternator, was used to charge Leyden jars. The jars were made to discharge over a single or multiple gap through the primary of the second coil. To insure the action of the gap, the arc was blown out by a magnet or air blast. To adjust the potential in the secondary a small oil condenser was used, or polished brass spheres of different sizes were screwed on the terminals and their distance adjusted.

When the conditions were carefully determined to suit each experiment, magnificent effects were obtained. Two wires, stretched through the room, each being connected to one of the terminals of the coil, emitted streams so powerful that the light from them allowed distinguishing the objects in the room; the wires became luminous even though covered with thick and most excellent insulation. When two straight wires, or two concentric circles of wire, are connected to the terminals, and set at the proper distance, a uniform luminous sheet is produced between them. It was possible in this way to cover an area of more than one meter square completely with the streams. By attaching to one terminal a large circle of wire and to the other terminal a small sphere, the streams are focused upon the sphere, produce a strongly lighted spot upon the same, and present the appearance of a luminous cone. A very thin wire glued upon a plate of hard rubber of great thickness, on the opposite side of which is fastened a tinfoil coating, is rendered intensely luminous when the coating is connected to the other terminal of the coil. Such an experiment can be performed also with low frequency currents, but much less satisfactorily.

When the terminals of such a coil, even of a very small one, are separated by a rubber or glass plate, the discharge spreads over the plate in the form of streams, threads or brilliant sparks, and affords a magnificent display, which cannot be equaled by the largest coil operated in the usual ways. By a simple adjustment it is possible to produce with the coil a succession of brilliant sparks, exactly as with a Holtz machine.

Under certain conditions, when the frequency of the oscillation is very great, white, phantom-like streams are seen to break forth from the terminals of the coil. The chief interesting feature about them is, that they stream freely against the outstretched hand or other conducting object without producing any sensation, and the hand may be approached very near to the terminal without a spark being induced to jump. This is due presumably to the fact that a considerable portion of the energy is carried away or dissipated in the streamers, and the difference of potential between the terminal and the hand is diminished.

It is found in such experiments that the frequency of the vibration and the quickness of succession of the sparks between the knobs affect to a marked degree the appearance of the streams. When the frequency is very low, the air gives way in more or less the same manner as by a steady difference of potential, and the streams consist of distinct threads, generally mingled with thin sparks, which probably correspond to the successive discharges occurring between the knobs. But when the frequency is very high, and the arc of the discharge produces a sound which is loud and smooth (which indicates both that oscillation takes place and that the sparks succeed each other with great rapidity), then the luminous streams formed are perfectly uniform. They are generally of a purplish hue, but when the molecular vibration is increased by raising the potential, they assume a white color.

The luminous intensity of the streams increases rapidly when the potential is increased; and with frequencies of only a few hundred thousand, could the coil be made to withstand a sufficiently high potential difference, there is no doubt that the space around a wire could be made to emit a strong light, merely by the agitation of the molecules of the air at ordinary pressure.

Such discharges of very high frequency which render luminous the air at ordinary pressure we have very likely occasion to witness in the aurora borealis. From many of these experiments it seems reasonable to infer that sudden cosmic disturbances, such as eruptions on the sun, set the electrostatic charge of the earth in an extremely rapid vibration, and produce the glow by the violent agitation of the air in the upper and even in the lower strata. It is thought that if the frequency were low, or even more so if the charge were not at all vibrating, the lower dense strata would break down as in a lightning discharge. Indications of such breaking down have been repeatedly observed, but they can be attributed to the fundamental disturbances, which are few in number, for the superimposed vibration would be so rapid as not to allow a disruptive break.

The study of these discharge phenomena has led Mr. Tesla to the recognition of some important facts. It was found, as already stated, that gaseous matter must be most carefully excluded from any dielectric which is subjected to great, rapidly changing electrostatic stresses. Since it is difficult to exclude the gas perfectly when solid insulators are used, it is necessary to resort to liquid dielectrics. When a solid dielectric is used, it matters little how thick and how good it is; if air be present, streamers form, which gradually heat the dielectric and impair its insulating power, and the discharge finally breaks through. Under ordinary conditions the best insulators are those which possess the highest specific inductive capacity, but such insulators are not the best to employ when working with these high frequency currents, for in most cases the higher specific inductive capacity is rather a disadvantage. The prime quality of the insulating medium for these currents is continuity. For this reason principally it is necessary to employ liquid insulators, such as oils. If two metal plates, connected to the terminals of the coil, are immersed in oil and set a distance apart, the coil may be kept working for any length of time without a break occurring, or without the oil being warmed, but if air bubbles are introduced, they become luminous; the air molecules, by their impact against the oil, heat it, and after some time cause the insulation to give way. If, instead of the oil, a solid plate of the best dielectric, even several times thicker than the oil intervening between the metal plates, is inserted between the latter, the air having free access to the charged surfaces, the dielectric invariably is warmed and breaks down.

The employment of oil is advisable or necessary even with low frequencies, if the potentials are such that streamers form, but only in such cases, as is evident from the theory of the action. If the potentials are so low that streamers do not form, then it is even disadvantageous to employ oil, for it may, principally by confining the heat, be the cause of the breaking down of the insulation.

The exclusion of gaseous matter is not only desirable on account of the safety of the apparatus, but also on account of economy, especially in a condenser, in which considerable waste of power may occur merely owing to the presence of air, if the electric density on the charged surfaces is great.

In the course of these investigations a phenomenon of special scientific interest was observed. It may be ranked among the brush phenomena, in fact it is a kind of brush which forms at, or near, a single terminal in high vacuum. In a bulb with a conducting electrode, even if the latter be of aluminum, the brush has only a very short existence, but it can be preserved for a considerable length of time in a bulb devoid of any conducting electrode. To observe the phenomenon it is found best to employ a large spherical bulb having in its center a small bulb supported on a tube sealed to the neck of the former. The large bulb being exhausted to a high degree, and the inside of the small bulb being connected to one of the terminals of the coil, under certain conditions there appears a misty haze around the small bulb, which, after passing through some stages, assumes the form of a brush, generally at right angles to the tube supporting the small bulb. When the brush assumes this form it may be brought to a state of extreme sensitiveness to electrostatic and magnetic influence. The bulb hanging straight down, and all objects being remote from it, the approach of the observer within a few paces will cause the brush to fly to the opposite side, and if he walks around the bulb it will always keep on the opposite side. It may begin to spin around the terminal long before it reaches that sensitive stage. When it begins to turn around, principally, but also before, it is affected by a magnet, and at a certain stage it is susceptible to magnetic influence to an astonishing degree. A small permanent magnet, with its poles at a distance of no more than two centimeters will affect it visibly at a distance of two meters, slowing down or accelerating the rotation according to how it is held relatively to the brush.

When the bulb hangs with the globe down, the rotation is always clockwise. In the southern hemisphere it would occur in the opposite direction, and on the (magnetic) equator the brush should not turn at all. The rotation may be reversed by a magnet kept at some distance. The brush rotates best, seemingly, when it is at right angles to the lines of force of the earth. It very likely rotates, when at its maximum speed, in synchronism with the alternations, say, 10,000 times a second. The rotation can be slowed down or accelerated by the approach or recession of the observer, or any conducting body, but it cannot be reversed by putting the bulb in any position. Very curious experiments may be performed with the brush when in its most sensitive state. For instance, the brush resting in one position, the experimenter may, by selecting a proper position, approach the hand at a certain considerable distance to the bulb, and he may cause the brush to pass off by merely stiffening the muscles of the arm, the mere change of configuration of the arm and the consequent imperceptible displacement being sufficient to disturb the delicate balance. When it begins to rotate slowly, and the hands are held at a proper distance, it is impossible to make even the slightest motion without producing a visible effect upon the brush. A metal plate connected to the other terminal of the coil affects it at a great distance, slowing down the rotation often to one turn a second.

Mr. Tesla hopes that this phenomenon will prove a valuable aid in the investigation of the nature of the forces acting in an electrostatic or magnetic field. If there is any motion which is measurable going on in the space, such a brush would be apt to reveal it. It is, so to speak, a beam of light, frictionless, devoid of inertia. On account of its marvellous sensitiveness to electrostatic or magnetic disturbances it may be the means of sending signals through submarine cables with any speed, and even of transmitting intelligence to a distance without wires.

In operating an induction coil with these rapidly alternating currents, it is astonishing to note, for the first time, the great importance of the relation of capacity, self-induction, and frequency as bearing upon the general result. The combined effect of these elements produces many curious effects. For instance, two metal plates are connected to the terminals and set at a small distance, so that an arc is formed between them. This arc *prevents* a strong current from flowing through the coil. If the arc be interrupted by the interposition of a glass plate, the capacity of the condenser obtained counteracts the self-induction, and a stronger current is made to pass. The effects of capacity are the most striking, for in these experiments, since the self-induction and frequency both are high, the critical capacity is very small, and need be but slightly varied to produce a very considerable change. The experimenter brings his body in contact with the terminals of the secondary of the coil, or attaches to one or both terminals insulated bodies of very small bulk, such as exhausted bulbs, and he produces a considerable rise or fall of potential on the secondary, and greatly affects the flow of the current through the primary coil.

In many of the phenomena observed, the presence of the air, or, generally speaking, of a medium of a gaseous nature (using this term not to imply specific properties, but in contradistinction to homogeneity or perfect continuity) plays an important part, as it allows energy to be dissipated by molecular impact or bombardment. The action is thus explained: When an insulated body connected to a terminal of the coil is suddenly charged to high potential, it acts inductively upon the surrounding air, or whatever gaseous medium there might be. The molecules or atoms which are near it are, of course, more attracted, and move through a greater distance than the further ones. When the nearest molecules strike the body they are repelled, and collisions occur at all distances within the inductive distance. It is now clear that, if the potential be steady, but little loss of energy can be caused in this way, for the molecules which are nearest to the body having had an additional charge imparted to them by contact, are not attracted until they have parted, if not with all, at least with most of the additional charge, which can be accomplished only after a great many collisions. This is inferred from the fact that with a steady potential there is but little loss in dry air. When the potential, instead of being steady, is alternating, the conditions are entirely different. In this case a rhythmical bombardment occurs, no matter whether the molecules after coming in contact with the body lose the imparted charge or not, and, what is more, if the charge is not lost, the impacts are all the more violent. Still, if the frequency of the impulses be very small, the loss caused by the impacts and collisions would not be serious unless the potential was excessive. But when extremely high frequencies and more or less high potentials are used, the loss may be very great. The total energy lost per unit of time is proportionate to the product of the number of impacts per second, or the frequency and the energy lost in each impact. But the energy of an impact must be proportionate to the square of the electric density of the body, on the assumption that the charge imparted to the molecule is proportionate to that density. It is concluded from this that the total energy lost must be proportionate to the product of the frequency and the square of the electric density; but this law needs experimental confirmation. Assuming the preceding considerations to be true, then, by rapidly alternating the potential of a body immersed in an insulating gaseous medium, any amount of energy may be dissipated into space. Most of that energy, then, is not dissipated in the form of long ether waves, propagated to considerable distance, as is thought most generally, but is consumed in impact and collisional losses—that is, heat vibrations—on the surface and in the vicinity of the body. To reduce the dissipation it is necessary to work with a small electric density—the smaller, the higher the frequency.

The behavior of a gaseous medium to such rapid alternations of potential makes it appear plausible that electrostatic disturbances of the earth, produced by cosmic events, may have great influence upon the meteorological conditions. When such disturbances occur both the frequency of the vibrations of the charge and the potential are in all probability excessive, and the energy converted into heat may be considerable. Since the density must be unevenly distributed, either in consequence of the irregularity of the earth's surface, or on account of the condition of the atmosphere in various places, the effect produced would accordingly vary from place to place. Considerable variations in the temperature and pressure of the atmosphere may in this manner be caused at any point of the surface of the earth. The variations may be gradual or very sudden, according to the nature of the original disturbance, and may produce rain and storms, or locally modify the weather in any way.

From many experiences gathered in the course of these investigations it appears certain that in lightning discharges the air is an element of importance. For instance, during a storm a stream may form on a nail or pointed projection of a building. If lightning strikes somewhere in the neighborhood, the harmless static discharge may, in consequence of the oscillations set up, assume the character of a high-frequency streamer, and the nail or projection may be brought to a high temperature by the violent impact of the air molecules. Thus, it is thought, a building may be set on fire without the lightning striking it. In like manner small metallic objects may be fused and volatilized— as frequently occurs in lightning discharges—merely because they are surrounded by air. Were they immersed in a practically continuous medium, such as oil, they would probably be safe, as the energy would have to spend itself elsewhere.

An instructive experience having a bearing on this subject is the following: A glass tube of an inch or so in diameter and several inches long is taken, and a platinum wire sealed into it, the wire running through the center of the tube from end to end. The tube is exhausted to a moderate degree. If a steady current is passed through the wire it is heated uniformly in all parts and the gas in the tube is of no consequence. But if high frequency discharges are directed through the wire, it is heated more on the ends than in the middle portion, and if the frequency, or rate of charge, is high enough, the wire might as well be cut in the middle as not, for most of the heating on the ends is due to the rarefied gas. Here the gas might only act as a conductor of no impedance, diverting the current from the wire as the impedance of the latter is enormously increased, and merely heating the ends of the wire by reason of their resistance to the passage of the discharge. But it is not at all necessary that the gas in the tube should be conducting; it might be at an extremely low pressure, still the ends of the wire would be heated; however, as is ascertained by experience, only the two ends would in such case not be electrically connected through the gaseous medium. Now, what with these frequencies and potentials occurs in an exhausted tube, occurs in the lightning discharge at ordinary pressure.

From the facility with which any amount of energy may be carried off through a gas, Mr. Tesla infers that the best way to render harmless a lightning discharge is to afford it in some way a passage through a volume of gas.

The recognition of some of the above facts has a bearing upon far-reaching scientific investigations in which extremely high frequencies and potentials are used. In such cases the air is an important factor to be considered. So, for instance, if two wires are attached to the terminals of the coil, and the streamers issue from them, there is dissipation of energy in the form of heat and light, and the wires behave like a condenser of larger capacity. If the wires be immersed in oil, the dissipation of energy is prevented, or at least reduced, and the apparent capacity is diminished. The action of the air would seem to make it very difficult to tell, from the measured or computed capacity of a condenser in which the air is acted upon, its actual capacity or vibration period, especially if the condenser is of very small surface and is charged to a very high potential. As many important results are dependant upon the correctness of the estimation of the vibration period, this subject demands the most careful scrutiny of investigators.

In Leyden jars the loss due to the presence of air is comparatively small, principally on account of the great surface of the coatings and the small external action, but if there are streamers on the top, the loss may be considerable, and the period of vibration is affected. In a resonator, the density is small, but the frequency is extreme, and may introduce a considerable error. It appears certain, at any rate, that the periods of vibration of a charged body in a gaseous and in a continuous medium, such as oil, are different, on account of the action of the former, as explained.

Another fact recognized, which is of some consequence, is, that in similar investigations the general considerations of static screening are not applicable when a gaseous medium is present. This is evident from the following experiment: A short and wide glass tube is taken and covered with a substantial coating of bronze powder, barely allowing the light to shine a little through. The tube is highly exhausted and suspended on a metallic clasp from the end of a wire. When the wire is connected with one of the terminals of the coil, the gas inside of the tube is lighted in spite of the metal coating. Here the metal evidently does not screen the gas inside as it ought to, even if it be very thin and poorly conducting. Yet, in a condition of rest the metal coating, however thin, screens the inside perfectly.

One of the most interesting results arrived at in pursuing these experiments, is the demonstration of the fact that a gaseous medium, upon which vibration is impressed by rapid changes of electrostatic potential, is rigid. In illustration of this result an experiment made by Mr. Tesla may be cited: A glass tube about one inch in diameter and three feet long, with outside condenser coatings on the ends, was exhausted to a certain point, when, the tube being suspended freely from a wire connecting the upper coating to one of the terminals of the coil, the discharge appeared in the form of a luminous thread passing through the axis of the tube. Usually the thread was sharply defined in the upper part of the tube and lost itself in the lower part. When a magnet or the finger was quickly passed near the upper part of the luminous thread, it was brought out of position by magnetic or electrostatic influence, and a transversal vibration like that of a suspended cord, with one or more distinct nodes, was set up, which lasted for a few minutes and gradually died out. By suspending from the lower condenser coating metal plates of different sizes, the speed of the vibration was varied. This vibration would seem to show beyond doubt that the thread possessed rigidity, at least to transversal displacements.

Many experiments were tried to demonstrate this property in air at ordinary pressure. Though no positive evidence has been obtained, it is thought, nevertheless, that a high frequency brush or streamer, if the frequency could be pushed far enough, would be decidedly rigid. A small sphere might then be moved within it quite freely, but if thrown against it the sphere would rebound. An ordinary flame cannot possess rigidity to a marked degree because the vibration is directionless; but an electric arc, it is believed, must possess that property more or less. A luminous band excited in a bulb by repeated discharges of a Leyden jar must also possess rigidity, and if deformed and suddenly released should vibrate.

From like considerations other conclusions of interest are reached. The most probable medium filling the space is one consisting of independent carriers immersed in an insulating fluid. If through this medium enormous electrostatic stresses are assumed to act, which vary rapidly in intensity, it would allow the motion of a body through it, yet it would be rigid and elastic, although the fluid itself might be devoid of these properties. Furthermore, on the assumption that the independent carriers are of any configuration such that the fluid resistance to motion in one direction is greater than in another, a stress of that nature would cause the carriers to arrange themselves in groups, since they would turn to each other their sides of the greatest electric density, in which position the fluid resistance to approach would be smaller than to receding. If in a medium of the above characteristics a brush would be formed by a steady potential, an exchange of the carriers would go on continually, and there would be less carriers per unit of volume in the brush than in the space at some distance from the electrode, this corresponding to rarefaction. If the potential were rapidly changing, the result would be very different; the higher the frequency of the pulses, the slower would be the exchange of the carriers; finally, the motion of translation through measurable space would cease, and, with a sufficiently high frequency and intensity of the stress, the carriers would be drawn towards the electrode, and compression would result.

An interesting feature of these high frequency currents is that they allow of operating all kinds of devices by connecting the device with only one leading wire to the electric source. In fact, under certain conditions it may be more economical to supply the electrical energy with one lead than with two.

An experiment of special interest shown by Mr. Tesla, is the running, by the use of only one insulated line, of a motor operating on the principle of the rotating magnetic field enunciated by Mr. Tesla. A simple form of such a motor is obtained by winding upon a laminated iron core a primary and close to it a secondary coil, closing the ends of the latter and placing a freely movable metal disc within the influence of the moving field. The secondary coil may, however, be omitted. When one of the ends of the primary coil of the motor is connected to one of the terminals of the high frequency coil and the other end to an insulated metal plate, which, it should be stated, is not absolutely necessary for the success of the experiment, the disc is set in rotation.

Experiments of this kind seem to bring it within possibility to operate a motor at any point of the earth's surface from a central source, without any connection to the same except through the earth. If, by means of powerful machinery, rapid variations of the earth's potential were produced, a grounded wire reaching up to some height would be traversed by a current which could be increased by connecting the free end of the wire to a body of some size. The current might be converted to low tension and used to operate a motor or other device. The experiment, which would be one of great scientific interest, would probably best succeed on a ship at sea. In this manner, even if it were not possible to operate machinery, intelligence might be transmitted quite certainly.

In the course of this experimental study special attention was devoted to the heating effects produced by these currents, which are not only striking, but open up the possibility of producing a more efficient illuminant. It is sufficient to attach to the coil terminal a thin wire or filament, to have the temperature of the latter perceptibly raised. If the wire or filament be enclosed in a bulb, the heating effect is increased by preventing the circulation of the air. If the air in the bulb be strongly compressed, the displacements are smaller, the impacts less violent, and the heating effect is diminished. On the contrary, if the air in the bulb be exhausted, an inclosed lamp filament is brought to incandescence, and any amount of light may thus be produced.

The heating of the inclosed lamp filament depends on so many things of a different nature, that it is difficult to give a generally applicable rule under which the maximum heating occurs. As regards the size of the bulb, it is ascertained that at ordinary or only slightly differing atmospheric pressures, when air is a good insulator, the filament is heated more in a small bulb, because of the better confinement of heat in this case. At lower pressures, when air becomes conducting, the heating effect is greater in a large bulb, but at excessively high degrees of exhaustion there seems to be, beyond a certain and rather small size of the vessel, no perceptible difference in the heating.

The shape of the vessel is also of some importance, and it has been found of advantage for reasons of economy to employ a spherical bulb with the electrode mounted in its center, where the rebounding molecules collide.

It is desirable on account of economy that all the energy supplied to the bulb from the source should reach without loss the body to be heated. The loss in conveying the energy from the source to the body may be reduced by employing thin wires heavily coated with insulation, and by the use of electrostatic screens. It is to be remarked, that the screen cannot be connected to the ground as under ordinary conditions.

In the bulb itself a large portion of the energy supplied may be lost by molecular bombardment against the wire connecting the body to be heated with the source. Considerable improvement was effected by covering the glass stem containing the wire with a closely fitting conducting tube. This tube is made to project a little above the glass, and prevents the cracking of the latter near the heated body. The effectiveness of the conducting tube is limited to very high degrees of exhaustion. It diminishes the energy lost in bombardment for two reasons; first, the charge given up by the atoms spreads over a greater area, and hence the electric density at any point is small, and the atoms are repelled with less energy than if they would strike against a good insulator; secondly, as the tube is electrified by the atoms which first come in contact with it, the progress of the following atoms against the tube is more or less checked by the repulsion which the electrified tube must exert upon the similarly electrified atoms. This, it is thought, explains why the discharge through a bulb is established with much greater facility when an insulator, than when a conductor, is present.

During the investigations a great many bulbs of different construction, with electrodes of different material, were experimented upon, and a number of observations of interest were made. Mr. Tesla has found that the deterioration of the electrode is the less, the higher the frequency. This was to be expected, as then the heating is effected by many small impacts, instead by fewer and more violent ones, which quickly shatter the structure. The deterioration is also smaller when the vibration is harmonic. Thus an electrode, maintained at a certain degree of heat, lasts much longer with currents obtained from an alternator, than with those obtained by means of a disruptive discharge. One of the most durable electrodes was obtained from strongly compressed carborundum, which is a kind of carbon recently produced by Mr. E. G. Acheson, of Monongahela City, Pa. From experience, it is inferred, that to be most durable, the electrode should be in the form of a sphere with a highly polished surface.

In some bulbs refractory bodies were mounted in a carbon cup and put under the molecular impact. It was observed in such experiments that the carbon cup was heated at first, until a higher temperature was reached; then most of the bombardment was directed against the refractory body, and the carbon was relieved. In general, when different bodies were mounted in the bulb, the hardest fusible would be relieved, and would remain at a considerably lower temperature. This was necessitated by the fact that most of the energy supplied would find its way through the body which was more easily fused or "evaporated."

Curiously enough it appeared in some of the experiments made, that a body was fused in a bulb under the molecular impact by evolution of less light than when fused by the application of heat in ordinary ways. This may be ascribed to a loosening of the structure of the body under the violent impacts and changing stresses.

Some experiments seem to indicate that under certain conditions a body, conducting or nonconducting, may, when bombarded, emit light, which to all appearances is due to phosphorescence, but may in reality be caused by the incandescence of an infinitesimal layer, the mean temperature of the body being comparatively small. Such might be the case if each single rhythmical impact were capable of instantaneously exciting the retina, and the rhythm were just high enough to cause a continuous impression in the eye. According to this view, a coil operated by disruptive discharge would be eminently adapted to produce such a result, and it is found by experience that its power of exciting phosphorescence is extraordinarily great. It is capable of exciting phosphorescence at comparatively low degrees of exhaustion, and also projects shadows at pressures far greater than those at which the mean free path is comparable to the dimensions of the vessel. The latter observation is of some importance, inasmuch as it may modify the generally accepted views in regard to the "radiant state" phenomena.

A thought which early and naturally suggested itself to Mr. Tesla, was to utilize the great inductive effects of high frequency currents to produce light in a sealed glass vessel without the use of leading in wires. Accordingly, many bulbs were constructed in which the energy necessary to maintain a button or filament at high incandescence, was supplied through the glass by either electrostatic or electrodynamic induction. It was easy to regulate the intensity of the light emitted by means of an externally applied condenser coating connected to an insulated plate, or simply by means of a plate attached to the bulb which at the same time performed the function of a shade.

A subject of experiment, which has been exhaustively treated in England by Prof. J. J. Thomson, has been followed up independently by Mr. Tesla from the beginning of this study, namely, to excite by electrodynamic induction a luminous band in a closed tube or bulb. In observing the behavior of gases, and the luminous phenomena obtained, the importance of the electrostatic effects was noted and it appeared desirable to produce enormous potential differences, alternating with extreme rapidity. Experiments in this direction led to some of the most interesting results arrived at in the course of these investigations. It was found that by rapid alternations of a high electrostatic potential, exhausted tubes could be lighted at considerable distances from a conductor connected to a properly constructed coil, and that it was practicable to establish with the coil an alternating electrostatic field, acting through the whole room and lighting a tube wherever it was placed within the four walls. Phosphorescent bulbs may be excited in such a field, and it is easy to regulate the effect by connecting to the bulb a small insulated metal plate. It was likewise possible to maintain a filament or button mounted in a tube at bright incandescence, and, in one experiment, a mica vane was spun by the incandescence of a platinum wire.

Coming now to the lecture delivered in Philadelphia and St. Louis, it may be remarked that to the superficial reader, Mr. Tesla's introduction, dealing with the importance of the eye, might appear as a digression, but the thoughtful reader will find therein much food for meditation and speculation. Throughout his discourse one can trace Mr. Tesla's effort to present in a popular way thoughts and views on the electrical phenomena which have in recent years captivated the scientific world, but of which the general public has even yet merely received an inkling. Mr. Tesla also dwells rather extensively on his well-known method of high-frequency conversion; and the large amount of detail information will be gratefully received by students and experimenters in this virgin field. The employment of apt analogies in explaining the fundamental principles involved makes it easy for all to gain a clear idea of their nature. Again, the ease with which, thanks to Mr. Tesla's efforts, these high-frequency currents may now be obtained from circuits carrying almost any kind of current, cannot fail to result in an extensive broadening of this field of research, which offers so many possibilities. Mr. Tesla, true philosopher as he is, does not hesitate to point out defects in some of his methods, and indicates the lines which to him seem the most promising. Particular stress is laid by him upon the employment of a medium in which the discharge electrodes should be immersed in order that this method of conversion may be brought to the highest perfection. He has evidently taken pains to give as much useful information as possible to those who wish to follow in his path, as he shows in detail the circuit arrangements to be adopted in all ordinary cases met with in practice, and although some of these methods were described by him two years before, the additional information is still timely and welcome.

In his experiments he dwells first on some phenomena produced by electrostatic force, which he considers in the light of modern theories to be the most important force in nature for us to investigate. At the very outset he shows a strikingly novel experiment illustrating the effect of a rapidly varying electrostatic force in a gaseous medium, by touching with one hand one of the terminals of a 200,000 volt transformer and bringing the other hand to the opposite terminal. The powerful streamers which issued from his hand and astonished his audiences formed a capital illustration of some of the views advanced, and afforded Mr. Tesla an opportunity of pointing out the true reasons why, with these currents, such an amount of energy can be passed through the body with impunity. He then showed by experiment the difference between a steady and a rapidly varying force upon the dielectric. This difference is most strikingly illustrated in the experiment in which a bulb attached to the end of a wire in connection with one of the terminals of the transformer is ruptured, although all extraneous bodies are remote from the bulb. He next illustrates how mechanical motions are produced by a varying electrostatic force acting through a gaseous medium. The importance of the action of the air is particularly illustrated by an interesting experiment.

Taking up another class of phenomena, namely, those of dynamic electricity, Mr. Tesla produced in a number of experiments a variety of effects by the employment of only a single wire with the evident intent of impressing upon his audience the idea that electric vibration or current can be transmitted with ease, without any return circuit; also how currents so transmitted can be converted and used for many practical purposes. A number of experiments are then shown, illustrating the effects of frequency, self-induction and capacity; then a number of ways of operating motive and other devices by the use of a single lead. A number of novel impedance phenomena are also shown which cannot fail to arouse interest.

Mr. Tesla next dwelt upon a subject which he thinks of great importance, that is, electrical resonance, which he explained in a popular way. He expressed his firm conviction that by observing proper conditions, intelligence, and possibly even power, can be transmitted through the medium or through the earth; and he considers this problem worthy of serious and immediate consideration.

Coming now to the light phenomena in particular, he illustrated the four distinct kinds of these phenomena in an original way, which to many must have been a revelation. Mr. Tesla attributes these light effects to molecular or atomic impacts produced by a varying electrostatic stress in a gaseous medium. He illustrated in a series of novel experiments the effect of the gas surrounding the conductor and shows beyond a doubt that with high frequency and high potential currents, the surrounding gas is of paramount importance in the heating of the conductor. He attributes the heating partially to a conduction current and partially to bombardment, and demonstrates that in many cases the heating may be practically due to the bombardment alone. He pointed out also that the skin effect is largely modified by the presence of the gas or of an atomic medium in general. He showed also some interesting experiments in which the effect of convection is illustrated. Probably one of the most curious experiments in this connection is that in which a thin platinum wire stretched along the axis of an exhausted tube is brought to incandescence at certain points corresponding to the position of the striae, while at others it remains dark. This experiment throws an interesting light upon the nature of the striae and may lead to important revelations.

Mr. Tesla also demonstrated the dissipation of energy through an atomic medium and dwelt upon the behavior of vacuous space in conveying heat, and in this connection showed the curious behavior of an electrode stream, from which he concludes that the molecules of a gas probably cannot be acted upon directly at measurable distances.

Mr. Tesla summarized the chief results arrived at in pursuing his investigations in a manner which will serve as a valuable guide to all who may engage in this work. Perhaps most interest will center on his general statements regarding the phenomena of phosphorescence, the most important fact revealed in this direction being that when exciting a phosphorescent bulb a certain definite potential gives the most economical result.

The lectures will now be presented in the order of their date of delivery.

CHAPTER XXVI

EXPERIMENTS WITH ALTERNATE CURRENTS OF VERY HIGH FREQUENCY AND THEIR APPLICATION TO METHODS OF ARTIFICIAL ILLUMINATION*

There is no subject more captivating, more worthy of study, than nature. To understand this great mechanism, to discover the forces which are active, and the laws which govern them, is the highest aim of the intellect of man.

Nature has stored up in the universe infinite energy. The eternal recipient and transmitter of this infinite energy is the ether. The recognition of the existence of ether, and of the functions it performs, is one of the most important results of modern scientific research. The mere abandoning of the idea of action at a distance, the assumption of a medium pervading all space and connecting all gross matter, has freed the minds of thinkers of an ever present doubt, and, by opening a new horizon—new and unforeseen possibilities—has given fresh interest to phenomena with which we are familiar of old. It has been a great step towards the understanding of the forces of nature and their multifold manifestations to our senses. It has been for the enlightened student of physics what the understanding of the mechanism of the firearm or of the steam engine is for the barbarian. Phenomena upon which we used to look as wonders baffling explanation, we now see in a different light. The spark of an induction coil, the glow of an incandescent lamp, the manifestations of the mechanical forces of currents and magnets are no longer beyond our grasp; instead of the incomprehensible, as before, their observation suggests now in our minds a simple mechanism, and although as to its precise nature all is still conjecture, yet we know that the truth cannot be much longer hidden, and instinctively we feel that the understanding is dawning upon us. We still admire these beautiful phenomena, these strange forces, but we are helpless no

---

* A lecture delivered beforee the American Institiute of Electrical Engineers, at Columbia College, N.Y., May 20, 1891.

longer; we can in a certain measure explain them, account for them, and we are hopeful of finally succeeding in unraveling the mystery which surrounds them.

In how far we can understand the world around us is the ultimate thought of every student of nature. The coarseness of our senses prevents us from recognizing the ulterior construction of matter, and astronomy, this grandest and most positive of natural sciences, can only teach us something that happens, as it were, in our immediate neighborhood: of the remoter portions of the boundless universe, with its numberless stars and suns, we know nothing. But far beyond the limit of perception of our senses the spirit still can guide us, and so we may hope that even these unknown worlds—infinitely small and great—may in a measure become known to us. Still, even if this knowledge should reach us, the searching mind will find a barrier, perhaps forever unsurpassable, to the *true* recognition of that which *seems* to be, the mere *appearance* of which is the only and slender basis of all our philosophy.

Of all the forms of nature's immeasurable, all-pervading energy, which ever and ever changing and moving, like a soul animates the inert universe, electricity and magnetism are perhaps the most fascinating. The effects of gravitation, of heat and light we observe daily, and soon we get accustomed to them, and soon they lose for us the character of the marvelous and wonderful; but electricity and magnetism, with their singular relationship, with their seemingly dual character, unique among the forces in nature, with their phenomena of attractions, repulsions and rotations, strange manifestations of mysterious agents, stimulate and excite the mind to thought and research. What is electricity, and what is magnetism? These questions have been asked again and again. The most able intellects have ceaselessly wrestled with the problem; still the question has not as yet been fully answered. But while we cannot even today state what these singular forces are, we have made good headway towards the solution of the problem. We are now confident that electric and magnetic phenomena are attributable to ether, and we are perhaps justified in saying that the effects of static electricity are effects of ether under strain, and those of dynamic electricity and electro-magnetism effects of ether in motion. But this still leaves the question, as to what electricity and magnetism are, unanswered.

First, we naturally inquire, What is electricity, and is there such a thing as electricity? In interpreting electric phenomena, we may speak of electricity or of an electric condition, state or effect. If we speak of electric effects we must distinguish two such effects, opposite in character and neutralizing each other, as observation shows that two such opposite effects exist. This is unavoidable, for in a medium of the properties of ether, we cannot possibly exert a strain, or produce a displacement or motion of any kind, without causing in the surrounding medium an equivalent and opposite effect. But if we speak of electricity, meaning a *thing*, we must, I think, abandon the idea of two electricities, as the existence of two such things is highly improbable. For how can we imagine that there should be two things, equivalent in amount, alike in their properties, but of opposite character, both clinging to matter, both attracting and completely neutralizing each other? Such an assumption, though suggested by many phenomena, though most convenient for explaining them, has little to commend it. If there *is* such a thing as electricity, there can be only one such thing, and excess and want of that one thing, possibly; but more probably its condition determines the positive and negative character. The old theory of Franklin, though falling short in some respects, is, from a certain point of view, after all, the most plausible one. Still, in spite of this, the theory of the two electricities is generally accepted, as it apparently explains electric phenomena in a more satisfactory manner. But a theory which better explains the facts is not necessarily true. Ingenious minds will invent theories to suit observation, and almost every independent thinker has his own views on the subject.

It is not with the object of advancing an opinion, but with the desire of acquainting you better with some of the results, which I will describe, to show you the reasoning I have followed, the departures I have made— that I venture to express, in a few words, the views and convictions which have led me to these results.

I adhere to the idea that there is a thing which we have been in the habit of calling electricity. The question is, What is that thing? or, What, of all things, the existence of which we know, have we the best reason to call electricity? We know that it acts like an incompressible fluid; that there must be a constant quantity of it in nature; that it can be neither produced nor destroyed; and, what is more important, the electro-magnetic theory of light and all facts observed teach us that electric and ether phenomena are identical. The idea at once suggests itself, therefore, that electricity might be called ether. In fact, this view has in a certain sense been advanced by Dr. Lodge. His interesting work has been read by everyone and many have been convinced by his arguments. His great ability and the interesting nature of the subject, keep the reader spellbound; but when the impressions fade, one realizes that he has to deal only with ingenious explanations. I must confess, that I cannot believe in two electricities, much less in a doubly-constituted ether. The puzzling behavior of the ether as a solid to waves of light and heat, and as a fluid to the motion of bodies through it, is certainly explained in the most natural and satisfactory manner by assuming it to be in motion, as Sir William Thomson has suggested; but regardless of this, there is nothing which would enable us to conclude with certainty that, while a fluid is not capable of transmitting transverse vibrations of a few hundred or thousand per second, it might not be capable of transmitting such vibrations when they range into hundreds of million millions per second. Nor can anyone prove that there are transverse ether waves emitted from an alternate current machine, giving a small number of alternations per second; to such slow disturbances, the ether, if at rest, may behave as a true fluid.

Returning to the subject, and bearing in mind that the existence of two electricities is, to say the least, highly improbable, we must remember, that we have no evidence of electricity, nor can we hope to get it, unless gross matter is present. Electricity, therefore, cannot be called ether in the broad sense of the term; but nothing would seem to stand in the way of calling electricity ether associated with matter, or bound ether; or, in other words, that the so-called static charge of the molecule is ether associated in some way with the molecule. Looking at it in that light, we would be justified in saying, that electricity is concerned in all molecular actions.

Now, precisely what the ether surrounding the molecules is, wherein it differs from ether in general, can only be conjectured. It cannot differ in density, ether being incompressible: it must, therefore, be under some strain or in motion, and the latter is the most probable. To understand its functions, it would be necessary to have an exact idea of the physical construction of matter, of which, of course, we can only form a mental picture.

But of all the views on nature, the one which assumes one matter and one force, and a perfect uniformity throughout, is the most scientific and most likely to be true. An infinitesimal world, with the molecules and their atoms spinning and moving in orbits, in much the same manner as celestial bodies, carrying with them and probably spinning with them ether, or in other words, carrying with them static charges, seems to my mind the most probable view, and one which, in a plausible manner, accounts for most of the phenomena observed. The spinning of the molecules and their ether sets up the ether tensions or electrostatic strains; the equalization of ether tensions sets up ether motions or electric currents, and the orbital movements produce the effects of electro and permanent magnetism.

About fifteen years ago, Prof. Rowland demonstrated a most interesting and important fact, namely, that a static charge carried around produces the effects of an electric current. Leaving out of consideration the precise nature of the mechanism, which produces the attraction and repulsion of currents, and conceiving the electrostatically charged molecules in motion, this experimental fact gives us a fair idea of magnetism. We can conceive lines or tubes of force which physically exist, being formed of rows of directed moving molecules; we can see that these lines must be closed, that they must tend to shorten and expand, etc. It likewise explains in a reasonable way, the most puzzling phenomenon of all, permanent magnetism, and, in general, has all the beauties of the Ampere theory without possessing the vital defect of the same, namely, the assumption of molecular currents. Without enlarging further upon the subject, I would say, that I look upon all electrostatic, current and magnetic phenomena as being due to electrostatic molecular forces.

The preceding remarks I have deemed necessary to a full understanding of the subject as it presents itself to my mind.

Of all these phenomena the most important to study are the current phenomena, on account of the already extensive and ever-growing use of currents for industrial purposes. It is now a century since the first practical source of current was produced, and, ever since, the phenomena which accompany the flow of currents have been diligently studied, and through the untiring efforts of scientific men the simple laws which govern them have been discovered. But these laws are found to hold good only when the currents are of a steady character. When the currents are rapidly varying in strength, quite different phenomena, often unexpected, present themselves, and quite different laws hold good, which even now have not been determined as fully as is desirable, though through the work, principally, of English scientists, enough knowledge has been gained on the subject to enable us to treat simple cases which now present themselves in daily practice.

The phenomena which are peculiar to the changing character of the currents are greatly exalted when the rate of change is increased, hence the study of these currents is considerably facilitated by the employment of properly constructed apparatus. It was with this and other objects in view that I constructed alternate current machines capable of giving more than two million reversals of current per minute, and to this circumstance it is principally due, that I am able to bring to your attention some of the results thus far reached, which I hope will prove to be a step in advance on account of their direct bearing upon one of the most important problems, namely, the production of a practical and efficient source of light.

The study of such rapidly alternating currents is very interesting. Nearly every experiment discloses something new. Many results may, of course, be predicted, but many more are unforeseen. The experimenter makes many interesting observations. For instance, we take a piece of iron and hold it against a magnet. Starting from low alternations and running up higher and higher we feel the impulses succeed each other faster and faster, get weaker and weaker, and finally disappear. We then observe a continuous pull; the pull, of course, is not continuous; it only appears so to us; our sense of touch is imperfect.

We may next establish an arc between the electrodes and observe, as the alternations rise, that the note which accompanies alternating arcs gets shriller and shriller, gradually weakens, and finally ceases. The air vibrations, of course, continue, but they are too weak to be perceived; our sense of hearing fails us.

We observe the small physiological effects, the rapid heating of the iron cores and conductors, curious inductive effects, interesting condenser phenomena, and still more interesting light phenomena with a high tension induction coil. All these experiments and observations would be of the greatest interest to the student, but their description would lead me too far from the principal subject. Partly for this reason, and partly on account of their vastly greater importance, I will confine myself to the description of the light effects produced by these currents.

In the experiments to this end a high tension induction coil or equivalent apparatus for converting currents of comparatively low into currents of high tension is used.

If you will be sufficiently interested in the results I shall describe as to enter into an experimental study of this subject; if you will be convinced of the truth of the arguments I shall advance—your aim will be to produce high frequencies and high potentials; in other words, powerful electrostatic effects. You will then encounter many difficulties, which, if completely overcome, would allow us to produce truly wonderful results.

First will be met the difficulty of obtaining the required frequencies by means of mechanical apparatus, and, if they be obtained otherwise, obstacles of a different nature will present themselves. Next it will be found difficult to provide the requisite insulation without considerably increasing the size of the apparatus, for the potentials required are high, and, owing to the rapidity of the alternations, the insulation presents peculiar difficulties. So, for instance, when a gas is present, the discharge may work, by the molecular bombardment of the gas and consequent heating, through as much as an inch of the best solid insulating material, such as glass, hard rubber, porcelain, sealing wax, etc.; in fact, through any known insulating substance. The chief requisite in the insulation of the apparatus is, therefore, the exclusion of any gaseous matter.

In general my experience tends to show that bodies which possess the highest specific inductive capacity, such as glass, afford a rather inferior insulation to others, which, while they are good insulators, have a much smaller specific inductive capacity, such as oils, for instance, the dielectric losses being no doubt greater in the former. The difficulty of insulating, of course, only exists when the potentials are excessively high, for with potentials such as a few thousand volts there is no particular difficulty encountered in conveying currents from a machine giving, say, 20,000 alternations per second, to quite a distance. This number of alternations, however, is by far too small for many purposes, though quite sufficient for some practical applications. This difficulty of insulating is fortunately not a vital drawback; it affects mostly the size of the apparatus, for, when excessively high potentials would be used, the light-giving devices would be located not far from the apparatus, and often they would be quite close to it. As the air-bombardment of the insulated wire is dependent on condenser action, the loss may be reduced to a trifle by using excessively thin wires heavily insulated.

Another difficulty will be encountered in the capacity and self-induction necessarily possessed by the coil. If the coil be large, that is, if it contain a great length of wire, it will be generally unsuited for excessively high frequencies; if it be small, it may be well adapted for such frequencies, but the potential might then not be as high as desired. A good insulator, and preferably one possessing a small specific inductive capacity, would afford a two-fold advantage. First, it would enable us to construct a very small coil capable of withstanding enormous differences of potential; and secondly, such a small coil, by reason of its smaller capacity and self-induction, would be capable of a quicker and more vigorous vibration. The problem then of constructing a coil or induction apparatus of any kind possessing the requisite qualities I regard as one of no small importance, and it has occupied me for a considerable time.

The investigator who desires to repeat the experiments which I will describe, with an alternate current machine, capable of supplying currents of the desired frequency, and an induction coil, will do well to take the primary coil out and mount the secondary in such a manner as to be able to look through the tube upon which the secondary is wound. He will then be able to observe the streams which pass from

Fig. 97.

the primary to the insulating tube, and from their intensity he will know how far he can strain the coil. Without this precaution he is sure to injure the insulation. This arrangement permits, however, an easy exchange of the primaries, which is desirable in these experiments.

The selection of the type of machine best suited for the purpose must be left to the judgment of the experimenter. There are here illustrated three distinct types of machines, which, besides others, I have used in my experiments.

Fig. 97 represents the machine used in my experiments before this Institute. The field magnet consists of a ring of wrought iron with 384 pole projections. The armature comprises a steel disc to which is fastened a thin, carefully welded rim of wrought iron. Upon the rim are wound several layers of fine, well annealed iron wire, which, when wound, is passed through shellac. The armature wires are wound around brass pins, wrapped with silk thread. The diameter of the armature wire in this type of machine should not be more than ⅛ of the thickness of the pole projections, else the local action will be considerable.

Fig. 98.

Fig. 98 represents a larger machine of a different type. The field magnet of this machine consists of two like parts which either enclose an exciting coil, or else are independently wound. Each part has 480 pole projections, the projections of one facing those of the other. The armature consists of a wheel of hard bronze, carrying the conductors which revolve between the projections of the field magnet. To wind the armature conductors, I have found it most convenient to proceed in the following manner. I construct a ring of hard bronze of the required size. This ring and the rim of the wheel are provided with the proper number of pins, and both fastened upon a plate. The armature conductors being wound, the pins are cut off and the ends of the conductors fastened by two rings which screw to the bronze ring and the rim of the wheel, respectively. The whole may then be taken off and forms a solid structure. The conductors in such a type of machine should consist of sheet copper, the thickness of which, of course, depends on the thickness of the pole projections; or else twisted thin wires should be employed.

Fig. 99.

Fig. 99 is a smaller machine, in many respects similar to the former, only here the armature conductors and the exciting coil are kept stationary, while only a block of wrought iron is revolved.

It would be uselessly lengthening this description were I to dwell more on the details of construction of these machines. Besides, they have been described somewhat more elaborately in *The Electrical Engineer*, of March 18, 1891. I deem it well, however, to call the attention of the investigator to two things, the importance of which, though self evident, he is nevertheless apt to underestimate; namely, to the local action in the conductors which must be carefully avoided, and to the clearance, which must be small. I may add, that since it is desirable to use very high peripheral speeds, the armature should be of very large diameter in order to avoid impracticable belt speeds. Of the several types of these machines which have been constructed by me, I have found that the type illustrated in Fig. 97 caused me the least trouble in construction, as well as in maintenance, and on the whole, it has been a good experimental machine.

In operating an induction coil with very rapidly alternating currents, among the first luminous phenomena noticed are naturally those presented by the high-tension discharge. As the number of alternations per second is increased, or as—the number being high—the current through the primary is varied, the discharge gradually changes in appearance. It would be difficult to describe the minor changes which occur, and the conditions which bring them about, but one may note five distinct forms of the discharge.

FIG. 100a.   FIG. 100b.

First, one may observe a weak, sensitive discharge in the form of a thin, feeble-colored thread. (Fig. 100a.) It always occurs when, the number of alternations per second being high, the current through the primary is very small. In spite of the excessively small current, the rate of change is great, and the difference of potential at the terminals of the secondary is therefore considerable, so that the arc is established at great distances; but the quantity of "electricity" set in motion is insignificant, barely sufficient to maintain a thin, threadlike arc. It is excessively sensitive and may be made so to such a degree that the mere act of breathing near the coil will affect it, and unless it is perfectly well protected from currents of air, it wriggles around constantly. Nevertheless, it is in this form excessively persistent, and when the terminals are approached to, say, one-third of the striking distance, it can be blown out only with difficulty. This exceptional persistency, when short, is largely due to the arc being excessively thin; presenting, therefore, a very small surface to the blast. Its great sensitiveness, when very long, is probably due to the motion of the particles of dust suspended in the air.

When the current through the primary is increased, the discharge gets broader and stronger, and the effect of the capacity of the coil becomes visible until, finally, under proper conditions, a white flaming arc, Fig. 100b, often as thick as one's finger, and striking across the whole coil, is produced. It develops remarkable heat, and may be further characterized by the absence of the high note which accompanies the less powerful discharges. To take a shock from the coil under these conditions would not be advisable, although under different conditions, the potential being much higher, a shock from the coil may be taken with impunity. To produce this kind of discharge the number of alternations per second must not be too great for the coil used; and, generally speaking, certain relations between capacity, self-induction and frequency must be observed.

The importance of these elements in an alternate current circuit is now well-known, and under ordinary conditions, the general rules are applicable. But in an induction coil exceptional conditions prevail. First, the self-induction is of little importance before the arc is established, when it asserts itself, but perhaps never as prominently as in ordinary alternate current circuits, because the capacity is distributed all along the coil, and by reason of the fact that the coil usually discharges through very great resistances; hence the currents are exceptionally small. Secondly, the capacity goes on increasing continually as the potential rises, in consequence of absorption which takes place to a considerable extent. Owing to this there exists no critical relationship between these quantities, and ordinary rules would not seem to be applicable. As the potential is increased either in consequence of the increased frequency or of the increased current through the primary, the amount of the energy stored becomes greater and greater, and the capacity gains more and more in importance. Up to a certain point the capacity is beneficial, but after that it begins to be an enormous drawback. It follows from this that each coil gives the best result with a given frequency and primary current. A very large coil, when operated with currents of very high frequency, may not give as much as ⅛ inch spark. By adding capacity to the terminals, the condition may be improved, but what the coil really wants is a lower frequency.

When the flaming discharge occurs, the conditions are evidently such that the greatest current is made to flow through the circuit. These conditions may be attained by varying the frequency within wide limits, but the highest frequency at which the flaming arc can still be produced, determines, for a given primary current, the maximum striking distance of the coil. In the flaming discharge the *eclat* effect of the capacity is not perceptible; the rate at which the energy is being stored then just equals the rate at which it can be disposed of through the circuit. This kind of discharge is the severest test for a coil; the break, when it occurs, is of the nature of that in an overcharged Leyden jar. To give a rough approximation I would state that, with an ordinary coil of, say 10,000 ohms resistance, the most powerful arc would be produced with about 12,000 alternations per second.

When the frequency is increased beyond that rate, the potential, of course, rises, but the striking distance may, nevertheless, diminish, paradoxical as it may seem. As the potential rises the coil attains more and more the properties of a static machine until, finally, one may observe the beautiful phenomenon of the streaming discharge, Fig. 101, which may be produced across the whole length of the coil. At that stage streams begin to issue freely from all points and projections. These streams will also be seen to pass in abundance in the space between the primary and the insulating tube. When the potential is excessively high they will always appear, even if the frequency be low, and even if the primary be surrounded by as much as an inch of wax, hard rubber, glass, or any other insulating substance. This limits greatly the output of the coil, but I will later show how I have been able to overcome to a considerable extent this disadvantage in the ordinary coil.

Besides the potential, the intensity of the streams depends on the frequency; but if the coil be very large they show themselves, no matter how low the frequencies used. For instance, in a very large coil of a resistance of 67,000 ohms, constructed by me some time ago, they appear with as low as 100 alternations per second and less, the insulation of the secondary being ¾ inch of ebonite. When very intense they produce a noise similar to that produced by the charging of a Holtz machine, but much more powerful, and they emit a strong smell of ozone. The lower the frequency, the more apt they are to suddenly injure the coil. With excessively high frequencies they may pass freely without producing any other effect than to heat the insulation slowly and uniformly.

The existence of these streams shows the importance of constructing an expensive coil so as to permit of one's seeing through the tube surrounding the primary, and the latter should be easily exchangeable; or else the space between the primary and secondary should be completely filled up with insulating material so as to exclude all air. The non-observance of this simple rule in the construction of commercial coils is responsible for the destruction of many an expensive coil.

FIG. 101.   FIG. 102.

At the stage when the streaming discharge occurs, or with somewhat higher frequencies, one may, by approaching the terminals quite nearly, and regulating properly the effect of capacity, produce a veritable spray of small silver-white sparks, or a bunch of excessively thin silvery threads (Fig. 102) amidst a powerful brush—each spark or thread possibly corresponding to one alternation. This, when produced under proper conditions, is probably the most beautiful discharge, and when an air blast is directed against it, it presents a singular appearance. The spray of sparks, when received through the body, causes some inconvenience, whereas, when the discharge simply streams, nothing at all is likely to be felt if large conducting objects are held in the hands to protect them from receiving small burns.

If the frequency is still more increased, then the coil refuses to give any spark unless at comparatively small distances, and the fifth typical form of discharge may be observed (Fig. 103). The tendency to stream out and dissipate is then so great that when the brush is produced at one terminal no sparking occurs, even if, as I have repeatedly tried, the hand, or any conducting object, is held within the stream; and, what is more singular, the luminous stream is not at all easily deflected by the approach of a conducting body.

FIG. 103.   FIG. 104.

At this stage the streams seemingly pass with the greatest freedom through considerable thicknesses of insulators, and it is particularly interesting to study their behavior. For this purpose it is convenient to connect to the terminals of the coil two metallic spheres which may be placed at any desired distance, Fig. 104. Spheres are preferable to plates, as the discharge can be better observed. By inserting dielectric bodies between the spheres, beautiful discharge phenomena may be observed. If the spheres be quite close and a spark be playing between them, by interposing a thin plate of ebonite between the spheres the spark instantly ceases and the discharge spreads into an intensely luminous circle several inches in diameter, provided the spheres are sufficiently large. The passage of the streams heats, and, after a while, softens, the rubber so much that two plates may be made to stick together in this manner. If the spheres are so far apart that no spark occurs, even if they are far beyond the striking distance, by inserting a thick plate of glass the discharge is instantly induced to pass from the spheres to the glass in the form of luminous streams. It appears almost as though these streams pass *through* the dielectric. In reality this is not the case, as the streams are due to the molecules of the air which are violently agitated in the space between the oppositely charged surfaces of the spheres. When no dielectric other than air is present, the bombardment goes on, but is too weak to be visible; by inserting a dielectric the inductive effect is much increased, and besides, the projected air molecules find an obstacle and the bombardment becomes so intense that the streams become luminous. If by any mechanical means we could effect such a violent agitation of the molecules we could produce the same phenomenon. A jet of air escaping through a small hole under enormous pressure and striking against an insulating substance, such as glass, may be luminous in the dark, and it might be possible to produce a phosphorescence of the glass or other insulators in this manner.

The greater the specific inductive capacity of the interposed dielectric, the more powerful the effect produced. Owing to this, the streams show themselves with excessively high potentials even if the glass be as much as one and one-half to two inches thick. But besides the heating due to bombardment, some heating goes on undoubtedly in the dielectric, being apparently greater in glass than in ebonite. I attribute this to the greater specific inductive capacity of the glass, in consequence of which, with the same potential difference, a greater amount of energy is taken up in it than in rubber. It is like connecting to a battery a copper and a brass wire of the same dimensions. The copper wire, though a more perfect conductor, would heat more by reason of its taking more current. Thus what is otherwise considered a virtue of the glass is here a defect. Glass usually gives way much quicker than ebonite; when it is heated to a certain degree, the discharge suddenly breaks through at one point, assuming then the ordinary form of an arc.

The heating effect produced by molecular bombardment of the dielectric would, of course, diminish as the pressure of the air is increased, and at enormous pressure it would be negligible, unless the frequency would increase correspondingly.

It will be often observed in these experiments that when the spheres are beyond the striking distance, the approach of a glass plate, for instance, may induce the spark to jump between the spheres. This occurs when the capacity of the spheres is somewhat below the critical value which gives the greatest difference of potential at the terminals of the coil. By approaching a dielectric, the specific inductive capacity of the space between the spheres is increased, producing the same effect as if the capacity of the spheres were increased. The potential at the terminals may then rise so high that the air space is cracked. The experiment is best performed with dense glass or mica.

Another interesting observation is that a plate of insulating material, when the discharge is passing through it, is strongly attracted by either of the spheres, that is by the nearer one, this being obviously due to the smaller mechanical effect of the bombardment on that side, and perhaps also to the greater electrification.

From the behavior of the dielectrics in these experiments, we may conclude that the best insulator for these rapidly alternating currents would be the one possessing the smallest specific inductive capacity and at the same time one capable of withstanding the greatest differences of potential; and thus two diametrically opposite ways of securing the required insulation are indicated, namely, to use either a perfect vacuum or a gas under great pressure; but the former would be preferable. Unfortunately neither of these two ways is easily carried out in practice.

It is especially interesting to note the behavior of an excessively high vacuum in these experiments. If a test tube, provided with external electrodes and exhausted to the highest possible degree, be connected to the terminals of the coil, Fig. 105, the electrodes of the tube are instantly brought to a high temperature and the glass at each end of the tube is rendered intensely phosphorescent, but the middle appears comparatively dark, and for a while remains cool.

When the frequency is so high that the discharge shown in Fig. 103 is observed, considerable dissipation no doubt occurs in the coil. Nevertheless the coil may be worked for a long time, as the heating is gradual.

In spite of the fact that the difference of potential may be enormous, little is felt when the discharge is passed through the body, provided the hands are armed. This is to some extent due to the higher frequency, but principally to the fact that less energy is available externally, when the difference of potential reaches an enormous value, owing to the circumstance that, with the rise of potential, the energy absorbed in the coil increases as the square of the potential. Up to a certain point the energy available externally increases with the rise of potential, then it begins to fall off rapidly. Thus, with the ordinary high tension induction coil, the curious paradox exists, that, while with a given current through the primary the shock might be fatal, with many times that current it might be perfectly harmless, even if the frequency be the same. With high frequencies and excessively high potentials when the terminals are not connected to bodies of some size, practically all the energy supplied to the primary is taken up by the coil. There is no breaking through, no local injury, but all the material, insulating and conducting, is uniformly heated.

To avoid misunderstanding in regard to the physiological effect of alternating currents of very high frequency, I think it necessary to state that, while it is an undeniable fact that they are incomparably less dangerous than currents of low frequencies, it should not be thought that they are altogether harmless. What has just been said refers only to currents from an ordinary high tension induction coil, which currents are necessarily very small; if received directly from a machine or from a secondary of low resistance, they produce more or less powerful effects, and may cause serious injury, especially when used in conjunction with condensers.

The streaming discharge of a high tension induction coil differs in many respects from that of a powerful static machine. In color it has neither the violet of the positive, nor the brightness of the negative, static discharge, but lies somewhere between, being, of course, alternatively positive and negative. But since the streaming is more powerful when the point or terminal is electrified positively, than when electrified negatively, it follows that the point of the brush is more like the positive, and the root more like the negative, static discharge. In the dark, when the brush is very powerful, the root may appear almost white. The wind produced by the escaping streams, though it may be very strong—often indeed to such a degree that it may be felt quite a distance from the coil—is, nevertheless, considering the quantity of the discharge, smaller than that produced by the positive brush of a static machine, and it affects the flame much less powerfully. From the nature of the phenomenon we can conclude that the higher the frequency, the smaller must, of course, be the wind produced by the streams, and with sufficiently high frequencies no wind at all would be produced at the ordinary atmospheric pressures. With frequencies obtainable by means of a machine, the mechanical effect is sufficiently great to revolve, with considerable speed, large pin-wheels, which in the dark present a beautiful appearance owing to the abundance of the streams (Fig. 106).

FIG. 105.   FIG. 106.

In general, most of the experiments usually performed with a static machine can be performed with an induction coil when operated with very rapidly alternating currents. The effects produced, however, are much more striking, being of incomparably greater power. When a small length of ordinary cotton covered wire, Fig. 107, is attached to one terminal of the coil, the streams issuing from all points of the wire may be so intense as to produce a considerable light effect. When the potentials and frequencies are very high, a wire insulated with gutta-percha or rubber and attached to one of the terminals, appears to be covered with a luminous film. A very thin bare wire when attached to a terminal emits powerful streams and vibrates continually to and fro or spins in a circle, producing a singular effect (Fig. 108). Some of these experiments have been described by me in *The Electrical World*, of February 21, 1891.

Another peculiarity of the rapidly alternating discharge of the induction coil is its radically different behavior with respect to points and rounded surfaces.

FIG. 107.   FIG. 108.

If a thick wire, provided with a ball at one end and with a point at the other, be attached to the positive terminal of a static machine, practically all the charge will be lost through the point, on account of the enormously greater tension, dependent on the radius of curvature. But if such a wire is attached to one of the terminals of the induction coil, it will be observed that with very high frequencies streams issue from the ball almost as copiously as from the point (Fig. 109).

FIG. 109.   FIG. 110.

It is hardly conceivable that we could produce such a condition to an equal degree in a static machine, for the simple reason, that the tension increases as the square of the density, which in turn is proportional to the radius of curvature; hence, with a steady potential an enormous charge would be required to make streams issue from a polished ball while it is connected with a point. But with an induction coil the discharge of which alternates with great rapidity it is different. Here we have to deal with two distinct tendencies. First, there is the tendency to escape which exists in a condition of rest, and which depends on the radius of curvature; second, there is the tendency to dissipate into the surrounding air by condenser action, which depends on the surface. When one of these tendencies is a maximum, the other is at a minimum. At the point the luminous stream is principally due to the air molecules coming bodily in contact with the point; they are attracted and repelled, charged and discharged, and, their atomic charges being thus disturbed, vibrate and emit light waves. At the ball, on the contrary, there is no doubt that the effect is to a great extent produced inductively, the air molecules not *necessarily* coming in contact with the ball, though they undoubtedly do so. To convince ourselves of this we only need to exalt the condenser action, for instance, by enveloping the ball, at some distance, by a better conductor than the surrounding medium, the conductor being, of course, insulated; or else by surrounding it with a better dielectric and approaching an insulated conductor; in both cases the streams will break forth more copiously. Also, the larger the ball with a given frequency, or the higher the frequency, the more will the ball have the advantage over the point. But, since a certain intensity of action is required to render the streams visible, it is obvious that in the experiment described the ball should not be taken too large.

In consequence of this two-fold tendency, it is possible to produce by means of points, effects identical to those produced by capacity. Thus, for instance, by attaching to one terminal of the coil a small length of soiled wire, presenting many points and offering great facility to escape, the potential of the coil may be raised to the same value as by attaching to the terminal a polished ball of a surface many times greater than that of the wire.

An interesting experiment, showing the effect of the points, may be performed in the following manner: Attach to one of the terminals of the coil a cotton covered wire about two feet in length, and adjust the conditions so that streams issue from the wire. In this experiment the primary coil should be preferably placed so that it extends only about half way into the secondary coil. Now touch the free terminal of the secondary with a conducting object held in the hand, or else connect it to an insulated body of some size. In this manner the potential on the wire may be enormously raised. The effect of this will be either to increase, or to diminish, the streams. If they increase, the wire is too short; if they diminish, it is too long. By adjusting the length of the wire, a point is found where the touching of the other terminal does not at all affect the streams. In this case the rise of potential is exactly counteracted by the drop through the coil. It will be observed that small lengths of wire produce considerable difference in the magnitude and luminosity of the streams. The primary coil is placed sidewise for two reasons: First, to increase the potential at the wire; and, second, to increase the drop through the coil. The sensitiveness is thus augmented.

There is still another and far more striking peculiarity of the brush discharge produced by very rapidly alternating currents. To observe this it is best to replace the usual terminals of the coil by two metal columns insulated with a good thickness of ebonite. It is also well to close all fissures and cracks with wax so that the brushes cannot form anywhere except at the tops of the columns. If the conditions are carefully adjusted—which, of course, must be left to the skill of the experimenter—so that the potential rises to an enormous value, one may produce two powerful brushes several inches long, nearly white at their roots, which in the dark bear a striking resemblance to two flames of a gas escaping under pressure (Fig. 110). But they do not only *resemble*, they *are* veritable flames, for they are hot. Certainly they are not as hot as a gas burner, *but they would be so if the frequency and the potential would be sufficiently high.* Produced with, say, twenty thousand alternations per second, the heat is easily perceptible even if the potential is not excessively high. The heat developed is, of course, due to the impact of the air molecules against the terminals and against each other. As, at the ordinary pressures, the mean free path is excessively small, it is possible that in spite of the enormous initial speed imparted to each molecule upon coming in contact with the terminal, its progress—by collision with other molecules—is retarded to such an extent, that it does not get away far from the terminal, but may strike the same many times in succession. The higher the frequency, the less the molecule is able to get away, and this the more so, as for a given effect the potential required is smaller;

and a frequency is conceivable—perhaps even obtainable—at which practically the same molecules would strike the terminal. Under such conditions the exchange of the molecules would be very slow, and the heat produced at, and very near, the terminal would be excessive. But if the frequency would go on increasing constantly, the heat produced would begin to diminish for obvious reasons. In the positive brush of a static machine the exchange of the molecules is very rapid, the stream is constantly of one direction, and there are fewer collisions; hence the heating effect must be very small. Anything that impairs the facility of exchange tends to increase the local heat produced. Thus, if a bulb be held over the terminal of the coil so as to enclose the brush, the air contained in the bulb is very quickly brought to a high temperature. If a glass tube be held over the brush so as to allow the draught to carry the brush upwards, scorching hot air escapes at the top of the tube. Anything held within the brush is, of course, rapidly heated, and the possibility of using such heating effects for some purpose or other suggests itself.

When contemplating this singular phenomenon of the hot brush, we cannot help being convinced that a similar process must take place in the ordinary flame, and it seems strange that after all these centuries past of familiarity with the flame, now, in this era of electric lighting and heating, we are finally led to recognize, that since time immemorial we have, after all, always had "electric light and heat" at our disposal. It is also of no little interest to contemplate, that we have a possible way of producing—by other than chemical means—a veritable flame, which would give light and heat without any material being consumed, without any chemical process taking place, and to accomplish this, we only need to perfect methods of producing enormous frequencies and potentials. I have no doubt that if the potential could be made to alternate with sufficient rapidity and power, the brush formed at the end of a wire would lose its electrical characteristics and would become flamelike. The flame must be due to electrostatic molecular action.

This phenomenon now explains in a manner which can hardly be doubted the frequent accidents occurring in storms. It is well known that objects are often set on fire without the lightning striking them. We shall presently see how this can happen. On a nail in a roof, for instance, or on a projection of any kind, more or less conducting, or rendered so by dampness, a powerful brush may appear. If the lightning strikes somewhere in the neighborhood the enormous potential may be made to alternate or fluctuate perhaps many million times a second. The air molecules are violently attracted and repelled, and by their impact produce such a powerful heating effect that a fire is started. It is conceivable that a ship at sea may, in this manner, catch fire at many points at once. When we consider, that even with the comparatively low frequencies obtained from a dynamo machine, and with potentials of no more than one or two hundred thousand volts, the heating effects are considerable, we may imagine how much more powerful they must be with frequencies and potentials many times greater; and the above explanation seems, to say the least, very probable. Similar explanations may have been suggested, but I am not aware that, up to the present, the heating effects of a brush produced by a rapidly alternating potential have been experimentally demonstrated, at least not to such a remarkable degree.

Fig. 111.

By preventing completely the exchange of the air molecules, the local heating effect may be so exalted as to bring a body to incandescence. Thus, for instance, if a small button, or preferably a very thin wire or filament be enclosed in an unexhausted globe and connected with the terminal of the coil, it may be rendered incandescent. The phenomenon is made much more interesting by the rapid spinning round in a circle of the top of the filament, thus presenting the appearance of a luminous funnel, Fig. 111, which widens when the potential is increased. When the potential is small the end of the filament may perform irregular motions, suddenly changing from one to the other, or it may describe an ellipse; but when the potential is very high it always spins in a circle; and so does generally a thin straight wire attached freely to the terminal of the coil. These motions are, of course, due to the impact of the molecules, and the irregularity in the distribution of the potential, owing to the roughness and dissymmetry of the wire or filament. With a perfectly symmetrical and polished wire such motions would probably not occur. That the motion is not likely to be due to other causes is evident from the fact that it is not of a definite direction, and that in a very highly exhausted globe it ceases altogether. The possibility of bringing a body to incandescence in an exhausted globe, or even when not at all enclosed, would seem to afford a possible way of obtaining light effects, which, in perfecting methods of producing rapidly alternating potentials, might be rendered available for useful purposes.

In employing a commercial coil, the production of very powerful brush effects is attended with considerable difficulties, for when these high frequencies and enormous potentials are used, the best insulation is apt to give way. Usually the coil is insulated well enough to stand the strain from convolution to convolution, since two double silk covered paraffined wires will withstand a pressure of several thousand volts; the difficulty lies principally in preventing the breaking through from the secondary to the primary, which is greatly facilitated by the streams issuing from the latter. In the coil, of course, the strain is greatest from section to section, but usually in a larger coil there are so many sections that the danger of a sudden giving way is not very great. No difficulty will generally be encountered in that direction, and besides, the liability of injuring the coil internally is very much reduced by the fact that the effect most likely to be produced is simply a gradual heating, which, when far enough advanced, could not fail to be observed. The principal necessity is then to prevent the streams between the primary and the tube, not only on account of the heating and possible injury, but also because the streams may diminish very considerably the potential difference available at the terminals. A few hints as to how this may be accomplished will probably be found useful in most of these experiments with the ordinary induction coil.

Fig. 112a.

One of the ways is to wind a short primary, Fig. 112a, so that the difference of potential is not at that length great enough to cause the breaking forth of the streams through the insulating tube. The length of the primary should be determined by experiment. Both the ends of the coil should be brought out on one end through a plug of insulating material fitting in the tube as illustrated. In such a disposition one terminal of the secondary is attached to a body, the surface of which is determined with the greatest care so as to produce the greatest rise in the potential. At the other terminal a powerful brush appears, which may be experimented upon.

The above plan necessitates the employment of a primary of comparatively small size, and it is apt to heat when powerful effects are desirable for a certain length of time. In such a case it is better to employ a larger coil, Fig. 112b, and introduce it from one side of the tube, until the streams begin to appear. In this case the nearest terminal of the secondary may be connected to the primary or to the ground, which is practically the same thing, if the primary is connected directly to the machine. In the case of ground connections it is well to determine experimentally the frequency which is best suited under the conditions of the test. Another way of obviating the streams, more or less, is to make the primary in sections and supply it from separate, well insulated sources.

FIG. 112b.

In many of these experiments, when powerful effects are wanted for a short time, it is advantageous to use iron cores with the primaries. In such case a very large primary coil may be wound and placed side by side with the secondary, and, the nearest terminal of the latter being connected to the primary, a laminated iron core is introduced through the primary into the secondary as far as the streams will permit. Under these conditions an excessively powerful brush, several inches long, which may be appropriately called "St. Elmo's hot fire," may be caused to appear at the other terminal of the secondary, producing striking effects. It is a most powerful ozonizer, so powerful indeed, that only a few minutes are sufficient to fill the whole room with the smell of ozone, and it undoubtedly possesses the quality of exciting chemical affinities.

For the production of ozone, alternating currents of very high frequency are eminently suited, not only on account of the advantages they offer in the way of conversion but also because of the fact, that the ozonizing action of a discharge is dependent on the frequency as well as on the potential, this being undoubtedly confirmed by observation.

In these experiments if an iron core is used it should be carefully watched, as it is apt to get excessively hot in an incredibly short time. To give an idea of the rapidity of the heating, I will state, that by passing a powerful current through a coil with many turns, the inserting within the same of a thin iron wire for no more than one second's time is sufficient to heat the wire to something like 100°C.

But this rapid heating need not discourage us in the use of iron cores in connection with rapidly alternating currents. I have for a long time been convinced that in the industrial distribution by means of transformers, some such plan as the following might be practicable. We may use a comparatively small iron core, subdivided, or perhaps not even subdivided. We may surround this core with a considerable thickness of material which is fire-proof and conducts the heat poorly, and on top of that we may place the primary and secondary windings. By using either higher frequencies or greater magnetizing forces, we may by hysteresis and eddy currents heat the iron core so far as to bring it nearly to its maximum permeability, which, as Hopkinson has shown, may be as much as sixteen times greater than that at ordinary temperatures. If the iron core were perfectly enclosed, it would not be deteriorated by the heat, and, if the enclosure of fire-proof material would be sufficiently thick, only a limited amount of energy could be radiated in spite of the high temperature. Transformers have been constructed by me on that plan, but for lack of time, no thorough tests have as yet been made.

Another way of adapting the iron core to rapid alternations, or, generally speaking, reducing the frictional losses, is to produce by continuous magnetization a flow of something like seven thousand or eight thousand lines per square centimeter through the core, and then work with weak magnetizing forces and preferably high frequencies around the point of greatest permeability. A higher efficiency of conversion and greater output are obtainable in this manner. I have also employed this principle in connection with machines in which there is no reversal of polarity. In these types of machines, as long as there are only few pole projections, there is no great gain, as the maxima and minima of magnetization are far from the point of maximum permeability; but when the number of the pole projections is very great, the required rate of change may be obtained, without the magnetization varying so far as to depart greatly from the point of maximum permeability, and the gain is considerable.

The above described arrangements refer only to the use of commercial coils as ordinarily constructed. If it is desired to construct a coil for the express purpose of performing with it such experiments as I have described, or, generally, rendering it capable of withstanding the greatest possible difference of potential, then a construction as indicated in

Fig. 113 will be found of advantage. The coil in this case is formed of two independent parts which are wound oppositely, the connection between both being made near the primary. The potential in the middle being zero, there is not much tendency to jump to the primary and not much insulation is required. In some cases the middle point may, however, be connected to the primary or to the ground. In such a coil the places of greatest difference of potential are far apart and the coil is capable of withstanding an enormous strain. The two parts may be movable so as to allow a slight adjustment of the capacity effect.

FIG. 113.

As to the manner of insulating the coil, it will be found convenient to proceed in the following way: First, the wire should be boiled in paraffin until all the air is out; then the coil is wound by running the wire through melted paraffin, merely for the purpose of fixing the wire. The coil is then taken off from the spool, immersed in a cylindrical vessel filled with pure melted wax and boiled for a long time until the bubbles cease to appear. The whole is then left to cool down thoroughly, and then the mass is taken out of the vessel and turned up in a lathe. A coil made in this manner and with care is capable of withstanding enormous potential differences.

It may be found convenient to immerse the coil in paraffin oil or some other kind of oil; it is a most effective way of insulating, principally on account of the perfect exclusion of air, but it may be found that, after all, a vessel filled with oil is not a very convenient thing to handle in a laboratory.

If an ordinary coil can be dismounted, the primary may be taken out of the tube and the latter plugged up at one end, filled with oil, and the primary reinserted. This affords an excellent insulation and prevents the formation of the streams.

Of all the experiments which may be performed with rapidly alternating currents the most interesting are those which concern the production of a practical illuminant. It cannot be denied that the present methods, though they were brilliant advances, are very wasteful. Some better methods must be invented, some more perfect apparatus devised. Modern research has opened new possibilities for the production of an efficient source of light, and the attention of all has been turned in the direction indicated by able pioneers. Many have been carried away by the enthusiasm and passion to discover, but in their zeal to reach results, some have been misled. Starting with the idea of producing electro-magnetic waves, they turned their attention, perhaps, too much to the study of electro-magnetic effects, and neglected the study of electrostatic phenomena. Naturally, nearly every investigator availed himself of an apparatus similar to that used in earlier experiments. But in those forms of apparatus, while the electro-magnetic inductive effects are enormous, the electrostatic effects are excessively small.

In the Hertz experiments, for instance, a high tension induction coil is short-circuited by an arc, the resistance of which is very small, the smaller, the more capacity is attached to the terminals; and the difference of potential at these is enormously diminished. On the other hand, when the discharge is not passing between the terminals, the static effects may be considerable, but only qualitatively so, not quantitatively, since their rise and fall is very sudden, and since their frequency is small. In neither case, therefore, are powerful electrostatic effects perceivable. Similar conditions exist when, as in some interesting experiments of Dr. Lodge, Leyden jars are discharged disruptively. It has been thought—and I believe asserted—that in such cases most of the energy is radiated into space. In the light of the experiments which I have described above, it will now not be thought so. I feel safe in asserting that in such cases most of the energy is partly taken up and converted into heat in the arc of the discharge and in the conducting and insulating material of the jar, some energy being, of course, given off by electrification of the air; but the amount of the directly radiated energy is very small.

When a high tension induction coil, operated by currents alternating only 20,000 times a second, has its terminals closed through even a very small jar, practically all the energy passes through the dielectric of the jar, which is heated, and the electrostatic effects manifest themselves outwardly only to a very weak degree. Now the external circuit of a Leyden jar, that is, the arc and the connections of the coatings, may be looked upon as a circuit generating alternating currents of excessively high frequency and fairly high potential, which is closed through the coatings and the dielectric between them, and from the above it is evident that the external electrostatic effects must be very small, even if a recoil circuit be used. These conditions make it appear that with the apparatus usually at hand, the observation of powerful electrostatic effects was impossible, and what experience has been gained in that direction is only due to the great ability of the investigators.

But powerful electrostatic effects are a *sine qua non* of light production on the lines indicated by theory. Electro-magnetic effects are primarily unavailable, for the reason that to produce the required effects we would have to pass current impulses through a conductor, which, long before the required frequency of the impulses could be reached, would cease to transmit them. On the other hand, electro-magnetic waves many times longer than those of light, and producible by sudden discharge of a condenser, could not be utilized, it would seem, except we avail ourselves of their effect upon conductors as in the present methods, which are wasteful. We could not affect by means of such waves the static molecular or atomic charges of a gas, cause them to vibrate and to emit light. Long transverse waves cannot, apparently, produce such effects, since excessively small electro-magnetic disturbances may pass readily through miles of air. Such dark waves, unless they are of the length of true light waves, cannot, it would seem, excite luminous radiation in a Geissler tube, and the luminous effects, which are producible by induction in a tube devoid of electrodes, I am inclined to consider as being of an electrostatic nature.

To produce such luminous effects, straight electrostatic thrusts are required; these, whatever be their frequency, may disturb the molecular charges and produce light. Since current impulses of the required frequency cannot pass through a conductor of measurable dimensions, we must work with a gas, and then the production of powerful electrostatic effects becomes an imperative necessity.

It has occurred to me, however, that electrostatic effects are in many ways available for the production of light. For instance, we may place a body of some refractory material in a closed, and preferably more or less exhausted, globe, connect it to a source of high, rapidly alternating potential, causing the molecules of the gas to strike it many times a second at enormous speeds, and in this manner, with trillions of invisible hammers, pound it until it gets incandescent; or we may place a body in a very highly exhausted globe, in a non-striking vacuum, and, by employing very high frequencies and potentials, transfer sufficient energy from it to other bodies in the vicinity, or in general to the surroundings, to maintain it at any degree of incandescence; or we may, by means of such rapidly alternating high potentials, disturb the ether carried by the molecules of a gas or their static charges, causing them to vibrate and to emit light.

But, electrostatic effects being dependent upon the potential and frequency, to produce the most powerful action it is desirable to increase both as far as practicable. It may be possible to obtain quite fair results by keeping either of these factors small, provided the other is sufficiently great; but we are limited in both directions. My experience demonstrates that we cannot go below a certain frequency, for, first, the potential then becomes so great that it is dangerous; and, secondly, the light production is less efficient.

I have found that, by using the ordinary low frequencies, the physiological effect of the current required to maintain at a certain degree of brightness a tube four feet long, provided at the ends with outside and inside condenser coatings, is so powerful that, I think, it might produce serious injury to those not accustomed to such shocks; whereas, with twenty thousand alternations per second, the tube may be maintained at the same degree of brightness without any effect being felt. This is due principally to the fact that a much smaller potential is required to produce the same light effect, and also to the higher efficiency in the light production. It is evident that the efficiency in such cases is the greater, the higher the frequency, for the quicker the process of charging and discharging the molecules, the less energy will be lost in the form of dark radiation. But, unfortunately, we cannot go beyond a certain frequency on account of the difficulty of producing and conveying the effects.

FIG. 114.　　　　　　　　　　　　　　　　FIG. 115.

I have stated above that a body inclosed in an unexhausted bulb may be intensely heated by simply connecting it with a source of rapidly alternating potential. The heating in such a case is, in all probability, due mostly to the bombardment of the molecules of the gas contained in the bulb. When the bulb is exhausted, the heating of the body is much more rapid, and there is no difficulty whatever in bringing a wire or filament to any degree of incandescence by simply connecting it to one terminal of a coil of the proper dimensions. Thus, if the well-known apparatus of Prof. Crookes, consisting of a bent platinum wire with vanes mounted over it (Fig. 114), be connected to one terminal of the coil—either one or both ends of the platinum wire being connected— the wire is rendered almost instantly incandescent, and the mica vanes are rotated as though a current from a battery were used. A thin carbon filament, or, preferably, a button of some refractory material (Fig. 115), even if it be a comparatively poor conductor, inclosed in an exhausted globe, may be rendered highly incandescent; and in this manner a simple lamp capable of giving any desired candle power is provided.

The success of lamps of this kind would depend largely on the selection of the light-giving bodies contained within the bulb. Since, under the conditions described, refractory bodies—which are very poor conductors and capable of withstanding for a long time excessively high degrees of temperature—may be used, such illuminating devices may be rendered successful.

It might be thought at first that if the bulb, containing the filament or button of refractory material, be perfectly well exhausted—that is, as far as it can be done by the use of the best apparatus—the heating would be much less intense, and that in a perfect vacuum it could not occur at all. This is not confirmed by my experience; quite the contrary, the better the vacuum the more easily the bodies are brought to incandescence. This result is interesting for many reasons.

At the outset of this work the idea presented itself to me, whether two bodies of refractory material enclosed in a bulb exhausted to such a degree that the discharge of a large induction coil, operated in the usual manner, cannot pass through, could be rendered incandescent by mere condenser action. Obviously, to reach this result enormous potential differences and very high frequencies are required, as is evident from a simple calculation.

But such a lamp would possess a vast advantage over an ordinary incandescent lamp in regard to efficiency. It is well-known that the efficiency of a lamp is to some extent a function of the degree of incandescence, and that, could we but work a filament at many times higher degrees of incandescence, the efficiency would be much greater. In an ordinary lamp this is impracticable on account of the destruction of the filament, and it has been determined by experience how far it is advisable to push the incandescence. It is impossible to tell how much higher efficiency could be obtained if the filament could withstand indefinitely, as the investigation to this end obviously cannot be carried beyond a certain stage; but there are reasons for believing that it would be very considerably higher. An improvement might be made in the ordinary lamp by employing a short and thick carbon; but then the leading-in wires would have to be thick, and, besides, there are many other considerations which render such a modification entirely impracticable. But in a lamp as above described, the leading-in wires may be very small, the incandescent refractory material may be in the shape of blocks offering a very small radiating surface, so that less energy would be required to keep them at the desired incandescence; and in addition to this, the refractory material need not be carbon, but may be manufactured from mixtures of oxides, for instance, with carbon or other material, or may be selected from bodies which are practically non-conductors, and capable of withstanding enormous degrees of temperature.

All this would point to the possibility of obtaining a much higher efficiency with such a lamp than is obtainable in ordinary lamps. In my experience it has been demonstrated that the blocks are brought to high degrees of incandescence with much lower potentials than those determined by calculation, and the blocks may be set at greater distances from each other. We may freely assume, and it is probable, that the molecular bombardment is an important element in the heating, even if the globe be exhausted with the utmost care, as I have done; for although the number of the molecules is, comparatively speaking, insignificant, yet on account of the mean free path being very great, there are fewer collisions, and the molecules may reach much higher speeds, so that the heating effect due to this cause may be considerable, as in the Crookes experiments with radiant matter.

But it is likewise possible that we have to deal here with an increased facility of losing the charge in very high vacuum, when the potential is rapidly alternating, in which case most of the heating would be directly due to the surging of the charges in the heated bodies. Or else the observed fact may be largely attributable to the effect of the points which I have mentioned above, in consequence of which the blocks or filaments contained in the vacuum are equivalent to condensers of many times greater surface than that calculated from their geometrical dimensions. Scientific men still differ in opinion as to whether a charge should, or should not, be lost in a perfect vacuum, or in other words, whether ether is, or is not, a conductor. If the former were the case, then a thin filament enclosed in a perfectly exhausted globe, and connected to a source of enormous, steady potential, would be brought to incandescence.

Various forms of lamps on the above described principle, with the refractory bodies in the form of filaments, Fig. 116, or blocks, Fig. 117, have been constructed and operated by me, and investigations are being carried on in this line. There is no difficulty in reaching such high degrees of incandescence that ordinary carbon is to all appearance melted and volatilized. If the vacuum could be made absolutely perfect, such a lamp, although inoperative with apparatus

FIG. 116.   FIG. 117.

ordinarily used, would, if operated with currents of the required character, afford an illuminant which would never be destroyed, and which would be far more efficient than an ordinary incandescent lamp. This perfection can, of course, never be reached, and a very slow destruction and gradual diminution in size always occurs, as in incandescent filaments; but there is no possibility of a sudden and premature disabling which occurs in the latter by the breaking of the filament, especially when the incandescent bodies are in the shape of blocks.

With these rapidly alternating potentials there is, however, no necessity of enclosing two blocks in a globe, but a single block, as in Fig. 115, or filament, Fig. 118, may be used. The potential in this case must of course be higher, but is easily obtainable, and besides it is not necessarily dangerous.

The facility with which the button or filament in such a lamp is brought to incandescence, other things being equal, depends on the size of the globe. If a perfect vacuum could be obtained, the size of the globe would not be of importance, for then the heating would be wholly due to the surging of the charges, and all the energy would be given off to the surroundings by radiation. But this can never occur in practice. There is always some gas left in the globe, and although the exhaustion may be carried to the highest degree, still the space inside of the bulb must be considered as conducting when such high potentials are used, and I assume that, in estimating the energy that may be given off from the filament to the surroundings, we may consider the inside surface of the bulb as one coating of a condenser, the air and other objects surrounding the bulb forming the other coating. When the alternations are very low there is no doubt that a considerable portion of the energy is given off by the electrification of the surrounding air.

FIG. 118.

In order to study this subject better, I carried on some experiments with excessively high potentials and low frequencies. I then observed that when the hand is approached to the bulb—the filament being connected with one terminal of the coil—a powerful vibration is felt, being due to the attraction and repulsion of the molecules of the air which are electrified by induction through the glass. In some cases when the action is very intense I have been able to hear a sound, which must be due to the same cause.

When the alternations are low, one is apt to get an excessively powerful shock from the bulb. In general, when one attaches bulbs or objects of some size to the terminals of the coil, one should look out for the rise of potential, for it may happen that by merely connecting a bulb or plate to the terminal, the potential may rise to many times its original value. When lamps are attached to the terminals, as illustrated in Fig. 119, then the capacity of the bulbs should be such as to give the maximum rise of potential under the existing conditions. In this manner one may obtain the required potential with fewer turns of wire.

FIG. 119.   FIG. 120.

The life of such lamps as described above depends, of course, largely on the degree of exhaustion, but to some extent also on the shape of the block of refractory material. Theoretically it would seem that a small sphere of carbon enclosed in a sphere of glass would not suffer deterioration from molecular bombardment, for, the matter in the globe being radiant, the molecules would move in straight lines, and would seldom strike the sphere obliquely. An interesting thought in connection with such a lamp is, that in it "electricity" and electrical energy apparently must move in the same lines.

The use of alternating currents of very high frequency makes it possible to transfer, by electrostatic or electromagnetic induction through the glass of a lamp, sufficient energy to keep a filament at incandescence and so do away with the leading-in wires. Such lamps have been proposed, but for want of proper apparatus they have not been successfully operated. Many forms of lamps on this principle with continuous and broken filaments have been constructed by me and experimented upon. When using a secondary enclosed within the lamp, a condenser is advantageously combined with the secondary.

When the transference is effected by electrostatic induction, the potentials used are, of course, very high with frequencies obtainable from a machine. For instance, with a condenser surface of forty square centimeters, which is not impracticably large, and with glass of good quality 1 mm. thick, using currents alternating twenty thousand times a second, the potential required is approximately 9,000 volts. This may seem large, but since each lamp may be included in the secondary of a transformer of very small dimensions, it would not be inconvenient, and, moreover, it would not produce fatal injury. The transformers would all be preferably in series. The regulation would offer no difficulties, as with currents of such frequencies it is very easy to maintain a constant current.

FIG. 121a.   FIG. 121b..

In the accompanying engravings some of the types of lamps of this kind are shown. Fig. 120 is such a lamp with a broken filament, and Figs. 121a and 121b one with a single outside and inside coating and a single filament. I have also made lamps with two outside and inside coatings and a continuous loop connecting the latter. Such lamps have been operated by me with current impulses of the enormous frequencies obtainable by the disruptive discharge of condensers.

The disruptive discharge of a condenser is especially suited for operating such lamps—with no outward electrical connections—by means of electromagnetic induction, the electromagnetic inductive effects being excessively high; and I have been able to produce the desired incandescence with only a few short turns of wire. Incandescence may also be produced in this manner in a simple closed filament.

Leaving now out of consideration the practicability of such lamps, I would only say that they possess a beautiful and desirable feature, namely, that they can be rendered, at will, more or less brilliant simply by altering the relative position of the outside and inside condenser coatings, or inducing and induced circuits.

Fig. 122.   Fig. 123.

When a lamp is lighted by connecting it to one terminal only of the source, this may be facilitated by providing the globe with an outside condenser coating, which serves at the same time as a reflector, and connecting this to an insulated body of some size. Lamps of this kind are illustrated in Fig. 122 and Fig. 123. Fig. 124 shows the plan of connection. The brilliancy of the lamp may, in this case, be regulated within wide limits by varying the size of the insulated metal plate to which the coating is connected.

It is likewise practicable to light with one leading wire lamps such as illustrated in Fig. 116 and Fig. 117, by connecting one terminal of the lamp to one terminal of the source, and the other to an insulated body of the required size. In all cases the insulated body serves to give off the energy into the surrounding space, and is equivalent to a return wire. Obviously, in the two last-named cases, instead of connecting the wires to an insulated body, connections may be made to the ground.

Fig. 124.

The experiments which will prove most suggestive and of most interest to the investigator are probably those performed with exhausted tubes. As might be anticipated, a source of such rapidly alternating potentials is capable of exciting the tubes at a considerable distance, and the light effects produced are remarkable.

During my investigations in this line I endeavored to excite tubes, devoid of any electrodes, by electromagnetic induction, making the tube the secondary of the induction device, and passing through the primary the discharges of a Leyden jar. These tubes were made of many shapes, and I was able to obtain luminous effects which I then thought were due wholly to electromagnetic induction. But on carefully investigating the phenomena I found that the effects produced were more of an electrostatic nature. It may be attributed to this circumstance that this mode of exciting tubes is very wasteful, namely, the primary circuit being closed, the potential, and consequently the electrostatic inductive effect, is much diminished.

When an induction coil, operated as above described, is used, there is no doubt that the tubes are excited by electrostatic induction, and that electromagnetic induction has little, if anything, to do with the phenomena.

This is evident from many experiments. For instance, if a tube be taken in one hand, the observer being near the coil, it is brilliantly lighted and remains so no matter in what position it is held relatively to the observer's body. Were the action electromagnetic, the tube could not be lighted when the observer's body is interposed between it and the coil, or at least its luminosity should be considerably diminished. When the tube is held exactly over the center of the coil—the latter being wound in sections and the primary placed symmetrically to the secondary—it may remain completely dark, whereas it is rendered intensely luminous by moving it slightly to the right or left from the center of the coil. It does not light because in the middle both halves of the coil neutralize each other, and the electric potential is zero. If the action were electromagnetic, the tube should light best in the plane through the center of the coil, since the electromagnetic effect there should be a maximum. When an arc is established between the terminals, the tubes and lamps in the vicinity of the coil go out, but light up again when the arc is broken, on account of the rise of potential. Yet the electromagnetic effect should be practically the same in both cases.

By placing a tube at some distance from the coil, and nearer to one terminal—preferably at a point on the axis of the coil—one may light it by touching the remote terminal with an insulated body of some size or with the hand, thereby raising the potential at that terminal nearer to the tube. If the tube is shifted nearer to the coil so that it is lighted by the action of the nearer terminal, it may be made to go out by holding, on an insulated support, the end of a wire connected to the remote terminal, in the vicinity of the nearer terminal, by this means counteracting the action of the latter upon the tube. These effects are evidently electrostatic. Likewise, when a tube is placed at a considerable distance from the coil, the observer may, standing upon an insulated support between coil and tube, light the latter by approaching the hand to it; or he may even render it luminous by simply stepping between it and the coil. This would be impossible with electro-magnetic induction, for the body of the observer would act as a screen.

When the coil is energized by excessively weak currents, the experimenter may, by touching one terminal of the coil with the tube, extinguish the latter, and may again light it by bringing it out of contact with the terminal and allowing a small arc to form. This is clearly due to the respective lowering and raising of the potential at that terminal. In the above experiment, when the tube is lighted through a small arc, it may go out when the arc is broken, because the electrostatic inductive effect alone is too weak, though the potential may be much higher; but when the arc is established, the electrification of the end of the tube is much greater, and it consequently lights.

If a tube is lighted by holding it near to the coil, and in the hand which is remote, by grasping the tube anywhere with the other hand, the part between the hands is rendered dark, and the singular effect of wiping out the light of the tube may be produced by passing the hand quickly along the tube and at the same time withdrawing it gently from the coil, judging properly the distance so that the tube remains dark afterwards.

If the primary coil is placed sidewise, as in Fig. 112b for instance, and an exhausted tube be introduced from the other side in the hollow space, the tube is lighted most intensely because of the increased condenser action, and in this position the striae are most sharply defined. In all these experiments described, and in many others, the action is clearly electrostatic.

The effects of screening also indicate the electrostatic nature of the phenomena and show something of the nature of electrification through the air. For instance, if a tube is placed in the direction of the axis of the coil, and an insulated metal plate be interposed, the tube will generally increase in brilliancy, or if it be too far from the coil to light, it may even be rendered luminous by interposing an insulated metal plate. The magnitude of the effects depends to some extent on the size of the plate. But if the metal plate be connected by a wire to the ground, its interposition will always make the tube go out even if it be very near the coil. In general, the interposition of a body between the coil and tube, increases or diminishes the brilliancy of the tube, or its facility to light up, according to whether it increases or diminishes the electrification. When experimenting with an insulated plate, the plate should not be taken too large, else it will generally produce a weakening effect by reason of its great facility for giving off energy to the surroundings.

If a tube be lighted at some distance from the coil, and a plate of hard rubber or other insulating substance be interposed, the tube may be made to go out. The interposition of the dielectric in this case only slightly increases the inductive effect, but diminishes considerably the electrification through the air.

In all cases, then, when we excite luminosity in exhausted tubes by means of such a coil, the effect is due to the rapidly alternating electrostatic potential;

and, furthermore, it must be attributed to the harmonic alternation produced directly by the machine, and not to any superimposed vibration which might be thought to exist. Such superimposed vibrations are impossible when we work with an alternate current machine. If a spring be gradually tightened and released, it does not perform independent vibrations; for this a sudden release is necessary. So with the alternate currents from a dynamo machine; the medium is harmonically strained and released, this giving rise to only one kind of waves; a sudden contact or break, or a sudden giving way of the dielectric, as in the disruptive discharge of a Leyden jar, are essential for the production of superimposed waves.

In all the last described experiments, tubes devoid of any electrodes may be used, and there is no difficulty in producing by their means sufficient light to read by. The light effect is, however, considerably increased by the use of phosphorescent bodies such as yttria, uranium glass, etc. A difficulty will be found when the phosphorescent material is used, for with these powerful effects, it is carried gradually away, and it is preferable to use material in the form of a solid.

Instead of depending on induction at a distance to light the tube, the same may be provided with an external—and, if desired, also with an internal—condenser coating, and it may then be suspended anywhere in the room from a conductor connected to one terminal of the coil, and in this manner a soft illumination may be provided.

The ideal way of lighting a hall or room would, however, be to produce such a condition in it that an illuminating device could be moved and put anywhere, and that it is lighted, no matter where it is put and without being electrically connected to anything. I have been able to produce such a condition by creating in the room a powerful, rapidly alternating electrostatic field. For this purpose I suspend a sheet of metal a distance from the ceiling on insulating cords and connect it to one terminal of the induction coil, the other terminal being preferably connected to the ground. Or else I suspend two sheets as illustrated in Fig. 125, each sheet being connected with one of the terminals of the coil, and their size being carefully determined. An exhausted tube may then be carried in the hand anywhere between the sheets or placed anywhere, even a certain distance beyond them; it remains always luminous.

Fig. 125.

In such an electrostatic field interesting phenomena may be observed, especially if the alternations are kept low and the potentials excessively high. In addition to the luminous phenomena mentioned, one may observe that any insulated conductor gives sparks when the hand or another object is approached to it, and the sparks may often be powerful. When a large conducting object is fastened on an insulating support, and the hand approached to it, a vibration, due to the rythmical motion of the air molecules is felt, and luminous streams may be perceived when the hand is held near a pointed projection. When a telephone receiver is made to touch with one or both of its terminals an insulated conductor of some size, the telephone emits a loud sound; it also emits a sound when a length of wire is attached to one or both terminals, and with very powerful fields a sound may be perceived even without any wire.

How far this principle is capable of practical application, the future will tell. It might be thought that electrostatic effects are unsuited for such action at a distance. Electromagnetic inductive effects, if available for the production of light, might be thought better suited. It is true the electrostatic effects diminish nearly with the cube of the distance from the coil, whereas the electromagnetic inductive effects diminish simply with the distance. But when we establish an electrostatic field of force, the condition is very different, for then, instead of the differential effect of both the terminals, we get their conjoint effect. Besides, I would call attention to the effect, that in an alternating electrostatic field, a conductor, such as an exhausted tube, for instance, tends to take up most of the energy, whereas in an electromagnetic alternating field the conductor tends to take up the least energy, the waves being reflected with but little loss. This is one reason why it is difficult to excite an exhausted tube, at a distance, by electromagnetic induction. I have wound coils of very large diameter and of many turns of wire, and connected a Geissler tube to the ends of the coil with the object of exciting the tube at a distance; but even with the powerful inductive effects producible by Leyden jar discharges, the tube could not be excited unless at a very small distance, although some judgment was used as to the dimensions of the coil. I have also found that even the most powerful Leyden jar discharges are capable of exciting only feeble luminous effects in a closed exhausted tube, and even these effects upon thorough examination I have been forced to consider of an electrostatic nature.

How then can we hope to produce the required effects at a distance by means of electromagnetic action, when even in the closest proximity to the source of disturbance, under the most advantageous conditions, we can excite but faint luminosity? It is true that when acting at a distance we have the resonance to help us out. We can connect an exhausted tube, or whatever the illuminating device may be, with an insulated system of the proper capacity, and so it may be possible to increase the effect qualitatively, and only qualitatively, for we would not get *more* energy through the device. So we may, by resonance effect, obtain the required electromotive force in an exhausted tube, and excite faint luminous effects, but we cannot get enough energy to render the light practically available, and a simple calculation, based on experimental results, shows that even if all the energy which a tube would receive at a certain distance from the source should be wholly converted into light, it would hardly satisfy the practical requirements. Hence the necessity of directing, by means of a conducting circuit, the energy to the place of transformation. But in so doing we cannot very sensibly depart from present methods, and all we could do would be to improve the apparatus.

From these considerations it would seem that if this ideal way of lighting is to be rendered practicable it will be only by the use of electrostatic effects. In such a case the most powerful electrostatic inductive effects are needed; the apparatus employed must, therefore, be capable of producing high electrostatic potentials changing in value with extreme rapidity. High frequencies are especially wanted, for practical considerations make it desirable to keep down the potential. By the employment of machines, or, generally speaking, of any mechanical apparatus, but low frequencies can be reached; recourse must, therefore, be had to some other means. The discharge of a condenser affords us a means of obtaining frequencies by far higher than are obtainable mechanically, and I have accordingly employed condensers in the experiments to the above end.

FIG. 126.

When the terminals of a high tension induction coil, Fig. 126, are connected to a Leyden jar, and the latter is discharging disruptively into a circuit, we may look upon the arc playing between the knobs as being a source of alternating, or generally speaking, undulating currents, and then we have to deal with the familiar system of a generator of such currents, a circuit connected to it, and a condenser bridging the circuit.

The condenser in such case is a veritable transformer, and since the frequency is excessive, almost any ratio in the strength of the currents in both the branches may be obtained. In reality the analogy is not quite complete, for in the disruptive discharge we have most generally a fundamental instantaneous variation of comparatively low frequency, and a superimposed harmonic vibration, and the laws governing the flow of currents are not the same for both.

In converting in this manner, the ratio of conversion should not be too great, for the loss in the arc between the knobs increases with the square of the current, and if the jar be discharged through very thick and short conductors, with the view of obtaining a very rapid oscillation, a very considerable portion of the energy stored is lost. On the other hand, too small ratios are not practicable for many obvious reasons.

FIG. 127.

As the converted currents flow in a practically closed circuit, the electrostatic effects are necessarily small, and I therefore convert them into currents or effects of the required character. I have effected such conversions in several ways. The preferred plan of connections is illustrated in Fig. 127. The manner of operating renders it easy to obtain by means of a small and inexpensive apparatus enormous differences of potential which have been usually obtained by means of large and expensive coils. For this it is only necessary to take an ordinary small coil, adjust to it a condenser and discharging circuit, forming the primary of an auxiliary small coil, and convert upward. As the inductive effect of the primary currents is excessively great, the second coil need have comparatively but very few turns. By properly adjusting the elements, remarkable results may be secured.

In endeavoring to obtain the required electrostatic effects in this manner, I have, as might be expected, encountered many difficulties which I have been gradually overcoming, but I am not as yet prepared to dwell upon my experiences in this direction.

I believe that the disruptive discharge of a condenser will play an important part in the future, for it offers vast possibilities, not only in the way of producing light in a more efficient manner and in the line indicated by theory, but also in many other respects.

For years the efforts of inventors have been directed towards obtaining electrical energy from heat by means of the thermopile. It might seem invidious to remark that but few know what is the real trouble with the thermopile. It is not the inefficiency or small output— though these are great drawbacks—but the fact that the thermopile has its phylloxera, that is, that by constant use it is deteriorated, which has thus far prevented its introduction on an industrial scale. Now that all modern research seems to point with certainty to the use of electricity of excessively high tension, the question must present itself to many whether it is not possible to obtain in a practicable manner this form of energy from heat. We have been used to look upon an electrostatic machine as a plaything, and somehow we couple with it the idea of the inefficient and impractical. But now we must think differently, for now we know that everywhere we have to deal with the same forces, and that it is a mere question of inventing proper methods or apparatus for rendering them available.

In the present systems of electrical distribution, the employment of the iron with its wonderful magnetic properties allows us to reduce considerably the size of the apparatus; but, in spite of this, it is still very cumbersome. The more we progress in the study of electric and magnetic phenomena, the more we become convinced that the present methods will be short-lived. For the production of light, at least, such heavy machinery would seem to be unnecessary. The energy required is very small, and if light can be obtained as efficiently as, theoretically, it appears possible, the apparatus need have but a very small output. There being a strong probability that the illuminating methods of the future will involve the use of very high potentials, it seems very desirable to perfect a contrivance capable of converting the energy of heat into energy of the requisite form. Nothing to speak of has been done towards this end, for the thought that electricity of some 50,000 or 100,000 volts pressure or more, even if obtained, would be unavailable for practical purposes, has deterred inventors from working in this direction.

In Fig. 126 a plan of connections is shown for converting currents of high, into currents of low, tension by means of the disruptive discharge of a condenser. This plan has been used by me frequently for operating a few incandescent lamps required in the laboratory. Some difficulties have been encountered in the arc of the discharge which I have been able to overcome to a great extent; besides this, and the adjustment necessary for the proper working, no other difficulties have been met with, and it was easy to operate ordinary lamps, and even motors, in this manner. The line being connected to the ground, all the wires could be handled with perfect impunity, no matter how high the potential at the terminals of the condenser. In these experiments a high tension induction coil, operated from a battery or from an alternate current machine, was employed to charge the condenser; but the induction coil might be replaced by an apparatus of a different kind, capable of giving electricity of such high tension. In this manner, direct or alternating currents may be converted, and in both cases the current-impulses may be of any desired frequency. When the currents charging the condenser are of the same direction, and it is desired that the converted currents should also be of one direction, the resistance of the discharging circuit should, of course, be so chosen that there are no oscillations.

In operating devices on the above plan I have observed curious phenomena of impedance which are of interest. For instance if a thick copper bar be bent, as indicated in Fig. 128, and shunted by ordinary incandescent lamps, then, by passing the discharge between the knobs, the lamps may be brought to incandescence although they are short-circuited. When a large induction coil is employed it is easy to obtain nodes on the bar, which are rendered evident by the different degree of brilliancy of the lamps, as shown roughly in Fig. 128. The nodes are never clearly defined, but they are simply maxima and minima of potentials along the bar. This is probably due to the irregularity of the arc between the knobs. In general when the above-described plan

FIG. 128.

of conversion from high to low tension is used, the behavior of the disruptive discharge may be closely studied. The nodes may also be investigated by means of an ordinary Cardew voltmeter which should be well insulated. Geissler tubes may also be lighted across the points of the bent bar; in this case, of course, it is better to employ smaller capacities. I have found it practicable to light up in this manner a lamp, and even a Geissler tube, shunted by a short, heavy block of metal, and this result seems at first very curious. In fact, the thicker the copper bar in Fig. 128, the better it is for the success of the experiments, as they appear more striking. When lamps with long slender filaments are used it will be often noted that the filaments are from time to time violently vibrated, the vibration being smallest at the nodal points.

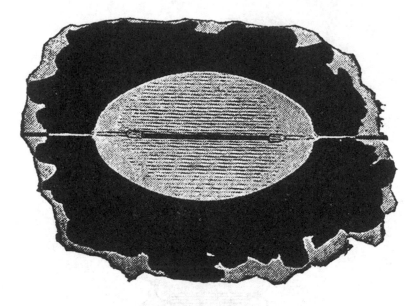

FIG. 129.

This vibration seems to be due to an electrostatic action between the filament and the glass of the bulb.

In some of the above experiments it is preferable to use special lamps having a straight filament as shown in Fig. 129. When such a lamp is used a still more curious phenomenon than those described may be observed. The lamp may be placed across the copper bar and lighted, and by using somewhat larger capacities, or, in other words, smaller frequencies or smaller impulsive impedances, the filament may be brought to any desired degree of incandescence. But when the impedance is increased, a point is reached when comparatively little current passes through the carbon, and most of it through the rarefied gas; or perhaps it may be more correct to state that the current divides nearly evenly through both, in spite of the enormous difference in the resistance, and this would be true unless the gas and the filament behave differently. It is then noted that the whole bulb is brilliantly illuminated, and the ends of the leading-in wires become incandescent and often throw off sparks in consequence of the violent bombardment, but the carbon filament remains dark. This is illustrated in Fig. 129. Instead of the filament a single wire extending through the whole bulb may be used, and in this case the phenomenon would seem to be still more interesting.

From the above experiment it will be evident, that when ordinary lamps are operated by the converted currents, those should be preferably taken in which the platinum wires are far apart, and the frequencies used should not be too great, else the discharge will occur at the ends of the filament or in the base of the lamp between the leading-in wires, and the lamp might then be damaged.

In presenting to you these results of my investigation on the subject under consideration, I have paid only a passing notice to facts upon which I could have dwelt at length, and among many observations I have selected only those which I thought most likely to interest you. The field is wide and completely unexplored, and at every step a new truth is gleaned, a novel fact observed.

How far the results here borne out are capable of practical applications will be decided in the future. As regards the production of light, some results already reached are encouraging and make me confident in asserting that the practical solution of the problem lies in the direction I have endeavored to indicate. Still, whatever may be the immediate outcome of these experiments I am hopeful that they will only prove a step in further development towards the ideal and final perfection. The possibilities which are opened by modern research are so vast that even the most reserved must feel sanguine of the future. Eminent scientists consider the problem of utilizing one kind of radiation without the others a rational one. In an apparatus designed for the production of light by conversion from any form of energy into that of light, such a result can never be reached, for no matter what the process of producing the required vibrations, be it electrical, chemical or any other, it will not be possible to obtain the higher light vibrations without going through the lower heat vibrations. It is the problem of imparting to a body a certain velocity without passing through all lower velocities. But there is a possibility of obtaining energy not only in the form of light, but motive power, and energy of any other form, in some more direct way from the medium. The time will be when this will be accomplished, and the time has come when one may utter such words before an enlightened audience without being considered a visionary. We are whirling through endless space with an inconceivable speed, all around us everything is spinning, everything is moving, everywhere is energy. There must be some way of availing ourselves of this energy more directly. Then, with the light obtained from the medium, with the power derived from it, with every form of energy obtained without effort, from the store forever inexhaustible, humanity will advance with giant strides. The mere contemplation of these magnificent possibilities expands our minds, strengthens our hopes and fills our hearts with supreme delight.

CHAPTER XXVII

EXPERIMENTS WITH ALTERNATE CURRENTS OF HIGH POTENTIAL AND HIGH FREQUENCY[*]

I cannot find words to express how deeply I feel the honor of addressing some of the foremost thinkers of the present time, and so many able scientific men, engineers and electricians, of the country greatest in scientific achievements.

The results which I have the honor to present before such a gathering I cannot call my own. There are among you not a few who can lay better claim than myself on any feature of merit which this work may contain. I need not mention many names which are world-known—names of those among you who are recognized as the leaders in this enchanting science; but one, at least, I must mention—a name which could not be omitted in a demonstration of this kind. It is a name associated with the most beautiful invention ever made: it is Crookes!

When I was at college, a good while ago, I read, in a translation (for then I was not familiar with your magnificent language), the description of his experiments on radiant matter. I read it only once in my life—that time—yet every detail about that charming work I can remember to this day. Few are the books, let me say, which can make such an impression upon the mind of a student.

But if, on the present occasion, I mention this name as one of many your Institution can boast of, it is because I have more than one reason to do so. For what I have to tell you and to show you this evening concerns, in a large measure, that same vague world which Professor Crookes has so ably explored; and, more than this, when I trace back the mental process which led me to these advances—which even by myself cannot be considered trifling, since they are so appreciated by you—I believe that their real origin, that which started me to work in this direction, and brought me to them, after a long period of constant thought, was that fascinating little book which I read many years ago.

---

[*]   Lecture delivered before the Institution of Electrical Engineers, London, February 1892.

And now that I have made a feeble effort to express my homage and acknowledge my indebtedness to him and others among you, I will make a second effort, which I hope you will not find so feeble as the first, to entertain you.

Give me leave to introduce the subject in a few words.

A short time ago I had the honor to bring before our American Institute of Electrical Engineers some results then arrived at by me in a novel line of work. I need not assure you that the many evidences which I have received that English scientific men and engineers were interested in this work have been for me a great reward and encouragement. I will not dwell upon the experiments already described, except with the view of completing, or more clearly expressing, some ideas advanced by me before, and also with the view of rendering the study here presented self-contained, and my remarks on the subject of this evening's lecture consistent.

This investigation, then, it goes without saying, deals with alternating currents, and to be more precise, with alternating currents of high potential and high frequency. Just in how much a very high frequency is essential for the production of the results presented is a question which, even with my present experience, would embarrass me to answer. Some of the experiments may be performed with low frequencies; but very high frequencies are desirable, not only on account of the many effects secured by their use, but also as a convenient means of obtaining, in the induction apparatus employed, the high potentials, which in their turn are necessary to the demonstration of most of the experiments here contemplated.

Of the various branches of electrical investigation, perhaps the most interesting and the most immediately promising is that dealing with alternating currents. The progress in this branch of applied science has been so great in recent years that it justifies the most sanguine hopes. Hardly have we become familiar with one fact, when novel experiences are met and new avenues of research are opened. Even at this hour possibilities not dreamed of before are, by the use of these currents, partly realized. As in nature all is ebb and tide, all is wave motion, so it seems that in all branches of industry alternating currents—electric wave motion—will have the sway.

One reason, perhaps, why this branch of science is being so rapidly developed is to be found in the interest which is attached to its

FIG. 130.   FIG. 131.

experimental study. We wind a simple ring of iron with coils; we establish the connections to the generator, and with wonder and delight we note the effects of strange forces which we bring into play, which allow us to transform, to transmit and direct energy at will. We arrange the circuits properly, and we see the mass of iron and wires behave as though it were endowed with life, spinning a heavy armature, through invisible connections, with great speed and power—with the energy possibly conveyed from a great distance. We observe how the energy of an alternating current traversing the wire manifests itself—not so much in the wire as in the surrounding space—in the most surprising manner, taking the forms of heat, light, mechanical energy, and, most surprising of all, even chemical affinity. All these observations fascinate us, and fill us with an intense desire to know more about the nature of these phenomena. Each day we go to our work in the hope of discovering—in the hope that someone, no matter who, may find a solution of one of the pending great problems—and each succeeding day we return to our task with renewed ardor; and even if we *are* unsuccessful, our work has not been in vain, for in these strivings, in these efforts, we have found hours of untold pleasure, and we have directed our energies to the benefit of mankind.

We may take—at random, if you choose—any of the many experiments which may be performed with alternating currents; a few of which only, and by no means the most striking, form the subject of this evening's demonstration; they are all equally interesting, equally inciting to thought.

Here is a simple glass tube from which the air has been partially exhausted. I take hold of it; I bring my body in contact with a wire conveying alternating currents of high potential, and the tube in my hand is brilliantly lighted. In whatever position I may put it, wherever I move it in space, as far as I can reach, its soft, pleasing light persists with undiminished brightness.

Here is an exhausted bulb suspended from a single wire. Standing on an insulated support, I grasp it, and a platinum button mounted in it is brought to vivid incandescence.

Here, attached to a leading wire, is another bulb, which, as I touch its metallic socket, is filled with magnificent colors of phosphorescent light.

Here still another, which by my fingers' touch casts a shadow—the Crookes shadow—of the stem inside of it.

Here, again, insulated as I stand on this platform, I bring my body in contact with one of the terminals of the secondary of this induction coil—with the end of a wire many miles long—and you see streams of light break forth from its distant end, which is set in violent vibration.

Here, once more, I attach these two plates of wire gauze to the terminals of the coil; I set them a distance apart, and I set the coil to work. You may see a small spark pass between the plates. I insert a thick plate of one of the best dielectrics between them, and instead of rendering altogether impossible, as we are used to expect, I *aid* the passage of the discharge, which, as I insert the plate, merely changes in appearance and assumes the form of luminous streams.

Is there, I ask, can there be, a more interesting study than that of alternating currents?

In all these investigations, in all these experiments, which are so very, very interesting, for many years past—ever since the greatest experimenter who lectured in this hall discovered its principle—we have had a steady companion, an appliance familiar to everyone, a plaything once, a thing of momentous importance now—the induction coil. There is no dearer appliance to the electrician. From the ablest among you, I dare say, down to the inexperienced student, to your lecturer, we all have passed many delightful hours in experimenting with the induction coil. We have watched its play, and thought and pondered over the beautiful phenomena which it disclosed to our ravished eyes. So well known is this apparatus, so familiar are these phenomena to everyone, that my courage nearly fails me when I think that I have ventured to address so able an audience, that I have ventured to entertain you with that same old subject. Here in reality is the same apparatus, and here are the same phenomena, only the apparatus is operated somewhat differently, the phenomena are presented in a different aspect. Some of the results we find as expected, others surprise us, but all captivate our attention, for in scientific investigation each novel result achieved may be the center of a new departure, each novel fact learned may lead to important developments.

Usually in operating an induction coil we have set up a vibration of moderate frequency in the primary, either by means of an interrupter or break, or by the use of an alternator. Earlier English investigators, to mention only Spottiswoode and J. E. H. Gordon, have used a rapid break in connection with the coil. Our knowledge and experience of today enables us to see clearly why these coils under the conditions of the test did not disclose any remarkable phenomena, and why able experimenters failed to perceive many of the curious effects which have since been observed.

In the experiments such as performed this evening, we operate the coil either from a specially constructed alternator capable of giving many thousands of reversals of current per second, or, by disruptively discharging a condenser through the primary, we set up a vibration in the secondary circuit of a frequency of many hundred thousand or millions per second, if we so desire; and in using either of these means we enter a field as yet unexplored.

It is impossible to pursue an investigation in any novel line without finally making some interesting observation or learning some useful fact. That this statement is applicable to the subject of this lecture the many curious and unexpected phenomena which we observe afford a convincing proof. By way of illustration, take for instance the most obvious phenomena, those of the discharge of the induction coil.

Here is a coil which is operated by currents vibrating with extreme rapidity, obtained by disruptively discharging a Leyden jar. It would not surprise a student were the lecturer to say that the secondary of this coil consists of a small length of comparatively stout wire; it would not surprise him were the lecturer to state that, in spite of this, the coil is capable of giving any potential which the best insulation of the turns is able to withstand; but although he may be prepared, and even be indifferent as to the anticipated result, yet the aspect of the discharge of the coil will surprise and interest him. Everyone is familiar with the discharge of an ordinary coil; it need not be reproduced here. But, by way of contrast, here is a form of discharge of a coil, the primary current of which is vibrating several hundred thousand times per second. The discharge of an ordinary coil appears as a simple line or band of light. The discharge of this coil appears in the form of powerful brushes and luminous streams issuing from all points of the two straight wires attached to the terminals of the secondary. (Fig. 130.)

Now compare this phenomenon which you have just witnessed with the discharge of a Holtz or Wimshurst machine—that other interesting appliance so dear to the experimenter. What a difference there is between these phenomena! And yet, had I made the necessary arrangements—which could have been made easily, were it not that they would interfere with other experiments—I could have produced with this coil sparks which, had I the coil hidden from your view and only two knobs exposed, even the keenest observer among you would find it difficult, if not impossible, to distinguish from those of an influence or friction machine. This may be done in many ways—for instance, by operating the induction coil which charges the condenser from an alternating-current machine of very low frequency, and preferably adjusting the discharge circuit so that there are no oscillations set up in it. We then obtain in the secondary circuit, if the knobs are of the required size and properly set, a more or less rapid succession of sparks of great intensity and small quantity, which possess the same brilliancy, and are accompanied by the same sharp crackling sound, as those obtained from a friction or influence machine.

Another way is to pass through two primary circuits, having a common secondary, two currents of a slightly different period, which produce in the secondary circuit sparks occurring at comparatively long intervals. But, even with the means at hand this evening, I may succeed in imitating the spark of a Holtz machine. For this purpose I establish between the terminals of the coil which charges the condenser a long, unsteady arc, which is periodically interrupted by the upward current of air produced by it. To increase the current of air I place on each side of the arc, and close to it, a large plate of mica. The condenser charged from this coil discharges into the primary circuit of a second coil through a small air gap, which is necessary to produce a sudden rush of current through the primary. The scheme of connections in the present experiment is indicated in Fig. 131.

G is an ordinarily constructed alternator, supplying the primary P of an induction coil, the secondary S of which charges the condensers or jars C C. The terminals of the secondary are connected to the inside coatings of the jars, the outer coatings being connected to the ends of the primary *p p* of a second induction coil. This primary *p p* has a small air gap *a b*.

The secondary S of this coil is provided with knobs or spheres K K of the proper size and set at a distance suitable for the experiment.

A long arc is established between the terminals A B of the first induction coil. M M are the mica plates.

Each time the arc is broken between A and B the jars are quickly charged and discharged through the primary *p p*, producing a snapping spark between the knobs K K. Upon the arc forming between A and B the potential falls, and the jars cannot be charged to such high potential as to break through the air gap *a b* until the arc is again broken by the draught.

In this manner sudden impulses, at long intervals, are produced in the primary *p p*, which in the secondary S give a corresponding number of impulses of great intensity. If the secondary knobs or spheres, K K, are of the proper size, the sparks show much resemblance to those of a Holtz machine.

But these two effects, which to the eye appear so very different, are only two of the many discharge phenomena. We only need to change the conditions of the test, and again we make other observations of interest.

When, instead of operating the induction coil as in the last two experiments, we operate it from a high frequency alternator, as in the next experiment, a systematic study of the phenomena is rendered much more easy. In such case, in varying the strength and frequency of the currents through the primary, we may observe five distinct forms of discharge, which I have described in my former paper on the subject before the American Institute of Electrical Engineers, May 20, 1891.

It would take too much time, and it would lead us too far from the subject presented this evening, to reproduce all these forms, but it seems to me desirable to show you one of them. It is a brush discharge, which is interesting in more than one respect. Viewed from a near position it resembles much a jet of gas escaping under great pressure. We know that the phenomenon is due to the agitation of the molecules near the terminal, and we anticipate that some heat must be developed by the impact of the molecules against the terminal or against each other. Indeed, we find that the brush is hot, and only a little thought leads us to the conclusion that, could we but reach sufficiently high frequencies, we could produce a brush which would give intense light and heat, and which would resemble in every particular an ordinary flame, save, perhaps, that both phenomena might not be due to the same agent—save, perhaps, that chemical affinity might not be *electrical* in its nature.

As the production of heat and light is here due to the impact of the molecules, or atoms of air, or something else besides, and, as we can augment the energy simply by raising the potential, we might, even with frequencies obtained from a dynamo machine, intensify the action to such a degree as to bring the terminal to melting heat. But with such low frequencies we would have to deal always with something of the nature of an *electric* current. If I approach a conducting object to the brush, a thin little spark passes, yet, even with the frequencies used this evening, the tendency to spark is not very great. So, for instance, if I hold a metallic sphere at some distance above the terminal, you may see the whole space between the terminal and sphere illuminated by the streams without the spark passing; and with the much higher frequencies obtainable by the disruptive discharge of a condenser, were it not for the sudden impulses, which are comparatively few in number, sparking would not occur even at very small distances. However, with incomparably higher frequencies, which we may yet find means to produce efficiently, and provided that electric impulses of such high frequencies could be transmitted through a conductor, the electrical characteristics of the brush discharge would completely vanish—no spark would pass, no shock would be felt—yet we would still have to deal with an electric phenomenon, but in the broad, modern interpretation of the word. In my first paper, before referred to, I have pointed out the curious properties of the brush, and described the best manner of producing it, but I have thought it worth while to endeavor to express myself more clearly in regard to this phenomenon, because of its absorbing interest.

When a coil is operated with currents of very high frequency, beautiful brush effects may be produced, even if the coil be of comparatively small dimensions. The experimenter may vary them in many ways, and, if it were for nothing else, they afford a pleasing sight. What adds to their interest is that they may be produced with one single terminal as well as with two—in fact, often better with one than with two.

But of all the discharge phenomena observed, the most pleasing to the eye, and the most instructive, are those observed with a coil which is operated by means of the disruptive discharge of a condenser. The power of the brushes, the abundance of the sparks, when the conditions are patiently adjusted, is often amazing. With even a very small coil, if it be so well insulated as to stand a difference of potential of several thousand volts per turn, the sparks may be so abundant that the whole coil may appear a complete mass of fire.

Curiously enough the sparks, when the terminals of the coil are set at a considerable distance, seem to dart in every possible direction as though the terminals were perfectly independent of each other. As the sparks would soon destroy the insulation, it is necessary to prevent them. This is best done by immersing the coil in a good liquid insulator, such as boiled-out oil. Immersion in a liquid may be considered almost an absolute necessity for the continued and successful working of such a coil.

It is, of course, out of the question, in an experimental lecture, with only a few minutes at disposal for the performance of each experiment, to show these discharge phenomena to advantage, as, to produce each phenomenon at its best, a very careful adjustment is required. But even if imperfectly produced, as they are likely to be this evening, they are sufficiently striking to interest an intelligent audience.

Before showing some of these curious effects I must, for the sake of completeness, give a short description of the coil and other apparatus used in the experiments with the disruptive discharge this evening.

It is contained in a box B (Fig. 132) of thick boards of hard wood, covered on the outside with a zinc sheet z, which is carefully soldered all around. It might be advisable, in a strictly scientific investigation, when accuracy is of great importance, to do away with the metal cover, as it might introduce many errors, principally on account of its complex action upon the coil, as a condenser of very small capacity and as an electrostatic and electromagnetic screen. When the coil is used for such experiments as are here contemplated, the employment of the metal cover offers some practical advantages, but these are not of sufficient importance to be dwelt upon.

Fig. 132.

The coil should be placed symmetrically to the metal cover, and the space between should, of course, not be too small, certainly not less than, say, five centimeters, but much more if possible; especially the two sides of the zinc box, which are at right angles to the axis of the coil, should be sufficiently remote from the latter, as otherwise they might impair its action and be a source of loss.

The coil consists of two spools of hard rubber R R, held apart at a distance of 10 centimeters by bolts c and nuts n, likewise of hard rubber. Each spool comprises a tube T of approximately 8 centimeters inside diameter, and 3 millimeters thick, upon which are screwed two flanges F F, 24 centimeters square, the space between the flanges being about 3 centimeters. The secondary, s s, of the best gutta percha–covered wire, has 26 layers, 10 turns in each, giving for each half a total of 260 turns. The two halves are wound oppositely and connected in series, the connection between both being made over the primary. This disposition, besides being convenient, has the advantage that when the coil is well balanced—that is, when both of its terminals T_1, T_2, are connected to bodies or devices of equal capacity—there is not much danger of breaking through to the primary, and the insulation between the primary and the secondary need not be thick. In using the coil it is advisable to attach to *both* terminals devices of nearly equal capacity, as, when the capacity of the terminals is not equal, sparks will be apt to pass to the primary. To avoid this, the middle point of the secondary may be connected to the primary, but this is not always practicable.

The primary P P is wound in two parts, and oppositely, upon a wooden spool w. and the four ends are led out of the oil through hard rubber tubes t t. The ends of the secondary T_1, T_1 are also led out of the oil through rubber tubes t_1, t_1 of great thickness. The primary and secondary layers are insulated by cotton cloth, the thickness of the insulation, of course, bearing some proportion to the difference of potential between the turns of the different layers. Each half of the primary has four layers, 24 turns in each, this giving a total of 96 turns. When both the parts are connected in series, this gives a ratio of conversion of about 1:2.7, and with the primaries in multiple, 1:5.4; but in operating with very rapidly alternating currents this ratio does not convey even an approximate idea of the ratio of the E. M. F's. in the primary and secondary circuits. The coil is held in position in the oil on wooden supports, there being about 5 centimeters thickness of oil all round. Where the oil is not specially needed, the space is filled with pieces of wood, and for this purpose principally the wooden box B surrounding the whole is used.

The construction here shown is, of course, not the best on general principles, but I believe it is a good and convenient one for the production of effects in which an excessive potential and a very small current are needed.

In connection with the coil I use either the ordinary form of discharger or a modified form. In the former I have introduced two changes which secure some advantages, and which are obvious. If they are mentioned, it is only in the hope that some experimenter may find them of use.

One of the changes is that the adjustable knobs A and B (Fig. 133), of the discharger are held in jaws of brass, J J, by spring pressure, this allowing of turning them successively into different positions, and so doing away with the tedious process of frequent polishing up.

The other change consists in the employment of a strong electromagnet N S, which is placed with its axis at right angles to the line joining the knobs A and B, and produces a strong magnetic field between them. The pole pieces of the magnet are movable and properly formed so as to protrude between the brass knobs, in order to make the field as intense as possible; but to prevent the discharge from jumping to the magnet the pole pieces are protected by a layer of mica, M M, of sufficient thickness; $s_1 s_1$ and $s_2 s_2$ are screws for fastening the wires. On each side one of the screws is for large and the other for small wires. L L are screws for fixing in position the rods R R, which support the knobs.

Fig. 133.

In another arrangement with the magnet I take the discharge between the rounded pole pieces themselves, which in such case are insulated and preferably provided with polished brass caps.

The employment of an intense magnetic field is of advantage principally when the induction coil or transformer which charges the condenser is operated by currents of very low frequency. In such a case the number of the fundamental discharges between the knobs may be so small as to render the currents produced in the secondary unsuitable for many experiments. The intense magnetic field then serves to blow out the arc between the knobs as soon as it is formed, and the fundamental discharges occur in quicker succession.

Instead of the magnet, a draught or blast of air may be employed with some advantage. In this case the arc is preferably established between the knobs A B, in Fig. 131 (the knobs *a b* being generally joined, or entirely done away with), as in this disposition the arc is long and unsteady, and is easily affected by the draught.

When a magnet is employed to break the arc, it is better to choose the connection indicated diagrammatically in Fig. 134, as in this case the currents forming the arc are much more powerful, and the magnetic field exercises a greater influence. The use of the magnet permits, however, of the arc being replaced by a vacuum tube, but I have encountered great difficulties in working with an exhausted tube.

FIG. 134.

The other form of discharger used in these and similar experiments is indicated in Figs. 135 and 136. It consists of a number of brass pieces *c c* (Fig. 135), each of which comprises a spherical middle portion *m* with an extension *e* below—which is merely used to fasten the piece in a lathe when polishing up the discharging surface—and a column above, which consists of a knurled flange *f* surmounted by a threaded stem *l* carrying a nut *n*, by means of which a wire is fastened to the column. The flange *f* conveniently serves for holding the brass piece when fastening the wire, and also for turning it in any position when it becomes necessary to present a fresh discharging surface. Two stout strips of hard rubber R R, with planed grooves *g g* (Fig. 136) to fit the middle portion of the pieces *c c*, serve to clamp the latter and hold them firmly in position by means of two bolts *c c* (of which only one is shown) passing through the ends of the strips.

Fig. 135.

Fig. 136.

In the use of this kind of discharger I have found three principal advantages over the ordinary form. First, the dielectric strength of a given total width of air space is greater when a great many small air gaps are used instead of one, which permits of working with a smaller length of air gap, and that means smaller loss and less deterioration of the metal; secondly, by reason of splitting the arc up into smaller arcs, the polished surfaces are made to last much longer; and, thirdly, the apparatus affords some gauge in the experiments. I usually set the pieces by putting between them sheets of uniform thickness at a certain very small distance which is known from the experiments of Sir William Thomson to require a certain electromotive force to be bridged by the spark.

It should, of course, be remembered that the sparking distance is much diminished as the frequency is increased. By taking any number of spaces the experimenter has a rough idea of the electromotive force, and he finds it easier to repeat an experiment, as he has not the trouble of setting the knobs again and again. With this kind of discharger I have been able to maintain an oscillating motion without any spark being visible with the naked eye between the knobs, and they would not show a very appreciable rise in temperature. This form of discharge also lends itself to many arrangements of condensers and circuits which are often very convenient and time-saving. I have used it preferably in a disposition similar to that indicated in Fig. 131, when the currents forming the arc are small.

I may here mention that I have also used dischargers with single or multiple air gaps, in which the discharge surfaces were rotated with great speed. No particular advantage was, however, gained by this method, except in cases where the currents from the condenser were large and the keeping cool of the surfaces was necessary, and in cases when, the discharge not being oscillating of itself, the arc as soon as established was broken by the air current, thus starting the vibration at intervals in rapid succession. I have also used mechanical interrupters in many ways. To avoid the difficulties with frictional contacts, the preferred plan adopted was to establish the arc and rotate through it at great speed a rim of mica provided with many holes and fastened to a steel plate. It is understood, of course, that the employment of a magnet, air current, or other interrupter, produces no effect worth noticing, unless the self-induction, capacity and resistance are so related that there are oscillations set up upon each interruption.

I will now endeavor to show you some of the most noteworthy of these discharge phenomena.

I have stretched across the room two ordinary cotton covered wires, each about seven meters in length. They are supported on insulating cords at a distance of about thirty centimeters. I attach now to each of the terminals of the coil one of the wires, and set the coil in action. Upon turning the lights off in the room you see the wires strongly illuminated by the streams issuing abundantly from their whole surface in spite of the cotton covering, which may even be very thick. When the experiment is performed under good conditions, the light from the wires is sufficiently intense to allow distinguishing the objects in a room. To produce the best result it is, of course, necessary to adjust carefully the capacity of the jars, the arc between the knobs and the length of the wires. My experience is that calculation of the length of the wires leads, in such case, to no result whatever. The experimenter will do best to take the wires at the start very long, and then adjust by cutting off first long pieces, and then smaller and smaller ones as he approaches the right length.

A convenient way is to use an oil condenser of very small capacity, consisting of two small adjustable metal plates, in connection with this and similar experiments. In such case I take wires rather short and at the beginning set the condenser plates at maximum distance. If the streams from the wires increase by approach of the plates, the length of the wires is about right; if they diminish, the wires are too long for that frequency and potential. When a condenser is used in connection with experiments with such a coil, it should be an oil condenser by all means, as in using an air condenser considerable energy might be wasted. The wires leading to the plates in the oil should be very thin, heavily coated with some insulating compound, and provided with a conducting covering—this preferably extending under the surface of the oil. The conducting cover should not be too near the terminals, or ends, of the wire, as a spark would be apt to jump from the wire to it. The conducting coating is used to diminish the air losses, in virtue of its action as an electrostatic screen. As to the size of the vessel containing the oil, and the size of the plates, the experimenter gains at once an idea from a rough trial. The size of the plates *in oil* is, however, calculable, as the dielectric losses are very small.

In the preceding experiment it is of considerable interest to know what relation the quantity of the light emitted bears to the frequency and potential of the electric impulses. My opinion is that the heat as well as light effects produced should be proportionate, under otherwise equal conditions of test, to the product of frequency and square of potential, but the experimental verification of the law, whatever it may be, would be exceedingly difficult. One thing is certain, at any rate, and that is, that in augmenting the potential and frequency we rapidly intensify the streams; and, though it may be very sanguine, it is surely not altogether hopeless to expect that we may succeed in producing a practical illuminant on these lines. We would then be simply using burners or flames, in which there would be no chemical process, no consumption of material, but merely a transfer of energy, and which would, in all probability, emit more light and less heat than ordinary flames.

The luminous intensity of the streams is, of course, considerably increased when they are focused upon a small surface. This may be shown by the following experiment:

FIG. 137.

I attach to one of the terminals of the coil a wire w (Fig. 137), bent in a circle of about 30 centimeters in diameter, and to the other terminal I fasten a small brass sphere s, the surface of the wire being preferably equal to the surface of the sphere, and the center of the latter being in a line at right angles to the plane of the wire circle and passing through its center. When the discharge is established under proper conditions, a luminous hollow cone is formed, and in the dark one-half of the brass sphere is strongly illuminated, as shown in the cut.

By some artifice or other it is easy to concentrate the streams upon small surfaces and to produce very strong light effects. Two thin wires may thus be rendered intensely luminous.

In order to intensify the streams the wires should be very thin and short; but as in this case their capacity would be generally too small for the coil—at least for such a one as the present—it is necessary to augment the capacity to the required value, while, at the same time, the surface of the wires remains very small. This may be done in many ways.

FIG. 138.

Here, for instance, I have two plates, R R, of hard rubber (Fig. 138), upon which I have glued two very thin wires *w w*, so as to form a name. The wires may be bare or covered with the best insulation—it is immaterial for the success of the experiment. Well insulated wires, if anything, are preferable. On the back of each plate, indicated by the shaded portion, is a tinfoil coating *t t*. The plates are placed in line at a sufficient distance to prevent a spark passing from one wire to the other. The two tinfoil coatings I have joined by a conductor c, and the two wires I presently connect to the terminals of the coil. It is now easy, by varying the strength and frequency of the currents through the primary, to find a point at which the capacity of the system is best suited to the conditions, and the wires become so strongly luminous that, when the light in the room is turned off the name formed by them appears in brilliant letters.

It is perhaps preferable to perform this experiment with a coil operated from an alternator of high frequency, as then, owing to the harmonic rise and fall, the streams are very uniform, though they are less abundant than when produced with such a coil as the present one. This experiment, however, may be performed with low frequencies, but much less satisfactorily.

FIG. 139.

When two wires, attached to the terminals of the coil, are set at the proper distance, the streams between them may be so intense as to produce a continuous luminous sheet. To show this phenomenon I have here two circles, c and c (Fig. 139), of rather stout wire, one being about 80 centimeters and the other 30 centimeters in diameter. To each of the terminals of the coil I attach one of the circles. The supporting wires are so bent that the circles may be placed in the same plane, coinciding as nearly as possible. When the light in the room is turned off and the coil set to work, you see the whole space between the wires uniformly filled with streams, forming a luminous disc, which could be seen from a considerable distance, such is the intensity of the streams. The outer circle could have been much larger than the present one; in fact, with this coil I have used much larger circles, and I have been able to produce a strongly luminous sheet, covering an area of more than one square meter, which is a remarkable effect with this very small coil. To avoid uncertainty, the circle has been taken smaller, and the area is now about 0.43 square meter.

The frequency of the vibration, and the quickness of succession of the sparks between the knobs, affect to a marked degree the appearance of the streams. When the frequency is very low, the air gives way in more or less the same manner, as by a steady difference of potential, and the streams consist of distinct threads, generally mingled with thin sparks, which probably correspond to the successive discharges occurring between the knobs. But when the frequency is extremely high, and the arc of the discharge produces a very *loud* but *smooth* sound— showing both that oscillation takes place and that the sparks succeed each other with great rapidity—then the luminous streams formed are perfectly uniform. To reach this result very small coils and jars of small capacity should be used. I take two tubes of thick Bohemian glass, about 5 centimeters in diameter and 20 centimeters long. In each of the tubes I slip a primary of very thick copper wire. On the top of each tube I wind a secondary of much thinner gutta-percha covered wire. The two secondaries I connect in series, the primaries preferably in multiple arc. The tubes are then placed in a large glass vessel, at a distance of 10 to 15 centimeters from each other, on insulating supports, and the vessel is filled with boiled-out oil, the oil reaching about an inch above the tubes. The free ends of the secondary are lifted out of the coil and placed parallel to each other at a distance of about ten centimeters. The ends which are scraped should be dipped in the oil. Two four-pint jars joined in series may be used to discharge through the primary. When the necessary adjustments in the length and distance of the wires above the oil and in the arc of discharge are made, a luminous sheet is produced between the wires which is perfectly smooth and textureless, like the ordinary discharge through a moderately exhausted tube.

I have purposely dwelt upon this apparently insignificant experiment. In trials of this kind the experimenter arrives at the startling conclusion that, to pass ordinary luminous discharges through gases, no particular degree of exhaustion is needed, but that the gas may be at ordinary or even greater pressure. To accomplish this, a very high frequency is essential; a high potential is likewise required, but this is merely an incidental necessity. These experiments teach us that, in endeavoring to discover novel methods of producing light by the agitation of atoms, or molecules, of a gas, we need not limit our research to the vacuum tube, but may look forward quite seriously to the possibility of obtaining the light effects without the use of any vessel whatever, with air at ordinary pressure.

Such discharges of very high frequency, which render luminous the air at ordinary pressures, we have probably occasion often to witness in Nature. I have no doubt that if, as many believe, the aurora borealis is produced by sudden cosmic disturbances, such as eruptions at the sun's surface, which set the electrostatic charge of the earth in an extremely rapid vibration, the red glow observed is not confined to the upper rarefied strata of the air, but the discharge traverses, by reason of its very high frequency, also the dense atmosphere in the form of a *glow*, such as we ordinarily produce in a slightly exhausted tube. If the frequency were very low, or even more so, if the charge were not at all vibrating, the dense air would break down as in a lightning discharge. Indications of such breaking down of the lower dense strata of the air have been repeatedly observed at the occurrence of this marvelous phenomenon; but if it does occur, it can only be attributed to the fundamental disturbances, which are few in number, for the vibration produced by them would be far too rapid to allow a disruptive break. It is the original and irregular impulses which affect the instruments; the superimposed vibrations probably pass unnoticed.

When an ordinary low frequency discharge is passed through moderately rarefied air, the air assumes a purplish hue. If by some means or other we increase the intensity of the molecular, or atomic, vibration, the gas changes to a white color. A similar change occurs at ordinary pressures with electric impulses of very high frequency. If the molecules of the air around a wire are moderately agitated, the brush formed is reddish or violet; if the vibration is rendered sufficiently intense, the streams become white. We may accomplish this in various ways. In the experiment before shown with the two wires across the room, I have endeavored to secure the result by pushing to a high value both the frequency and potential; in the experiment with the thin wires glued on the rubber plate I have concentrated the action upon a very small surface—in other words, I have worked with a great electric density.

A most curious form of discharge is observed with such a coil when the frequency and potential are pushed to the extreme limit. To perform the experiment, every part of the coil should be heavily insulated, and only two small spheres—or, better still, two sharp-edged metal discs (*d d*, Fig. 140) of no more than a few centimeters in diameter—should be exposed to the air. The coil here used is immersed in oil, and the ends of the secondary reaching out of the oil are covered with an air-tight cover of hard rubber of great thickness. All cracks, if there are any, should be carefully stopped up, so that the brush discharge cannot form anywhere except on the small spheres or plates which are exposed to the air. In this case, since there are no large plates or other bodies of capacity attached to the terminals, the coil is capable of an extremely rapid vibration. The potential may be raised by increasing, as far as the experimenter judges proper, the rate of change of the primary current. With a coil not widely differing from the present, it is best to connect the two primaries in multiple arc; but if the secondary should have a much greater number of turns the primaries should preferably be used in series, as otherwise the vibration might be too fast for the secondary. It occurs under these conditions that misty white streams break forth from the edges of the discs and spread out phantom-like into space. With this coil, when fairly well produced, they are about 25 to 30 centimeters long. When the hand is held against them no sensation is produced, and a spark, causing a shock, jumps from the terminal only upon the hand being brought much nearer. If the oscillation of the primary current is rendered intermittent by some means or other, there is a corresponding throbbing of the streams, and now the hand or other conducting object may be brought in still greater proximity to the terminal without a spark being caused to jump.

Fig. 140.

Among the many beautiful phenomena which may be produced with such a coil, I have here selected only those which appear to possess some features of novelty, and lead us to some conclusions of interest. One will not find it at all difficult to produce in the laboratory, by means of it, many other phenomena which appeal to the eye even more than these here shown, but present no particular feature of novelty.

Early experimenters describe the display of sparks produced by an ordinary large induction coil upon an insulating plate separating the terminals. Quite recently Siemens performed some experiments in which fine effects were obtained, which were seen by many with interest. No doubt large coils, even if operated with currents of low frequencies, are capable of producing beautiful effects. But the largest coil ever made could not, by far, equal the magnificent display of streams and sparks obtained from such a disruptive discharge coil when properly adjusted. To give an idea, a coil such as the present one will cover easily a plate of one meter in diameter completely with the streams. The best way to perform such experiments is to take a very thin rubber or a glass plate and glue on one side of it a narrow ring of tinfoil of very large diameter, and on the other a circular washer, the center of the latter coinciding with that of the ring, and the surfaces of both being preferably equal, so as to keep the coil well balanced. The washer and ring should be connected to the terminals by heavily insulated thin wires. It is easy in observing the effect of the capacity to produce a sheet of uniform streams, or a fine network of thin silvery threads, or a mass of loud brilliant sparks, which completely cover the plate.

Since I have advanced the idea of the conversion by means of the disruptive discharge, in my paper before the American Institute of Electrical Engineers at the beginning of the past year, the interest excited in it has been considerable. It affords us a means for producing any potentials by the aid of inexpensive coils operated from ordinary systems of distribution, and—what is perhaps more appreciated—it enables us to convert currents of any frequency into currents of any other lower or higher frequency. But its chief value will perhaps be found in the help which it will afford us in the investigations of the phenomena of phosphorescence, which a disruptive discharge coil is capable of exciting in innumerable cases where ordinary coils, even the largest, would utterly fail.

Considering its probable uses for many practical purposes, and its possible introduction into laboratories for scientific research, a few additional remarks as to the construction of such a coil will perhaps not be found superfluous.

It is, of course, absolutely necessary to employ in such a coil wires provided with the best insulation.

Good coils may be produced by employing wires covered with several layers of cotton, boiling the coil a long time in pure wax, and cooling under moderate pressure. The advantage of such a coil is that it can be easily handled, but it cannot probably give as satisfactory results as a coil immersed in pure oil. Besides, it seems that the presence of a large body of wax affects the coil disadvantageously, whereas this does not seem to be the case with oil. Perhaps it is because the dielectric losses in the liquid are smaller.

I have tried at first silk and cotton covered wires with oil immersions, but I have been gradually led to use gutta-percha covered wires, which proved most satisfactory. Gutta-percha insulation adds, of course, to the capacity of the coil, and this, especially if the coil be large, is a great disadvantage when extreme frequencies are desired; but, on the other hand, gutta-percha will withstand much more than an equal thickness of oil, and this advantage should be secured at any price. Once the coil has been immersed, it should never be taken out of the oil for more than a few hours, else the gutta-percha will crack up and the coil will not be worth half as much as before. Gutta-percha is probably slowly attacked by the oil, but after an immersion of eight to nine months I have found no ill effects.

I have obtained two kinds of gutta-percha wire known in commerce: in one the insulation sticks tightly to the metal, in the other it does not. Unless a special method is followed to expel all air, it is much safer to use the first kind. I wind the coil within an oil tank so that all interstices are filled up with the oil. Between the layers I use cloth boiled out thoroughly in oil, calculating the thickness according to the difference of potential between the turns. There seems not to be a very great difference whatever kind of oil is used; I use paraffin or linseed oil.

To exclude more perfectly the air, an excellent way to proceed, and easily practicable with small coils, is the following: Construct a box of hardwood of very thick boards which have been for a long time boiled in oil. The boards should be so joined as to safely withstand the external air pressure. The coil being placed and fastened in position within the box, the latter is closed with a strong lid, and covered with closely fitting metal sheets, the joints of which are soldered very carefully. On the top two small holes are drilled, passing through the metal sheet and the wood, and in these holes two small glass tubes are inserted and the joints made air-tight. One of the tubes is connected to a vacuum pump, and the other with a vessel containing a sufficient quantity of boiled-out oil. The latter tube has a very small hole at the bottom, and is provided with a stopcock. When a fairly good vacuum has been obtained, the stopcock is opened and the oil slowly fed in. Proceeding in this manner, it is impossible that any big bubbles, which are the principal danger, should remain between the turns. The air is most completely excluded, probably better than by boiling out, which, however, when gutta-percha coated wires are used, is not practicable.

For the primaries I use ordinary line wire with a thick cotton coating. Strands of very thin insulated wires properly interlaced would, of course, be the best to employ for the primaries, but they are not to be had.

In an experimental coil the size of the wires is not of great importance. In the coil here used the primary is No. 12 and the secondary No. 24 Brown & Sharpe gauge wire; but the sections may be varied considerably. It would only imply different adjustments; the results aimed at would not be materially affected.

I have dwelt at some length upon the various forms of brush discharge because, in studying them, we not only observe phenomena which please our eye, but also afford us food for thought, and lead us to conclusions of practical importance. In the use of alternating currents of very high tension, too much precaution cannot be taken to prevent the brush discharge. In a main conveying such currents, in an induction coil or transformer, or in a condenser, the brush discharge is a source of great danger to the insulation. In a condenser, especially, the gaseous matter must be most carefully expelled, for in it the charged surfaces are near each other, and if the potentials are high, just as sure as a weight will fall if let go, so the insulation will give way if a single gaseous bubble of some size be present, whereas, if all gaseous matter were carefully excluded, the condenser would safely withstand a much higher difference of potential. A main conveying alternating currents of very high tension may be injured merely by a blowhole or small crack in the insulation, the more so as a blowhole is apt to contain gas at low pressure; and as it appears almost impossible to completely obviate such little imperfections, I am led to believe that in our future distribution of electrical energy by currents of very high tension, liquid insulation will be used. The cost is a great drawback, but if we employ an oil as an insulator the distribution of electrical energy with something like 100,000 volts, and even more, becomes, at least with higher frequencies, so easy that it could be hardly called an engineering feat. With oil insulation and alternate current motors, transmissions of power can be affected with safety and upon an industrial basis at distances of as much as a thousand miles.

A peculiar property of oils, and liquid insulation in general, when subjected to rapidly changing electric stresses, is to disperse any gaseous bubbles which may be present, and diffuse them through its mass, generally long before any injurious break can occur. This feature may be easily observed with an ordinary induction coil by taking the primary out, plugging up the end of the tube upon which the secondary is wound, and filling it with some fairly transparent insulator, such as paraffin oil. A primary of a diameter something like six millimeters smaller than the inside of the tube may be inserted in the oil. When the coil is set to work one may see, looking from the top through the oil, many luminous points—air bubbles which are caught by inserting the primary, and which are rendered luminous in consequence of the violent bombardment. The occluded air, by its impact against the oil, heats it; the oil begins to circulate, carrying some of the air along with it, until the bubbles are dispersed and the luminous points disappear. In this manner, unless large bubbles are occluded in such way that circulation is rendered impossible, a damaging break is averted, the only effect being a moderate warming up of the oil. If, instead of the liquid, a solid insulation, no matter how thick, were used, a breaking through and injury of the apparatus would be inevitable.

The exclusion of gaseous matter from any apparatus in which the dielectric is subjected to more or less rapidly changing electric forces is, however, not only desirable in order to avoid a possible injury of the apparatus, but also on account of economy. In a condenser, for instance, as long as only a solid or only a liquid dielectric is used, the loss is small; but if a gas under ordinary or small pressure be present the loss may be very great. Whatever the nature of the force acting in the dielectric may be, it seems that in a solid or liquid the molecular displacement produced by the force is small: hence the product of force and displacement is insignificant, unless the frce be very great; but in a gas the displacement, and therefore this product, is considerable; the molecules are free to move, they reach high speeds, and the energy of their impact is lost in heat or otherwise. If the gas be strongly compressed, the displacement due to the force is made smaller, and the losses are reduced.

In most of the succeeding experiments I prefer, chiefly on account of the regular and positive action, to employ the alternator before referred to. This is one of the several machines constructed by me for the purpose of these investigations. It has 384 pole projections, and is capable of giving currents of a frequency of about 10,000 per second. This machine has been illustrated and briefly described in my first paper before the American Institute of Electrical Engineers, May 20th, 1891, to which I have already referred. A more detailed description, sufficient to enable any engineer to build a similar machine, will be found in several electrical journals of that period.

The induction coils operated from the machine are rather small, containing from 5,000 to 15,000 turns in the secondary. They are immersed in boiled-out linseed oil, contained in wooden boxes covered with zinc sheet.

I have found it advantageous to reverse the usual position of the wires, and to wind, in these coils, the primaries on the top; thus allowing the use of a much larger primary, which, of course, reduces the danger of overheating and increases the output of the coil. I make the primary on each side at least one centimeter shorter than the secondary, to prevent the breaking through on the ends, which would surely occur unless the insulation on the top of the secondary be very thick, and this, of course, would be disadvantageous.

When the primary is made movable, which is necessary in some experiments, and many times convenient for the purposes of adjustment, I cover the secondary with wax, and turn it off in a lathe to a diameter slightly smaller than the inside of the primary coil. The latter I provide with a handle reaching out of the oil, which serves to shift it in any position along the secondary.

I will now venture to make, in regard to the general manipulation of induction coils, a few observations bearing upon points which have not been fully appreciated in earlier experiments with such coils, and are even now often overlooked.

The secondary of the coil possesses usually such a high self-induction that the current through the wire is inappreciable, and may be so even when the terminals are joined by a conductor of small resistance. If capacity is added to the terminals, the self-induction is counteracted, and a stronger current is made to flow through the secondary, though its terminals are insulated from each other. To one entirely unacquainted with the properties of alternating currents nothing will look more puzzling. This feature was illustrated in the experiment performed at the beginning with the top plates of wire gauze attached to the terminals and the rubber plate. When the plates of wire gauze were close together, and a small arc passed between them, the arc *prevented* a strong current from passing through the secondary, because it did away with the capacity on the terminals; when the rubber plate was inserted between, the capacity of the condenser formed counteracted the self-induction of the secondary, a stronger current passed now, the coil performed more work, and the discharge was by far more powerful.

The first thing, then, in operating the induction coil is to combine capacity with the secondary to overcome the self-induction. If the frequencies and potentials are very high, gaseous matter should be carefully kept away from the charged surfaces. If Leyden jars are used, they should be immersed in oil, as otherwise considerable dissipation may occur if the jars are greatly strained. When high frequencies are used, it is of equal importance to combine a condenser with the primary. One may use a condenser connected to the ends of the primary or to the terminals of the alternator, but the latter is not to be recommended, as the machine might be injured. The best way is undoubtedly to use the condenser in series with the primary and with the alternator, and to adjust its capacity so as to annul the self-induction of both the latter. The condenser should be adjustable by very small steps, and for a finer adjustment a small oil condenser with movable plates may be used conveniently.

I think it best at this juncture to bring before you a phenomenon, observed by me some time ago, which to the purely scientific investigator may perhaps appear more interesting than any of the results which I have the privilege to present to you this evening.

It may be quite properly ranked among the brush phenomena—in fact, it is a brush, formed at, or near, a single terminal in high vacuum.

In bulbs provided with a conducting terminal, though it be of aluminum, the brush has but an ephemeral existence, and cannot, unfortunately, be indefinitely preserved in its most sensitive state, even in a bulb devoid of any conducting electrode. In studying the phenomenon, by all means a bulb having no leading-in wire should be used. I have found it best to use bulbs constructed as indicated in Figs. 141 and 142.

FIG. 141.   FIG. 142.

In Fig. 141 the bulb comprises an incandescent lamp globe L, in the neck of which is sealed a barometer tube b, the end of which is blown out to form a small sphere s. This sphere should be sealed as closely as possible in the center of the large globe. Before sealing, a thin tube t, of aluminum sheet, may be slipped in the barometer tube, but it is not important to employ it.

The small hollow sphere s is filled with some conducting powder, and a wire w is cemented in the neck for the purpose of connecting the conducting powder with the generator.

The construction shown in Fig. 142 was chosen in order to remove from the brush any conducting body which might possibly affect it. The bulb consists in this case of a lamp globe L, which has a neck n, provided with a tube b and small sphere s, sealed to it, so that two entirely independent compartments are formed, as indicated in the drawing. When the bulb is in use the neck n is provided with a tinfoil coating, which is connected to the generator and acts inductively upon the moderately rarefied and highly conducted gas inclosed in the neck. From there the current passes through the tube *b* into the small sphere *s*, to act by induction upon the gas contained in the globe L.

It is of advantage to make the tube *t* very thick, the hole through it very small, and to blow the sphere *s* very thin. It is of the greatest importance that the sphere *s* be placed in the center of the globe L.

FIG. 143.

Figs. 143, 144 and 145 indicate different forms, or stages, of the brush. Fig. 143 shows the brush as it first appears in a bulb provided with a conducting terminal; but, as in such a bulb it very soon disappears— often after a few minutes—I will confine myself to the description of the phenomenon as seen in a bulb without conducting electrode. It is observed under the following conditions:

When the globe L (Figs. 141 and 142) is exhausted to a very high degree, generally the bulb is not excited upon connecting the wire *w* (Fig. 141) or the tinfoil coating of the bulb (Fig. 142) to the terminal of the induction coil. To excite it, it is usually sufficient to grasp the globe L with the hand. An intense phosphorescence then spreads at first over the globe, but soon gives place to a white, misty light. Shortly afterward one may notice that the luminosity is unevenly distributed in the globe, and after passing the current for some time the bulb appears as in Fig. 144. From this stage the phenomenon will gradually pass to that indicated in Fig. 145, after some minutes, hours, days or weeks, according as the bulb is worked. Warming the bulb or increasing the potential hastens the transit.

FIG. 144.

FIG. 145.

When the brush assumes the form indicated in Fig. 145, it may be brought to a state of extreme sensitiveness to electrostatic and magnetic influence. The bulb hanging straight down from a wire, and all objects being remote from it, the approach of the observer at a few paces from the bulb will cause the brush to fly to the opposite side, and if he walks around the bulb it will always keep on the opposite side. It may begin to spin around the terminal long before it reaches that sensitive stage. When it begins to turn around, principally, but also before, it is affected by a magnet, and at a certain stage it is susceptible to magnetic influence to an astonishing degree. A small permanent magnet, with its poles at a distance of no more than two centimetres, will affect it visibly at a distance of two metres, slowing down or accelerating the rotation according to how it is held relatively to the brush. I think I have observed that at the stage when it is most sensitive to magnetic, it is not most sensitive to electrostatic, influence. My explanation is, that the electrostatic attraction between the brush and the glass of the bulb, which retards the rotation, grows much quicker than the magnetic influence when the intensity of the stream is increased.

When the bulb hangs with the globe l down, the rotation is always clockwise. In the southern hemisphere it would occur in the opposite direction and on the equator the brush should not turn at all. The rotation may be reversed by a magnet kept at some distance. The brush rotates best, seemingly, when it is at right angles to the lines of force of the earth. It very likely rotates, when at its maximum speed, in synchronism with the alternations, say, 10,000 times a second. The rotation can be slowed down or accelerated by the approach or receding of the observer, or any conducting body, but it cannot be reversed by putting the bulb in any position. When it is in the state of the highest sensitiveness and the potential or frequency be varied, the sensitiveness is rapidly diminished. Changing either of these but little will generally stop the rotation. The sensitiveness is likewise affected by the variations of temperature. To attain great sensitiveness it is necessary to have the small sphere s in the center of the globe L, as otherwise the electrostatic action of the glass of the globe will tend to stop the rotation. The sphere s should be small and of uniform thickness; any dissymmetry of course has the effect to diminish the sensitiveness.

The fact that the brush rotates in a definite direction in a permanent magnetic field seems to show that in alternating currents of very high frequency the positive and negative impulses are not equal, but that one always preponderates over the other.

Of course, this rotation in one direction may be due to the action of the two elements of the same current upon each other, or to the action of the field produced by one of the elements upon the other, as in a series motor, without necessarily one impulse being stronger than the other. The fact that the brush turns, as far as I could observe, in any position, would speak for this view. In such case it would turn at any point of the earth's surface. But, on the other hand, it is then hard to explain why a permanent magnet should reverse the rotation, and one must assume the preponderance of impulses of one kind.

As to the causes of the formation of the brush or stream, I think it is due to the electrostatic action of the globe and the dissymmetry of the parts. If the small bulb s and the globe L were perfect concentric spheres, and the glass throughout of the same thickness and quality, I think the brush would not form, as the tendency to pass would be equal on all sides. That the formation of the stream is due to an irregularity is apparent from the fact that it has the tendency to remain in one position, and rotation occurs most generally only when it is brought out of this position by electrostatic or magnetic influence. When in an extremely sensitive state it rests in one position, most curious experiments may be performed with it. For instance, the experimenter may, by selecting a proper position, approach the hand at a certain considerable distance to the bulb, and he may cause the brush to pass off by merely stiffening the muscles of the arm. When it begins to rotate slowly, and the hands are held at a proper distance, it is impossible to make even the slightest motion without producing a visible effect upon the brush. A metal plate connected to the other terminal of the coil affects it at a great distance, slowing down the rotation often to one turn a second.

I am firmly convinced that such a brush, when we learn how to produce it properly, will prove a valuable aid in the investigation of the nature of the forces acting in an electrostatic or magnetic field. If there is any motion which is measurable going on in the space, such a brush ought to reveal it. It is, so to speak, a beam of light, frictionless, devoid of inertia.

I think that it may find practical applications in telegraphy. With such a brush it would be possible to send dispatches across the Atlantic, for instance, with any speed, since its sensitiveness may be so great that the slightest changes will affect it. If it were possible to make the stream more intense and very narrow, its deflections could be easily photographed.

I have been interested to find whether there is a rotation of the stream itself, or whether there is simply a stress traveling around the bulb. For this purpose I mounted a light mica fan so that its vanes were in the path of the brush. If the stream itself was rotating the fan would be spun around. I could produce no distinct rotation of the fan, although I tried the experiment repeatedly; but as the fan exerted a noticeable influence on the stream, and the apparent rotation of the latter was, in this case, never quite satisfactory, the experiment did not appear to be conclusive.

I have been unable to produce the phenomenon with the disruptive discharge coil, although every other of these phenomena can be well produced by it—many, in fact, much better than with coils operated from an alternator.

It may be possible to produce the brush by impulses of one direction, or even by a steady potential, in which case it would be still more sensitive to magnetic influence.

In operating an induction coil with rapidly alternating currents, we realize with astonishment, for the first time, the great importance of the relation of capacity, self-induction and frequency as regards the general results. The effects of capacity are the most striking, for in these experiments, since the self-induction and frequency both are high, the critical capacity is very small, and need be but slightly varied to produce a very considerable change. The experimenter may bring his body in contact with the terminals of the secondary of the coil, or attach to one or both terminals insulated bodies of very small bulk, such as bulbs, and he may produce a considerable rise or fall of potential, and greatly affect the flow of the current through the primary. In the experiment before shown, in which a brush appears at a wire attached to one terminal, and the wire is vibrated when the experimenter brings his insulated body in contact with the other terminal of the coil, the sudden rise of potential was made evident.

I may show you the behavior of the coil in another manner which possesses a feature of some interest. I have here a little light fan of aluminum sheet, fastened to a needle and arranged to rotate freely in a metal piece screwed to one of the terminals of the coil. When the coil is set to work, the molecules of the air are rhythmically attracted and repelled. As the force with which they are repelled is greater than that with which they are attracted, it results that there is a repulsion exerted on the surfaces of the fan. If the fan were made simply of a metal sheet, the repulsion would be equal on the opposite sides, and would produce no effect. But if one of the opposing surfaces is screened, or if, generally speaking, the bombardment on this side is weakened in some way or other, there remains the repulsion exerted upon the other, and the fan is set in rotation. The screening is best effected by fastening upon one of the opposing sides of the fan insulated conducting coatings, or, if the fan is made in the shape of an ordinary propeller screw, by fastening on one side, and close to it, an insulated metal plate. The static screen may, however, be omitted, and simply a thickness of insulating material fastened to one of the sides of the fan.

To show the behavior of the coil, the fan may be placed upon the terminal and it will readily rotate when the coil is operated by currents of very high frequency. With a steady potential, of course, and even with alternating currents of very low frequency, it would not turn, because of the very slow exchange of air and, consequently, smaller bombardment; but in the latter case it might turn if the potential were excessive. With a pin wheel, quite the opposite rule holds good; it rotates best with a steady potential, and the effort is the smaller the higher the frequency. Now, it is very easy to adjust the conditions so that the potential is normally not sufficient to turn the fan, but that by connecting the other terminal of the coil with an insulated body it rises to a much greater value, so as to rotate the fan, and it is likewise possible to stop the rotation by connecting to the terminal a body of different size, thereby diminishing the potential.

Instead of using the fan in this experiment, we may use the "electric" radiometer with similar effect. But in this case it will be found that the vanes will rotate only at high exhaustion or at ordinary pressures; they will not rotate at moderate pressures, when the air is highly conducting. This curious observation was made conjointly by Professor Crookes and myself. I attribute the result to the high conductivity of the air, the molecules of which then do not act as independent carriers of electric charges, but act all together as a single conducting body. In such case, of course, if there is any repulsion at all of the molecules from the vanes, it must be very small. It is possible, however, that the result is in part due to the fact that the greater part of the discharge passes from the leading-in wire through the highly conducting gas, instead of passing off from the conducting vanes.

In trying the preceding experiment with the electric radiometer the potential should not exceed a certain limit, as then the electrostatic attraction between the vanes and the glass of the bulb may be so great as to stop the rotation.

A most curious feature of alternate currents of high frequencies and potentials is that they enable us to perform many experiments by the use of one wire only. In many respects this feature is of great interest.

In a type of alternate current motor invented by me some years ago I produced rotation by inducing, by means of a single alternating current passed through a motor circuit, in the mass or other circuits of the motor, secondary currents, which, jointly with the primary or inducing current, created a moving field of force. A simple but crude form of such a motor is obtained by winding upon an iron core a primary, and close to it a secondary coil, joining the ends of the latter and placing a freely movable metal disc within the influence of the field produced by both. The iron core is employed for obvious reasons, but it is not essential to the operation. To improve the motor, the iron core is made to encircle the armature. Again to improve, the secondary coil is made to partly overlap the primary, so that it cannot free itself from a strong inductive action of the latter, repel its lines as it may. Once more to improve, the proper difference of phase is obtained between the primary and secondary currents by a condenser, self-induction, resistance or equivalent windings.

FIG. 146.

I had discovered, however, that rotation is produced by means of a single coil and core; my explanation of the phenomenon, and leading thought in trying the experiment, being that there must be a true time lag in the magnetization of the core. I remember the pleasure I had when, in the writings of Professor Ayrton, which came later to my hand, I found the idea of the time lag advocated. Whether there is a true time lag, or whether the retardation is due to eddy currents circulating in minute paths, must remain an open question, but the fact is that a coil wound upon an iron core and traversed by an alternating current creates a moving field of force, capable of setting an armature in rotation. It is of some interest, in conjunction with the historical Arago experiment, to mention that in lag or phase motors I have produced rotation in the opposite direction to the moving field, which means that in that experiment the magnet may not rotate, or may even rotate in the opposite direction to the moving disc. Here, then, is a motor (diagrammatically illustrated in Fig. 146), comprising a coil and iron core, and a freely movable copper disc in proximity to the latter.

To demonstrate a novel and interesting feature, I have, for a reason which I will explain, selected this type of motor. When the ends of the coil are connected to the terminals of an alternator the disc is set in rotation. But it is not this experiment, now well known, which I desire to perform. What I wish to show you is that this motor rotates with *one single* connection between it and the generator; that is to say, one terminal of the motor is connected to one terminal of the generator—in this case the secondary of a high-tension induction coil—the other terminals of motor and generator being insulated in space. To produce rotation it is generally (but not absolutely) necessary to connect the free end of the motor coil to an insulated body of some size. The experimenter's body is more than sufficient. If he touches the free terminal with an object held in the hand, a current passes through the coil and the copper disc is set in rotation. If an exhausted tube is put in series with the coil, the tube lights brilliantly, showing the passage of a strong current. Instead of the experimenter's body, a small metal sheet suspended on a cord may be used with the same result. In this case the plate acts as a condenser in series with the coil. It counteracts the self-induction of the latter and allows a strong current to pass. In such a combination, the greater the self-induction of the coil the smaller need be the plate, and this means that a lower frequency, or eventually a lower potential, is required to operate the motor. A single coil wound upon a core has a high self-induction; for this reason, principally, this type of motor was chosen to perform the experiment. Were a secondary closed coil wound upon the core, it would tend to diminish the self-induction, and then it would be necessary to employ a much higher frequency and potential. Neither would be advisable, for a higher potential would endanger the insulation of the small primary coil, and a higher frequency would result in a materially diminished torque.

It should be remarked that when such a motor with a closed secondary is used, it is not at all easy to obtain rotation with excessive frequencies, as the secondary cuts off almost completely the lines of the primary— and this, of course, the more, the higher the frequency—and allows the passage of but a minute current. In such a case, unless the secondary is closed through a condenser, it is almost essential, in order to produce rotation, to make the primary and secondary coils overlap each other more or less.

But there is an additional feature of interest about this motor, namely, it is not necessary to have even a single connection between the motor and generator, except, perhaps, through the ground; for not only is an insulated plate capable of giving off energy into space, but it is likewise capable of deriving it from an alternating electrostatic field, though in the latter case the available energy is much smaller. In this instance one of the motor terminals is connected to the insulated plate or body located within the alternating electrostatic field, and the other terminal preferably to the ground.

It is quite possible, however, that such "no wire" motors, as they might be called, could be operated by conduction through the rarefied air at considerable distances. Alternate currents, especially of high frequencies, pass with astonishing freedom through even slightly rarefied gases. The upper strata of the air are rarefied. To reach a number of miles out into space requires the overcoming of difficulties of a merely mechanical nature. There is no doubt that with the enormous potentials obtainable by the use of high frequencies and oil insulation, luminous discharges might be passed through many miles of rarefied air, and that, by thus directing the energy of many hundreds or thousands of horsepower, motors or lamps might be operated at considerable distances from stationary sources. But such schemes are mentioned merely as possibilities. We shall have no need to transmit power in this way. We shall have no need to *transmit* power at all. Ere many generations pass, our machinery will be driven by a power obtainable at any point of the universe. This idea is not novel. Men have been led to it long ago by instinct or reason. It has been expressed in many ways, and in many places, in the history of old and new. We find it in the delightful myth of Antheus, who derives power from the earth; we find it among the subtle speculations of one of your splendid mathematicians, and in many hints and statements of thinkers of the present time. Throughout space there is energy. Is this energy static or kinetic? If static our hopes are in vain; if kinetic—and this we know it is, for certain—then it is a mere question of time when men will succeed in attaching their machinery to the very wheelwork of nature. Of all, living or dead, Crookes came nearest to doing it. His radiometer will turn in the light of day and in the darkness of the night; it will turn everywhere where there is heat, and heat is everywhere. But, unfortunately, this beautiful little machine, while it goes down to posterity as the most interesting, must likewise be put on record as the most inefficient machine ever invented!

The preceding experiment is only one of many equally interesting experiments which may be performed by the use of only one wire with alternations of high potential and frequency. We may connect an insulated line to a source of such currents, we may pass an inappreciable current over the line, and on any point of the same we are able to obtain a heavy current, capable of fusing a thick copper wire. Or we may, by the help of some artifice, decompose a solution in any electrolytic cell by connecting only one pole of the cell to the line or source of energy. Or we may, by attaching to the line, or only bringing into its vicinity, light up an incandescent lamp, an exhausted tube, or a phosphorescent bulb.

However impracticable this plan of working may appear in many cases, it certainly seems practicable, and even recommendable, in the production of light. A perfected lamp would require but little energy, and if wires were used at all we ought to be able to supply that energy without a return wire.

It is now a fact that a body may be rendered incandescent or phosphorescent by bringing it either in single contact or merely in the vicinity of a source of electric impulses of the proper character, and that in this manner a quantity of light sufficient to afford a practical illuminant may be produced. It is, therefore, to say the least, worth while to attempt to determine the best conditions and to invent the best appliances for attaining this object.

Some experiences have already been gained in this direction, and I will dwell on them briefly, in the hope that they might prove useful.

The heating of a conducting body inclosed in a bulb, and connected to a source of rapidly alternating electric impulses, is dependent on so many things of a different nature, that it would be difficult to give a generally applicable rule under which the maximum heating occurs. As regards the size of the vessel, I have lately found that at ordinary or only slightly differing atmospheric pressures, when air is a good insulator, and hence practically the same amount of energy by a certain potential and frequency is given off from the body, whether the bulb be small or large, the body is brought to a higher temperature if enclosed in a small bulb, because of the better confinement of heat in this case.

At lower pressures, when air becomes more or less conducting, or if the air be sufficiently warmed to become conducting, the body is rendered more intensely incandescent in a large bulb, obviously because, under otherwise equal conditions of test, more energy may be given off from the body when the bulb is large.

At very high degrees of exhaustion, when the matter in the bulb becomes "radiant," a large bulb has still an advantage, but a comparatively slight one, over the small bulb.

Finally, at excessively high degrees of exhaustion, which cannot be reached except by the employment of special means, there seems to be, beyond a certain and rather small size of vessel, no perceptible difference in the heating.

These observations were the result of a number of experiments, of which one, showing the effect of the size of the bulb at a high degree of exhaustion, may be described and shown here, as it presents a feature of interest. Three spherical bulbs of 2 inches, 3 inches and 4 inches diameter were taken, and in the center of each was mounted an equal length of an ordinary incandescent lamp filament of uniform thickness. In each bulb the piece of filament was fastened to the leading-in wire of platinum, contained in a glass stem sealed in the bulb; care being taken, of course, to make everything as nearly alike as possible. On each glass stem in the inside of the bulb was slipped a highly polished tube made of aluminum sheet, which fitted the stem and was held on it by spring pressure. The function of this aluminum tube will be explained subsequently. In each bulb an equal length of filament protruded above the metal tube. It is sufficient to say now that under these conditions equal lengths of filament of the same thickness—in other words, bodies of equal bulk— were brought to incandescence. The three bulbs were sealed to a glass tube, which was connected to a Sprengel pump. When a high vacuum had been reached, the glass tube carrying the bulbs was sealed off. A current was then turned on successively on each bulb, and it was found that the filaments came to about the same brightness, and, if anything, the smallest bulb, which was placed midway between the two larger ones, may have been slightly brighter. This result was expected, for when either of the bulbs was connected to the coil the luminosity spread through the other two, hence the three bulbs constituted really one vessel. When all the three bulbs were connected in multiple arc to the coil, in the largest of them the filament glowed brightest, in the next smaller it was a little less bright, and in the smallest it only came to redness. The bulbs were then sealed off and separately tried. The brightness of the filaments was now such as would have been expected on the supposition that the energy given off was proportionate to the surface of the bulb, this surface in each case representing one of the coatings of a condenser. Accordingly, there was less difference between the largest and the middle sized than between the latter and the smallest bulb.

An interesting observation was made in this experiment. The three bulbs were suspended from a straight bare wire connected to a terminal of a coil, the largest bulb being placed at the end of the wire, at some distance from it the smallest bulb, and at an equal distance from the latter the middle-sized one. The carbons glowed then in both the larger bulbs about as expected, but the smallest did not get its share by far. This observation led me to exchange the position of the bulbs, and I then observed that whichever of the bulbs was in the middle was by far less bright than it was in any other position. This mystifying result was, of course, found to be due to the electrostatic action between the bulbs. When they were placed at a considerable distance, or when they were attached to the corners of an equilateral triangle of copper wire, they glowed in about the order determined by their surfaces.

As to the shape of the vessel, it is also of some importance, especially at high degrees of exhaustion. Of all the possible constructions, it seems that a spherical globe with the refractory body mounted in its center is the best to employ. By experience it has been demonstrated that in such a globe a refractory body of a given bulk is more easily brought to incandescence than when differently shaped bulbs are used. There is also an advantage in giving to the incandescent body the shape of a sphere, for self-evident reasons. In any case the body should be mounted in the center, where the atoms rebounding from the glass collide. This object is best attained in the spherical bulb; but it is also attained in a cylindrical vessel with one or two straight filaments coinciding with its axis, and possibly also in parabolical or spherical bulbs with refractory body or bodies placed in the focus or foci of the same; though the latter is not probable, as the electrified atoms should in all cases rebound normally from the surface they strike, unless the speed were excessive, in which case they *would* probably follow the general law of reflection. No matter what shape the vessel may have, if the exhaustion be low, a filament mounted in the globe is brought to the same degree of incandescence in all parts; but if the exhaustion be high and the bulb be spherical or pear-shaped, as usual, focal points form and the filament is heated to a higher degree at or near such points.

To illustrate the effect, I have here two small bulbs which are alike, only one is exhausted to a low and the other to a very high degree. When connected to the coil, the filament in the former glows uniformly throughout all its length; whereas in the latter, that portion of the filament which is in the center of the bulb glows far more intensely than the rest. A curious point is that the phenomenon occurs even if two filaments are mounted in a bulb, each being connected to one terminal of the coil, and, what is still more curious, if they be very near together, provided the vacuum be very high. I noted in experiments with such bulbs that the filaments would give way usually at a certain point, and in the first trials I attributed it to a defect in the carbon. But when the phenomenon occurred many times in succession I recognized its real cause.

In order to bring a refractory body inclosed in a bulb to incandescence, it is desirable, on account of economy, that all the energy supplied to the bulb from the source should reach without loss the body to be heated; from there, and from nowhere else, it should be radiated. It is, of course, out of the question to reach this theoretical result, but it is possible by a proper construction of the illuminating device to approximate it more or less.

For many reasons, the refractory body is placed in the center of the bulb, and it is usually supported on a glass stem containing the leading-in wire. As the potential of this wire is alternated, the rarefied gas surrounding the stem is acted upon inductively, and the glass stem is violently bombarded and heated. In this manner by far the greater portion of the energy supplied to the bulb—especially when exceedingly high frequencies are used—may be lost for the purpose contemplated. To obviate this loss, or at least to reduce it to a minimum, I usually screen the rarefied gas surrounding the stem from the inductive action of the leading-in wire by providing the stem with a tube or coating of conducting material. It seems beyond doubt that the best among metals to employ for this purpose is aluminum, on account of its many remarkable properties. Its only fault is that it is easily fusible, and, therefore, its distance from the incandescing body should be properly estimated. Usually, a thin tube, of a diameter somewhat smaller than that of the glass stem, is made of the finest aluminum sheet, and slipped on the stem. The tube is conveniently prepared by wrapping around a rod fastened in a lathe a piece of aluminum sheet of proper size, grasping the sheet firmly with clean chamois leather or blotting paper, and spinning the rod very fast. The sheet is wound tightly around the rod, and a highly polished tube of one or three layers of the sheet is obtained. When slipped on the stem, the pressure is generally sufficient to prevent it from slipping off, but, for safety, the lower edge of the sheet may be turned inside. The upper inside corner of the sheet—that is, the one which is nearest to the refractory incandescent body—should be cut out diagonally, as it often happens that, in consequence of the intense heat, this corner turns toward the inside and comes very near to, or in contact with, the wire, or filament, supporting the refractory body. The greater part of the energy supplied to the bulb is then used up in heating the metal tube, and the bulb is rendered useless for the purpose. The aluminum sheet should project above the glass stem more or less— one inch or so—or else, if the glass be too close to the incandescing body, it may be strongly heated and become more or less conducting, whereupon it may be ruptured, or may, by its conductivity, establish a good electrical connection between the metal tube and the leading-in wire, in which case, again, most of the energy will be lost in heating the former. Perhaps the best way is to make the top of the glass tube, for about an inch, of a much smaller diameter. To still further reduce the danger arising from the heating of the glass stem, and also with the view of preventing an electrical connection between the metal tube and the electrode, I preferably wrap the stem with several layers of thin mica, which extends at least as far as the metal tube. In some bulbs I have also used an outside insulating cover.

The preceding remarks are only made to aid the experimenter in the first trials, for the difficulties which he encounters he may soon find means to overcome in his own way.

To illustrate the effect of the screen, and the advantage of using it, I have here two bulbs of the same size, with their stems, leading-in wires and incandescent lamp filaments tied to the latter, as nearly alike as possible. The stem of one bulb is provided with an aluminum tube, the stem of the other has none. Originally the two bulbs were joined by a tube which was connected to a Sprengel pump. When a high vacuum had been reached, first the connecting tube, and then the bulbs, were sealed off; they are therefore of the same degree of exhaustion. When they are separately connected to the coil giving a certain potential, the carbon filament in the bulb provided with the aluminum screen is rendered highly incandescent, while the filament in the other bulb may, with the same potential, not even come to redness, although in reality the latter bulb takes generally more energy than the former. When they are both connected together to the terminal, the difference is even more apparent, showing the importance of the screening. The metal tube placed on the stem containing the leading-in wire performs really two distinct functions: First, it acts more or less as an electrostatic screen, thus economizing the energy supplied to the bulb; and, second, to whatever extent it may fail to act electrostatically, it acts mechanically, preventing the bombardment, and consequently intense heating and possible deterioration of the *slender* support of the refractory incandescent body, or of the glass stem containing the leading-in wire. I say slender support, for it is evident that in order to confine the heat more completely to the incandescing body its support should be very thin, so as to carry away the smallest possible amount of heat by conduction. Of all the supports used I have found an ordinary incandescent lamp filament to be the best, principally because among conductors it can withstand the highest degree of heat.

The effectiveness of the metal tube as an electrostatic screen depends largely on the degree of exhaustion.

At excessively high degrees of exhaustion—which are reached by using great care and special means in connection with the Sprengel pump—when the matter in the globe is in the ultra-radiant state, it acts most perfectly. The shadow of the upper edge of the tube is then sharply defined upon the bulb.

At a somewhat lower degree of exhaustion, which is about the ordinary "non-striking" vacuum, and generally as long as the matter moves predominantly in straight lines, the screen still does well. In elucidation of the preceding remark it is necessary to state that what is a "non-striking" vacuum for a coil operated as ordinarily, by impulses, or currents, of low frequency, is not so, by far, when the coil is operated by currents of very high frequency. In such case the discharge may pass with great freedom through the rarefied gas through which a low frequency discharge may not pass, even though the potential be much higher. At ordinary atmospheric pressures just the reverse rule holds good: the higher the frequency, the less the spark discharge is able to jump between the terminals, especially if they are knobs or spheres of some size.

Finally, at very low degrees of exhaustion, when the gas is well conducting, the metal tube not only does not act as an electrostatic screen, but even is a drawback, aiding to a considerable extent the dissipation of the energy laterally from the leading-in wire. This, of course, is to be expected. In this case, namely, the metal tube is in good electrical connection with the leading-in wire, and most of the bombardment is directed upon the tube. As long as the electrical connection is not good, the conducting tube is always of some advantage, for although it may not greatly economize energy, still it protects the support of the refractory button, and is the means of concentrating more energy upon the same.

To whatever extent the aluminum tube performs the function of a screen, its usefulness is therefore limited to very high degrees of exhaustion when it is insulated from the electrode—that is, when the gas as a whole is non-conducting, and the molecules, or atoms, act as independent carriers of electric charges.

In addition to acting as a more or less effective screen, in the true meaning of the word, the conducting tube or coating may also act, by reason of its conductivity, as a sort of equalizer or dampener of the bombardment against the stem. To be explicit, I assume the action to be as follows: Suppose a rhythmical bombardment to occur against the conducting tube by reason of its imperfect action as a screen, it certainly must happen that some molecules, or atoms, strike the tube sooner than others. Those which come first in contact with it give up their superfluous charge, and the tube is electrified, the electrification instantly spreading over its surface. But this must diminish the energy lost in the bombardment, for two reasons: first, the charge given up by the atoms spreads over a great area, and hence the electric density at any point is small, and the atoms are repelled with less energy than they would be if they struck against a good insulator; secondly, as the tube is electrified by the atoms which first come in contact with it, the progress of the following atoms against the tube is more or less checked by the repulsion which the electrified tube must exert upon the similarly electrified atoms. This repulsion may perhaps be sufficient to prevent a large portion of the atoms from striking the tube, but at any rate it must diminish the energy of their impact. It is clear that when the exhaustion is very low, and the rarefied gas well conducting, neither of the above effects can occur, and, on the other hand, the fewer the atoms, with the greater freedom they move; in other words, the higher the degree of exhaustion, up to a limit, the more telling will be both the effects.

What I have just said may afford an explanation of the phenomenon observed by Prof. Crookes, namely, that a discharge through a bulb is established with much greater facility when an insulator than when a conductor is present in the same. In my opinion, the conductor acts as a dampener of the motion of the atoms in the two ways pointed out; hence, to cause a visible discharge to pass through the bulb, a much higher potential is needed if a conductor, especially of much surface, be present.

For the sake of elucidating of some of the remarks before made, I must now refer to Figs. 147, 148 and 149, which illustrate various arrangements with a type of bulb most generally used.

FIG. 147.

FIG. 148.

Fig. 147 is a section through a spherical bulb L, with the glass stem *s*, contains the leading-in wire *w*, which has a lamp filament *l* fastened to it, serving to support the refractory button *m* in the center. M is a sheet of thin mica wound in several layers around the stem *s*, and *a* is the aluminum tube.

Fig. 148 illustrates such a bulb in a somewhat more advanced stage of perfection. A metallic tube *s* is fastened by means of some cement to the neck of the tube. In the tube is screwed a plug P, of insulating material, in the center of which is fastened a metallic terminal *t*, for the connection to the leading-in wire *w*. This terminal must be well insulated from the metal tube *s*; therefore, if the cement used is conducting—and most generally it is sufficiently so—the space between the plug P and the neck of the bulb should be filled with some good insulating material, such as mica powder.

FIG. 149.　　　　　　　　　FIG. 150.

Fig. 149 shows a bulb made for experimental purposes. In this bulb the aluminum tube is provided with an external connection, which serves to investigate the effect of the tube under various conditions. It is referred to chiefly to suggest a line of experiment followed.

Since the bombardment against the stem containing the leading-in wire is due to the inductive action of the latter upon the rarefied gas, it is of advantage to reduce this action as far as practicable by employing a very thin wire, surrounded by a very thick insulation of glass or other material, and by making the wire passing through the rarefied gas as short as practicable. To combine these features I employ a large tube T

(Fig. 150), which protrudes into the bulb to some distance, and carries on the top a very short glass stem *s*, into which is sealed the leading-in wire *w*, and I protect the top of the glass stem against the heat by a small aluminum tube *a* and a layer of mica underneath the same, as usual. The wire *w*, passing through the large tube to the outside of the bulb, should be well insulated—with a glass tube, for instance—and the space between ought to be filled out with some excellent insulator. Among many insulating powders I have found that mica powder is the best to employ. If this precaution is not taken, the tube T, protruding into the bulb, will surely be cracked in consequence of the heating by the brushes which are apt to form in the upper part of the tube, near the exhausted globe, especially if the vacuum be excellent, and therefore the potential necessary to operate the lamp be very high.

Fig. 151 illustrates a similar arrangement, with a large tube T protruding into the part of the bulb containing the refractory button *m*. In this case the wire leading from the outside into the bulb is omitted, the energy required being supplied through condenser coatings *c c*. The insulating packing P should in this construction be tightly fitting to the glass, and rather wide, or otherwise the discharge might avoid passing through the wire *w*, which connects the inside condenser coating to the incandescent button *m*.

FIG. 151.

FIG. 152.

The molecular bombardment against the glass stem in the bulb is a source of great trouble. As an illustration I will cite a phenomenon only too frequently and unwillingly observed. A bulb, preferably a large one, may be taken, and a good conducting body, such as a piece of carbon, may be mounted in it upon a platinum wire sealed in the glass stem. The bulb may be exhausted to a fairly high degree, nearly to the point when phosphorescence begins to appear. When the bulb is connected with the coil, the piece of carbon, if small, may become highly incandescent at first, but its brightness immediately diminishes, and then the discharge may break through the glass somewhere in the middle of the stem, in the form of bright sparks, in spite of the fact that the platinum wire is in good electrical connection with the rarefied gas through the piece of carbon or metal at the top. The first sparks are singularly bright, recalling those drawn from a clear surface of mercury. But, as they heat the glass rapidly, they, of course, lose their brightness, and cease when the glass at the ruptured place becomes incandescent, or generally sufficiently hot to conduct. When observed for the first time the phenomenon must appear very curious, and shows in a striking manner how radically different alternate currents, or impulses, of high frequency behave, as compared with steady currents, or currents of low frequency. With such currents—namely, the latter—the phenomenon would of course not occur. When frequencies such as are obtained by mechanical means are used, I think that the rupture of the glass is more or less the consequence of the bombardment, which warms it up and impairs its insulating power; but with frequencies obtainable with condensers I have no doubt that the glass may give way without previous heating. Although this appears most singular at first, it is in reality what we might expect to occur. The energy supplied to the wire leading into the bulb is given off partly by direct action through the carbon button, and partly by inductive action through the glass surrounding the wire. The case is thus analogous to that in which a condenser shunted by a conductor of low resistance is connected to a source of alternating current. As long as the frequencies are low, the conductor gets the most and the condenser is perfectly safe; but when the frequency becomes excessive, the *role* of the conductor may become quite insignificant. In the latter case the difference of potential at the terminals of the condenser may become so great as to rupture the dielectric, notwithstanding the fact that the terminals are joined by a conductor of low resistance.

It is, of course, not necessary, when it is desired to produce the incandescence of a body inclosed in a bulb by means of these currents, that the body should be a conductor, for even a perfect non-conductor may be quite as readily heated. For this purpose it is sufficient to surround a conducting electrode with a non-conducting material, as, for instance, in the bulb described before in Fig. 150, in which a thin incandescent lamp filament is coated with a non-conductor, and supports a button of the same material on the top. At the start the bombardment goes on by inductive action through the non-conductor, until the same is sufficiently heated to become conducting, when the bombardment continues in the ordinary way.

A different arrangement used in some of the bulbs constructed is illustrated in Fig. 152. In this instance a non-conductor m is mounted in a piece of common arc light carbon so as to project some small distance above the latter. The carbon piece is connected to the leading-in wire passing through a glass stem, which is wrapped with several layers of mica. An aluminum tube a is employed as usual for screening. It is so arranged that it reaches very nearly as high as the carbon and only the non-conductor m projects a little above it. The bombardment goes at first against the upper surface of carbon, the lower parts being protected by the aluminum tube. As soon, however, as the non-conductor m is heated it is rendered good conducting, and then it becomes the center of the bombardment, being most exposed to the same.

I have also constructed during these experiments many such single-wire bulbs with or without internal electrode, in which the radiant matter was projected against, or focused upon, the body to be rendered incandescent. Fig. 153 (page 516) illustrates one of the bulbs used. It consists of a spherical globe L, provided with a long neck n, on top, for increasing the action in some cases by the application of an external conducting coating. The globe L is blown out on the bottom into a very small bulb b, which serves to hold it firmly in a socket s of insulating material into which it is cemented. A fine lamp filament f, supported on a wire w, passes through the center of the globe L. The filament is rendered incandescent in the middle portion, where the bombardment proceeding from the lower inside surface of the globe is most intense. The lower portion of the globe, as far as the socket s reaches, is rendered conducting, either by a tinfoil coating or otherwise, and the external electrode is connected to a terminal of the coil.

The arrangement diagrammatically indicated in Fig. 153 was found to be an inferior one when it was desired to render incandescent a filament or button supported in the center of the globe, but it was convenient when the object was to excite phosphorescence.

In many experiments in which bodies of different kind were mounted in the bulb as, for instance, indicated in Fig. 152, some observations of interest were made.

It was found, among other things, that in such cases, no matter where the bombardment began, just as soon as a high temperature was reached there was generally one of the bodies which seemed to take most of the bombardment upon itself, the other, or others, being thereby relieved. The quality appeared to depend principally on the point of fusion, and on the facility with which the body was "evaporated," or, generally speaking, disintegrated—meaning by the latter term not only the throwing off of atoms, but likewise of large lumps. The observation made was in accordance with generally accepted notions. In a highly exhausted bulb, electricity is carried off from the electrode by independent carriers, which are partly the atoms, or molecules, of the residual atmosphere, and partly the atoms, molecules, or lumps thrown off from the electrode. If the electrode is composed of bodies of different character, and if one of these is more easily disintegrated than the other, most of the electricity supplied is carried off from that body, which is then brought to a higher temperature than the others, and this the more, as upon an increase of the temperature the body is still more easily disintegrated.

It seems to me quite probable that a similar process takes place in the bulb even with a homogeneous electrode, and I think it to be the principal cause of the disintegration. There is bound to be some irregularity, even if the surface is highly polished, which, of course, is impossible with most of the refractory bodies employed as electrodes. Assume that a point of the electrode gets hotter; instantly most of the discharge passes through that point, and a minute patch it probably fused and evaporated. It is now possible that in consequence of the violent disintegration the spot attacked sinks in temperature, or that a counter force is created, as in an arc; at any rate, the local tearing off meets with the limitations incident to the experiment, whereupon the same process occurs on another place.

To the eye the electrode appears uniformly brilliant, but there are upon it points constantly shifting and wandering around, of a temperature far above the mean, and this materially hastens the process of deterioration. That some such thing occurs, at least when the electrode is at a lower temperature, sufficient experimental evidence can be obtained in the following manner: Exhaust a bulb to a very high degree, so that with a fairly high potential the discharge cannot pass—that is, not a *luminous* one, for a weak invisible discharge occurs always, in all probability. Now raise slowly and carefully the potential, leaving the primary current on no more than for an instant. At a certain point, two, three, or half a dozen phosphorescent spots will appear on the globe. These places of the glass are evidently more violently bombarded than others, this being due to the unevenly distributed electric density, necessitated, of course, by sharp projections, or, generally speaking, irregularities of the electrode. But the luminous patches are constantly changing in position, which is especially well observable if one manages to produce very few, and this indicates that the configuration of the electrode is rapidly changing.

From experiences of this kind I am led to infer that, in order to be most durable, the refractory button in the bulb should be in the form of a sphere with a highly polished surface. Such a small sphere could be manufactured from a diamond or some other crystal, but a better way would be to fuse, by the employment of extreme degrees of temperature, some oxide—as, for instance, zirconia—into a small drop, and then keep it in the bulb at a temperature somewhat below its point of fusion.

Interesting and useful results can, no doubt, be reached in the direction of extreme degrees of heat. How can such high temperatures be arrived at? How are the highest degrees of heat reached in nature? By the impact of stars, by high speeds and collisions. In a collision any rate of heat generation may be attained. In a chemical process we are limited. When oxygen and hydrogen combine, they fall, metaphorically speaking, from a definite height. We cannot go very far with a blast, nor by confining heat in a furnace, but in an exhausted bulb we can concentrate any amount of energy upon a minute button. Leaving practicability out of consideration, this, then, would be the means which, in my opinion, would enable us to reach the highest temperature. But a great difficulty when proceeding in this way is encountered, namely, in most cases the body is carried off before it can fuse and form a drop. This difficulty exists principally with an oxide, such as zirconia, because it cannot be compressed in so hard a cake that it would not be carried off quickly. I have endeavored repeatedly to fuse zirconia, placing it in a cup of arc light carbon, as indicated in Fig. 152. It glowed with a most intense light, and the stream of the particles projected out of the carbon cup was of a vivid white; but whether it was compressed in a cake or made into a paste with carbon, it was carried off before it could be fused. The carbon cup, containing zirconia, had to be mounted very low in the neck of a large bulb, as the heating of the glass by the projected particles of the oxide was so rapid that in the first trial the bulb was cracked almost in an instant, when the current was turned on. The heating of the glass by the projected particles was found to be always greater when the carbon cup contained a body which was rapidly carried off—I presume, because in such cases, with the same potential, higher speeds were reached, and also because, per unit of time, more matter was projected—that is, more particles would strike the glass.

The before-mentioned difficulty did not exist, however, when the body mounted in the carbon cup offered great resistance to deterioration. For instance, when an oxide was first fused in an oxygen blast, and then mounted in the bulb, it melted very readily into a drop.

Generally, during the process of fusion, magnificent light effects were noted, of which it would be difficult to give an adequate idea. Fig. 152 is intended to illustrate the effect observed with a ruby drop. At first one may see a narrow funnel of white light projected against the top of the globe, where it produces an irregularly outlined phosphorescent patch. When the point of the ruby fuses, the phosphorescence becomes very powerful; but as the atoms are projected with much greater speed from the surface of the drop, soon the glass gets hot and "tired," and now only the outer edge of the patch glows. In this manner an intensely phosphorescent, sharply defined line, l, corresponding to the outline of the drop, is produced, which spreads slowly over the globe as the drop gets larger. When the mass begins to boil, small bubbles and cavities are formed, which cause dark colored spots to sweep across the globe. The bulb may be turned downward without fear of the drop falling off, as the mass possesses considerable viscosity.

I may mention here another feature of some interest, which I believe to have noted in the course of these experiments, though the observations do not amount to a certitude. It *appeared* that under the molecular impact caused by the rapidly alternating potential, the body was fused and maintained in that state at a lower temperature in a highly exhausted bulb than was the case at normal pressure and application of heat in the ordinary way—that is, at least, judging from the quantity of the light emitted. One of the experiments performed may be mentioned here by way of illustration. A small piece of pumice stone was stuck on a platinum wire, and first melted to it in a gas burner. The wire was next placed between two pieces of charcoal, and a burner applied, so as to produce an intense heat, sufficient to melt down the pumice stone into a small glass-like button. The platinum wire had to be taken of sufficient thickness, to prevent its melting in the fire. While in the charcoal fire, or when held in a burner to get a better idea of the degree of heat, the button glowed with great brilliancy. The wire with the button was then mounted in a bulb, and upon exhausting the same to a high degree, the current was turned on slowly, so as to prevent the cracking of the button. The button was heated to the point of fusion, and when it melted, it did not, apparently, glow with the same brilliancy as before, and this would indicate a lower temperature. Leaving out of consideration the observer's possible, and even probable, error, the question is, can a body under these conditions be brought from a solid to a liquid state with the evolution of *less* light?

When the potential of a body is rapidly alternated, it is certain that the structure is jarred. When the potential is very high, although the vibrations may be few—say 20,000 per second—the effect upon the structure may be considerable. Suppose, for example, that a ruby is melted into a drop by a steady application of energy. When it forms a drop, it will emit visible and invisible waves, which will be in a definite ratio, and to the eye the drop will appear to be of a certain brilliancy. Next, suppose we diminish to any degree we choose the energy steadily supplied, and, instead, supply energy which rises and falls according to a certain law. Now, when the drop is formed, there will be emitted from it three different kinds of vibrations—the ordinary visible, and two kinds of invisible waves: that is, the ordinary dark waves of all lengths, and, in addition, waves of a well defined character. The latter would not exist by a steady supply of the energy; still they help to jar and loosen the structure. If this really be the case, then the ruby drop will emit relatively less visible and more invisible waves than before. Thus it would seem that when a platinum wire, for instance, is fused by currents alternating with extreme rapidity, it emits at the point of fusion less light and more visible radiation than it does when melted by a steady current, though the total energy used up in the process of fusion is the same in both cases. Or, to cite another example, a lamp filament is not capable of withstanding as long with currents of extreme frequency as it does with steady currents, assuming that it be worked at the same luminous intensity. This means that for rapidly alternating currents the filament should be shorter and thicker. The higher the frequency—that is, the greater the departure from the steady flow—the worse it would be for the filament. But if the truth of this remark were demonstrated, it would be erroneous to conclude that such a refractory button as used in these bulbs would be deteriorated quicker by currents of extremely high frequency than by steady or low frequency currents. From experience I may say that just the opposite holds good: the button withstands the bombardment better with currents of very high frequency. But this is due to the fact that a high frequency discharge passes through a rarefied gas with much greater freedom than a steady or low frequency discharge, and this will mean that with the former we can work with a lower potential or with a less violent impact. As long, then, as the gas is of no consequence, a steady or low frequency current is better; but as soon as the action of the gas is desired and important, high frequencies are preferable.

In the course of these experiments a great many trials were made with all kinds of carbon buttons. Electrodes made of ordinary carbon buttons were decidedly more durable when the buttons were obtained by the application of enormous pressure. Electrodes prepared by depositing carbon in well known ways did not show up well; they blackened the globe very quickly. From many experiences I conclude that lamp filaments obtained in this manner can be advantageously used only with low potentials and low frequency currents. Some kinds of carbon withstand so well that, in order to bring them to the point of fusion, it is necessary to employ very small buttons. In this case the observation is rendered very difficult on account of the intense heat produced. Nevertheless there can be no doubt that all kinds of carbon are fused under the molecular bombardment, but the liquid state must be one of great instability. Of all the bodies tried there were two which withstood best—diamond and carborundum. These two showed up about equally, but the latter was preferable for many reasons. As it is more than likely that this body is not yet generally known, I will venture to call your attention to it.

It has been recently produced by Mr. E. G. Acheson, of Monongahela City, Pa., U.S.A. It is intended to replace ordinary diamond powder for polishing precious stones, etc., and I have been informed that it accomplishes this object quite successfully. I do not know why the name "carborundum" has been given to it, unless there is something in the process of its manufacture which justifies this selection. Through the kindness of the inventor, I obtained a short while ago some samples which I desired to test in regard to their qualities of phosphorescence and capability of withstanding high degrees of heat.

Carborundum can be obtained in two forms—in the form of "crystals" and of powder. The former appear to the naked eye dark colored, but are very brilliant; the latter is of nearly the same color as ordinary diamond powder, but very much finer. When viewed under a microscope the samples of crystals given to me did not appear to have any definite form, but rather resembled pieces of broken up egg coal of fine quality. The majority were opaque, but there were some which were transparent and colored. The crystals are a kind of carbon containing some impurities; they are extremely hard, and withstand for a long time even an oxygen blast. When the blast is directed against them they at first form a cake of some compactness, probably in consequence of the fusion of impurities they contain. The mass withstands for a very long time the blast without further fusion; but a slow carrying off, or burning, occurs, and, finally, a small quantity of a glass-like residue is left, which, I suppose, is melted alumina. When compressed strongly they conduct very well, but not as well as ordinary carbon. The powder, which is obtained from the crystals in some way, is practically non-conducting. It affords a magnificent polishing material for stones.

The time has been too short to make a satisfactory study of the properties of this product, but enough experience has been gained in a few weeks I have experimented upon it to say that it does possess some remarkable properties in many respects. It withstands excessively high degrees of heat, it is little deteriorated by molecular bombardment, and it does not blacken the globe as ordinary carbon does. The only difficulty which I have experienced in its use in connection with these experiments was to find some binding material which would resist the heat and the effect of the bombardment as successfully as carborundum itself does.

I have here a number of bulbs which I have provided with buttons of carborundum. To make such a button of carborundum crystals I proceed in the following manner: I take an ordinary lamp filament and dip its point in tar, or some other thick substance or paint which may be readily carbonized. I next pass the point of the filament through the crystals, and then hold it vertically over a hot plate. The tar softens and forms a drop on the point of the filament, the crystals adhering to the surface of the drop. By regulating the distance from the plate the tar is slowly dried out and the button becomes solid. I then once more dip the button in tar and hold it again over a plate until the tar is evaporated, leaving only a hard mass which firmly binds the crystals. When a larger button is required I repeat the process several times, and I generally also cover the filament a certain distance below the button with crystals. The button being mounted in a bulb, when a good vacuum has been reached, first a weak and then a strong discharge is passed through the bulb to carbonize the tar and expel all gases, and later it is brought to a very intense incandescence.

When the powder is used I have found it best to proceed as follows: I make a thick paint of carborundum and tar, and pass a lamp filament through the paint. Taking then most of the paint off by rubbing the filament against a piece of chamois leather, I hold it over a hot plate until the tar evaporates and the coating becomes firm. I repeat this process as many times as it is necessary to obtain a certain thickness of coating. On the point of the coated filament I form a button in the same manner.

There is no doubt that such a button—properly prepared under great pressure—of carborundum, especially of powder of the best quality, will withstand the effect of the bombardment fully as well as anything we know. The difficulty is that the binding material gives way, and the carborundum is slowly thrown off after some time. As it does not seem to blacken the globe in the least, it might be found useful for coating the filaments of ordinary incandescent lamps, and I think that it is even possible to produce thin threads or sticks of carborundum which will replace the ordinary filaments in an incandescent lamp. A carborundum coating seems to be more durable than other coatings, not only because the carborundum can withstand high degrees of heat, but also because it seems to unite with the carbon better than any other material I have tried. A coating of zirconia or any other oxide, for instance, is far more quickly destroyed. I prepared buttons of diamond dust in the same manner as of carborundum, and these came in durability nearest to those prepared of carborundum, but the binding paste gave way much more quickly in the diamond buttons; this, however, I attributed to the size and irregularity of the grains of the diamond.

It was of interest to find whether carborundum possesses the quality of phosphorescence. One is, of course, prepared to encounter two difficulties: first, as regards the rough product, the "crystals," they are good conducting, and it is a fact that conductors do not phosphoresce; second, the powder, being exceedingly fine, would not be apt to exhibit very prominently this quality, since we know that when crystals, even such as diamond or ruby, are finely powdered, they lose the property of phosphorescence to a considerable degree.

The question presents itself here, can a conductor phosphoresce? What is there in such a body as a metal, for instance, that would deprive it of the quality of phosphoresence, unless it is that property which characterizes it as a conductor? For it is a fact that most of the phosphorescent bodies lose that quality when they are sufficiently heated to become more or less conducting. Then, if a metal be in a large measure, or perhaps entirely, deprived of that property, it should be capable of phosphoresence. Therefore it is quite possible that at some extremely high frequency, when behaving practically as a non-conductor, a metal or any other conductor might exhibit the quality of phosphoresence, even though it be entirely incapable of phosphorescing under the impact of a low-frequency discharge. There is, however, another possible way how a conductor might at least *appear* to phosphoresce.

Considerable doubt still exists as to what really is phosphorescence, and as to whether the various phenomena comprised under this head are due to the same causes. Suppose that in an exhausted bulb, under the molecular impact, the surface of a piece of metal or other conductor is rendered strongly luminous, but at the same time it is found that it remains comparatively cool, would not this luminosity be called phosphorescence? Now such a result, theoretically at least, is possible, for it is a mere question of potential or speed. Assume the potential of the electrode, and consequently the speed of the projected atoms, to be sufficiently high, the surface of the metal piece, against which the atoms are projected, would be rendered highly incandescent, since the process of heat generation would be incomparably faster than that of radiating or conducting away from the surface of the collision. In the eye of the observer a single impact of the atoms would cause an instantaneous flash, but if the impacts were repeated with sufficient rapidity, they would produce a continuous impression upon his retina. To him then the surface of the metal would appear continuously incandescent and of constant luminous intensity, while in reality the light would be either intermittent, or at least changing periodically in intensity. The metal piece would rise in temperature until equilibrium was attained—that is, until the energy continuously radiated would equal that intermittently supplied. But the supplied energy might under such conditions not be sufficient to bring the body to any more than a very moderate mean temperature, especially if the frequency of the atomic impacts be very low—just enough that the fluctuation of the intensity of the light emitted could not be detected by the eye. The body would now, owing to the manner in which the energy is supplied, emit a strong light, and yet be at a comparatively very low mean temperature. How should the observer name the luminosity thus produced? Even if the analysis of the light would teach him something definite, still he *would* probably rank it under the phenomena of phosphorescence. It is conceivable that in such a way both conducting and non-conducting bodies may be maintained at a certain luminous intensity, but the energy required would very greatly vary with the nature and properties of the bodies.

These and some foregoing remarks of a speculative nature were made merely to bring out curious features of alternate currents or electric impulses. By their help we may cause a body to emit *more* light, while at a certain mean temperature, than it would emit if brought to that temperature by a steady supply; and, again, we may bring a body to the point of fusion, and cause it to emit *less* light than when fused by the application of energy in ordinary ways. It all depends on how we supply the energy, and what kind of vibrations we set up; in one case the vibrations are more, in the other less, adapted to affect our sense of vision.

Some effects, which I had not observed before, obtained with carborundum in the first trials, I attributed to phosphorescence, but in subsequent experiments it appeared that it was devoid of that quality. The crystals possess a noteworthy feature. In a bulb provided with a single electrode in the shape of a small circular metal disc, for instance, at a certain degree of exhaustion the electrode is covered with a milky film, which is separated by a dark space from the glow filling the bulb. When the metal disc is covered with carborundum crystals, the film is far more intense, and snow-white. This I found later to be merely an effect of the bright surface of the crystals, for when an aluminum electrode was highly polished, it exhibited more or less the same phenomenon. I made a number of experiments with the samples of crystals obtained, principally because it would have been of special interest to find that they are capable of phosphorescence, on account of their being conducting. I could not produce phosphorescence distinctly, but I must remark that a decisive opinion cannot be formed until other experimenters have gone over the same ground.

The powder behaved in some experiments as though it contained alumina, but it did not exhibit with sufficient distinctness the red of the latter. Its dead color brightens considerably under the molecular impact, but I am now convinced it does not phosphoresce. Still, the tests with the powder are not conclusive, because powdered carborundum probably does not behave like a phosphorescent sulphide, for example, which could be finely powdered without impairing the phosphorescence, but rather like powdered ruby or diamond, and therefore it would be necessary, in order to make a decisive test, to obtain it in a large lump and polish up the surface.

If the carborundum proves useful in connection with these and similar experiments, its chief value will be found in the production of coatings, thin conductors, buttons, or other electrodes capable of withstanding extremely high degrees of heat.

The production of a small electrode, capable of withstanding enormous temperatures, I regard as of the greatest importance in the manufacture of light. It would enable us to obtain, by means of currents of very high frequencies, certainly 20 times, if not more, the quantity of light which is obtained in the present incandescent lamp by the same expenditure of energy. This estimate may appear to many exaggerated, but in reality I think it is far from being so. As this statement might be misunderstood, I think it is necessary to expose clearly the problem with which, in this line of work, we are confronted, and the manner in which, in my opinion, a solution will be arrived at.

Anyone who begins a study of the problem will be apt to think that what is wanted in a lamp with an electrode is a very high degree of incandescence of the electrode. There he will be mistaken. The high incandescence of the button is a necessary evil, but what is really wanted is the high incandescence of the gas surrounding the button. In other words, the problem in such a lamp is to bring a mass of gas to the highest possible incandescence. The higher the incandescence, the quicker the mean vibration, the greater is the economy of the light production. But to maintain a mass of gas at a high degree of incandescence in a glass vessel, it will always be necessary to keep the incandescent mass away from the glass; that is, to confine it as much as possible to the central portion of the globe.

In one of the experiments this evening a brush was produced at the end of a wire. The brush was a flame, a source of heat and light. It did not emit much perceptible heat, nor did it glow with an intense light; but is it the less a flame because it does not scorch my hand? Is it the less a flame because it does not hurt my eyes by its brilliancy? The problem is precisely to produce in the bulb such a flame, much smaller in size, but incomparably more powerful. Were there means at hand for producing electric impulses of a sufficiently high frequency, and for transmitting them, the bulb could be done away with, unless it were used to protect the electrode, or to economize the energy by confining the heat. But as such means are not at disposal, it becomes necessary to place the terminal in the bulb and rarefy the air in the same. This is done merely to enable the apparatus to perform the work which it is not capable of performing at ordinary air pressure. In the bulb we are able to intensify the action to any degree—so far that the brush emits a powerful light.

The intensity of the light emitted depends principally on the frequency and potential of the impulses, and on the electric density on the surface of the electrode. It is of the greatest importance to employ the smallest possible button, in order to push the density very far. Under the violent impact of the molecules of the gas surrounding it, the small electrode is of course brought to an extremely high temperature, but around it is a mass of highly incandescent gas, a flame photosphere, many hundred times the volume of the electrode. With a diamond, carborundum or zirconia button the photosphere can be as much as one thousand times the volume of the button. Without much reflection one would think that in pushing so far the incandescence of the electrode it would be instantly volatilized. But after a careful consideration one would find that, theoretically, it should not occur, and in this fact—which, moreover, is experimentally demonstrated—lies principally the future value of such a lamp.

At first, when the bombardment begins, most of the work is performed on the surface of the button, but when a highly conducting photosphere is formed the button is comparatively relieved. The higher the incandescence of the photosphere, the more it approaches in conductivity to that of the electrode, and the more, therefore, the solid and the gas form one conducting body. The consequence is that the further the incandescence is forced the more work, comparatively, is performed on the gas, and the less on the electrode. The formation of a powerful photosphere is consequently the very means for protecting the electrode. This protection, of course, is a relative one, and it should not be thought that by pushing the incandescence higher the electrode is actually less deteriorated. Still, theoretically, with extreme frequencies, this result must be reached, but probably at a temperature too high for most of the refractory bodies known. Given, then, an electrode which can withstand to a very high limit the effect of the bombardment and outward strain, it would be safe, no matter how much it was forced beyond that limit. In an incandescent lamp quite different considerations apply. There the gas is not at all concerned; the whole of the work is performed on the filament; and the life of the lamp diminishes so rapidly with the increase of the degree of incandescence that economical reasons compel us to work it at a low incandescence. But if an incandescent lamp is operated with currents of very high frequency, the action of the gas cannot be neglected, and the rules for the most economical working must be considerably modified.

In order to bring such a lamp with one or two electrodes to a great perfection, it is necessary to employ impulses of very high frequency. The high frequency secures, among others, two chief advantages, which have a most important bearing upon the economy of the light production.

First, the deterioration of the electrode is reduced by reason of the fact that we employ a great many small impacts, instead of a few violent ones, which quickly shatter the structure; secondly, the formation of a large photosphere is facilitated.

In order to reduce the deterioration of the electrode to the minimum, it is desirable that the vibration be harmonic, for any suddenness hastens the process of destruction. An electrode lasts much longer when kept at incandescence by currents, or impulses, obtained from a high frequency alternator, which rise and fall more or less harmonically, than by impulses obtained from a disruptive discharge coil. In the latter case there is no doubt that most of the damage is done by the fundamental sudden discharges.

One of the elements of loss in such a lamp is the bombardment of the globe. As the potential is very high, the molecules are projected with great speed; they strike the glass, and usually excite a strong phosphorescence. The effect produced is very pretty, but for economical reasons it would be perhaps preferable to prevent, or at least reduce to a minimum, the bombardment against the globe, as in such case it is, as a rule, not the object to excite phosphorescence, and as some loss of energy results from the bombardment. This loss in the bulb is principally dependent on the potential of the impulses and on the electric density on the surface of the electrode. In employing very high frequencies the loss of energy by the bombardment is greatly reduced, for, first, the potential needed to perform a given amount of work is much smaller; and, secondly, by producing a highly conducting photosphere around the electrode, the same result is obtained as though the electrode were much larger, which is equivalent to a smaller electric density. But be it by the diminution of the maximum potential or of the density, the gain is effected in the same manner, namely, by avoiding violent shocks, which strain the glass much beyond its limit of elasticity. If the frequency could be brought high enough, the loss due to the imperfect elasticity of the glass would be entirely negligible. The loss due to bombardment of the globe may, however, be reduced by using two electrodes instead of one. In such case each of the electrodes may be connected to one of the terminals; or else, if it is preferable to use only one wire, one electrode may be connected to one terminal and the other to the ground or to an insulated body of some surface, as, for instance, a shade on the lamp. In the latter case, unless some judgment is used, one of the electrodes might glow more intensely than the other.

But on the whole I find it preferable, when using such high frequencies, to employ only one electrode and one connecting wire. I am convinced that the illuminating device of the near future will not require for its operation more than one lead, and, at any rate, it will have no leading-in wire, since the energy required can be as well transmitted through the glass. In experimental bulbs the leading-in wire is not generally used on account of convenience, as in employing condenser coatings in the manner indicated in Fig. 151, for example, there is some difficulty in fitting the parts, but these difficulties would not exist if a great many bulbs were manufactured; otherwise the energy can be conveyed through the glass as well as through a wire, and with these high frequencies the losses are very small. Such illustrating devices will necessarily involve the use of very high potentials, and this, in the eyes of practical men, might be an objectionable feature. Yet, in reality, high potentials are not objectionable—certainly not in the least so far as the safety of the devices is concerned.

There are two ways of rendering an electric appliance safe. One is to use low potentials, the other is to determine the dimensions of the apparatus so that it is safe, no matter how high a potential is used. Of the two, the latter seems to me the better way, for then the safety is absolute, unaffected by any possible combination of circumstances which might render even a low-potential appliance dangerous to life and property. But the practical conditions require not only the judicious determination of the dimensions of the apparatus; they likewise necessitate the employment of energy of the proper kind. It is easy, for instance, to construct a transformer capable of giving, when operated from an ordinary alternate current machine of low tension, say 50,000 volts, which might be required to light a highly exhausted phosphorescent tube, so that, in spite of the high potential, it is perfectly safe, the shock from it producing no inconvenience. Still such a transformer would be expensive, and in itself inefficient; and, besides, what energy was obtained from it would not be economically used for the production of light. The economy demands the employment of energy in the form of extremely rapid vibrations. The problem of producing light has been likened to that of maintaining a certain high-pitch note by means of a bell. It should be said a *barely audible* note; and even these words would not express it, so wonderful is the sensitiveness of the eye. We may deliver powerful blows at long intervals, waste a good deal of energy, and still not get what we want; or we may keep up the note by delivering frequent taps, and get nearer to the object sought by the expenditure of much less energy. In the production of light, as far as the illuminating device is concerned, there can be only one rule—that is, to use as high frequencies as can be obtained; but the means for the production and conveyance of impulses of such character impose, at present at least, great limitations. Once it is decided to use very high frequencies, the return wire becomes unnecessary, and all the appliances are simplified. By the use of obvious means the same result is obtained as though the return wire were used. It is sufficient for this purpose to bring in contact with the bulb, or merely in the vicinity of the same, an insulated body of some surface. The surface need, of course, be the smaller, the higher the frequency and potential used, and necessarily, also, the higher the economy of the lamp or other device.

This plan of working has been resorted to on several occasions this evening. So, for instance, when the incandescence of a button was produced by grasping the bulb with the hand, the body of the experimenter merely served to intensify the action. The bulb used was similar to that illustrated in Fig. 148, and the coil was excited to a small potential, not sufficient to bring the button to incandescence when the bulb was hanging from the wire; and incidentally, in order to perform the experiment in a more suitable manner, the button was taken so large that a perceptible time had to elapse before, upon grasping the bulb, it could be rendered incandescent. The contact with the bulb was, of course, quite unnecessary. It is easy, by using a rather large bulb with an exceedingly small electrode, to adjust the conditions so that the latter is brought to bright incandescence by the mere approach of the experimenter within a few feet of the bulb, and that the incandescence subsides upon his receding.

In another experiment, when phosphorescence was excited, a similar bulb was used. Here again, originally, the potential was not sufficient to excite phosphorescence until the action was intensified—in this case, however, to present a different feature, by touching the socket with a metallic object held in the hand. The electrode in the bulb was a carbon button so large that it could not be brought to incandescence, and thereby spoil the effect produced by phosphorescence.

Again, in another of the early experiments, a bulb was used, as illustrated in Fig. 141. In this instance, by touching the bulb with one or two fingers, one or two shadows of the stem inside were projected against the glass, the touch of the finger producing the same results as the application of an external negative electrode under ordinary circumstances.

In all these experiments the action was intensified by augmenting the capacity at the end of the lead connected to the terminal. As a rule, it is not necessary to resort to such means, and would be quite unnecessary with still higher frequencies; but when it *is* desired, the bulb, or tube, can be easily adapted to the purpose.

Fig. 153.

Fig. 154.

In Fig. 153, for example, an experimental bulb, L, is shown, which is provided with a neck, *n*, on the top, for the application of an external tinfoil coating, which may be connected to a body of larger surface. Such a lamp as illustrated in Fig. 154 may also be lighted by connecting the tinfoil coating on the neck *n* to the terminal, and the leading-in wire, *w*, to an insulated plate. If the bulb stands in a socket upright, as shown in the cut, a shade of conducting material may be slipped in the neck, *n*, and the action thus magnified.

A more perfected arrangement used in some of these bulbs is illustrated in Fig. 155. In this case the construction of the bulb is as shown and described before, when reference was made to Fig. 148. A zinc sheet, z, with a tubular extension, т, is applied over the metallic socket, s. The bulb hangs downward from the terminal, t, the zinc sheet, z, performing the double office of intensifier and reflector. The reflector is separated from the terminal, t, by an extension of the insulating plug, P.

FIG. 155.

A similar disposition with a phosphorescent tube is illustrated in Fig. 156. The tube, т, is prepared from two short tubes of different diameter, which are sealed on the ends. On the lower end is placed an inside conducting coating, c, which connects to the wire w. The wire has a hook on the upper end for suspension, and passes through the center of the inside tube, which is filled with some good and tightly packed insulator. On the outside of the upper end of the tube, т, is another conducting coating, c_1, upon which is slipped a metallic reflector z, which should be separated by a thick insulation from the end of wire w.

The economical use of such a reflector or intensifier would require that all energy supplied to an air condenser should be recoverable, or, in other words, that there should not be any losses, neither in the gaseous medium nor through its action elsewhere. This is far from being so, but, fortunately, the losses may be reduced to anything desired. A few remarks are necessary on this subject, in order to make the experiences gathered in the course of these investigations perfectly clear.

FIG. 156.

Suppose a small helix with many well insulated turns, as in experiment Fig. 146, has one of its ends connected to one of the terminals of the induction coil, and the other to a metal plate, or, for the sake of simplicity, a sphere, insulated in space. When the coil is set to work, the potential of the sphere is alternated, and a small helix now behaves as though its free end were connected to the other terminal of the induction coil. If an iron rod be held within a small helix, it is quickly brought to a high temperature, indicating the passage of a strong current through the helix. How does the insulated sphere act in this case? It can be a condenser, storing and returning the energy supplied to it, or it can be a mere sink of energy, and the conditions of the experiment determine whether it is rather one than the other. The sphere being charged to a high potential, it acts inductively upon the surrounding air, or whatever gaseous medium there might be. The molecules, or atoms, which are near the sphere, are of course more attracted, and move through a greater distance than the farther ones. When the nearest molecules strike the sphere, they are repelled, and collisions occur at all distances within the inductive action of the sphere. It is now clear that, if the potential be steady, but little loss of energy can be caused in this way, for the molecules which are nearest to the sphere, having had an additional charge imparted to them by contact, are not attracted until they have parted, if not with all, at least with most of the additional charge, which can be accomplished only after a great many collisions. From the fact, that with a steady potential there is but little loss in dry air, one must come to such a conclusion. When the potential of a sphere, instead of being steady, is alternating, the conditions are entirely different. In this case a rhythmical bombardment occurs, no matter whether the molecules, after coming in contact with the sphere, lose the imparted charge or not; what is more, if the charge is not lost, the impacts are only the more violent. Still, if the frequency of the impulses be very small, the loss caused by the impacts and collisions would not be serious, unless the potential were excessive. But when extremely high frequencies and more or less high potentials are used, the loss may be very great. The total energy lost per unit of time is proportionate to the product of the number of impacts per second, or the frequency and the energy lost in each impact. But the energy of an impact must be proportionate to the square of the electric density of the sphere, since the charge imparted to the molecule is proportionate to that density. I conclude from this that the total energy lost must be proportionate to the product of the frequency and the square of the electric density; but this law needs experimental confirmation. Assuming the preceding considerations to be true, then, by rapidly alternating the potential of a body immersed in an insulating gaseous medium, any amount of energy may be dissipated into space. Most of that energy then, I believe, is not dissipated in the form of long ether waves, propagated to considerable distance, as is thought most generally, but is consumed—in the case of an insulated sphere, for example—in impact and collisional losses—that is, heat vibrations—on the surface and in the vicinity of the sphere. To reduce the dissipation, it is necessary to work with a small electric density—the smaller, the higher the frequency.

But since, on the assumption before made, the loss is diminished with the square of the density, and since currents of very high frequencies involve considerable waste when transmitted through conductors, it follows that, on the whole, it is better to employ one wire than two. Therefore, if motors, lamps, or devices of any kind are perfected, capable of being advantageously operated by currents of extremely high frequency, economical reasons will make it advisable to use only one wire, especially if the distances are great.

When energy is absorbed in a condenser, the same behaves as though its capacity were increased. Absorption always exists more or less, but generally it is small and of no consequence as long as the frequencies are not very great. In using extremely high frequencies, and, necessarily in such case, also high potentials, the absorption—or, what is here meant more particularly by this term, the loss of energy due to the presence of a gaseous medium—is an important factor to be considered, as the energy absorbed in the air condenser may be any fraction of the supplied energy. This would seem to make it very difficult to tell from the measured or computed capacity of an air condenser its actual capacity or vibration period, especially if the condenser is of very small surface and is charged to a very high potential. As many important results are dependent upon the correctness of the estimation of the vibration period, this subject demands the most careful scrutiny of other investigators. To reduce the probable error as much as possible in experiments of the kind alluded to, it is advisable to use spheres or plates of large surface, so as to make the density exceedingly small. Otherwise, when it is practicable, an oil condenser should be used in preference. In oil or other liquid dielectrics there are seemingly no such losses as in gaseous media. It being impossible to exclude entirely the gas in condensers with solid dielectrics, such condensers should be immersed in oil, for economical reasons, if nothing else; they can then be strained to the utmost, and will remain cool. In Leyden jars the loss due to air is comparatively small, as the tinfoil coatings are large, close together, and the charged surfaces not directly exposed; but when the potentials are very high, the loss may be more or less considerable at, or near, the upper edge of the foil, where the air is principally acted upon. If the jar be immersed in boiled-out oil, it will be capable of performing four times the amount of work which it can for any length of time when used in the ordinary way, and the loss will be inappreciable.

It should not be thought that the loss in heat in an air condenser is necessarily associated with the formation of *visible* streams or brushes. If a small electrode, inclosed in an unexhausted bulb, is connected to one of the terminals of the coil, streams can be seen to issue from the electrode, and the air in the bulb is heated; if instead of a small electrode a large sphere is inclosed in the bulb, no streams are observed, still the air is heated.

Nor should it be thought that the temperature of an air condenser would give even an approximate idea of the loss in heat incurred, as in such case heat must be given off much more quickly, since there is, in addition to the ordinary radiation, a very active carrying away of heat by independent carriers going on, and since not only the apparatus, but the air at some distance from it is heated in consequence of the collisions which must occur.

Owing to this, in experiments with such a coil, a rise of temperature can be distinctly observed only when the body connected to the coil is very small. But with apparatus on a larger scale, even a body of considerable bulk would be heated, as, for instance, the body of a person; and I think that skilled physicians might make observations of utility in such experiments, which, if the apparatus were judiciously designed, would not present the slightest danger.

A question of some interest, principally to meteorologists, presents itself here. How does the earth behave? The earth is an air condenser, but is it a perfect or a very imperfect one—a mere sink of energy? There can be little doubt that to such small disturbance as might be caused in an experiment, the earth behaves as an almost perfect condenser. But it might be different when its charge is set in vibration by some sudden disturbance occurring in the heavens. In such case, as before stated, probably only little of the energy of the vibrations set up would be lost into space in the form of long ether radiations, but most of the energy, I think, would spend itself in molecular impacts and collisions, and pass off into space in the form of short heat, and possibly light, waves. As both the frequency of the vibrations of the charge and the potential are in all probability excessive, the energy converted into heat may be considerable. Since the density must be unevenly distributed, either in consequence of the irregularity of the earth's surface, or on account of the condition of the atmosphere in various places, the effect produced would accordingly vary from place to place. Considerable variations in the temperature and pressure of the atmosphere may in this manner be caused at any point of the surface of the earth. The variations may be gradual or very sudden, according to the nature of the general disturbance, and may produce rain and storms, or locally modify the weather in any way.

From the remarks before made, one may see what an important factor of loss the air in the neighborhood of a charged surface becomes when the electric density is great and the frequency of the impulses excessive. But the action, as explained, implies that the air is insulating—that is, that it is composed of independent carriers immersed in an insulating medium. This is the case only when the air is at something like ordinary or greater, or at extremely small, pressure. When the air is slightly rarefied and conducting, then true conduction losses occur also. In such case, of course, considerable energy may be dissipated into space even with a steady potential, or with impulses of low frequency, if the density is very great.

When the gas is at very low pressure, an electrode is heated more because higher speeds can be reached. If the gas around the electrode is strongly compressed, the displacements, and consequently the speeds, are very small, and the heating is insignificant. But if in such case the frequency could be sufficiently increased, the electrode would be brought to a high temperature as well as if the gas were at very low pressure; in fact, exhausting the bulb is only necessary because we cannot produce (and possibly not convey) currents of the required frequency.

Returning to the subject of electrode lamps, it is obviously of advantage in such a lamp to confine as much as possible the heat to the electrode by preventing the circulation of the gas in the bulb. If a very small bulb be taken, it would confine the heat better than a large one, but it might not be of sufficient capacity to be operated from the coil, or, if so, the glass might get too hot. A simple way to improve in this direction is to employ a globe of the required size, but to place a small bulb, the diameter of which is properly estimated, over the refractory button contained in the globe. This arrangement is illustrated in Fig. 157.

The globe L has in this case a large neck n, allowing the small bulb b to slip through. Otherwise the construction is the same as shown in Fig. 147, for example. The small bulb is conveniently supported upon the stem s, carrying the refractory button m. It is separated from the aluminum tube a by several layers of mica M, in order to prevent the cracking of the neck by the rapid heating of the aluminum tube upon a sudden turning on of the current. The inside bulb should be as small as possible when it is desired to obtain light only by incandescence of the electrode. If it is desired to produce phosphorescence, the bulb should be larger, else it would be apt to get too hot, and the phosphorescence would cease. In this arrangement usually only the small bulb shows phosphorescence, as there is practically no bombardment against the outer globe. In some of these bulbs constructed as illustrated in Fig. 157, the small tube was coated with phosphorescent paint, and beautiful effects were obtained. Instead of making the inside bulb large, in order to avoid undue heating, it answers the purpose to make the electrode m larger. In this case the bombardment is weakened by reason of the smaller electric density.

Many bulbs were constructed on the plan illustrated in Fig. 158. Here a small bulb b, containing the refractory button m, upon being exhausted to a very high degree was sealed in a large globe L, which was then moderately exhausted and sealed off. The principal advantage of this construction was that it allowed of reaching extremely high vacua, and, at the same time of using a large bulb. It was found, in the course of experiments with bulbs such as illustrated in Fig. 158, that it was well to make the stem s, near the seal at e, very thick, and the leading-in wire w

FIG. 157.

FIG. 158.

thin, as it occurred sometimes that the stem at e was heated and the bulb was cracked. Often the outer globe L was exhausted only just enough to allow the discharge to pass through, and the space between the bulbs appeared crimson, producing a curious effect. In some cases, when the exhaustion in globe L was very low, and the air good conducting, it was found necessary, in order to bring the button m to high incandescence, to place, preferably on the upper part of the neck of the globe, a tinfoil coating which was connected to an insulated body, to the ground, or to the other terminal of the coil, as the highly conducting air weakened the effect somewhat, probably by being acted upon inductively from the wire *w*, where it entered the bulb at *e*. Another difficulty—which, however, is always present when the refractory button is mounted in a very small bulb—existed in the construction illustrated in Fig. 158, namely, the vacuum in the bulb *b* would be impaired in a comparatively short time.

The chief idea in the two last described constructions was to confine the heat to the central portion of the globe by preventing the exchange of air. An advantage is secured, but owing to the heating of the inside bulb and slow evaporation of the glass, the vacuum is hard to maintain, even if the construction illustrated in Fig. 157 be chosen, in which both bulbs communicate.

But by far the better way—the ideal way—would be to reach sufficiently high frequencies. The higher the frequency, the slower would be the exchange of the air, and I think that a frequency may be reached, at which there would be no exchange whatever of the air molecules around the terminal. We would then produce a flame in which there would be no carrying away of material, and a queer flame it would be, for it would be rigid! With such high frequencies the inertia of the particles would come into play. As the brush, or flame, would gain rigidity in virtue of the inertia of the particles, the exchange of the latter would be prevented. This would necessarily occur, for, the number of impulses being augmented, the potential energy of each would diminish, so that finally only atomic vibrations could be set up, and the motion of translation through measurable space would cease. Thus an ordinary gas burner connected to a source of rapidly alternating potential might have its efficiency augmented to a certain limit, and this for two reasons—because of the additional vibration imparted, and because of a slowing down of the process of carrying off. But the renewal being rendered difficult, a renewal being necessary to maintain the *burner*, a continued increase of the frequency of the impulses, assuming they could be transmitted to and impressed upon the flame, would result in the "extinction" of the latter, meaning by this term only the cessation of the chemical process.

I think, however, that in the case of an electrode immersed in a fluid insulating medium, and surrounded by independent carriers of electric charges, which can be acted upon inductively, a sufficient high frequency of the impulses would probably result in a gravitation of the gas all around toward the electrode. For this it would be only necessary to assume that the independent bodies are irregularly shaped; they would then turn toward the electrode their side of the greatest electric density, and this would be a position in which the fluid resistance to approach would be smaller than that offered to the receding.

The general opinion, I do not doubt, is that it is out of the question to reach any such frequencies as might—assuming some of the views before expressed to be true—produce any of the results which I have pointed out as mere possibilities. This may be so, but in the course of these investigations, from the observation of many phenomena, I have gained the conviction that these frequencies would be much lower than one is apt to estimate at first. In a flame we set up light vibrations by causing molecules, or atoms, to collide. But what is the ratio of the frequency of the collisions and that of the vibrations set up? Certainly it must be incomparably smaller than that of the strokes of the bell and the sound vibrations, or that of the discharges and the oscillations of the condenser. We may cause the molecules of the gas to collide by the use of alternate electric impulses of high frequency, and so we may imitate the process in a flame; and from experiments with frequencies which we are now able to obtain, I think that the result is producible with impulses which are transmissible through a conductor.

In connection with thoughts of a similar nature, it appeared to me of great interest to demonstrate the rigidity of a vibrating gaseous column. Although with such low frequencies as, say 10,000 per second, which I was able to obtain without difficulty from a specially constructed alternator, the task looked discouraging at first, I made a series of experiments. The trials with air at ordinary pressure led to no result, but with air moderately rarefied I obtain what I think to be an unmistakable experimental evidence of the property sought for. As a result of this kind might lead able investigators to conclusions of importance, I will describe one of the experiments performed.

It is well known that when a tube is slightly exhausted, the discharge may be passed through it in the form of a thin luminous thread. When produced with currents of low frequency, obtained from a coil operated as usual, this thread is inert. If a magnet be approached to it, the part near the same is attracted or repelled, according to the direction of the lines of force of the magnet. It occurred to me that if such a thread would be produced with currents of very high frequency, it should be more or less rigid, and as it was visible it could be easily studied. Accordingly I prepared a tube about one inch in diameter and one meter long, with outside coating at each end. The tube was exhausted to a point at which, by a little working, the thread discharge could be obtained. It must be remarked here that the general aspect of the tube, and the degree of exhaustion, are quite other than when ordinary low frequency currents are used. As it was found preferable to work with one terminal, the tube prepared was suspended from the end of a wire connected to the terminal, the tinfoil coating being connected to the wire, and to the lower coating sometimes a small insulated plate was attached. When the thread was formed, it extended through the upper part of the tube and lost itself in the lower end. If it possessed rigidity it resembled, not exactly an elastic cord stretched tight between two supports, but a cord suspended from a height with a small weight attached at the end. When the finger or a small magnet was approached to the upper end of the luminous thread, it could be brought locally out of position by electrostatic or magnetic action; and when the disturbing object was very quickly removed, an analogous result was produced, as though a suspended cord would be displaced and quickly released near the point of suspension. In doing this the luminous thread was set in vibration, and two very sharply marked nodes, and a third indistinct one, were formed. The vibration, once set up, continued for fully eight minutes, dying gradually out. The speed of the vibration often varied perceptibly, and it could be observed that the electrostatic attraction of the glass affected the vibrating thread; but it was clear that the electrostatic action was not the cause of the vibration, for the thread was most generally stationary, and could always be set in vibration by passing the finger quickly near the upper part of the tube. With a magnet the thread could be split in two and both parts vibrated. By approaching the hand to the lower coating of the tube, or insulation plate if attached, the vibration was quickened; also, as far as I could see, by raising the potential or frequency. Thus, either increasing the frequency or passing a stronger discharge of the same frequency corresponded to a tightening of the cord. I did not obtain any experimental evidence with condenser discharges. A luminous band excited in the bulb by repeated discharges of a Leyden jar must possess rigidity, and if deformed and suddenly released, should vibrate. But probably the amount of vibrating matter is so small that in spite of the extreme speed, the inertia cannot prominently assert itself. Besides, the observation in such a case is rendered extremely difficult on account of the fundamental vibration.

The demonstration of the fact—which still needs better experimental confirmation—that a vibrating gaseous column possesses rigidity, might greatly modify the views of thinkers. When with low frequencies and insignificant potentials indications of that property may be noted, how must a gaseous medium behave under the influence of enormous electrostatic stresses which may be active in the interstellar space, and which may alternate with inconceivable rapidity? The existence of such an electrostatic, rhythmically throbbing force—of a vibrating electrostatic field—would show a possible way how solids might have formed from the ultra-gaseous uterus, and how transverse and all kinds of vibrations may be transmitted through a gaseous medium filling all space. Then, ether might be a true fluid, devoid of rigidity, and at rest, it being merely necessary as a connecting link to enable interaction. What determines the rigidity of a body? It must be the speed and the amount of motive matter. In a gas the speed maybe considerable, but the density is exceedingly small; in a liquid the speed would be likely to be small, though the density may be considerable; and in both cases the inertia resistance offered to displacement is practically *nil*. But place a gaseous (or liquid) column in an intense, rapidly alternating electrostatic field, set the particles vibrating with enormous speeds, then the inertia resistance asserts itself. A body might move with more or less freedom through the vibrating mass, but as a whole it would be rigid.

There is a subject which I must mention in connection with these experiments: it is that of high vacua. This is a subject, the study of which is not only interesting, but useful, for it may lead to results of great practical importance. In commercial apparatus, such as incandescent lamps, operated from ordinary systems of distribution, a much higher vacuum than is obtained at present would not secure a very great advantage. In such a case the work is performed on the filament, and the gas is little concerned; the improvement, therefore, would be but trifling. But when we begin to use very high frequencies and potentials, the action of the gas becomes all important, and the degree of exhaustion materially modifies the results. As long as ordinary coils, even very large ones, were used, the study of the subject was limited, because just at a point when it became most interesting it had to be interrupted on account of the "non-striking" vacuum being reached. But at present we are able to obtain from a small disruptive discharge coil potentials much higher than even the largest coil was capable of giving, and, what is more, we can make the potential alternate with great rapidity. Both of these results enable us now to pass a luminous discharge through almost any vacua obtainable, and the field of our investigations is greatly extended. Think we as we may, of all the possible directions to develop a practical illuminant, the line of high vacua seems to be the most promising at present. But to reach extreme vacua the appliances must be much more improved, and ultimate perfection will not be attained until we shall have discharged the mechanical and perfected an *electrical* vacuum pump. Molecules and atoms can be thrown out of a bulb under the action of an enormous potential: *this* will be the principle of the vacuum pump of the future. For the present, we must secure the best results we can with mechanical appliances. In this respect, it might not be out of the way to say a few words about the method of, and apparatus for, producing excessively high degrees of exhaustion of which I have availed myself in the course of these investigations. It is very probable that other experimenters have used similar arrangements; but as it is possible that there may be an item of interest in their description, a few remarks, which will render this investigation more complete, might be permitted.

The apparatus is illustrated in a drawing shown in Fig. 159. s represents a Sprengel pump, which has been specially constructed to better suit the work required. The stopcock which is usually employed has been omitted, and instead of it a hollow stopper s has been fitted in the neck of the reservoir R. This stopper has a small hole h, through which the mercury descends; the size of the outlet o being properly determined with respect to the section of the fall tube t, which is sealed to the reservoir instead of being connected to it in the usual manner. This arrangement overcomes the imperfections and troubles which often arise from the use of the stopcock on the reservoir and the connections of the latter with the fall tube.

The pump is connected through a U-shaped tube t to a very large reservoir R_1. Especial care was taken in fitting the grinding surfaces of the stoppers p and p_1, and both of these and the mercury caps above them were made exceptionally long. After the U-shaped tube was fitted and put in place, it was heated, so as to soften and take off the strain resulting from imperfect fitting. The U-shaped tube was provided with a stopcock C, and two ground connections g and g_1—one for a small bulb b, usually containing caustic potash, and the other for the receiver r, to be exhausted.

Fig. 159.

The reservoir R_1, was connected by means of a rubber tube to a slightly larger reservoir R_2, each of the two reservoirs being provided with a stopcock c_1 and c_2, respectively. The reservoir R_2 could be raised and lowered by a wheel and rack, and the range of its motion was so determined that when it was filled with mercury and the stopcock c_2 closed, so as to form a Torricellian vacuum in it when raised, it could be lifted so high that the reservoir R_1 would stand a little above stopcock c_1; and when this stopcock was closed and the reservoir R_2 descended, so as to form a Torricellian vacuum in reservoir R_1, it could be lowered so far as to completely empty the latter, the mercury filling the reservoir R_2 up to a little above stopcock c_2.

The capacity of the pump and of the connections was taken as small as possible relatively to the volume of reservoir, R_1, since, of course, the degree of exhaustion depended upon the ratio of these quantities.

With this apparatus I combined the usual means indicated by former experiments for the production of very high vacua. In most of the experiments it was most convenient to use caustic potash. I may venture to say, in regard to its use, that much time is saved and a more perfect action of the pump insured by fusing and boiling the potash as soon as, or even before, the pump settles down. If this course is not followed, the sticks, as ordinarily employed, may give off moisture at a certain very slow rate, and the pump may work for many hours without reaching a very high vacuum. The potash was heated either by a spirit lamp or by passing a discharge through it, or by passing a current through a wire contained in it. The advantage in the latter case was that the heating could be more rapidly repeated.

Generally the process of exhaustion was the following: At the start, the stopcocks c and c_1 being open, and all other connections closed, the reservoir R_2 was raised so far that the mercury filled the reservoir R_1 and a part of the narrow connecting U-shaped tube. When the pump was set to work, the mercury would, of course, quickly rise in the tube, and reservoir R_2 was lowered, the experimenter keeping the mercury at about the same level. The reservoir R_2 was balanced by a long spring which facilitated the operation, and the friction of the parts was generally sufficient to keep it in almost any position. When the Sprengel pump had done its work, the reservoir R_2 was further lowered and the mercury descended in R_1 and filled R_2, whereupon stopcock c_2 was closed. The air adhering to the walls of R_1 and that absorbed by the mercury was carried off, and to free the mercury of all air the reservoir R_2 was for a long time worked up and down. During this process some air, which would gather below stopcock c_2, was expelled from R_2 by lowering it far enough and opening the stopcock, closing the latter again before raising the reservoir. When all the air had been expelled from the mercury, and no air would gather in R_2 when it was lowered, the caustic potash was resorted to. The reservoir R_2 was now again raised until the mercury in R_1, stood above stopcock c_1. The caustic potash was fused and boiled, and moisture partly carried off by the pump and partly re-absorbed; and this process of heating and cooling was repeated many times, and each time, upon the moisture being absorbed or carried off, the reservoir R_2 was for a long time raised and lowered. In this manner all the moisture was carried off from the mercury, and both the reservoirs were in proper condition to be used. The reservoir R_2 was then again raised to the top, and the pump was kept working for a long time. When the highest vacuum obtainable with the pump had been reached, the potash bulb was usually wrapped with cotton which was sprinkled with ether so as to keep the potash at a very low temperature, then the reservoir R_2 was lowered, and upon reservoir R_1 being emptied the receiver was quickly sealed up.

When a new bulb was put on, the mercury was always raised above stopcock c_1, which was closed, so as to always keep the mercury and both the reservoirs in fine condition, and the mercury was never withdrawn from R_1 except when the pump had reached the highest degree of exhaustion. It is necessary to observe this rule if it is desired to use the apparatus to advantage.

By means of this arrangement I was able to proceed very quickly, and when the apparatus was in perfect order it was possible to reach the phosphorescent stage in a small bulb in less than fifteen minutes, which is certainly very quick work for a small laboratory arrangement requiring all in all about 100 pounds of mercury. With ordinary small bulbs the ratio of the capacity of the pump, receiver, and connections, and that of reservoir R was about 1 to 20, and the degrees of exhaustion reached were necessarily very high, though I am unable to make a precise and reliable statement how far the exhaustion was carried.

What impresses the investigator most in the course of these experiences is the behavior of gases when subjected to great rapidly alternating electrostatic stresses. But he must remain in doubt as to whether the effects observed are due wholly to the molecules, or atoms, of the gas which chemical analysis discloses to us, or whether there enters into play another medium of a gaseous nature, comprising atoms, or molecules, immersed in a fluid pervading the space. Such a medium surely must exist, and I am convinced that, for instance, even if air were absent, the surface and neighborhood of a body in space would be heated by rapidly alternating the potential of the body; but no such heating of the surface or neighborhood could occur if all free atoms were removed and only a homogeneous, incompressible, and elastic fluid—such as ether is supposed to be— would remain, for then there would be no impacts, no collisions. In such a case, as far as the body itself is concerned, only frictional losses in the inside could occur.

It is a striking fact that the discharge through a gas is established with ever-increasing freedom as the frequency of the impulses is augmented. It behaves in this respect quite contrarily to a metallic conductor. In the latter the impedance enters prominently into play as the frequency is increased, but the gas acts much as a series of condensers would; the facility with which the discharge passes through, seems to depend on the rate of change of potential. If it acts so, then in a vacuum tube even of great length, and no matter how strong the current, self-induction could not assert itself to any appreciable degree. We have, then, as far as we can now see, in the gas a conductor which is capable of transmitting electric impulses of any frequency which we may be able to produce. Could the frequency be brought high enough, then a queer system of electric distribution, which would be likely to interest gas companies, might be realized: metal pipes filled with gas—the metal being the insulator, the gas the conductor—supplying phosphorescent bulbs, or perhaps devices as yet uninvented. It is certainly possible to take a hollow core of copper, rarefy the gas in the same, and by passing impulses of sufficiently high frequency through a circuit around it, bring the gas inside to a high degree of incandescence; but as to the nature of the forces there would be considerable uncertainty, for it would be doubtful whether with such impulses the copper core would act as a static screen. Such paradoxes and apparent impossibilities we encounter at every step in this line of work, and therein lies, to a great extent, the charm of the study.

I have here a short and wide tube which is exhausted to a high degree and covered with a substantial coating of bronze, the coating barely allowing the light to shine through. A metallic cap, with a hook for suspending the tube, is fastened around the middle portion of the latter, the clasp being in contact with the bronze coating. I now want to light the gas inside by suspending the tube on a wire connected to the coil. Anyone who would try the experiment for the first time, not having any previous experience, would probably take care to be quite alone when making the trial, for fear that he might become the joke of his assistants. Still, the bulb lights in spite of the metal coating, and the light can be distinctly perceived through the latter. A long tube covered with aluminum bronze lights when held in one hand—the other touching the terminal of the coil—quite powerfully. It might be objected that the coatings are not sufficiently conducting; still, even if they were highly resistant, they ought to screen the gas. They certainly screen it perfectly in a condition of rest, but far from perfectly when the charge is surging in the coating. But the loss of energy which occurs within the tube, notwithstanding the screen, is occasioned principally by the presence of the gas. Were we to take a large hollow metallic sphere and fill it with a perfect, incompressible, fluid dielectric, there would be no loss inside of the sphere, and consequently the inside might be considered as perfectly screened, though the potential be very rapidly alternating. Even were the sphere filled with oil, the loss would be incomparably smaller than when the fluid is replaced by a gas, for in the latter case the force produces displacements; that means impact and collisions in the inside.

No matter what the pressure of the gas may be, it becomes an important factor in the heating of a conductor when the electric density is great and the frequency very high. That in the heating of conductors by lightning discharges, air is an element of great importance, is almost as certain as an experimental fact. I may illustrate the action of the air by the following experiment: I take a short tube which is exhausted to a moderate degree and has a platinum wire running through the middle from one end to the other. I pass a steady or low frequency current through the wire, and it is heated uniformly in all parts. The heating here is due to conduction, or frictional losses, and the gas around the wire has—as far as we can see—no function to perform. But now let me pass sudden discharges, or high frequency currents, through the wire. Again the wire is heated, this time principally on the ends and least in the middle portion; and if the frequency of the impulses, or the rate of change, is high enough, the wire might as well be cut in the middle as not, for practically all heating is due to the rarefied gas. Here the gas might only act as a conductor of no impedance diverting the current from the wire as the impedance of the latter is enormously increased, and merely heating the ends of the wire by reason of their resistance to the passage of the discharge. But it is not at all necessary that the gas in the tube should be conducting; it might be at an extremely low pressure, still the ends of the wire would be heated—as, however, is ascertained by experience—only the two ends would in such case not be electrically connected through the gaseous medium. Now what with these frequencies and potentials occurs in an exhausted tube, occurs in the lightning discharges at ordinary pressure. We only need remember one of the facts arrived at in the course of these investigations, namely, that to impulses of very high frequency the gas at ordinary pressure behaves much in the same manner as though it were at moderately low pressure. I think that in lightning discharges frequently wires or conducting objects are volatilized merely because air is present, and that, were the conductor immersed in an insulating liquid, it would be safe, for then the energy would have to spend itself somewhere else. From the behavior of gases under sudden impulses of high potential, I am led to conclude that there can be no surer way of diverting a lightning discharge than by affording it a passage through a volume of gas, if such a thing can be done in a practical manner.

There are two more features upon which I think it necessary to dwell in connection with these experiments—the "radiant state" and the "non-striking vacuum."

Anyone who has studied Crookes' work must have received the impression that the "radiant state" is a property of the gas inseparably connected with an extremely high degree of exhaustion. But it should be remembered that the phenomena observed in an exhausted vessel are limited to the character and capacity of the apparatus which is made use of. I think that in a bulb a molecule, or atom, does not precisely move in a straight line because it meets no obstacle, but because the velocity imparted to it is sufficient to propel it in a sensibly straight line. The mean free path is one thing, but the velocity—the energy associated with the moving body—is another, and under ordnary circumstances I believe that it is a mere question of potential or speed. A disruptive discharge coil, when the potential is pushed very far, excites phosphorescence and projects shadows, at comparatively low degrees of exhaustion. In a lightning discharge, matter moves in straight lines at ordinary pressure when the mean free path is exceedingly small, and frequently images of wires or other metallic objects have been produced by the particles thrown off in straight lines.

Fig. 160.

I have prepared a bulb to illustrate by an experiment the correctness of these assertions. In a globe L, Fig. 160, I have mounted upon a lamp filament *f* a piece of lime *l*. The lamp filament is connected with a wire which leads into the bulb, and the general construction of the latter is as indicated in Fig. 148, before described. The bulb being suspended from a wire connected to the terminal of the coil, and the latter being set to work, the lime piece *l* and the projecting parts of the filament *f* are bombarded. The degree of exhaustion is just such that with the potential the coil is capable of giving, phosphorescence of the glass is produced, but disappears as soon as the vacuum is impaired. The lime containing moisture, and moisture being given off as soon as heating occurs, the phosphorescence lasts only for a few moments. When the lime has been sufficiently heated, enough moisture has been given off to impair materially the vacuum of the bulb. As the bombardment goes on, one point of the lime piece is more heated than other points, and the result is that finally practically all the discharge passes through that point which is intensely heated, and a white stream of lime particles (Fig. 160) then breaks forth from that point. This stream is composed of "radiant" matter, yet the degree of exhaustion is low. But the particles move in straight lines because the velocity imparted to them is great, and this is due to three causes—to the great electric density, the high temperature of the small point, and the fact that the particles of the lime are easily torn and thrown off—far more easily than those of carbon. With frequencies such as we are able to obtain, the particles are bodily thrown off and projected to a considerable distance; but with sufficiently high frequencies no such thing would occur; in such case only a stress would spread or a vibration would be propagated through the bulb. It would be out of the question to reach any such frequency on the assumption that the atoms move with the speed of light; but I believe that such a thing is impossible; for this an enormous potential would be required. With potentials which we are able to obtain, even with a disruptive discharge coil, the speed must be quite insignificant.

As to the "non-striking vacuum," the point to be noted is, that it can occur only with low frequency impulses, and it is necessitated by the impossibility of carrying off enough energy with such impulses in high vacuum, since the few atoms which are around the terminal upon coming in contact with the same, are repelled and kept at a distance for a comparatively long period of time, and not enough work can be performed to render the effect perceptible to the eye. If the difference of potential between the terminals is raised, the dielectric breaks down. But with very high frequency impulses there is no necessity for such breaking down, since any amount of work can be performed by continually agitating the atoms in the exhausted vessel, provided the frequency is high enough. It is easy to reach—even with frequencies obtained from an alternator as here used—a stage at which the discharge does not pass between two electrodes in a narrow tube, each of these being connected to one of the terminals of the coil, but it is difficult to reach a point at which a luminous discharge would not occur around each electrode.

A thought which naturally presents itself in connection with high frequency currents, is to make use of their powerful electrodynamic inductive action to produce light effects in a sealed glass globe. The leading-in wire is one of the defects of the present incandescent lamp, and if no other improvement were made, that imperfection at least should be done away with. Following this thought, I have carried on experiments in various directions, of which some were indicated in my former paper. I may here mention one or two more lines of experiment which have been followed up.

FIG. 161.

FIG. 162.

Many bulbs were constructed as shown in Fig. 161 and Fig. 162.

In Fig. 161, a wide tube, T, was sealed to a smaller W-shaped tube U, of phosphorescent glass. In the tube T, was placed a coil C, of aluminum wire, the ends of which were provided with small spheres, t and t_1, of aluminum, and reached into the U tube. The tube T was slipped into a socket containing a primary coil, through which usually the discharges of Leyden jars were directed, and the rarefied gas in the small U tube was excited to strong luminosity by the high-tension current induced in the coil C. When Leyden jar discharges were used to induce currents in the coil C, it was found necessary to pack the tube T tightly with insulating powder, as a discharge would occur frequently between the turns of the coil, especially when the primary was thick and the air gap, through which the jars discharged, large, and no little trouble was experienced in this way.

In Fig. 162 is illustrated another form of the bulb constructed. In this case a tube T is sealed to a globe L. The tube contains a coil C, the ends of which pass through two small glass tubes t and t_1, which are sealed to the tube T. Two refractory buttons m and m_1, are mounted on lamp filaments which are fastened to the ends of the wires passing through the glass tubes

t and t_1. Generally in bulbs made on this plan the globe L communicated with the tube T. For this purpose the ends of the small tubes t and t_1 were heated just a trifle in the burner, merely to hold the wires, but not to interfere with the communication. The tube T, with the small tubes, wires through the same, and the refractory buttons m and m_1, were first prepared, and then sealed to globe L, whereupon the coil C was slipped in and the connections made to its ends. The tube was then packed with insulating powder, jamming the latter as tight as possible up to very nearly the end; then it was closed and only a small hole left through which the remainder of the powder was introduced, and finally the end of the tube was closed. Usually in bulbs constructed as shown in Fig. 162 an aluminum tube a was fastened to the upper end s of each of the tubes t and t_1 in order to protect that end against the heat. The buttons m and m_1 could be brought to any degree of incandescence by passing the discharges of Leyden jars around the coil C. In such bulbs with two buttons a very curious effect is produced by the formation of the shadows of each of the two buttons.

Another line of experiment, which has been assiduously followed, was to induce by electro-dynamic induction a current or luminous discharge in an exhausted tube or bulb. This matter has received such able treatment at the hands of Prof. J. J. Thomson, that I could add but little to what he has made known, even had I made it the special subject of this lecture. Still, since experiments in this line have gradually led me to the present views and results, a few words must be devoted here to this subject.

It has occurred, no doubt, to many that as a vacuum tube is made longer, the electromotive force per unit length of the tube, necessary to pass a luminous discharge through the latter, becomes continually smaller; therefore, if the exhausted tube be made long enough, even with low frequencies a luminous discharge could be induced in such a tube closed upon itself. Such a tube might be placed around a hall or on a ceiling, and at once a simple appliance capable of giving considerable light would be obtained. But this would be an appliance hard to manufacture and extremely unmanageable. It would not do to make the tube up of small lengths, because there would be with ordinary frequencies considerable loss in the coatings, and besides, if coatings were used, it would be better to supply the current directly to the tube by connecting the coatings to a transformer. But even if all objections of such nature were removed, with low frequencies the light conversion itself would be inefficient, as I have before stated. In using extremely high frequencies the length of the secondary—in other words, the size of the vessel—can be reduced as much as desired, and the efficiency of the light conversion is increased, provided that means are invented for efficiently obtaining such high frequencies. Thus one is led, from theoretical and practical considerations, to the use of high frequencies, and this means high electromotive forces and small currents in the primary. When one works with condenser charges—and they are the only means up to the present known for reaching these extreme frequencies—one gets to electromotive forces of several thousands of volts per turn of the primary. We cannot multiply the electro-dynamic inductive effect by taking more turns in the primary, for we arrive at the conclusion that the best way is to work with one single turn—though we must sometimes depart from this rule—and we must get along with whatever inductive effect we can obtain with one turn. But before one has long experimented with the extreme frequencies required to set up in a small bulb an electromotive force of several thousands of volts, one realizes the great importance of electrostatic effects, and these effects grow relatively to the electro-dynamic in significance as the frequency is increased.

Now, if anything is desirable in this case, it is to increase the frequency, and this would make it still worse for the electrodynamic effects. On the other hand, it is easy to exalt the electrostatic action as far as one likes by taking more turns on the secondary, or combining self-induction and capacity to raise the potential. It should also be remembered that, in reducing the current to the smallest value and increasing the potential, the electric impulses of high frequency can be more easily transmitted through a conductor.

These and similar thoughts determined me to devote more attention to the electrostatic phenomena, and to endeavor to produce potentials as high as possible, and alternating as fast as they could be made to alternate. I then found that I could excite vacuum tubes at considerable distance from a conductor connected to a properly constructed coil, and that I could, by converting the oscillatory current of a conductor to a higher potential, establish electrostatic alternating fields which acted through the whole extent of the room, lighting up a tube no matter where it was held in space. I thought I recognized that I had made a step in advance, and I have persevered in this line; but I wish to say that I share with all lovers of science and progress the one and only desire—to reach a result of utility to men in any direction to which thought or experiment may lead me. I think that this departure is the right one, for I cannot see, from the observation of the phenomena which manifest themselves as the frequency is increased, what there would remain to act between two circuits conveying, for instance, impulses of several hundred millions per second, except electrostatic forces. Even with such trifling frequencies the energy would be practically all potential, and my conviction has grown strong that, to whatever kind of motion light may be due, it is produced by tremendous electrostatic stresses vibrating with extreme rapidity.

Of all these phenomena observed with currents, or electric impulses, of high frequency, the most fascinating for an audience are certainly those which are noted in an electrostatic field acting through considerable distance; and the best an unskilled lecturer can do is to begin and finish with the exhibition of these singular effects. I take a tube in my hand and move it about, and it is lighted wherever I may hold it; throughout space the invisible forces act. But I may take another tube and it might not light, the vacuum being very high. I excite it by means of a disruptive discharge coil, and now it will light in the electrostatic field. I may put it away for a few weeks or months, still it retains the faculty of being excited. What change have I produced in the tube in the act of exciting it? If a motion imparted to atoms, it is difficult to perceive how it can persist so long without being arrested by frictional losses; and if a strain exerted in the dielectric, such as a simple electrification would produce, it is easy to see how it may persist indefinitely, but very difficult to understand why such a condition should aid the excitation when we have to deal with potentials which are rapidly alternating.

Since I have exhibited these phenomena for the first time, I have obtained some other interesting effects. For instance, I have produced the incandescence of a button, filament, or wire enclosed in a tube. To get to this result it was necessary to economize the energy which is obtained from the field, and direct most of it on the small body to be rendered incandescent. At the beginning the task appeared difficult, but the experiences gathered permitted me to reach the result easily. In Fig. 163 and Fig. 164, two such tubes are illustrated, which are prepared for the occasion. In Fig. 163 a short tube T_1, sealed to another long tube

T, is provided with a stem *s*, with a platinum wire sealed in the latter. A very thin lamp filament *l*, is fastened to this wire and connection to the outside is made through a thin copper wire *w*. The tube is provided with outside and inside coatings, C and C₁, respectively, and is filled as far as the coatings reach with conducting, and the space above with insulating, powder. These coatings are merely used to enable me to perform two experiments with the tube—namely, to produce the effect desired either by direct connection of the body of the experimenter or of another body to the wire *w*, or by acting inductively through the glass. The stem *s* is provided with an aluminum tube *a*, for purposes before explained, and only a small part of the filament reaches out of this tube. By holding the tube T₁ anywhere in the electrostatic field, the filament is rendered incandescent.

FIG. 163.   FIG. 164.

A more interesting piece of apparatus is illustrated in Fig. 164. The construction is the same as before, only instead of the lamp filament a small platinum wire p, sealed in a stem s, and bent above it in a circle, is connected to the copper wire w, which is joined to an inside coating c. A small stem s_1 is provided with a needle, on the point of which is arranged, to rotate very freely, a very light fan of mica v. To prevent the fan from falling out, a thin stem of glass g, is bent properly and fastened to the aluminum tube. When the glass tube is held anywhere in the electrostatic field the platinum wire becomes incandescent, and the mica vanes are rotated very fast.

Intense phosphorescence may be excited in a bulb by merely connecting it to a plate within the field, and the plate need not be any larger than an ordinary lamp shade. The phosphorescence excited with these currents is incomparably more powerful than with ordinary apparatus. A small phosphorescent bulb, when attached to a wire connected to a coil, emits sufficient light to allow reading ordinary print at a distance of five to six paces. It was of interest to see how some of the phosphorescent bulbs of Professor Crookes would behave with these currents, and he has had the kindness to lend me a few for the occasion. The effects produced are magnificent, especially by the sulphide of calcium and sulphide of zinc. With the disruptive discharge coil they glow intensely merely by holding them in the hand and connecting the body to the terminal of the coil.

To whatever results investigations of this kind may lead, the chief interest lies, for the present, in the possibilities they offer for the production of an efficient illuminating device. In no branch of electric industry is an advance more desired than in the manufacture of light. Every thinker, when considering the barbarous methods employed, the deplorable losses incurred in our best systems of light production, must have asked himself, What is likely to be the light of the future? Is it to be an incandescent solid, as in the present lamp, or an incandescent gas, or a phosphorescent body, or something like a burner, but incomparably more efficient?

There is little chance to perfect a gas burner; not, perhaps, because human ingenuity has been bent upon that problem for centuries without a radical departure having been made—though the argument is not devoid of force—but because in a burner the highest vibrations can never be reached, except by passing through all the low ones. For how is a flame to proceed unless by a fall of lifted weights? Such process cannot be maintained without renewal, and renewal is repeated passing from low to high vibrations. One way only seems to be open to improve a burner, and that is by trying to reach higher degrees of incandescence. Higher incandescence is equivalent to a quicker vibration: that means more light from the same material, and that again, means more economy. In this direction some improvements have been made, but the progress is hampered by many limitations. Discarding, then, the burner, there remains the three ways first mentioned, which are essentially electrical.

Suppose the light of the immediate future to be a solid, rendered incandescent by electricity. Would it not seem that it is better to employ a small button than a frail filament? From many considerations it certainly must be concluded that a button is capable of a higher economy, assuming, of course, the difficulties connected with the operation of such a lamp to be effectively overcome. But to light such a lamp we require a high potential; and to get this economically, we must use high frequencies.

Such considerations apply even more to the production of light by the incandescence of a gas, or by phosphorescence. In all cases we require high frequencies and high potentials. These thoughts occurred to me a long time ago.

Incidentally we gain, by the use of high frequencies, many advantages, such as higher economy in the light production, the possibility of working with one lead, the possibility of doing away with the leading-in wire, etc.

The question is, how far can we go with frequencies? Ordinary conductors rapidly lose the facility of transmitting electric impulses when the frequency is greatly increased. Assume the means for the production of impulses of very great frequency brought to the utmost perfection, everyone will naturally ask how to transmit them when the necessity arises. In transmitting such impulses through conductors we must remember that we have to deal with *pressure* and *flow*, in the ordinary interpretation of these terms. Let the pressure increase to an enormous value, and let the flow correspondingly diminish, then such impulses— variations merely of pressure, as it were—can no doubt be transmitted through a wire even if their frequency be many hundreds of millions per second. It would, of course, be out of question to transmit such impulses through a wire immersed in a gaseous medium, even if the wire were provided with a thick and excellent insulation, for most of the energy would be lost in molecular bombardment and consequent heating. The end of the wire connected to the source would be heated, and the remote end would receive but a trifling part of the energy supplied. The prime necessity, then, if such electric impulses are to be used, is to find means to reduce as much as possible the dissipation.

The first thought is, to employ the thinnest possible wire surrounded by the thickest practicable insulation. The next thought is to employ electrostatic screens. The insulation of the wire may be covered with a thin conducting coating and the latter connected to the ground. But this would not do, as then all the energy would pass through the conducting coating to the ground and nothing would get to the end of the wire. If a ground connection is made it can only be made through a conductor offering an enormous impedance, or through a condenser of extremely small capacity. This, however, does not do away with other difficulties.

If the wave length of the impulses is much smaller than the length of the wire, then corresponding short waves will be set up in the conducting coating, and it will be more or less the same as though the coating were directly connected to earth. It is therefore necessary to cut up the coating in sections much shorter than the wave length. Such an arrangement does not still afford a perfect screen, but it is ten thousand times better than none. I think it preferable to cut up the conducting coating in small sections, even if the current waves be much longer than the coating.

If a wire were provided with a perfect electrostatic screen, it would be the same as though all objects were removed from it at infinite distance. The capacity would then be reduced to the capacity of the wire itself, which would be very small. It would then be possible to send over the wire current vibrations of very high frequencies at enormous distances, without affecting greatly the character of the vibrations. A perfect screen is of course out of the question, but I believe that with a screen such as I have just described telephony could be rendered practicable across the Atlantic. According to my ideas, the gutta-percha covered wire should be provided with a third conducting coating subdivided in sections. On the top of this should be again placed a layer of gutta-percha and other insulation, and on the top of the whole the armor. But such cables will not be constructed, for ere long intelligence—transmitted without wires—will throb through the earth like a pulse through a living organism. The wonder is that, with the present state of knowledge and the experiences gained, no attempt is being made to disturb the electrostatic or magnetic condition of the earth, and transmit, if nothing else, intelligence.

It has been my chief aim in presenting these results to point out phenomena or features of novelty, and to advance ideas which I am hopeful will serve as starting points of new departures. It has been my chief desire this evening to entertain you with some novel experiments. Your applause, so frequently and generously accorded, has told me that I have succeeded.

In conclusion, let me thank you most heartily for your kindness and attention, and assure you that the honor I have had in addressing such a distinguished audience, the pleasure I have had in presenting these results to a gathering of so many able men—and among them also some of those in whose work for many years past I have found enlightenment and constant pleasure—I shall never forget.

CHAPTER XXVIII

ON LIGHT AND OTHER HIGH FREQUENCY PHENOMENA*

INTRODUCTORY—SOME THOUGHTS ON THE EYE

When we look at the world around us, on Nature, we are impressed with its beauty and grandeur. Each thing we perceive, though it may be vanishingly small, is in itself a world, that is, like the whole of the universe, matter and force governed by law—a world, the contemplation of which fills us with feelings of wonder and irresistibly urges us to ceaseless thought and inquiry. But in all this vast world, of all objects our senses reveal to us, the most marvellous, the most appealing to our imagination, appears no doubt a highly developed organism, a thinking being. If there is anything fitted to make us admire Nature's handiwork, it is certainly this inconceivable structure, which performs its innumerable motions of obedience to external influence. To understand its workings, to get a deeper insight into this Nature's masterpiece, has ever been for thinkers a fascinating aim, and after many centuries of arduous research men have arrived at a fair understanding of the functions of its organs and senses. Again, in all the perfect harmony of its parts, of the parts which constitute the material or tangible of our being, of all its organs and senses, the eye is the most wonderful. It is the most precious, the most indispensable of our perceptive or directive organs, it is the great gateway through which all knowledge enters the mind. Of all our organs, it is the one, which is in the most intimate relation with that which we call intellect. So intimate is this relation, that it is often said, the very soul shows itself in the eye.

It can be taken as a fact, which the theory of the action of the eye implies, that for each external impression, that is, for each image produced upon the retina, the ends of the visual nerves, concerned in the conveyance of the impression to the mind, must be under a peculiar stress or in a vibratory state. It now does not seem improbable that,

---

* A lecture delivered before the Franklin Institute, at Philadelphia, Feb. 24, 1893, and before the National Electric Light Association, at St. Louis, Mo., March 1, 1893.

when by the power of thought an image is evoked, a distinct reflex action, no matter how weak, is exerted upon certain ends of the visual nerves, and therefore upon the retina. Will it ever be within human power to analyze the condition of the retina when disturbed by thought or reflex action, by the help of some optical or other means of such sensitiveness, that a clear idea of its state might be gained at any time? If this were possible, then the problem of reading one's thoughts with precision, like the characters of an open book, might be much easier to solve than many problems belonging to the domain of positive physical science, in the solution of which many, if not the majority, of scientific men implicitly believe. Helmholtz has shown that the fundi of the eye are themselves, luminous, and he was able to *see*, in total darkness, the movement of his arm by the light of his own eyes. This is one of the most remarkable experiments recorded in the history of science, and probably only a few men could satisfactorily repeat it, for it is very likely, that the luminosity of the eyes is associated with uncommon activity of the brain and great imaginative power. It is fluorescence of brain action, as it were.

Another fact having a bearing on this subject which has probably been noted by many, since it is stated in popular expressions, but which I cannot recollect to have found chronicled as a positive result of observation is, that at times, when a sudden idea or image presents itself to the intellect, there is a distinct and sometimes painful sensation of luminosity produced in the eye, observable even in broad daylight.

The saying then, that the soul shows itself in the eye, is deeply founded, and we feel that it expresses a great truth. It has a profound meaning even for one who, like a poet or artist, only following his inborn instinct or love for Nature, finds delight in aimless thoughts and in the mere contemplation of natural phenomena, but a still more profound meaning for one who, in the spirit of positive scientific investigation, seeks to ascertain the causes of the effects. It is principally the natural philosopher, the physicist, for whom the eye is the subject of the most intense admiration.

Two facts about the eye must forcibly impress the mind of the physicist, notwithstanding he may think or say that it is an imperfect optical instrument, forgetting, that the very conception of that which is perfect or seems so to him, has been gained through this same instrument. First, the eye is, as far as our positive knowledge goes, the only organ which is *directly* affected by that subtile medium, which as science teaches us, must fill all space; secondly, it is the most sensitive of our organs, incomparably more sensitive to external impressions than any other.

The organ of hearing implies the impact of ponderable bodies, the organ of smell the transference of detached material particles, and the organs of taste, and of touch or force, the direct contact, or at least some interference of ponderable matter, and this is true even in those instances of animal organisms, in which some of these organs are developed to a degree of truly marvelous perfection. This being so, it seems wonderful that the organ of sight solely should be capable of being stirred by that, which all our other organs are powerless to detect, yet which plays an essential part in all natural phenomena, which transmits all energy and sustains all motion and, that most intricate of all, life, but which has properties such that even a scientifically trained mind cannot help drawing a distinction between it and all that is called matter. Considering merely this, and the fact that the eye, by its marvelous power, widens our otherwise very narrow range of perception far beyond the limits of the small world which is our own, to embrace myriads of other worlds, suns and stars in the infinite depths of the universe, would make it justifiable to assert, that it is an organ of a higher order. Its performances are beyond comprehension. Nature as far as we know never produced anything more wonderful. We can get barely a faint idea of its prodigious power by analyzing what it does and by comparing. When ether waves impinge upon the human body, they produce the sensations of warmth or cold, pleasure or pain, or perhaps other sensations of which we are not aware, and any degree or intensity of these sensations, which degrees are infinite in number, hence an infinite number of distinct sensations. But our sense of touch, or our sense of force, cannot reveal to us these differences in degree or intensity, unless they are very great. Now we can readily conceive how an organism, such as the human, in the eternal process of evolution, or more philosophically speaking, adaptation to Nature, being constrained to the use of only the sense of touch or force, for instance, might develop this sense to such a degree of sensitiveness or perfection, that it would be capable of distinguishing the minutest differences in the temperature of a body even at some distance, to a hundredth, or thousandth, or millionth part of a degree. Yet, even this apparently impossible performance would not begin to compare with that of the eye, which is capable of distinguishing and conveying to the mind in a single instant innumerable peculiarities of the body, be it in form, or color, or other respects. This power of the eye rests upon two things, namely, the rectilinear propagation of the disturbance by which it is effected, and upon its sensitiveness. To say that the eye is sensitive is not saying anything. Compared with it, all other organs are monstrously crude. The organ of smell which guides a dog on the trail of a deer, the organ of touch or force which guides an insect in its wanderings, the organ of hearing, which is affected by the slightest disturbances of the air, are sensitive organs, to be sure, but what are they compared with the human eye! No doubt it responds to the faintest echoes or reverberations of the medium; no doubt, it brings us tidings from other worlds, infinitely remote, but in a language we cannot as yet always understand. And why not? Because we live in a medium filled with air and other gases, vapors and a dense mass of solid particles flying about. These play an important part in many phenomena; they fritter away the energy of the vibrations before they can reach the eye; they too, are the carriers of germs of destruction, they get into our lungs and other organs, clog up the channels and imperceptibly, yet inevitably, arrest the stream of life. Could we but do away with all ponderable matter in the line of sight of the telescope, it would reveal to us undreamt of marvels. Even the unaided eye, I think, would be capable of distinguishing in the pure medium, small objects at distances measured probably by hundreds or perhaps thousands of miles.

But there is something else about the eye which impresses us still more than these wonderful features which we observed, viewing it from the standpoint of a physicist, merely as an optical instrument—something which appeals to us more than its marvelous faculty of being directly affected by the vibrations of the medium, without interference of gross matter, and more than its inconceivable sensitiveness and discerning power. It is its significance in the processes of life. No matter what one's views on nature and life may be, he must stand amazed when, for the first time in his thoughts, he realizes the importance of the eye in the physical processes and mental performances of the human organism. And how could it be otherwise, when he realizes, that the eye is the means through which the human race has acquired the entire knowledge it possesses, that it controls all our motions, more still, all our actions.

There is no way of acquiring knowledge except through the eye. What is the foundation of all philosophical systems of ancient and modern times, in fact, of all the philosophy of man? *I am, I think; I think, therefore I am.* But how could I think and how would I know that I exist, if I had not the eye? For knowledge involves consciousness; consciousness involves ideas, conceptions; conceptions involve pictures or images, and images the sense of vision, and therefore the organ of sight. But how about blind men, will be asked? Yes, a blind man may depict in magnificent poems, forms and scenes from real life, from a world he physically does not see. A blind man may touch the keys of an instrument with unerring precision, may model the fastest boat, may discover and invent, calculate and construct, may do still greater wonders—but all the blind men who have done such things have descended from those who had seeing eyes. Nature may reach the same result in many ways. Like a wave in the physical world, in the infinite ocean of the medium which pervades all, so in the world of organisms, in life, an impulse started proceeds onward, at times, may be, with the speed of light, at times, again, so slowly that for ages and ages it seems to stay, passing through processes of a complexity inconceivable to men, but in all its forms, in all its stages, its energy ever and ever integrally present. A single ray of light from a distant star falling upon the eye of a tyrant in bygone times, may have altered the course of his life, may have changed the destiny of nations, may have transformed the surface of the globe, so intricate, so inconceivably complex are the processes in Nature. In no way can we get such an overwhelming idea of the grandeur of Nature, as when we consider, that in accordance with the law of the conservation of energy, throughout the infinite, the forces are in a perfect balance, and hence the energy of a single thought may determine the motion of a Universe. It is not necessary that every individual, not even that every generation or many generations, should have the physical instrument of sight, in order to be able to form images and to think, that is, form ideas or conceptions; but sometime or other, during the process of evolution, the eye certainly must have existed, else thought, as we understand it, would be impossible; else conceptions, like spirit, intellect, mind, call it as you may, could not exist. It is conceivable, that in some other world, in some other beings, the eye is replaced by a different organ, equally or more perfect, but these beings cannot be men.

Now what prompts us all to voluntary motions and actions of any kind? Again the eye. If I am conscious of the motion, I must have an idea or conception, that is, an image, therefore the eye. If I am not precisely conscious of the motion, it is, because the images are vague or indistinct, being blurred by the superimposition of many. But when I perform the motion, does the impulse which prompts me to the action come from within or from without? The greatest physicists have not disdained to endeavor to answer this and similar questions and have at times abandoned themselves to the delights of pure and unrestrained thought. Such questions are generally considered not to belong to the realm of positive physical science, but will before long be annexed to its domain. Helmholtz has probably thought more on life than any modern scientist. Lord Kelvin expressed his belief that life's process is electrical and that there is a force inherent to the organism and determining its motions. Just as much as I am convinced of any physical truth I am convinced that the motive impulse must come from the outside. For, consider the lowest organism we know—and there are probably many lower ones—an aggregation of a few cells only. If it is capable of voluntary motion it can perform an infinite number of motions, all definite and precise. But now a mechanism consisting of a finite number of parts and few at that, cannot perform an infinite number of definite motions, hence the impulses which govern its movements must come from the environment. So, the atom, the ulterior element of the Universe's structure, is tossed about in space, eternally, a play to external influences, like a boat in a troubled sea. Were it to stop its motion *it would die*. Matter at rest, if such a thing could exist, would be matter dead. Death of matter! Never has a sentence of deeper philosophical meaning been uttered. This is the way in which Prof. Dewar forcibly expresses it in the description of his admirable experiments, in which liquid oxygen is handled as one handles water, and air at ordinary pressure is made to condense and even to solidify by the intense cold. Experiments, which serve to illustrate, in his language, the last feeble manifestations of life, the last quiverings of matter about to die. But human eyes shall not witness such death. There is no death of matter, for throughout the infinite universe, all has to move, to vibrate, that is, to live.

I have made the preceding statements at the peril of treading upon metaphysical ground, in my desire to introduce the subject of this lecture in a manner not altogether uninteresting, I may hope, to an audience such as I have the honor to address. But now, then, returning to the subject, this divine organ of sight, this indispensable instrument for thought and all intellectual enjoyment, which lays open to us the marvels of this universe, through which we have acquired what knowledge we possess, and which prompts us to, and controls, all our physical and mental activity. By what is it affected? By light! What is light?

We have witnessed the great strides which have been made in all departments of science in recent years. So great have been the advances that we cannot refrain from asking ourselves, Is this all true, or is it but a dream? Centuries ago men have lived, have thought, discovered, invented, and have believed that they were soaring, while they were merely proceeding at a snail's pace. So we too may be mistaken. But taking the truth of the observed events as one of the implied facts of science, we must rejoice in the immense progress already made and still more in the anticipation of what must come, judging from the possibilities opened up by modern research. There is, however, an advance which we have been witnessing, which must be particularly gratifying to every lover of progress. It is not a discovery, or an invention, or an achievement in any particular direction. It is an advance in all directions of scientific thought and experiment. I mean the generalization of the natural forces and phenomena, the looming up of a certain broad idea on the scientific horizon. It is this idea which has, however, long ago taken possession of the most advanced minds, to which I desire to call your attention, and which I intend to illustrate in a general way, in these experiments, as the first step in answering the question "What is light?" and to realize the modern meaning of this word.

It is beyond the scope of my lecture to dwell upon the subject of light in general, my object being merely to bring presently to your notice a certain class of light effects and a number of phenomena observed in pursuing the study of these effects. But to be consistent in my remarks it is necessary to state that, according to that idea, now accepted by the majority of scientific men as a positive result of theoretical and experimental investigation, the various forms or manifestations of energy which were generally designated as "electric" or more precisely "electromagnetic" are energy manifestations of the same nature as those of radiant heat and light. Therefore the phenomena of light and heat and others besides these, may be called electrical phenomena. Thus electrical science has become the mother science of all and its study has become all important. The day when we shall know exactly what "electricity" is, will chronicle an event probably greater, more important than any other recorded in the history of the human race. The time will come when the comfort, the very existence, perhaps, of man will depend upon that wonderful agent. For our existence and comfort we require heat, light and mechanical power. How do we now get all these? We get them from fuel, we get them by consuming material. What will man do when the forests disappear, when the coal fields are exhausted? Only one thing, according to our present knowledge will remain; that is, to transmit power at great distances. Men will go to the waterfalls, to the tides, which are the stores of an infinitesimal part of Nature's immeasurable energy. There will they harness the energy and transmit the same to their settlements, to warm their homes by, to give them light, and to keep their obedient slaves, the machines, toiling. But how will they transmit this energy if not by electricity? Judge then, if the comfort, nay, the very existence, of man will not depend on electricity. I am aware that this view is not that of a practical engineer, but neither is it that of an illusionist, for it is certain, that power transmission, which at present is merely a stimulus to enterprise, will someday be a dire necessity.

It is more important for the student, who takes up the study of light phenomena, to make himself thoroughly acquainted with certain modern views, than to peruse entire books on the subject of light itself, as disconnected from these views. Were I therefore to make these demonstrations before students seeking information—and for the sake of the few of those who may be present, give me leave to so assume—it would be my principal endeavor to impress these views upon their minds in this series of experiments.

It might be sufficient for this purpose to perform a simple and well-known experiment. I might take a familiar appliance, a Leyden jar, charge it from a frictional machine, and then discharge it. In explaining to you its permanent state when charged, and its transitory condition when discharging, calling your attention to the forces which enter into play and to the various phenomena they produce, and pointing out the relation of the forces and phenomena, I might fully succeed in illustrating that modern idea. No doubt, to the thinker, this simple experiment would appeal as much as the most magnificent display. But this is to be an experimental demonstration, and one which should possess, besides instructive, also entertaining features and as such, a simple experiment, such as the one cited, would not go very far towards the attainment of the lecturer's aim. I must therefore choose another way of illustrating, more spectacular certainly, but perhaps also more instructive. Instead of the frictional machine and Leyden jar, I shall avail myself in these experiments, of an induction coil of peculiar properties, which was described in detail by me in a lecture before the London Institution of Electrical Engineers, in Feb., 1892. This induction coil is capable of yielding currents of enormous potential differences, alternating with extreme rapidity. With this apparatus I shall endeavor to show you three distinct classes of effects, or phenomena, and it is my desire that each experiment, while serving for the purposes of illustration, should at the same time teach us some novel truth, or show us some novel aspect of this fascinating science. But before doing this, it seems proper and useful to dwell upon the apparatus employed, and method of obtaining the high potentials and high-frequency currents which are made use of in these experiments.

On the Apparatus and Method of Conversion

These high-frequency currents are obtained in a peculiar manner. The method employed was advanced by me about two years ago in an experimental lecture before the American Institute of Electrical Engineers. A number of ways, as practiced in the laboratory, of obtaining these currents either from continuous or low frequency alternating currents, is diagrammatically indicated in Fig. 165, which will be later described in detail. The general plan is to charge condensers, from a direct or alternate-current source, preferably of high-tension, and to discharge them disruptively while observing well-known conditions necessary to maintain the oscillations of the current. In view of the general interest taken in high-frequency currents and effects producible by them, it seems to me advisable to dwell at some length upon this method of conversion. In order to give you a clear idea of the action, I will suppose that a continuous-current generator is employed, which is often very convenient. It is desirable that the generator should possess such high tension as

FIG. 165.

to be able to break through a small air space. If this is not the case, then auxiliary means have to be resorted to, some of which will be indicated subsequently. When the condensers are charged to a certain potential, the air, or insulating space, gives way and a disruptive discharge occurs. There is then a sudden rush of current and generally a large portion of accumulated electrical energy spends itself. The condensers are thereupon quickly charged and the same process is repeated in more or less rapid succession. To produce such sudden rushes of current it is necessary to observe certain conditions. If the rate at which the condensers are discharged is the same as that at which they are charged, then, clearly, in the assumed case the condensers do not come into play. If the rate of discharge be smaller than the rate of charging, then, again, the condensers cannot play an important part. But if, on the contrary, the rate of discharging is greater than that of charging, then a succession of rushes of current is obtained. It is evident that, if the rate at which the energy is dissipated by the discharge is very much greater than the rate of supply to the condensers, the sudden rushes will be comparatively few, with long-time intervals between. This always occurs when a condenser of considerable capacity is charged by means of a comparatively small machine. If the rates of supply and dissipation are not widely different, then the rushes of current will be in quicker succession, and this the more, the more nearly equal both the rates are, until limitations incident to each case and depending upon a number of causes are reached. Thus we are able to obtain from a continuous-current generator as rapid a succession of discharges as we like. Of course, the higher the tension of the generator, the smaller need be the capacity of the condensers, and for this reason, principally, it is of advantage to employ a generator of very high tension. Besides, such a generator permits the attaining of greater rates of vibration.

The rushes of current may be of the same direction under the conditions before assumed, but most generally there is an oscillation superimposed upon the fundamental vibration of the current. When the conditions are so determined that there are no oscillations, the current impulses are unidirectional and thus a means is provided of transforming a continuous current of high tension, into a direct current of lower tension, which I think may find employment in the arts.

This method of conversion is exceedingly interesting and I was much impressed by its beauty when I first conceived it. It is ideal in certain respects. It involves the employment of no mechanical devices of any kind, and it allows of obtaining currents of any desired frequency from an ordinary circuit, direct or alternating. The frequency of the fundamental discharges depending on the relative rates of supply and dissipation can be readily varied within wide limits, by simple adjustments of these quantities, and the frequency of the superimposed vibration by the determination of the capacity, self-induction and resistance of the circuit. The potential of the currents, again, may be raised as high as any insulation is capable of withstanding safely by combining capacity and self-induction or by induction in a secondary, which need have but comparatively few turns.

As the conditions are often such that the intermittence or oscillation does not readily establish itself, especially when a direct current source is employed, it is of advantage to associate an interrupter with the arc, as I have, some time ago, indicated the use of an air-blast or magnet, or other such device readily at hand. The magnet is employed with special advantage in the conversion of direct currents, as it is then very effective. If the primary source is an alternate current generator, it is desirable, as I have stated on another occasion, that the frequency should be low, and that the current forming the arc be large, in order to render the magnet more effective.

FIG. 166.

A form of such discharger with a magnet which has been found convenient, and adopted after some trials, in the conversion of direct currents particularly, is illustrated in Fig. 166. N S are the pole pieces of a very strong magnet which is excited by a coil C. The pole pieces are slotted for adjustment and can be fastened in any position by screws $s\ s_1$. The discharge rods $d\ d_1$, thinned down on the ends in order to allow a closer approach of the magnetic pole pieces, pass through the columns of brass $b\ b_1$ and are fastened in position by screws $s_2\ s_2$. Springs $r\ r_1$ and collars $c\ c_1$ are slipped on the rods, the latter serving to set the points of the rods at a certain suitable distance by means of screws $s_3\ s_3$, and the former to draw the points apart. When it is desired to start the arc, one of the large rubber handles $h\ h_1$ is tapped quickly with the hand, whereby the points of the rods are brought in contact but are instantly separated by the springs $r\ r_1$. Such an arrangement has been found to be often necessary, namely in cases when the E. M. F. was not large enough to cause the discharge to break through the gap, and also when it was desirable to avoid short circuiting of the generator by the metallic contact of the rods. The rapidity of the interruptions of the current with a magnet depends on the intensity of the magnetic field and on the potential difference at the end of the arc. The interruptions are generally in such quick succession as to produce a musical sound. Years ago it was observed that when a powerful induction coil is discharged between the poles of a strong magnet, the discharge produces a loud noise, not unlike a small pistol shot. It was vaguely stated that the spark was intensified by the presence of the magnetic field. It is now clear that the discharge current, flowing for some time, was interrupted a great number of times by the magnet, thus producing the sound. The phenomenon is especially marked when the field circuit of a large magnet or dynamo is broken in a powerful magnetic field.

When the current through the gap is comparatively large, it is of advantage to slip on the points of the discharge rods pieces of very hard carbon and let the arc play between the carbon pieces. This preserves the rods, and besides has the advantage of keeping the air space hotter, as the heat is not conducted away as quickly through the carbons, and the result is that a smaller E. M. F. in the arc gap is required to maintain a succession of discharges.

FIG. 167.

Another form of discharger, which may be employed with advantage in some cases, is illustrated in Fig. 167. In this form the discharge rods $d\,d_1$ pass through perforations in a wooden box B, which is thickly coated with mica on the inside, as indicated by the heavy lines. The perforations are provided with mica tubes m m_1 of some thickness, which are preferably not in contact with the rods d d_1. The box has a cover c which is a little larger and descends on the outside of the box. The spark gap is warmed by a small lamp l contained in the box. A plate p above the lamp allows the draught to pass only through the chimney e of the lamp, the air entering through holes o o in or near the bottom of the box and following the path indicated by the arrows. When the discharger is in operation, the door of the box is closed so that the light of the arc is not visible outside. It is desirable to exclude the light as perfectly as possible, as it interferes with some experiments. This form of discharger is simple and very effective when properly manipulated. The air being warmed to a certain temperature, has its insulating power impaired; it becomes dielectrically weak, as it were, and the consequence is that the arc can be established at much greater distance. The arc should, of course, be sufficiently insulating to allow the discharge to pass through the gap *disruptively.* The arc formed under such conditions, when long, may be made extremely sensitive, and the weak draught through the lamp chimney c is quite sufficient to produce rapid interruptions. The adjustment is made by regulating the temperature and velocity of the draught. Instead of using the lamp, it answers the purpose to provide for a draught of warm air in other ways. A very simple way which has been practiced is to enclose the arc in a long vertical tube, with plates on the top and bottom for regulating the temperature and velocity of the air current. Some provision had to be made for deadening the sound.

The air may be rendered dielectrically weak also by rarefaction. Dischargers of this kind have likewise been used by me in connection with a magnet. A large tube is for this purpose provided with heavy electrodes of carbon or metal, between which the discharge is made to pass, the tube being placed in a powerful magnetic field. The exhaustion of the tube is carried to a point at which the discharge breaks through easily, but the pressure should be more than 75 millimeters, at which the ordinary thread discharge occurs. In another form of discharger, combining the features before mentioned, the discharge was made to pass between two adjustable magnetic pole pieces, the space between them being kept at an elevated temperature.

It should be remarked here that when such, or interrupting devices of any kind, are used and the currents are passed through the primary of a disruptive discharge coil, it is not, as a rule, of advantage to produce a number of interruptions of the current per second greater than the natural frequency of vibration of the dynamo supply circuit, which is ordinarily small. It should also be pointed out here, that while the devices mentioned in connection with the disruptive discharge are advantageous under certain conditions, they may be sometimes a source of trouble, as they produce intermittences and other irregularities in the vibration which it would be very desirable to overcome.

There is, I regret to say, in this beautiful method of conversion a defect, which fortunately is not vital, and which I have been gradually overcoming. I will best call attention to this defect and indicate a fruitful line of work, by comparing the electrical process with its mechanical analogue. The process may be illustrated in this manner. Imagine a tank with a wide opening at the bottom, which is kept closed by spring pressure, but so that it snaps off *suddenly* when the liquid in the tank has reached a certain height. Let the fluid be supplied to the tank by means of a pipe feeding at a certain rate. When the critical height of the liquid is reached, the spring gives way and the bottom of the tank drops out. Instantly the liquid falls through the wide opening, and the spring, reasserting itself, closes the bottom again. The tank is now filled, and after a certain time interval the same process is repeated. It is clear, that if the pipe feeds the fluid quicker than the bottom outlet is capable of letting it pass through, the bottom will remain off and the tank will still overflow. If the rates of supply are exactly equal, then the bottom lid will remain partially open and no vibration of the same and of the liquid column will generally occur, though it might, if started by some means. But if the inlet pipe does not feed the fluid fast enough for the outlet, then there will be always vibration. Again, in such case, each time the bottom flaps up or down, the spring and the liquid column, if the pliability of the spring and the inertia of the moving parts are properly chosen, will perform independent vibrations. In this analogue the fluid may be likened to electricity or electrical energy, the tank to the condenser, the spring to the dielectric, and the pipe to the conductor through which electricity is supplied to the condenser. To make this analogy quite complete it is necessary to make the assumption, that the bottom, each time it gives way, is knocked violently against a non-elastic stop, this impact involving some loss of energy; and that, besides, some dissipation of energy results due to frictional losses. In the preceding analogue the liquid is supposed to be under a steady pressure. If the presence of the fluid be assumed to vary rhythmically, this may be taken as corresponding to the case of an alternating current. The process is then not quite as simple to consider, but the action is the same in principle.

It is desirable, in order to maintain the vibration economically, to reduce the impact and frictional losses as much as possible. As regards the latter, which in the electrical analogue correspond to the losses due to the resistance of the circuits, it is impossible to obviate them entirely, but they can be reduced to a minimum by a proper selection of the dimensions of the circuits and by the employment of thin conductors in the form of strands. But the loss of energy caused by the first breaking through of the dielectric—which in the above example corresponds to the violent knock of the bottom against the inelastic stop—would be more important to overcome. At the moment of the breaking through, the air space has a very high resistance, which is probably reduced to a very small value when the current has reached some strength, and the space is brought to a high temperature. It would materially diminish the loss of energy if the space were always kept at an extremely high temperature, but then there would be no disruptive break. By warming the space moderately by means of a lamp or otherwise, the economy as far as the arc is concerned is sensibly increased. But the magnet or other interrupting device does not diminish the loss in the arc. Likewise, a jet of air only facilitates the carrying off of the energy. Air, or a gas in general, behaves curiously in this respect. When two bodies charged to a very high potential, discharge disruptively through an air space, any amount of energy may be carried off by the air. This energy is evidently dissipated by bodily carriers, in impact and collisional losses of the molecules. The exchange of the molecules in the space occurs with inconceivable rapidity. A powerful discharge taking place between two electrodes, they may remain entirely cool, and yet the loss in the air may represent any amount of energy. It is perfectly practicable, with very great potential differences in the gap, to dissipate several horsepower in the arc of the discharge without even noticing a small increase in the temperature of the electrodes. All the frictional losses occur then practically in the air. If the exchange of the air molecules is prevented, as by enclosing the air hermetically, the gas inside of the vessel is brought quickly to a high temperature, even with a very small discharge. It is difficult to estimate how much of the energy is lost in sound waves, audible or not, in a powerful discharge. When the currents through the gap are large, the electrodes may become rapidly heated, but this is not a reliable measure of the energy wasted in the arc, as the loss through the gap itself may be comparatively small. The air or a gas in general is, at ordinary pressure at least, clearly not the best medium through which a disruptive discharge should occur. Air or other gas under great pressure is of course a much more suitable medium for the discharge gap. I have carried on long-continued experiments in this direction, unfortunately less practicable on account of the difficulties and expense in getting air under great pressure. But even if the medium in the discharge space is solid or liquid, still the same losses take place, though they are generally smaller, for just as soon as the arc is established, the solid or liquid is volatilized. Indeed, there is no body known which would not be disintegrated by the arc, and it is an open question among scientific men, whether an arc discharge could occur at all in the air itself without the particles of the electrodes being torn off. When the current through the gap is very small and the arc very long, I believe that a relatively considerable amount of heat is taken up in the disintegration of the electrodes, which partially on this account may remain quite cold.

The ideal medium for a discharge gap should only *crack*, and the ideal electrode should be of some material which cannot be disintegrated. With small currents through the gap it is best to employ aluminum, but not when the currents are large. The disruptive break in the air, or more or less in any ordinary medium, is not of the nature of a crack, but it is rather comparable to the piercing of innumerable bullets through a mass offering great frictional resistances to the motion of the bullets, this involving considerable loss of energy. A medium which would merely crack when strained electrostatically—and this possibly might be the case with a perfect vacuum, that is, pure ether—would involve a very small loss in the gap, so small as to be entirely negligible, at least theoretically, because a crack may be produced by an infinitely small displacement. In exhausting an oblong bulb provided with two aluminum terminals, with the greatest care, I have succeeded in producing such a vacuum that the secondary discharge of a disruptive discharge coil would break disruptively through the bulb in the form of fine spark streams.

The curious point was that the discharge would completely ignore the terminals and start far behind the two aluminum plates which served as electrodes. This extraordinary high vacuum could only be maintained for a very short while. To return to the ideal medium, think, for the sake of illustration, of a piece of glass or similar body clamped in a vice, and the latter tightened more and more. At a certain point a minute increase of the pressure will cause the glass to crack. The loss of energy involved in splitting the glass may be practically nothing, for though the force is great, the displacement need be but extremely small. Now imagine that the glass would possess the property of closing again perfectly the crack upon a minute diminution of the pressure. This is the way the dielectric in the discharge space should behave. But inasmuch as there would be always some loss in the gap, the medium, which should be continuous, should exchange through the gap at a rapid rate. In the preceding example, the glass being perfectly closed, it would mean that the dielectric in the discharge space possesses a great insulating power; the glass being cracked, it would signify that the medium in the space is a good conductor. The dielectric should vary enormously in resistance by minute variations of the E. M. F. across the discharge space. This condition is attained, but in an extremely imperfect manner, by warming the air space to a certain critical temperature, dependent on the E. M. F. across the gap, or by otherwise impairing the insulating power of the air. But as a matter of fact the air does never break down *disruptively*, if this term be rigorously interpreted, for before the sudden rush of the current occurs, there is always a weak current preceding it, which rises first gradually and then with comparative suddenness. That is the reason why the rate of change is very much greater when glass, for instance, is broken through, than when the break takes place through an air space of equivalent dielectric strength. As a medium for the discharge space, a solid, or even a liquid, would be preferable therefor. It is somewhat difficult to conceive of a solid body which would possess the property of closing instantly after it has been cracked. But a liquid, especially under great pressure, behaves practically like a solid, while it possesses the property of closing the crack. Hence it was thought that a liquid insulator might be more suitable as a dielectric than air. Following out this idea, a number of different forms of dischargers in which a variety of such insulators, sometimes under great pressure, were employed, have been experimented upon. It is thought sufficient to dwell in a few words upon one of the forms experimented upon. One of these dischargers is illustrated in Figs. 168a and 168b.

FIG. 168a.   FIG. 168b.

A hollow metal pulley p (Fig. 168a), was fastened upon an arbor a, which by suitable means was rotated at a considerable speed. On the inside of the pulley, but disconnected from the same, was supported a thin disc h (which is shown thick for the sake of clearness), of hard rubber in which there were embedded two metal segments s s with metallic extensions e e into which were screwed conducting terminals t t covered with thick tubes of hard rubber t1 t1. The rubber disc h with its metallic segments s s, was finished in a lathe, and its entire surface highly polished so as to offer the smallest possible frictional resistance to the motion through a fluid. In the hollow of the pulley an insulating liquid such as a thin oil was poured so as to reach very nearly to the opening left in the flange f, which was screwed tightly on the front side of the pulley. The terminals t t, were connected to the opposite coatings of a battery of condensers so that the discharge occurred through the liquid. When the pulley was rotated, the liquid was forced against the rim of the pulley and considerable fluid pressure resulted. In this simple way the discharge gap was filled with a medium which behaved practically like a solid, which possessed the quality of closing instantly upon the occurrence of the break, and which moreover was circulating through the gap at a rapid rate. Very powerful effects were produced by discharges of this kind with liquid interrupters, of which a number of different forms were made. It was found that, as expected, a longer spark for a given length of wire was obtainable in this way than by using air as an interrupting device. Generally the speed, and therefore also the fluid pressure, was limited by reason of the fluid friction, in the form of discharger described, but the practically obtainable speed was more than sufficient to produce a number of breaks suitable for the circuits ordinarily used. In such instances the metal pulley p was provided with a few projections inwardly, and a definite number of breaks was then produced which could be computed from the speed of rotation of the pulley. Experiments were also carried on with liquids of different insulating power with the view of reducing the loss in the arc. When an insulating liquid is moderately warmed, the loss in the arc is diminished.

A point of some importance was noted in experiments with various discharges of this kind. It was found, for instance, that whereas the conditions maintained in these forms were favorable for the production of a great spark length, the current so obtained was not best suited to the production of light effects. Experience undoubtedly has shown, that for such purposes a harmonic rise and fall of the potential is preferable. Be it that a solid is rendered incandescent, or phosphorescent, or be it that energy is transmitted by condenser coating through the glass, it is quite certain that a harmonically rising and falling potential produces less destructive action, and that the vacuum is more permanently maintained. This would be easily explained if it were ascertained that the process going on in an exhausted vessel is of an electrolytic nature.

In the diagrammatical sketch, Fig. 165, which has been already referred to, the cases which are most likely to be met with in practice are illustrated. One has at his disposal either direct or alternating currents from a supply station. It is convenient for an experimenter in an isolated laboratory to employ a machine G, such as illustrated, capable of giving both kinds of currents. In such case it is also preferable to use a machine with multiple circuits, as in many experiments it is useful and convenient to have at one disposal currents of different phases. In the sketch, D represents the direct and A the alternating circuit. In each of these, three branch circuits are shown, all of which are provided with double line switches s s s s s. Consider first the direct current conversion; Ia represents the simplest case. If the E. M. F. of the generator is sufficient to break through a small air space, at least when the latter is warmed or otherwise rendered poorly insulating, there is no difficulty in maintaining a vibration with fair economy by judicious adjustment of the capacity, self-induction and resistance of the circuit L containing the devices *l l m.* The magnet N, S, can be in this case advantageously combined with the air space. The discharger *d d* with the magnet may be placed either way, as indicated by the full or by the dotted lines. The circuit Ia with the connections and devices is supposed to possess dimensions such as are suitable for the maintenance of a vibration. But usually the E. M. F. on the circuit or branch ı*a* will be something like a 100 volts or so, and in this case it is not sufficient to break through the gap. Many different means may be used to remedy this by raising the E. M. F. across the gap. The simplest is probably to insert a large self-induction coil in series with the circuit L. When the arc is established, as by the discharger illustrated in Fig. 166, the magnet blows the arc out the instant it is formed. Now the extra current of the break, being of high E. M. F., breaks through the gap, and a path of low resistance for the dynamo current being again provided, there is a sudden rush of current from the dynamo upon the weakening or subsidence of the extra current. This process is repeated in rapid succession, and in this manner I have maintained oscillation with as low as 50 volts, or even less, across the gap. But conversion on this plan is not to be recommended on account of the too heavy currents through the gap and consequent heating of the electrodes; besides, the frequencies obtained in this way are low, owing to the high self-induction necessarily associated with the circuit. It is very desirable to have the E. M. F. as high as possible, first, in order to increase the economy of the conversion, and, secondly, to obtain high frequencies. The difference of potential in this electric oscillation is, of course, the equivalent of the stretching force in the mechanical vibration of the spring. To obtain very rapid vibration in a circuit of some inertia, a great stretching force or difference of potential is necessary. Incidentally, when the E. M. F. is very great, the condenser which is usually employed in connection with the circuit need but have a small capacity, and many other advantages are gained. With a view of raising the E. M. F. to a many times greater value than obtainable from ordinary distribution circuits, a rotating transformer *g* is used, as indicated at ıı*a*, Fig. 165, or else a separate high potential machine is driven by means of a motor operated from the generator G. The latter plan is in fact preferable, as changes are easier made. The connections from the high tension winding are quite similar to those in branch ı*a* with the exception that a condenser C, which should be adjustable, is connected to the high tension circuit. Usually, also, an adjustable self-induction coil in series with the circuit has been employed in these experiments. When the tension of the currents is very high, the magnet ordinarily used in connection with the discharger is of comparatively small value, as it is quite easy to adjust the dimensions of the circuit so that oscillation is maintained. The employment of a steady E. M. F. in the high frequency conversion affords some advantages over the employment of alternating E. M. F., as the adjustments are much simpler and the action can be easier controlled. But unfortunately one is limited by the obtainable potential difference. The winding also breaks down easily in consequence of the sparks which form between the sections of the armature or commutator when a vigorous oscillation takes place. Besides, these transformers are expensive to build. It has been found by experience that it is best to follow the plan illustrated at IIIa. In this arrangement a rotating transformer g, is employed to convert the low tension direct currents into low frequency alternating currents, preferably also of small tension. The tension of the currents is then raised in a stationary transformer T. The secondary S of this transformer is connected to an adjustable condenser C which discharges through the gap or discharger d d, placed in either of the ways indicated, through the primary P of a disruptive discharge coil, the high frequency current being obtained from the secondary S of this coil, as described on previous occasions. This will undoubtedly be found the cheapest and most convenient way of converting direct currents.

The three branches of the circuit A represent the usual cases met in practice when alternating currents are converted. In Fig. 1b a condenser C, generally of large capacity, is connected to the circuit L containing the devices l l, m m. The devices m m are supposed to be of high self-induction so as to bring the frequency of the circuit more or less to that of the dynamo. In this instance the discharger d d should best have a number of makes and breaks per second equal to twice the frequency of the dynamo. If not so, then it should have at least a number equal to a multiple or even fraction of the dynamo frequency. It should be observed, referring to 1b, that the conversion to a high potential is also effected when the discharger d d, which is shown in the sketch, is omitted. But the effects which are produced by currents which rise instantly to high values, as in a disruptive discharge, are entirely different from those produced by dynamo currents which rise and fall harmonically. So, for instance, there might be in a given case a number of makes and breaks at $d\,d$ equal to just twice the frequency of the dynamo, or in other words, there may be the same number of fundamental oscillations as would be produced without the discharge gap, and there might even not be any quicker superimposed vibration; yet the differences of potential at the various points of the circuit, the impedance and other phenomena, dependent upon the rate of change, will bear no similarity in the two cases. Thus, when working with currents discharging disruptively, the element chiefly to be considered is not the frequency, as a student might be apt to believe, but the rate of change per unit of time. With low frequencies in a certain measure the same effects may be obtained as with high frequencies, provided the rate of change is sufficiently great. So if a low frequency current is raised to a potential of, say, 75,000 volts, and the high tension current passed through a series of high resistance lamp filaments, the importance of the rarefied gas surrounding the filament is clearly noted, as will be seen later; or, if a low frequency current of several thousand amperes is passed through a metal bar, striking phenomena of impedance are observed, just as with currents of high frequencies. But it is, of course, evident that with low frequency currents it is impossible to obtain such rates of change per unit of time as with high frequencies, hence the effects produced by the latter are much more prominent. It is deemed advisable to make the preceding remarks, inasmuch as many more recently described effects have been unwittingly identified with high frequencies. Frequency alone in reality does not mean anything, except when an undisturbed harmonic oscillation is considered.

In the branch IIIb a similar disposition to that in Ib is illustrated, with the difference that the currents discharging through the gap $d\,d$ are used to induce currents in the secondary s of a transformer T. In such case the secondary should be provided with an adjustable condenser for the purpose of tuning it to the primary.

IIb illustrates a plan of alternate current high frequency conversion which is most frequently used and which is found to be most convenient. This plan has been dwelt upon in detail on previous occasions and need not be described here.

Some of these results were obtained by the use of a high frequency alternator. A description of such machines will be found in my original paper before the American Institute of Electrical Engineers, and in periodicals of that period, notably in *The Electrical Engineer* of March 18, 1891.

I will now proceed with the experiments.

ON PHENOMENA PRODUCED BY ELECTROSTATIC FORCE

The first class of effects I intend to show you are effects produced by electrostatic force. It is the force which governs the motion of the atoms, which causes them to collide and develop the life-sustaining energy of heat and light, and which causes them to aggregate in an infinite variety of ways, according to Nature fanciful designs, and to form all these wondrous structures we perceive around us; it is, in fact, if our present views be true, the most important force for us to consider in Nature. As the term *electrostatic* might imply a steady electric condition, it should be remarked, that in these experiments the force is not constant, but varies at a rate which may be considered moderate, about one million times a second, or thereabouts. This enables me to produce many effects which are not producible with an unvarying force.

When two conducting bodies are insulated and electrified, we say that an electrostatic force is acting between them. This force manifests itself in attractions, repulsions and stresses in the bodies and space or medium without. So great may be the strain exerted in the air, or whatever separates the two conducting bodies, that it may break down, and we observe sparks or bundles of light or streamers, as they are called. These streamers form abundantly when the force through the air is rapidly varying. I will illustrate this action of electrostatic force in a novel experiment in which I will employ the induction coil before referred to. The coil is contained in a trough filled with oil, and placed under the table. The two ends of the secondary wire pass through the two thick columns of hard rubber which protrude to some height above the table. It is necessary to insulate the ends or terminals of the secondary heavily with hard rubber, because even dry wood is by far too poor an insulator for these currents of enormous potential differences. On one of the terminals of the coil, I have placed a large sphere of sheet brass, which is connected to a larger insulated brass plate, in order to enable me to perform the experiments under conditions, which, as you will see, are more suitable for this experiment. I now set the coil to work and approach the free terminal with a metallic object held in my hand, this simply to avoid burns. As I approach the metallic object to a distance of eight or ten inches, a torrent of furious sparks breaks forth from the end of the secondary wire, which passes through the rubber column. The sparks cease when the metal in my hand touches the wire. My arm is now traversed by a powerful electric current, vibrating at about the rate of one million times a second. All around me the electrostatic force makes itself felt, and the air molecules and particles of dust flying about are acted upon and are hammering violently against my body. So great is this agitation of the particles, that when the lights are turned out you may see streams of feeble light appear on some parts of my body. When such a streamer breaks out on any part of the body, it produces a sensation like the pricking of a needle. Were the potentials sufficiently high and the frequency of the vibration rather low, the skin would probably be ruptured under the tremendous strain, and the blood would rush out with great force in the form of fine spray or jet so thin as to be invisible, just as oil will when placed on the positive terminal of a Holtz machine. The breaking through of the skin though it may seem impossible at first, would perhaps occur, by reason of the tissues under the skin being incomparably better conducting. This, at least, appears plausible, judging from some observations.

I can make these streams of light visible to all, by touching with the metallic object one of the terminals as before, and approaching my free hand to the brass sphere, which is connected to the second terminal of the coil. As the hand is approached, the air between it and the sphere, or in the immediate neighborhood, is more violently agitated, and you see streams of light now break forth from my finger tips and from the whole hand (Fig. 169). Were I to approach the hand closer, powerful sparks would jump from the brass sphere to my hand, which might be injurious. The streamers offer no particular inconvenience, except that in the ends of the finger tips a burning sensation is felt. They should not be confounded with those produced by an influence machine, because in many respects they behave differently. I have attached the brass sphere and plate to one of the terminals in order to prevent the formation of visible streamers on that terminal, also in order to prevent sparks from jumping at a considerable distance. Besides, the attachment is favorable for the working of the coil.

Fig. 169.

The streams of light which you have observed issuing from my hand are due to a potential of about 200,000 volts, alternating in rather irregular intervals, sometimes like a million times a second. A vibration of the same amplitude, but four times as fast, to maintain which over 3,000,000 volts would be required, would be more than sufficient to envelop my body in a complete sheet of flame. But this flame would not burn me up; quite contrarily, the probability is that I would not be injured in the least. Yet a hundredth part of that energy, otherwise directed, would be amply sufficient to kill a person.

The amount of energy which may thus be passed into the body of a person depends on the frequency and potential of the currents, and by making both of these very great, a vast amount of energy may be passed into the body without causing any discomfort, except perhaps, in the arm, which is traversed by a true conduction current. The reason why no pain in the body is felt, and no injurious effect noted, is that everywhere, if a current be imagined to flow through the body, the direction of its flow would be at right angles to the surface; hence the body of the experimenter offers an enormous section to the current, and the density is very small, with the exception of the arm, perhaps, where the density may be considerable. But if only a small fraction of that energy would be applied in such a way that a current would traverse the body in the same manner as a low frequency current, a shock would be received which might be fatal. A direct or low frequency alternating current is fatal, I think, principally because its distribution through the body is not uniform, as it must divide itself in minute streamlets of great density, whereby some organs are vitally injured. That such a process occurs I have not the least doubt, though no evidence might apparently exist, or be found upon examination. The surest to injure and destroy life, is a continuous current, but the most painful is an alternating current of very low frequency. The expression of these views, which are the result of long continued experiment and observation, both with steady and varying currents, is elicited by the interest which is at present taken in this subject, and by the manifestly erroneous ideas which are daily propounded in journals on this subject.

I may illustrate an effect of the electrostatic force by another striking experiment, but before, I must call your attention to one or two facts. I have said before, that when the medium between two oppositely electrified bodies is strained beyond a certain limit it gives way and, stated in popular language, the opposite electric charges unite and neutralize each other. This breaking down of the medium occurs principally when the force acting between the bodies is steady, or varies at a moderate rate. Were the variation sufficiently rapid, such a destructive break would not occur, no matter how great the force, for all the energy would be spent in radiation, convection and mechanical and chemical action. Thus the *spark* length, or greatest distance which a *spark* will jump between the electrified bodies is the smaller, the greater the variation or time rate of change. But this rule may be taken to be true only in a general way, when comparing rates which are widely different.

I will show you by an experiment the difference in the effect produced by a rapidly varying and a steady or moderately varying force. I have here two large circular brass plates p p (Fig. 170a and Fig. 170b), supported on movable insulating stands on the table, connected to the ends of the secondary of a coil similar to the one used before. I place the plates ten or twelve inches apart and set the coil to work. You see the whole space between the plates, nearly two cubic feet, filled with uniform light, Fig. 170a. This light is due to the streamers you have seen in the first experiment, which are now much more intense. I have already pointed out the importance of these streamers in commercial apparatus and their still greater importance in some purely scientific investigations. Often they are too weak to be visible, but they always exist, consuming energy and modifying the action of the apparatus. When intense, as they are at present, they produce ozone in great quantity, and also, as Professor Crookes has pointed out, nitrous acid. So quick is the chemical action that if a coil, such as this one, is worked for a very long time it will make the atmosphere of a small room unbearable, for the eyes and throat are attacked. But when moderately produced, the streamers refresh the atmosphere wonderfully, like a thunder-storm, and exercises unquestionably a beneficial effect.

FIG. 170a. FIG. 170b.

In this experiment the force acting between the plates changes in intensity and direction at a very rapid rate. I will now make the rate of change per unit time much smaller. This I effect by rendering the discharges through the primary of the induction coil less frequent, and also by diminishing the rapidity of the vibration in the secondary. The former result is conveniently secured by lowering the E. M. F. over the air gap in the primary circuit, the latter by approaching the two brass plates to a distance of about three or four inches. When the coil is set to work, you see no streamers or light between the plates, yet the medium between them is under a tremendous strain. I still further augment the strain by raising the E. M. F. in the primary circuit, and soon you see the air give way and the hall is illuminated by a shower of brilliant and noisy sparks, Fig. 170b. These sparks could be produced also with unvarying force; they have been for many years a familiar phenomenon, though they were usually obtained from an entirely different apparatus. In describing these two phenomena so radically different in appearance, I have advisedly spoken of a "force" acting between the plates. It would be in accordance with accepted views to say, that there was an "alternating E. M. F." acting between the plates. This term is quite proper and applicable in all cases where there is evidence of at least a possibility of an essential inter-dependence of the electric state of the plates, or electric action in their neighborhood. But if the plates were removed to an infinite distance, or if at a finite distance, there is no probability or necessity whatever for such dependence. I prefer to use the term "electrostatic force," and to say that such a force is acting around each plate or electrified insulated body in general. There is an inconvenience in using this expression as the term incidentally means a steady electric condition; but a proper nomenclature will eventually settle this difficulty.

Fig. 171.

Fig. 172a.

Fig. 172b.

I now return to the experiment to which I have already alluded, and with which I desire to illustrate a striking effect produced by a rapidly varying electrostatic force. I attach to the end of the wire, l (Fig. 171), which is in connection with one of the terminals of the secondary of the induction coil, an exhausted bulb b. This bulb contains a thin carbon filament f, which is fastened to a platinum wire w, sealed in the glass and leading outside of the bulb, where it connects to the wire l. The bulb may be exhausted to any degree attainable with ordinary apparatus. Just a moment before, you have witnessed the breaking down of the air between the charged brass plates. You know that a plate of glass, or any other insulating material, would break down in like manner. Had I therefore a metallic coating attached to the outside of the bulb, or placed near the same, and were this coating connected to the other terminal of the coil, you would be prepared to see the glass give way if the strain were sufficiently increased. Even were the coating not connected to the other terminal, but to an insulated plate, still, if you have followed recent developments, you would naturally expect a rupture of the glass.

But it will certainly surprise you to note that under the action of the varying electrostatic force, the glass gives way when all other bodies are removed from the bulb. In fact, all the surrounding bodies we perceive might be removed to an infinite distance without affecting the result in the slightest. When the coil is set to work, the glass is invariably broken through at the seal, or other narrow channel, and the vacuum is quickly impaired. Such a damaging break would not occur with a steady force, even if the same were many times greater. The break is due to the agitation of the molecules of the gas within the bulb, and outside of the same. This agitation, which is generally most violent in the narrow pointed channel near the seal, causes a heating and rupture of the glass. This rupture, would, however, not occur, not even with a varying force, if the medium filling the inside of the bulb, and that surrounding it, were perfectly homogeneous. The break occurs much quicker if the top of the bulb is drawn out into a fine fibre. In bulbs used with these coils such narrow, pointed channels must therefore be avoided.

When a conducting body is immersed in air, or similar insulating medium, consisting of, or containing, small freely movable particles capable of being electrified, and when the electrification of the body is made to undergo a very rapid change—which is equivalent to saying that the electrostatic force acting around the body is varying in intensity— the small particles are attracted and repelled, and their violent impacts against the body may cause a mechanical motion of the latter. Phenomena of this kind are noteworthy, inasmuch as they have not been observed before with apparatus such as has been commonly in use. If a very light conducting sphere be suspended on an exceedingly fine wire, and charged to a steady potential, however high, the sphere will remain at rest. Even if the potential would be rapidly varying, provided that the small particles of matter, molecules or atoms, are evenly distributed, no motion of the sphere should result. But if one side of the conducting sphere is covered with a thick insulating layer, the impacts of the particles will cause the sphere to move about, generally in irregular curves, Fig. 172a. In like manner, as I have shown on a previous occasion, a fan of sheet metal, Fig. 172b, covered partially with insulating material as indicated, and placed upon the terminal of the coil so as to turn freely on it, is spun around.

FIG. 173.   FIG. 174.

All these phenomena you have witnessed and others which will be shown later, are due to the presence of a medium like air, and would not occur in a continuous medium. The action of the air may be illustrated still better by the following experiment. I take a glass tube t, Fig. 173, of about an inch in diameter, which has a platinum wire w sealed in the lower end, and to which is attached a thin lamp filament f. I connect the wire with the terminal of the coil and set the coil to work. The platinum wire is now electrified positively and negatively in rapid succession and the wire and air inside of the tube is rapidly heated by the impacts of the particles, which may be so violent as to render the filament incandescent. But if I pour oil in the tube, just as soon as the wire is covered with the oil, all action apparently ceases and there is no marked evidence of heating. The reason of this is that the oil is a practically continuous medium. The displacements in such a continuous medium are, with these frequencies, to all appearance incomparably smaller than in air, hence the work performed in such a medium is insignificant. But oil would behave very differently with frequencies many times as great, for even though the displacements be small, if the frequency were much greater, considerable work might be performed in the oil.

The electrostatic attractions and repulsions between bodies of measurable dimensions are, of all the manifestations of this force, the first so-called *electrical* phenomena noted. But though they have been known to us for many centuries, the precise nature of the mechanism concerned in these actions is still unknown to us, and has not been even quite satisfactorily explained. What kind of mechanism must that be? We cannot help wondering when we observe two magnets attracting and repelling each other with a force of hundreds of pounds with apparently nothing between them. We have in our commercial dynamos magnets capable of sustaining in mid-air tons of weight. But what are even these forces acting between magnets when compared with the tremendous attractions and repulsions produced by electrostatic force, to which there is apparently no limit as to intensity. In lightning discharges bodies are often charged to so high a potential that they are thrown away with inconceivable force and torn asunder or shattered into fragments. Still even such effects cannot compare with the attractions and repulsions which exist between charged molecules or atoms, and which are sufficient to project them with speeds of many kilometers a second, so that under their violent impact bodies are rendered highly incandescent and are volatilized. It is of special interest for the thinker who inquires into the nature of these forces to note that whereas the actions between individual molecules or atoms occur seemingly under any conditions, the attractions and repulsions of bodies of measurable dimensions imply a medium possessing insulating properties. So, if air, either by being rarefied or heated, is rendered more or less conducting, these actions between two electrified bodies practically cease, while the actions between the individual atoms continue to manifest themselves.

An experiment may serve as an illustration and as a means of bringing out other features of interest. Some time ago I showed that a lamp filament or wire mounted in a bulb and connected to one of the terminals of a high tension secondary coil is set spinning, the top of the filament generally describing a circle. This vibration was very energetic when the air in the bulb was at ordinary pressure and became less energetic when the air in the bulb was strongly compressed. It ceased altogether when the air was exhausted so as to become comparatively good conducting. I found at that time that no vibration took place when the bulb was very highly exhausted. But I conjectured that the vibration which I ascribed to the electrostatic action between the walls of the bulb and the filament should take place also in a highly exhausted bulb. To test this under conditions which were more favorable, a bulb like the one in Fig. 174, was constructed. It comprised a globe b, in the neck of which was sealed a platinum wire w carrying a thin lamp filament f. In the lower part of the globe a tube t was sealed so as to surround the filament. The exhaustion was carried as far as it was practicable with the apparatus employed.

This bulb verified my expectation, for the filament was set spinning when the current was turned on, and became incandescent. It also showed another interesting feature, bearing upon the preceding remarks, namely, when the filament had been kept incandescent some time, the narrow tube and the space inside were brought to an elevated temperature, and as the gas in the tube then became conducting, the electrostatic attraction between the glass and the filament became very weak or ceased, and the filament came to rest. When it came to rest it would glow far more intensely. This was probably due to its assuming the position in the center of the tube where the molecular bombardment was most intense, and also partly to the fact that the individual impacts were more violent and that no part of the supplied energy was converted into mechanical movement. Since, in accordance with accepted views, in this experiment the incandescence must be attributed to the impacts of the particles, molecules or atoms in the heated space, these particles must therefore, in order to explain such action, be assumed to behave as independent carriers of electric charges immersed in an insulating medium; yet there is no attractive force between the glass tube and the filament because the space in the tube is, as a whole, conducting.

It is of some interest to observe in this connection that whereas the attraction between two electrified bodies may cease owing to the impairing of the insulating power of the medium in which they are immersed, the repulsion between the bodies may still be observed. This may be explained in a plausible way. When the bodies are placed at some distance in a poorly conducting medium, such as slightly warmed or rarefied air, and are suddenly electrified, opposite electric charges being imparted to them, these charges equalize more or less by leakage through the air. But if the bodies are similarly electrified, there is less opportunity afforded for such dissipation, hence the repulsion observed in such case is greater than the attraction. Repulsive actions in a gaseous medium are however, as Prof. Crookes has shown, enhanced by molecular bombardment.

On Current or Dynamic Electricity Phenomena

So far, I have considered principally effects produced by a varying electrostatic force in an insulating medium, such as air. When such a force is acting upon a conducting body of measurable dimensions, it causes within the same, or on its surface, displacements of the electricity and gives rise to electric currents, and these produce another kind of phenomena, some of which I shall presently endeavor to illustrate. In presenting this second class of electrical effects, I will avail myself principally of such as are producible without any return circuit, hoping to interest you the more by presenting these phenomena in a more or less novel aspect.

It has been a long time customary, owing to the limited experience with vibratory currents, to consider an electric current as something circulating in a closed conducting path. It was astonishing at first to realize that a current may flow through the conducting path even if the latter be interrupted, and it was still more surprising to learn, that sometimes it may be even easier to make a current flow under such conditions than through a closed path. But that old idea is gradually disappearing, even among practical men, and will soon be entirely forgotten.

If I connect an insulated metal plate P, Fig. 175, to one of the terminals T of the induction coil by means of a wire, though this plate be very well insulated, a current passes through the wire when the coil is set to work. First I wish to give you evidence that there *is* a current passing through the connecting wire. An obvious way of demonstrating this is to insert between the terminal of the coil and the insulated plate a very thin platinum or german silver wire *w* and bring the latter to incandescence or fusion by the current. This requires a rather large plate or else current impulses of very

FIG. 175.

high potential and frequency. Another way is to take a coil c, Fig. 175, containing many turns of thin insulated wire and to insert the same in the path of the current to the plate. When I connect one of the ends of the coil to the wire leading to another insulated plate P_1, and its other end to the terminal T_1 of the induction coil, and set the latter to work, a current passes through the inserted coil c and the existence of the current may be made manifest in various ways. For instance, I insert an iron core i within the coil. The current being one of very high frequency, will, if it be of some strength, soon bring the iron core to a noticeably higher temperature, as the hysteresis and current losses are great with such high frequencies. One might take a core of some size, laminated or not, it would matter little; but ordinary iron wire $1/16$th or $1/8$th of an inch thick is suitable for the purpose. While the induction coil is working, a current traverses the inserted coil and only a few moments are sufficient to bring the iron wire i to an elevated temperature sufficient to soften the sealing-wax s, and cause a paper washer p fastened by it to the iron wire to fall off. But with the apparatus

FIG. 176.

such as I have here, other, much more interesting, demonstrations of this kind can be made. I have a secondary s, Fig. 176, of coarse wire, wound upon a coil similar to the first. In the preceding experiment the current through the coil c, Fig. 175, was very small, but there being many turns a strong heating effect was, nevertheless, produced in the iron wire. Had I passed that current through a conductor in order to show the heating of the latter, the current might have been too small to produce the effect desired. But with this coil provided with a secondary winding, I can now transform the feeble current of high tension which passes through the primary P into a strong secondary current of low tension, and this current will quite certainly do what I expect. In a small glass tube (*t*, Fig. 176), I have enclosed a coiled platinum wire, *w*, this merely in order to protect the wire. On each end of the glass tube is sealed a terminal of stout wire to which one of the ends of the platinum wire *w*, is connected. I join the terminals of the secondary coil to these terminals and insert the primary *p*, between the insulated plate P_1, and the terminal T_1, of the induction coil as before. The latter being set to work, instantly the platinum wire *w* is rendered incandescent and can be fused, even if it be very thick.

FIG. 177.

Instead of the platinum wire I now take an ordinary 50-volt 16 c. p. lamp. When I set the induction coil in operation the lamp filament is brought to high incandescence. It is, however, not necessary to use the insulated plate, for the lamp (*l*, Fig. 177) is rendered incandescent even if the plate P_1 be disconnected. The secondary may also be connected to the primary as indicated by the dotted line in Fig. 177, to do away more or less with the electrostatic induction or to modify the action otherwise.

I may here call attention to a number of interesting observations with the lamp. First, I disconnect one of the terminals of the lamp from the secondary s. When the induction coil plays, a glow is noted which fills the whole bulb. This glow is due to electrostatic induction. It increases when the bulb is grasped with the hand, and the capacity of the experimenter's body thus added to the secondary circuit. The secondary, in effect, is equivalent to a metallic coating, which would be placed near the primary. If the secondary, or its equivalent, the coating, were placed symmetrically to the primary, the electrostatic induction would be nil under ordinary conditions, that is, when a primary return circuit is used, as both halves would neutralize each other. The secondary *is* in fact placed symmetrically to the primary, but the action of both halves of the latter, when only one of its ends is connected to the induction coil, is not exactly equal; hence electrostatic induction takes place, and hence the glow in the bulb. I can nearly equalize the action of both halves of the primary by connecting the other, free end of the same to the insulated plate, as in the preceding experiment. When the plate is connected, the glow disappears. With a smaller plate it would not entirely disappear and then it would contribute to the brightness of the filament when the secondary is closed, by warming the air in the bulb.

FIG. 178a.

FIG. 178b.

To demonstrate another interesting feature, I have adjusted the coils used in a certain way. I first connect both the terminals of the lamp to the secondary, one end of the primary being connected to the terminal T_1 of the induction coil and the other to the insulated plate P_1 as before. When the current is turned on, the lamp glows brightly, as shown in Fig. 178b, in which C is a fine wire coil and S a coarse wire secondary wound upon it. If the insulated plate P_1 is disconnected, leaving one of the ends a of the primary insulated, the filament becomes dark or generally it diminishes in brightness (Fig. 178a). Connecting again the

FIG. 179a.

FIG. 179b.

plate P_1 and raising the frequency of the current, I make the filament quite dark or barely red (Fig. 179b). Once more I will disconnect the plate. One will of course infer that when the plate is disconnected, the current through the primary will be weakened, that therefore the E. M. F. will fall in the secondary S, and that the brightness of the lamp will diminish. This might be the case and the result can be secured by an easy adjustment of the coils; also by varying the frequency and potential of the currents. But it is perhaps of greater interest to note, that the lamp increases in brightness when the plate is disconnected (Fig. 179a). In this case all the energy the primary receives is now sunk into it, like the charge of a battery in an ocean cable, but most of that energy is recovered through the secondary and used to light the lamp. The current traversing the primary is strongest at the end b which is connected to the terminal T_1 of the induction coil, and diminishes in strength towards the remote end a. But the dynamic inductive effect exerted upon the secondary s is now greater than before, when the suspended plate was connected to the primary. These results might have been produced by a number of causes. For instance, the plate P_1 being connected, the reaction from the coil c may be such as to diminish the potential at the terminal T_1 of the induction coil, and therefore weaken the current through the primary of the coil c. Or the disconnecting of the plate may diminish the capacity effect with relation to the primary of the latter coil to such an extent that the current through it is diminished, though the potential at the terminal T_1 of the induction coil may be the same or even higher. Or the result might have been produced by the change of phase of the primary and secondary currents and consequent reaction. But the chief determining factor is the relation of the self-induction and capacity of coil c and plate P_1 and the frequency of the currents. The greater brightness of the filament in Fig. 179a, is, however, in part due to the heating of the rarefied gas in the lamp by electrostatic induction, which, as before remarked, is greater when the suspended plate is disconnected.

Still another feature of some interest I may here bring to your attention. When the insulated plate is disconnected and the secondary of the coil opened, by approaching a small object to the secondary, but very small sparks can be drawn from it, showing that the electrostatic induction is small in this case. But upon the secondary being closed upon itself or through the lamp, the filament glowing brightly, strong sparks are obtained from the secondary. The electrostatic induction is now much greater, because the closed secondary determines a greater flow of current through the primary and principally through that half of it which is connected to the induction coil. If now the bulb be grasped with the hand, the capacity of the secondary with reference to the primary is augmented by the experimenter's body and the luminosity of the filament is increased, the incandescence now being due partly to the flow of current through the filament and partly to the molecular bombardment of the rarefied gas in the bulb.

The preceding experiments will have prepared one for the next following results of interest, obtained in the course of these investigations. Since I can pass a current through an insulated wire merely by connecting one of its ends to the source of electrical energy, since I can induce by it another current, magnetize an iron core, and, in short, perform all operations as though a return circuit were used, clearly I can also drive a motor by the aid of only one wire. On a former occasion I have described a simple form of motor comprising a single exciting coil, an iron core and disc. Fig. 180 illustrates a modified way of operating such an alternate current motor by currents induced in a transformer connected to one lead, and several other arrangements of circuits for operating a certain class of alternating motors founded on the action of currents of differing phase. In view of the present state of the art it is thought sufficient to describe these arrangements in a few words only. The diagram, Fig. 180 II, shows a primary coil P, connected with one of its ends to the line L leading from a high tension transformer terminal T_1. In inductive relation to this primary P is a secondary S of coarse wire in the circuit of which is a coil c. The currents induced in the secondary energize the iron core i, which is preferably, but not necessarily, subdivided, and set the metal disc d in rotation. Such a motor M_2 as diagrammatically shown in Fig. 180 II, has been called a "magnetic lag motor," but this expression may be objected to by those who attribute the rotation of the disc to eddy currents circulating in minute paths when the core i is finally subdivided. In order to operate such a motor effectively on the plan indicated, the frequencies should not be too high, not more than four or five thousand, though the rotation is produced even with ten thousand per second, or more.

In Fig. 180 I, a motor M_1 having two energizing circuits, A and B, is diagrammatically indicated. The circuit A is connected to the line L and in series with it is a primary P, which may have its free end connected to an insulated plate P_1, such connection being indicated by the dotted lines. The other motor circuit B is connected to the secondary S which is in inductive relation to the primary P. When the transformer terminal T_1 is alternately electrified, currents traverse the open line L and also circuit

FIG. 180.

FIG. 181.

FIG. 182.

A and primary P. The currents through the latter induce secondary currents in the circuit s, which pass through the energizing coil B of the motor. The currents through the secondary s and those through the primary P differ in phase 90 degrees, or nearly so, and are capable of rotating an armature placed in inductive relation to the circuits A and B.

In Fig. 180 III, a similar motor M_3 with two energizing circuits A_1 and B_1 is illustrated. A primary P, connected with one of its ends to the line L has a secondary s, which is preferably wound for a tolerably high E. M. F., and to which the two energizing circuits of the motor are connected, one directly to the ends of the secondary and the other through a condenser c, by the action of which the currents traversing the circuit A_1 and B_1 are made to differ in phase.

In Fig. 180 IV, still another arrangement is shown. In this case two primaries P_1 and P_2 are connected to the line L, one through a condenser c of small capacity, and the other directly. The primaries are provided with secondaries s_1 and s_2 which are in series with the energizing circuits, A_2 and B_2 and a motor M_3, the condenser c again serving to produce the requisite difference in the phase of the currents traversing the motor circuits. As such phase motors with two or more circuits are now well known in the art, they have been here illustrated diagrammatically. No difficulty whatever is found in operating a motor in the manner indicated, or in similar ways; and although such experiments up to this day present only scientific interest, they may at a period not far distant, be carried out with practical objects in view.

It is thought useful to devote here a few remarks to the subject of operating devices of all kinds by means of only one leading wire. It is quite obvious, that when high-frequency currents are made use of, ground connections are—at least when the E. M. F. of the currents is great—better than a return wire. Such ground connections are objectionable with steady or low frequency currents on account of destructive chemical actions of the former and disturbing influences exerted by both on the neighboring circuits; but with high frequencies these actions practically do not exist. Still, even ground connections become superfluous when the E. M. F. is very high, for soon a condition is reached, when the current may be passed more economically through open, than through closed, conductors. Remote as might seem an industrial application of such single wire transmission of energy to one not experienced in such lines of experiment, it will not seem so to anyone who for some time has carried on investigations of such nature. Indeed I cannot see why such a plan should not be practicable. Nor should it be thought that for carrying out such a plan currents of very high frequency are expressly required, for just as soon as potentials of say 30,000 volts are used, the single wire transmission may be effected with low frequencies, and experiments have been made by me from which these inferences are made.

When the frequencies are very high it has been found in laboratory practice quite easy to regulate the effects in the manner shown in diagram Fig. 181. Here two primaries P and P_1 are shown, each connected with one of its ends to the line L and with the other end to the condenser plates c and c, respectively. Near these are placed other condenser plates c_1 and c_1, the former being connected to the line L and the latter to an insulated larger plate P_2. On the primaries are wound secondaries s and s_1, of coarse wire, connected to the devices *d* and *l* respectively. By varying the distances of the condenser plates c and c_1, and c and c_1 the currents through the secondaries s and s_1 are varied in intensity. The curious feature is the great sensitiveness, the slightest change in the distance of the plates producing considerable variations in the intensity or strength of the currents. The sensitiveness may be rendered extreme by making the frequency such, that the primary itself, without any plate attached to its free end, satisfies, in conjunction with the closed secondary, the condition of resonance. In such condition an extremely small change in the capacity of the free terminal produces great variations. For instance, I have been able to adjust the conditions so that the mere approach of a person to the coil produces a considerable change in the brightness of the lamps attached to the secondary. Such observations and experiments possess, of course, at present, chiefly scientific interest, but they may soon become of practical importance.

Very high frequencies are of course not practicable with motors on account of the necessity of employing iron cores. But one may use sudden discharges of low frequency and thus obtain certain advantages of high-frequency currents without rendering the iron core entirely incapable of following the changes and without entailing a very great expenditure of energy in the core. I have found it quite practicable to operate with such low frequency disruptive discharges of condensers, alternating-current motors. A certain class of such motors which I advanced a few years ago, which contain closed secondary circuits, will rotate quite vigorously when the discharges are directed through the exciting coils. One reason that such a motor operates so well with these discharges is that the difference of phase between the primary and secondary currents is 90 degrees, which is generally not the case with harmonically rising and falling currents of low frequency. It might not be without interest to show an experiment with a simple motor of this kind, inasmuch as it is commonly thought that disruptive discharges are unsuitable for such purposes. The motor is illustrated in Fig. 182. It comprises a rather large iron core i with slots on the top into which are embedded thick copper washers $c\ c$. In proximity to the core is a freely-movable metal disc D. The core is provided with a primary exciting coil c_1 the ends a and b of which are connected to the terminals of the secondary s of an ordinary transformer, the primary P of the latter being connected to an alternating distribution circuit or generator G of low or moderate frequency. The terminals of the secondary s are attached to a condenser c which discharges through an air gap $d\ d$ which may be placed in series or shunt to the coil c_1. When the conditions are properly chosen the disc D rotates with considerable effort and the iron core i does not get very perceptibly hot. With currents from a high-frequency alternator, on the contrary, the core gets rapidly hot and the disc rotates with a much smaller effort. To perform the experiment properly it should be first ascertained that the disc D is not set in rotation when the discharge is not occurring at $d\ d$. It is preferable to use a large iron core and a condenser of large capacity so as to bring the superimposed quicker oscillation to a very low pitch or to do away with it entirely. By observing certain elementary rules I have also found it practicable to operate ordinary series or shunt direct-current motors with such disruptive discharges, and this can be done with or without a return wire.

IMPEDANCE PHENOMENA

Among the various current phenomena observed, perhaps the most interesting are those of impedance presented by conductors to currents varying at a rapid rate. In my first paper before the American Institute of Electrical Engineers, I have described a few striking observations of this kind. Thus I showed that when such currents or sudden discharges are passed through a thick metal bar there may be points on the bar only a few inches apart, which have a sufficient potential difference between them to maintain at bright incandescence an ordinary filament lamp. I have also described the curious behavior of rarefied gas surrounding a conductor, due to such sudden rushes of current. These phenomena have since been more carefully studied and one or two novel experiments of this kind are deemed of sufficient interest to be described here.

Fig. 183a, 183b, and 183c.

Referring to Fig. 183a, B and B_1 are very stout copper bars connected at their lower ends to plates C and C_1, respectively, of a condenser, the opposite plates of the latter being connected to the terminals of the secondary s of a high-tension transformer, the primary p of which is supplied with alternating currents from an ordinary low-frequency dynamo g or distribution circuit. The condenser discharges through an adjustable gap $d\,d$ as usual. By establishing a rapid vibration it was found quite easy to perform the following curious experiment. The bars b and b_1 were joined at the top by a low-voltage lamp l_3; a little lower was placed by means of clamps $c\,c$, a 50-volt lamp l_2; and still lower another 100-volt lamp l_1; and finally, at a certain distance below the latter lamp, an exhausted tube t. By carefully determining the positions of these devices it was found practicable to maintain them all at their proper illuminating power. Yet they were all connected in multiple arc to the two stout copper bars and required widely different pressures. This experiment requires of course some time for adjustment but is quite easily performed.

In Figs. 183b and 183c, two other experiments are illustrated which, unlike the previous experiment, do not require very careful adjustments. In Fig. 183b, two lamps, l_1 and l_2, the former a 100-volt and the latter a 50-volt are placed in certain positions as indicated, the 100-volt lamp being below the 50-volt lamp. When the arc is playing at $d\,d$ and the sudden discharges are passed through the bars b b_1, the 50-volt lamp will, as a rule, burn brightly, or at least this result is easily secured, while the 100-volt lamp will burn very low or remain quite dark, Fig. 183b. Now the bars b b_1 may be joined at the top by a thick cross bar b_2 and it is quite easy to maintain the 100-volt lamp at full candle-power while the 50-volt lamp remains dark, Fig. 183c. These results, as I have pointed out previously, should not be considered to be due exactly to frequency but rather to the time rate of change which may be great, even with low frequencies. A great many other results of the same kind, equally interesting, especially to those who are only used to manipulate steady currents, may be obtained and they afford precious clues in investigating the nature of electric currents.

In the preceding experiments I have already had occasion to show some light phenomena and it would now be proper to study these in particular; but to make this investigation more complete I think it necessary to make first a few remarks on the subject of electrical resonance which has to be always observed in carrying out these experiments.

On Electrical Resonance

The effects of resonance are being more and more noted by engineers and are becoming of great importance in the practical operation of apparatus of all kinds with alternating currents. A few general remarks may therefore be made concerning these effects. It is clear, that if we succeed in employing the effects of resonance practically in the operation of electric devices the return wire will, as a matter of course, become unnecessary, for the electric vibration may be conveyed with one wire just as well as, and sometimes even better than, with two. The question first to answer is, then, whether pure resonance effects are producible. Theory and experiment both show that such is impossible in Nature, for as the oscillation becomes more and more vigorous, the losses in the vibrating bodies and environing media rapidly increase and necessarily check the vibration which otherwise would go on increasing forever. It is a fortunate circumstance that pure resonance is not producible, for if it were there is no telling what dangers might not lie in wait for the innocent experimenter. But to a certain degree resonance is producible, the magnitude of the effects being limited by the imperfect conductivity and imperfect elasticity of the media or, generally stated, by frictional losses. The smaller these losses, the more striking are the effects. The same is the case in mechanical vibration. A stout steel bar may be set in vibration by drops of water falling upon it at proper intervals; and with glass, which is more perfectly elastic, the resonance effect is still more remarkable, for a goblet may be burst by singing into it a note of the proper pitch. The electrical resonance is the more perfectly attained, the smaller the resistance or the impedance of the conducting path and the more perfect the dielectric. In a Leyden jar discharging through a short stranded cable of thin wires these requirements are probably best fulfilled, and the resonance effects are therefore very prominent. Such is not the case with dynamo machines, transformers and their circuits, or with commercial apparatus in general in which the presence of iron cores complicates the action or renders it impossible. In regard to Leyden jars with which resonance effects are frequently demonstrated, I would say that the effects observed are often *attributed* but are seldom *due* to true resonance, for an error is quite easily made in this respect.

This may be undoubtedly demonstrated by the following experiment. Take, for instance, two large insulated metallic plates or spheres which I shall designate A and B; place them at a certain small distance apart and charge them from a frictional or influence machine to a potential so high that just a slight increase of the difference of potential between them will cause the small air or insulating space to break down. This is easily reached by making a few preliminary trials. If now another plate—fastened on an insulating handle and connected by a wire to one of the terminals of a high tension secondary of an induction coil, which is maintained in action by an alternator (preferably high frequency)—is approached to one of the charged bodies A or B, so as to be nearer to either one of them, the discharge will invariably occur between them; at least it will, if the potential of the coil in connection with the plate is sufficiently high. But the explanation of this will soon be found in the fact that the approached plate acts inductively upon the bodies A and B and causes a spark to pass between them. When this spark occurs, the charges which were previously imparted to these bodies from the influence machine, must needs be lost, since the bodies are brought in electrical connection through the arc formed. Now this arc is formed whether there be resonance or not. But even if the spark would not be produced, still there is an alternating E. M. F. set up between the bodies when the plate is brought near one of them; therefore the approach of the plate, if it *does* not always actually, will, at any rate, *tend* to break down the air space by inductive action. Instead of the spheres or plates A and B we may take the coatings of a Leyden jar with the same result, and in place of the machine—which is a high frequency alternator preferably, because it is more suitable for the experiment and also for the argument—we may take another Leyden jar or battery of jars. When such jars are discharging through a circuit of low resistance the same is traversed by currents of very high frequency. The plate may now be connected to one of the coatings of the second jar, and when it is brought near to the first jar just previously charged to a high potential from an influence machine, the result is the same as before, and the first jar will discharge through a small air space upon the second being caused to discharge. But both jars and their circuits need not be tuned any closer than a basso profundo is to the note produced by a mosquito, as small sparks will be produced through the air space, or at least the latter will be considerably more strained owing to the setting up of an alternating E. M. F. by induction, which takes place when one of the jars begins to discharge. Again another error of a similar nature is quite easily made. If the circuits of the two jars are run parallel and close together, and the experiment has been performed of discharging one by the other, and now a coil of wire be added to one of the circuits whereupon the experiment does not succeed, the conclusion that this is due to the fact that the circuits are now not tuned, would be far from being safe. For the two circuits act as condenser coatings and the addition of the coil to one of them is equivalent to bridging them, at the point where the coil is placed, by a small condenser, and the effect of the latter might be to prevent the spark from jumping through the discharge space by diminishing the alternating E. M. F. acting across the same. All these remarks, and many more which might be added but for fear of wandering too far from the subject, are made with the pardonable intention of cautioning the unsuspecting student, who might gain an entirely unwarranted opinion of his skill at seeing every experiment succeed; but they are in no way thrust upon the experienced as novel observations.

In order to make reliable observations of electric resonance effects it is very desirable, if not necessary, to employ an alternator giving currents which rise and fall harmonically, as in working with make and break currents the observations are not always trustworthy, since many phenomena, which depend on the rate of change, may be produced with widely different frequencies. Even when making such observations with an alternator one is apt to be mistaken. When a circuit is connected to an alternator there are an indefinite number of values for capacity and self-induction which, in conjunction, will satisfy the condition of resonance. So there are in mechanics an infinite number of tuning forks which will respond to a note of a certain pitch, or loaded springs which have a definite period of vibration. But the resonance will be most perfectly attained in that case in which the motion is effected with the greatest freedom. Now in mechanics, considering the vibration in the common medium—that is, air—it is of comparatively little importance whether one tuning fork be somewhat larger than another, because the losses in the air are not very considerable. One may, of course, enclose a tuning fork in an exhausted vessel and by thus reducing the air resistance to a minimum obtain better resonant action. Still the difference would not be very great. But it would make a great difference if the tuning fork were immersed in mercury. In the electrical vibration it is of enormous importance to arrange the conditions so that the vibration is effected with the greatest freedom. The magnitude of the resonance effect depends, under otherwise equal conditions, on the quantity of electricity set in motion or on the strength of the current driven through the circuit. But the circuit opposes the passage of the currents by reason of its impedance and therefore, to secure the best action it is necessary to reduce the impedance to a minimum. It is impossible to overcome it entirely, but merely in part, for the ohmic resistance cannot be overcome. But when the frequency of the impulses is very great, the flow of the current is practically determined by self-induction. Now self-induction can be overcome by combining it with capacity. If the relation between these is such, that at the frequency used they annul each other, that is, have such values as to satisfy the condition of resonance, and the greatest quantity of electricity is made to flow through the external circuit, then the best result is obtained. It is simpler and safer to join the condenser in series with the self-induction. It is clear that in such combinations there will be, for a given frequency, and considering only the fundamental vibration, values which will give the best result, with the condenser in shunt to the self-induction coil; of course more such values than with the condenser in series. But practical conditions determine the selection. In the latter case in performing the experiments one may take a small self-induction and a large capacity or a small capacity and a large self-induction, but the latter is preferable, because it is inconvenient to adjust a large capacity by small steps. By taking a coil with a very large self-induction the critical capacity is reduced to a very small value, and the capacity of the coil itself may be sufficient. It is easy, especially by observing certain artifices, to wind a coil through which the impedance will be reduced to the value of the ohmic resistance only; and for any coil there is, of course, a frequency at which the maximum current will be made to pass through the coil. The observation of the relation between self-induction, capacity and frequency is becoming important in the operation of alternate current apparatus, such as transformers or motors, because by a judicious determination of the elements the employment of an expensive condenser becomes unnecessary. Thus it is possible to pass through the coils of an alternating current motor under the normal working conditions the required current with a low E. M. F. and do away entirely with the false current, and the larger the motor, the easier such a plan becomes practicable; but it is necessary for this to employ currents of very high potential or high frequency.

FIG. 184.

In Fig. 184 I is shown a plan which has been followed in the study of the resonance effects by means of a high frequency alternator. c_1 is a coil of many turns, which is divided into small separate sections for the purpose of adjustment. The final adjustment was made sometimes with a few thin iron wires (though this is not always advisable) or with a closed secondary. The coil c_1 is connected with one of its ends to the line L from the alternator G and with the other end to one of the plates c of a condenser $c\,c_1$, the plate (c_1) of the latter being connected to a much larger plate P_1. In this manner both capacity and self-induction were adjusted to suit the dynamo frequency.

As regards the rise of potential through resonant action, of course, theoretically, it may amount to anything since it depends on self-induction and resistance and since these may have any value. But in practice one is limited in the selection of these values and besides these, there are other limiting causes. One may start with, say, 1,000 volts and raise the E. M. F. to 50 times that value, but one cannot start with 100,000 and raise it to ten times that value because of the losses in the media which are great, especially if the frequency is high. It should be possible to start with, for instance, two volts from a high or low frequency circuit of a dynamo and raise the E. M. F. to many hundred times that value. Thus coils of the proper dimensions might be connected each with only one of its ends to the mains from a machine of low E. M. F., and though the circuit of the machine would not be closed in the ordinary acceptance of the term, yet the machine might be burned out if a proper resonance effect would be obtained. I have not been able to produce, nor have I observed with currents from a dynamo machine, such great rises of potential. It is possible, if not probable, that with currents obtained from apparatus containing iron the disturbing influence of the latter is the cause that these theoretical possibilities cannot be realized. But if such is the case I attribute it solely to the hysteresis and Foucault current losses in the core. Generally it was necessary to transform upward, when the E. M. F. was very low, and usually an ordinary form of induction coil was employed, but sometimes the arrangement illustrated in Fig. 184 II, has been found to be convenient. In this case a coil c is made in a great many sections, a few of these being used as a primary. In this manner both primary and secondary are adjustable. One end of the coil is connected to the line L_1 from the alternator, and the other line L is connected to the intermediate point of the coil. Such a coil with adjustable primary and secondary will be found also convenient in experiments with the disruptive discharge. When true resonance is obtained the top of the wave must of course be on the free end of the coil as, for instance, at the terminal of the phosphorescence bulb B. This is easily recognized by observing the potential of a point on the wire w near to the coil.

In connection with resonance effects and the problem of transmission of energy over a single conductor which was previously considered, I would say a few words on a subject which constantly fills my thoughts and which concerns the welfare of all. I mean the transmission of intelligible signals or perhaps even power to any distance without the use of wires. I am becoming daily more convinced of the practicability of the scheme; and though I know full well that the great majority of scientific men will not believe that such results can be practically and immediately realized, yet I think that all consider the developments in recent years by a number of workers to have been such as to encourage thought and experiment in this direction. My conviction has grown so strong, that I no longer look upon this plan of energy or intelligence transmission as a mere theoretical possibility, but as a serious problem in electrical engineering, which must be carried out some day. The idea of transmitting intelligence without wires is the natural outcome of the most recent results of electrical investigations. Some enthusiasts have expressed their belief that telephony to any distance by induction through the air is possible. I cannot stretch my imagination so far, but I do firmly believe that it is practicable to disturb by means of powerful machines the electrostatic condition of the earth and thus transmit intelligible signals and perhaps power. In fact, what is there against the carrying out of such a scheme? We now know that electric vibration may be transmitted through a single conductor. Why then not try to avail ourselves of the earth for this purpose? We need not be frightened by the idea of distance. To the weary wanderer counting the mile-posts the earth may appear very large, but to that happiest of all men, the astronomer, who gazes at the heavens and by their standard judges the magnitude of our globe, it appears very small. And so I think it must seem to the electrician, for when he considers the speed with which an electric disturbance is propagated through the earth all his ideas of distance must completely vanish.

A point of great importance would be first to know what is the capacity of the earth? and what charge does it contain if electrified? Though we have no positive evidence of a charged body existing in space without other oppositely electrified bodies being near, there is a fair probability that the earth is such a body, for by whatever process it was separated from other bodies—and this is the accepted view of its origin—it must have retained a charge, as occurs in all processes of mechanical separation. If it be a charged body insulated in space its capacity should be extremely small, less than one-thousandth of a farad. But the upper strata of the air are conducting, and so, perhaps, is the medium in free space beyond the atmosphere, and these may contain an opposite charge. Then the capacity might be incomparably greater. In any case it is of the greatest importance to get an idea of what quantity of electricity the earth contains. It is difficult to say whether we shall ever acquire this necessary knowledge, but there is hope that we may, and that is, by means of electrical resonance. If ever we can ascertain at what period the earth's charge, when disturbed, oscillates with respect to an oppositely electrified system or known circuit, we shall know a fact possibly of the greatest importance to the welfare of the human race. I propose to seek for the period by means of an electrical oscillator, or a source of alternating electric currents. One of the terminals of the source would be connected to earth as, for instance, to the city water mains, the other to an insulated body of large surface. It is possible that the outer conducting air strata, or free space, contain an opposite charge and that, together with the earth, they form a condenser of very large capacity. In such case the period of vibration may be very low and an alternating dynamo machine might serve for the purpose of the experiment. I would then transform the current to a potential as high as it would be found possible and connect the ends of the high tension secondary to the ground and to the insulated body. By varying the frequency of the currents and carefully observing the potential of the insulated body and watching for the disturbance at various neighboring points of the earth's surface resonance might be detected. Should, as the majority of scientific men in all probability believe, the period be extremely small, then a dynamo machine would not do and a proper electrical oscillator would have to be produced and perhaps it might not be possible to obtain such rapid vibrations. But whether this be possible or not, and whether the earth contains a charge or not, and whatever may be its period of vibration, it certainly is possible—for of this we have daily evidence—to produce some electrical disturbance sufficiently powerful to be perceptible by suitable instruments at any point of the earth's surface.

FIG. 185.

Assume that a source of alternating current s be connected, as in Fig. 185, with one of its terminals to earth (conveniently to the water mains) and with the other to a body of large surface P. When the electric oscillation is set up there will be a movement of electricity in and out of P, and alternating currents will pass through the earth, converging to, or diverging from, the point c where the ground connection is made. In this manner neighboring points on the earth's surface within a certain radius will be disturbed. But the disturbance will diminish with the distance, and the distance at which the effect will still be perceptible will depend on the quantity of electricity set in motion. Since the body P is insulated, in order to displace a considerable quantity, the potential of the source must be excessive, since there would be limitations as to the surface of P. The conditions might be adjusted so that the generator or source s will set up the same electrical movement as though its circuit were closed. Thus it is certainly practicable to impress an electric vibration at least of a certain low period upon the earth by means of proper machinery. At what distance such a vibration might be made perceptible can only be conjectured. I have on another occasion considered the question how the earth might behave to electric disturbances. There is no doubt that, since in such an experiment the electrical density at the surface could be but extremely small considering the size of the earth, the air would not act as a very disturbing factor, and there would be not much energy lost through the action of the air, which would be the case if the density were great. Theoretically, then, it could not require a great amount of energy to produce a disturbance perceptible at great distance, or even all over the surface of the globe. Now, it is quite certain that at any point within a certain radius of the source s a properly adjusted self-induction and capacity device can be set in action by resonance. But not only can this be done, but another source s_1, Fig. 185, similar to s, or any number of such sources, can be set to work in synchronism with the latter, and the vibration thus intensified and spread over a large area, or a flow of electricity produced to or from the source s_1 if the same be of opposite phase to the source s. I think that beyond doubt it is possible to operate electrical devices in a city through the ground or pipe system by resonance from an electrical oscillator located at a central point. But the practical solution of this problem would be of incomparably smaller benefit to man than the realization of the scheme of transmitting intelligence, or perhaps power, to any distance through the earth or environing medium. If this is at all possible, distance does not mean anything. Proper apparatus must first be produced by means of which the problem can be attacked and I have devoted much thought to this subject. I am firmly convinced that it can be done and hope that we shall live to see it done.

On the Light Phenomena Produced by High-Frequency Currents of High Potential and General Remarks Relating to the Subject

Returning now to the light effects which it has been the chief object to investigate, it is thought proper to divide these effects into four classes: 1. Incandescence of a solid. 2. Phosphorescence. 3. Incandescence or phosphorescence of a rarefied gas; and 4. Luminosity produced in a gas at ordinary pressure. The first question is: How are these luminous effects produced? In order to answer this question as satisfactorily as I am able to do in the light of accepted views and with the experience acquired, and to add some interest to this demonstration, I shall dwell here upon a feature which I consider of great importance, inasmuch as it promises, besides, to throw a better light upon the nature of most of the phenomena produced by high-frequency electric currents. I have on other occasions pointed out the great importance of the presence of the rarefied gas, or atomic medium in general, around the conductor through which alternate currents of high frequency are passed, as regards the heating of the conductor by the currents. My experiments, described some time ago, have shown that, the higher the frequency and potential difference of the currents, the more important becomes the rarefied gas in which the conductor is immersed, as a factor of the heating. The potential difference, however, is, as I then pointed out, a more important element than the frequency. When both of these are sufficiently high, the heating may be almost entirely due to the presence of the rarefied gas. The experiments to follow will show the importance of the rarefied gas, or, generally, of gas at ordinary or other pressure as regards the incandescence or other luminous effects produced by currents of this kind.

I take two ordinary 50-volt 16 c. p. lamps which are in every respect alike, with the exception, that one has been opened at the top and the air has filled the bulb, while the other is at the ordinary degree of exhaustion of commercial lamps. When I attach the lamp which is exhausted to the terminal of the secondary of the coil, which I have already used, as in experiments illustrated in Fig. 179a for instance, and turn on the current, the filament, as you have before seen, comes to high incandescence. When I attach the second lamp, which is filled with air, instead of the former, the filament still glows, but much less brightly.

This experiment illustrates only in part the truth of the statements before made. The importance of the filament's being immersed in rarefied gas is plainly noticeable but not to such a degree as might be desirable. The reason is that the secondary of this coil is wound for low tension, having only 150 turns, and the potential difference at the terminals of the lamp is therefore small. Were I to take another coil with many more turns in the secondary, the effect would be increased, since it depends partially on the potential difference, as before remarked. But since the effect likewise depends on the frequency, it maybe properly stated that it depends on the time rate of the variation of the potential difference. The greater this variation, the more important becomes the gas as an element of heating. I can produce a much greater rate of variation in another way, which, besides, has the advantage of doing away with the objections, which might be made in the experiment just shown, even if both the lamps were connected in series or multiple arc to the coil, namely, that in consequence of the reactions existing between the primary and secondary coil the conclusions are rendered uncertain. This result I secure by charging, from an ordinary transformer which is fed from the alternating current supply station, a battery of condensers, and discharging the latter directly through a circuit of small self-induction, as before illustrated in Figs. 183a, 183b, and 183c.

FIG. 186a.   FIG. 186b.   FIG. 186c.

In Figs. 186a, 186b and 186c, the heavy copper bars B B, are connected to the opposite coatings of a battery of condensers, or generally in such way, that the high frequency or sudden discharges are made to traverse them. I connect first an ordinary 50-volt incandescent lamp to the bars by means of the clamps *c c.* The discharges being passed through the lamp, the filament is rendered incandescent, though the current through it is very small, and would not be nearly sufficient to produce a visible effect under the conditions of ordinary use of the lamp. Instead of this I now attach to the bars another lamp exactly like the first, but with the seal broken off, the bulb being therefore filled with air at ordinary pressure. When the discharges are directed through the filament, as before, it does not become incandescent. But the result might still be attributed to one of the many possible reactions. I therefore connect both the lamps in multiple arc as illustrated in Fig. 186a. Passing the discharges through both the lamps, again the filament in the exhausted lamp l glows very brightly while that in the non-exhausted lamp l_1 remains dark, as previously. But it should not be thought that the latter lamp is taking only a small fraction of the energy supplied to both the lamps; on the contrary, it may consume a considerable portion of the energy and it may become even hotter than the one which burns brightly. In this experiment the potential difference at the terminals of the lamps varies in sign theoretically three to four million times a second. The ends of the filaments are correspondingly electrified, and the gas in the bulbs is violently agitated and a large portion of the supplied energy is thus converted into heat. In the non-exhausted bulb, there being a few million times more gas molecules than in the exhausted one, the bombardment, which is most violent at the ends of the filament, in the neck of the bulb, consumes a large portion of the energy without producing any visible effect. The reason is that, there being many molecules, the bombardment is quantitatively considerable, but the individual impacts are not very violent, as the speeds of the molecules are comparatively small owing to the small free path. In the exhausted bulb, on the contrary, the speeds are very great, and the individual impacts are violent and therefore better adapted to produce a visible effect. Besides, the convection of heat is greater in the former bulb. In both the bulbs the current traversing the filaments is very small, incomparably smaller than that which they require on an ordinary low-frequency circuit. The potential difference, however, at the ends of the filaments is very great and might be possibly 20,000 volts or more, if the filaments were straight and their ends far apart. In the ordinary lamp a spark generally occurs between the ends of the filament or between the platinum wires outside, before such a difference of potential can be reached.

It might be objected that in the experiment before shown the lamps, being in multiple arc, the exhausted lamp might take a much larger current and that the effect observed might not be exactly attributable to the action of the gas in the bulbs. Such objections will lose much weight if I connect the lamps in series, with the same result. When this is done and the discharges are directed through the filaments, it is again noted that the filament in the non-exhausted bulb l_1 remains dark, while that in the exhausted one (l) glows even more intensely than under its normal conditions of working, Fig. 186b. According to general ideas the current through the filaments should now be the same, were it not modified by the presence of the gas around the filaments.

At this juncture I may point out another interesting feature, which illustrates the effect of the rate of change of potential of the currents. I will leave the two lamps connected in series to the bars B B$_1$, as in the previous experiment, Fig. 186b, but will presently reduce considerably the frequency of the currents, which was excessive in the experiment just before shown. This I may do by inserting a self-induction coil in the path of the discharges, or by augmenting the capacity of the condensers. When I now pass these low-frequency discharges through the lamps, the exhausted lamp l again is as bright as before, but it is noted also that the non-exhausted lamp l_1 glows, though not quite as intensely as the other. Reducing the current through the lamps, I may bring the filament in the latter lamp to redness, and, though the filament in the exhausted lamp l is bright, Fig. 186c, the degree of its incandescence is much smaller than in Fig. 186b, when the currents were of a much higher frequency.

In these experiments the gas acts in two opposite ways in determining the degree of the incandescence of the filaments, that is, by convection and bombardment. The higher the frequency and potential of the currents, the more important becomes the bombardment. The convection on the contrary should be the smaller, the higher the frequency. When the currents are steady there is practically no bombardment, and convection may therefore with such currents also considerably modify the degree of incandescence and produce results similar to those just before shown. Thus, if two lamps exactly alike, one exhausted and one not exhausted, are connected in multiple arc or series to a direct-current machine, the filament in the non-exhausted lamp will require a considerably greater current to be rendered incandescent. This result is entirely due to convection, and the effect is the more prominent the thinner the filament. Professor Ayrton and Mr. Kilgour some time ago published quantitative results concerning the thermal emissivity by radiation and convection in which the effect with thin wires was clearly shown. This effect may be strikingly illustrated by preparing a number of small, short, glass tubes, each containing through its axis the thinnest obtainable platinum wire. If these tubes be highly exhausted, a number of them may be connected in multiple arc to a direct-current machine and all of the wires may be kept at incandescence with a smaller current than that required to render incandescent a single one of the wires if the tube be not exhausted. Could the tubes be so highly exhausted that convection would be nil, then the relative amounts of heat given off by convection and radiation could be determined without the difficulties attending thermal quantitative measurements. If a source of electric impulses of high frequency and very high potential is employed, a still greater number of the tubes may be taken and the wires rendered incandescent by a current not capable of warming perceptibly a wire of the same size immersed in air at ordinary pressure, and conveying the energy to all of them.

I may here describe a result which is still more interesting, and to which I have been led by the observation of these phenomena. I noted that small differences in the density of the air produced a considerable difference in the degree of incandescence of the wires, and I thought that, since in a tube, through which a luminous discharge is passed, the gas is generally not of uniform density, a very thin wire contained in the tube might be rendered incandescent at certain places of smaller density of the gas, while it would remain dark at the places of greater density, where the convection would be greater and the bombardment less intense. Accordingly a tube t was prepared, as illustrated in Fig. 187, which contained through the middle a very fine platinum wire w. The tube was exhausted to a moderate degree and it was found that when it was attached to the terminal of a high-frequency coil the platinum wire w would indeed, become incandescent in patches, as illustrated in Fig. 187. Later a number of these tubes with one or more wires were prepared, each showing this result. The effect was best noted when the striated discharge occurred in the tube, but was also produced when the striae were not visible, showing that, even then, the gas in the tube was not of uniform density. The position of the striae was generally such, that the rarefactions corresponded to the places of incandescence or greater brightness on the wire w. But in a few instances it was noted,

Fig. 187.

Fig. 188.

that the bright spots on the wire were covered by the dense parts of the striated discharge as indicated by l in Fig. 187, though the effect was barely perceptible. This was explained in a plausible way by assuming that the convection was not widely different in the dense and rarefied places, and that the bombardment was greater on the dense places of the striated discharge. It is, in fact, often observed in bulbs, that under certain conditions a thin wire is brought to higher incandescence when the air is not too highly rarefied. This is the case when the potential of the coil is not high enough for the vacuum, but the result may be attributed to many different causes. In all cases this curious phenomenon of incandescence disappears when the tube, or rather the wire, acquires throughout a uniform temperature.

Disregarding now the modifying effect of convection there are then two distinct causes which determine the incandescence of a wire or filament with varying currents, that is, conduction current and bombardment. With steady currents we have to deal only with the former of these two causes, and the heating effect is a minimum, since the resistance is least to steady flow. When the current is a varying one the resistance is greater, and hence the heating effect is increased. Thus if the rate of change of the current is very great, the resistance may increase to such an extent that the filament is brought to incandescence with inappreciable currents, and we are able to take a short and thick block of carbon or other material and bring it to bright incandescence with a current incomparably smaller than that required to bring to the same degree of incandescence an ordinary thin lamp filament with a steady or low frequency current. This result is important, and illustrates how rapidly our views on these subjects are changing, and how quickly our field of knowledge is extending. In the art of incandescent lighting, to view this result in one aspect only, it has been commonly considered as an essential requirement for practical success, that the lamp filament should be thin and of high resistance. But now we know that the resistance of the filament to the steady flow does not mean anything; the filament might as well be short and thick; for if it be immersed in rarefied gas it will become incandescent by the passage of a small current. It all depends on the frequency and potential of the currents. We may conclude from this, that it would be of advantage, so far as the lamp is considered, to employ high frequencies for lighting, as they allow the use of short and thick filaments and smaller currents.

If a wire or filament be immersed in a homogeneous medium, all the heating is due to true conduction current, but if it be enclosed in an exhausted vessel the conditions are entirely different. Here the gas begins to act and the heating effect of the conduction current, as is shown in many experiments, may be very small compared with that of the bombardment. This is especially the case if the circuit is not closed and the potentials are of course very high. Suppose that a fine filament enclosed in an exhausted vessel be connected with one of its ends to the terminal of a high tension coil and with its other end to a large insulated plate. Though the circuit is not closed, the filament, as I have before shown, is brought to incandescence. If the frequency and potential be comparatively low, the filament is heated by the current passing *through it*. If the frequency and potential, and principally the latter, be increased, the insulated plate need be but very small, or may be done away with entirely; still the filament will become incandescent, practically all the heating being then due to the bombardment. A practical way of combining both the effects of conduction currents and bombardment is illustrated in Fig. 188, in which an ordinary lamp is shown provided with a very thin filament which has one of the ends of the latter connected to a shade serving the purpose of the insulated plate, and the other end to the terminal of a high tension source. It should not be thought that only rarefied gas is an important factor in the heating of a conductor by varying currents, but gas at ordinary pressure may become important, if the potential difference and frequency of the currents is excessive. On this subject I have already stated, that when a conductor is fused by a stroke of lightning, the current through it may be exceedingly small, not even sufficient to heat the conductor perceptibly, were the latter immersed in a homogeneous medium.

From the preceding it is clear that when a conductor of high resistance is connected to the terminals of a source of high frequency currents of high potential, there may occur considerable dissipation of energy, principally at the ends of the conductor, in consequence of the action of the gas surrounding the conductor. Owing to this, the current through a section of the conductor at a point midway between its ends may be much smaller than through a section near the ends. Furthermore, the current passes principally through the outer portions of the conductor, but this effect is to be distinguished from the skin effect as ordinarily interpreted, for the latter would, or should, occur also in a continuous incompressible medium. If a great many incandescent lamps are connected in series to a source of such currents, the lamps at the ends may burn brightly, whereas those in the middle may remain entirely dark. This is due principally to bombardment, as before stated. But even if the currents be steady, provided the difference of potential is very great, the lamps at the end will burn more brightly than those in the middle. In such case there is no rhythmical bombardment, and the result is produced entirely by leakage. This leakage or dissipation into space when the tension is high, is considerable when incandescent lamps are used, and still more considerable with arcs, for the latter act like flames. Generally, of course, the dissipation is much smaller with steady, than with varying, currents.

I have contrived an experiment which illustrates in an interesting manner the effect of lateral diffusion. If a very long tube is attached to the terminal of a high frequency coil, the luminosity is greatest near the terminal and falls off gradually towards the remote end. This is more marked if the tube is narrow.

FIG. 189.   FIG. 190.

A small tube about one-half inch in diameter and twelve inches long (Fig. 189), has one of its ends drawn out into a fine fibre f nearly three feet long. The tube is placed in a brass socket T which can be screwed on the terminal T_1 of the induction coil. The discharge passing through the tube first illuminates the bottom of the same, which is of comparatively large section; but through the long glass fibre the discharge cannot pass. But gradually the rarefied gas inside becomes warmed and more conducting and the discharge spreads into the glass fibre. This spreading is so slow, that it may take half a minute or more until the discharge has worked through up to the top of the glass fibre, then presenting the appearance of a strongly luminous thin thread. By adjusting the potential at the terminal the light may be made to travel upwards at any speed. Once, however, the glass fibre is heated, the discharge breaks through its entire length instantly. The interesting point to be noted is that, the higher the frequency of the currents, or in other words, the greater relatively the lateral dissipation, at a slower rate may the light be made to propagate through the fibre. This experiment is best performed with a highly exhausted and freshly made tube. When the tube has been used for some time the experiment often fails. It is possible that the gradual and slow impairment of the vacuum is the cause. This slow propagation of the discharge through a very narrow glass tube corresponds exactly to the propagation of heat through a bar warmed at one end. The quicker the heat is carried away laterally the longer time it will take for the heat to warm the remote end. When the current of a low frequency coil is passed through the fibre from end to end, then the lateral dissipation is small and the discharge instantly breaks through almost without exception.

After these experiments and observations which have shown the importance of the discontinuity or atomic structure of the medium and which will serve to explain, in a measure at least, the nature of the four kinds of light effects producible with these currents, I may now give you an illustration of these effects. For the sake of interest I may do this in a manner which to many of you might be novel. You have seen before that we may now convey the electric vibration to a body by means of a single wire or conductor of any kind. Since the human frame is conducting I may convey the vibration through my body.

First, as in some previous experiments, I connect my body with one of the terminals of a high-tension transformer and take in my hand an exhausted bulb which contains a small carbon button mounted upon a platinum wire leading to the outside of the bulb, and the button is rendered incandescent as soon as the transformer is set to work (Fig. 190). I may place a conducting shade on the bulb which serves to intensify the action, but is not necessary. Nor is it required that the button should be in conducting connection with the hand through a wire leading through the glass, for sufficient energy may be transmitted through the glass itself by inductive action to render the button incandescent.

FIG. 191.   FIG. 192.

Next I take a highly exhausted bulb containing a strongly phosphorescent body, above which is mounted a small plate of aluminum on a platinum wire leading to the outside, and the currents flowing through my body excite intense phosphorescence in the bulb (Fig. 191). Next again I take in my hand a simple exhausted tube, and in the same manner the gas inside the tube is rendered highly incandescent or phosphorescent (Fig. 192). Finally, I may take in my hand a wire, bare or covered with thick insulation, it is quite immaterial; the electrical vibration is so intense as to cover the wire with a luminous film (Fig. 193).

FIG. 193.   FIG. 194.   FIG. 195.

A few words must now be devoted to each of these phenomena. In the first place, I will consider the incandescence of a button or of a solid in general, and dwell upon some facts which apply equally to all these phenomena. It was pointed out before that when a thin conductor, such as a lamp filament, for instance, is connected with one of its ends to the terminal of a transformer of high tension the filament is brought to incandescence partly by a conduction current and partly by bombardment. The shorter and thicker the filament the more important becomes the latter, and finally, reducing the filament to a mere button, all the heating must practically be attributed to the bombardment. So in the experiment before shown, the button is rendered incandescent by the rhythmical impact of freely movable small bodies in the bulb. These bodies may be the molecules of the residual gas, particles of dust or lumps torn from the electrode; whatever they are, it is certain that the heating of the button is essentially connected with the pressure of such freely movable particles, or of atomic matter in general in the bulb. The heating is the more intense the greater the number of impacts per second and the greater the energy of each impact. Yet the button would be heated also if it were connected to a source of a steady potential. In such a case electricity would be carried away from the button by the freely movable carriers or particles flying about, and the quantity of electricity thus carried away might be sufficient to bring the button to incandescence by its passage through the latter. But the bombardment could not be of great importance in such case. For this reason it would require a comparatively very great supply of energy to the button to maintain it at incandescence with a steady potential. The higher the frequency of the electric impulses the more economically can the button be maintained at incandescence. One of the chief reasons why this is so, is, I believe, that with impulses of very high frequency there is less exchange of the freely movable carriers around the electrode and this means, that in the bulb the heated matter is better confined to the neighborhood of the button. If a double bulb, as illustrated in Fig. 194 be made, comprising a large globe B and a small one b, each containing as usual a filament f mounted on a platinum wire w and w_1, it is found, that if the filaments ff be exactly alike, it requires less energy to keep the filament in the globe b at a certain degree of incandescence, than that in the globe B. This is due to the confinement of the movable particles around the button. In this case it is also ascertained, that the filament in the small globe *b* is less deteriorated when maintained a certain length of time at incandescence. This is a necessary consequence of the fact that the gas in the small bulb becomes strongly heated and therefore a very good conductor, and less work is then performed on the button, since the bombardment becomes less intense as the conductivity of the gas increases. In this construction, of course, the small bulb becomes very hot and when it reaches an elevated temperature the convection and radiation on the outside increase. On another occasion I have shown bulbs in which this drawback was largely avoided. In these instances a very small bulb, containing a refractory button, was mounted in a large globe and the space between the walls of both was highly exhausted. The outer large globe remained comparatively cool in such constructions. When the large globe was on the pump and the vacuum between the walls maintained permanent by the continuous action of the pump, the outer globe would remain quite cold, while the button in the small bulb was kept at incandescence. But when the seal was made, and the button in the small bulb maintained incandescent some length of time, the large globe too would become warmed. From this I conjecture that if vacuous space (as Prof. Dewar finds) cannot convey heat, it is so merely in virtue of our rapid motion through space or, generally speaking, by the motion of the medium relatively to us, for a permanent condition could not be maintained without the medium being constantly renewed. A vacuum cannot, according to all evidence, be permanently maintained around a hot body.

In these constructions, before mentioned, the small bulb inside would, at least in the first stages, prevent all bombardment against the outer large globe. It occurred to me then to ascertain how a metal sieve would behave in this respect, and several bulbs, as illustrated in Fig. 195, were prepared for this purpose. In a globe *b*, was mounted a thin filament *f* (or button) upon a platinum wire *w* passing through a glass stem and leading to the outside of the globe. The filament *f* was surrounded by a metal sieve *s*. It was found in experiments with such bulbs that a sieve with wide meshes apparently did not in the slightest affect the bombardment against the globe *b*. When the vacuum was high, the shadow of the sieve was clearly projected against the globe and the latter would get hot in a short while. In some bulbs the sieve *s* was connected to a platinum wire sealed in the glass. When this wire was connected to the other terminal of the induction coil (the E. M. F. being kept low in this case), or to an insulated plate, the bombardment against the outer globe *b* was diminished. By taking a sieve with fine meshes the bombardment against the globe *b* was always diminished, but even then if the exhaustion was carried very far, and when the potential of the transformer was very high, the globe *b* would be bombarded and heated quickly, though no shadow of the sieve was visible, owing to the smallness of the meshes. But a glass tube or other continuous body mounted so as to surround the filament, did entirely cut off the bombardment and for a while the outer globe *b* would remain perfectly cold. Of course when the glass tube was sufficiently heated the bombardment against the outer globe could be noted at once. The experiments with these bulbs seemed to show that the speeds of the projected molecules or particles must be considerable (though quite insignificant when compared with that of light), otherwise it would be difficult to understand how they could traverse a fine metal sieve without being affected, unless it were found that such small particles or atoms cannot be acted upon directly at measurable distances. In regard to the speed of the projected atoms, Lord Kelvin has recently estimated it at about one kilometer a second or thereabouts in an ordinary Crookes bulb. As the potentials obtainable with a disruptive discharge coil are much higher than with ordinary coils, the speeds must, of course, be much greater when the bulbs are lighted from such a coil. Assuming the speed to be as high as five kilometers and uniform through the whole trajectory, as it should be in a very highly exhausted vessel, then if the alternate electrifications of the electrode would be of a frequency of five million, the greatest distance a particle could get away from the electrode would be one millimeter, and if it could be acted upon directly at that distance, the exchange of electrode matter or of the atoms would be very slow and there would be practically no bombardment against the bulb. This at least should be so, if the action of an electrode upon the atoms of the residual gas would be such as upon electrified bodies which we can perceive. A hot body enclosed in an exhausted bulb produces always atomic bombardment, but a hot body has no definite rhythm, for its molecules perform vibrations of all kinds.

If a bulb containing a button or filament be exhausted as high as is possible with the greatest care and by the use of the best artifices, it is often observed that the discharge cannot, at first, break through, but after some time, probably in consequence of some changes within the bulb, the discharge finally passes through and the button is rendered incandescent. In fact, it appears that the higher the degree of exhaustion the easier is the incandescence produced. There seem to be no other causes to which the incandescence might be attributed in such case except to the bombardment or similar action of the residual gas, or of particles of matter in general. But if the bulb be exhausted with the greatest care can these play an important part? Assume the vacuum in the bulb to be tolerably perfect, the great interest then centers in the question: Is the medium which pervades all space continuous or atomic? If atomic, then the heating of a conducting button or filament in an exhausted vessel might be due largely to ether bombardment, and then the heating of a conductor in general through which currents of high frequency or high potential are passed must be modified by the behavior of such medium; then also the skin effect, the apparent increase of the ohmic resistance, etc., admit, partially at least, of a different explanation.

It is certainly more in accordance with many phenomena observed with high-frequency currents to hold that all space is pervaded with free atoms, rather than to assume that it is devoid of these, and dark and cold, for so it must be, if filled with a continuous medium, since in such there can be neither heat nor light. Is then energy transmitted by independent carriers or by the vibration of a continuous medium? This important question is by no means as yet positively answered. But most of the effects which are here considered, especially the light effects, incandescence, or phosphorescence, involve the presence of free atoms and would be impossible without these.

In regard to the incandescence of a refractory button (or filament) in an exhausted receiver, which has been one of the subjects of this investigation, the chief experiences, which may serve as a guide in constructing such bulbs, may be summed up as follows: 1. The button should be as small as possible, spherical, of a smooth or polished surface, and of refractory material which withstands evaporation best. 2. The support of the button should be very thin and screened by an aluminum and mica sheet, as I have described on another occasion. 3. The exhaustion of the bulb should be as high as possible. 4. The frequency of the currents should be as high as practicable. 5. The currents should be of a harmonic rise and fall, without sudden interruptions. 6. The heat should be confined to the button by inclosing the same in a small bulb or otherwise. 7. The space between the walls of the small bulb and the outer globe should be highly exhausted.

Most of the considerations which apply to the incandescence of a solid just considered may likewise be applied to phosphorescence. Indeed, in an exhausted vessel the phosphorescence is, as a rule, primarily excited by the powerful beating of the electrode stream of atoms against the phosphorescent body. Even in many cases, where there is no evidence of such a bombardment, I think that phosphorescence is excited by violent impacts of atoms, which are not necessarily thrown off from the electrode but are acted upon from the same inductively through the medium or through chains of other atoms. That mechanical shocks play an important part in exciting phosphorescence in a bulb may be seen from the following experiment. If a bulb, constructed as that illustrated in Fig. 174, be taken and exhausted with the greatest care so that the discharge cannot pass, the filament f acts by electrostatic induction upon the tube t and the latter is set in vibration. If the tube o be rather wide, about an inch or so, the filament may be so powerfully vibrated that whenever it hits the glass tube it excites phosphorescence. But the phosphorescence ceases when the filament comes to rest. The vibration can be arrested and again started by varying the frequency of the currents. Now the filament has its own period of vibration, and if the frequency of the currents is such that there is resonance, it is easily set vibrating, though the potential of the currents be small. I have often observed that the filament in the bulb is destroyed by such mechanical resonance. The filament vibrates as a rule so rapidly that it cannot be seen and the experimenter may at first be mystified. When such an experiment as the one described is carefully performed, the potential of the currents need be extremely small, and for this reason I infer that the phosphorescence is then due to the mechanical shock of the filament against the glass, just as it is produced by striking a loaf of sugar with a knife. The mechanical shock produced by the projected atoms is easily noted when a bulb containing a button is grasped in the hand and the current turned on suddenly. I believe that a bulb could be shattered by observing the conditions of resonance.

In the experiment before cited it is, of course, open to say, that the glass tube, upon coming in contact with the filament, retains a charge of a certain sign upon the point of contact. If now the filament again touches the glass at the same point while it is oppositely charged, the charges equalize under evolution of light. But nothing of importance would be gained by such an explanation. It is unquestionable that the initial charges given to the atoms or to the glass play some part in exciting phosphorescence. So, for instance, if a phosphorescent bulb be first excited by a high frequency coil by connecting it to one of the terminals of the latter and the degree of luminosity be noted, and then the bulb be highly charged from a Holtz machine by attaching it preferably to the positive terminal of the machine, it is found that when the bulb is again connected to the terminal of the high frequency coil, the phosphorescence is far more intense. On another occasion I have considered the possibility of some phosphorescent phenomena in bulbs being produced by the incandescence of an infinitesimal layer on the surface of the phosphorescent body. Certainly the impact of the atoms is powerful enough to produce intense incandescence by the collisions, since they bring quickly to a high temperature a body of considerable bulk. If any such effect exists, then the best appliance for producing phosphorescence in a bulb, which we know so far, is a disruptive discharge coil giving an enormous potential with but few fundamental discharges, say 25–30 per second, just enough to produce a continuous impression upon the eye. It is a fact that such a coil excites phosphorescence under almost any condition and at all degrees of exhaustion, and I have observed effects which appear to be due to phosphorescence even at ordinary pressures of the atmosphere, when the potentials are extremely high. But if phosphorescent light is produced by the equalization of charges of electrified atoms (whatever this may mean ultimately), then the higher the frequency of the impulses or alternate electrifications, the more economical will be the light production. It is a long known and noteworthy fact that all the phosphorescent bodies are poor conductors of electricity and heat, and that all bodies cease to emit phosphorescent light when they are brought to a certain temperature. Conductors on the contrary do not possess this quality. There are but few exceptions to the rule. Carbon is one of them. Becquerel noted that carbon phosphoresces at a certain elevated temperature preceding the dark red. This phenomenon may be easily observed in bulbs provided with a rather large carbon electrode (say, a sphere of six millimeters diameter). If the current is turned on after a few seconds, a snow white film covers the electrode, just before it gets dark red. Similar effects are noted with other conducting bodies, but many scientific men will probably not attribute them to true phosphorescence. Whether true incandescence has anything to do with phosphorescence excited by atomic impact or mechanical shocks still remains to be decided, but it is a fact that all conditions, which tend to localize and increase the heating effect at the point of impact, are almost invariably the most favorable for the production of phosphorescence. So, if the electrode be very small, which is equivalent to saying in general, that the electric density is great; if the potential be high, and if the gas be highly rarefied, all of which things imply high speed of the projected atoms, or matter, and consequently violent impacts—the phosphorescence is very intense. If a bulb provided with a large and small electrode be attached to the terminal of an induction coil, the small electrode excites phosphorescence while the large one may not do so, because of the smaller electric density and hence smaller speed of the atoms. A bulb provided with a large electrode may be grasped with the hand while the electrode is connected to the terminal of the coil and it may not phosphoresce; but if instead of grasping the bulb with the hand, the same be touched with a pointed wire, the phosphorescence at once spreads through the bulb, because of the great density at the point of contact. With low frequencies it seems that gases of great atomic weight excite more intense phosphorescence than those of smaller weight, as for instance, hydrogen. With high frequencies the observations are not sufficiently reliable to draw a conclusion. Oxygen, as is well-known, produces exceptionally strong effects, which may be in part due to chemical action. A bulb with hydrogen residue seems to be most easily excited. Electrodes which are most easily deteriorated produce more intense phosphorescence in bulbs, but the condition is not permanent because of the impairment of the vacuum and the deposition of the electrode matter upon the phosphorescent surfaces. Some liquids, as oils, for instance, produce magnificent effects of phosphorescence (or fluorescence?), but they last only a few seconds. So if a bulb has a trace of oil on the walls and the current is turned on, the phosphorescence only persists for a few moments until the oil is carried away. Of all bodies so far tried, sulphide of zinc seems to be the most susceptible to phosphorescence. Some samples, obtained through the kindness of Prof. Henry in Paris, were employed in many of these bulbs. One of the defects of this sulphide is, that it loses its quality of emitting light when brought to a temperature which is by no means high. It can therefore, be used only for feeble intensities. An observation which might deserve notice is, that when violently bombarded from an aluminum electrode it assumes a black color, but singularly enough, it returns to the original condition when it cools down.

The most important fact arrived at in pursuing investigations in this direction is, that in all cases it is necessary, in order to excite phosphorescence with a minimum amount of energy, to observe certain conditions. Namely, there is always, no matter what the frequency of the currents, degree of exhaustion and character of the bodies in the bulb, a certain potential (assuming the bulb excited from one terminal) or potential difference (assuming the bulb to be excited with both terminals) which produces the most economical result. If the potential be increased, considerable energy may be wasted without producing any more light, and if it be diminished, then again the light production is not as economical. The exact condition under which the best result is obtained seems to depend on many things of a different nature, and it is to be yet investigated by other experimenters, but it will certainly have to be observed when such phosphorescent bulbs are operated, if the best results are to be obtained.

Coming now to the most interesting of these phenomena, the incandescence or phosphorescence of gases, at low pressures or at the ordinary pressure of the atmosphere, we must seek the explanation of these phenomena in the same primary causes, that is, in shocks or impacts of the atoms. Just as molecules or atoms beating upon a solid body excite phosphorescence in the same or render it incandescent, so when colliding among themselves they produce similar phenomena. But this is a very insufficient explanation and concerns only the crude mechanism. Light is produced by vibrations which go on at a rate almost inconceivable. If we compute, from the energy contained in the form of known radiations in a definite space the force which is necessary to set up such rapid vibrations, we find, that though the density of the ether be incomparably smaller than that of any body we know, even hydrogen, the force is something surpassing comprehension. What is this force, which in mechanical measure may amount to thousands of tons per square inch? It is electrostatic force in the light of modern views. It is impossible to conceive how a body of measurable dimensions could be charged to so high a potential that the force would be sufficient to produce these vibrations. Long before any such charge could be imparted to the body it would be shattered into atoms. The sun emits light and heat, and so does an ordinary flame or incandescent filament, but in neither of these can the force be accounted for if it be assumed that it is associated with the body as a whole. Only in one way may we account for it, namely, by identifying it with the atom. An atom is so small, that if it be charged by coming in contact with an electrified body and the charge be assumed to follow the same law as in the case of bodies of measurable dimensions, it must retain a quantity of electricity which is fully capable of accounting for these forces and tremendous rates of vibration. But the atom behaves singularly in this respect—it always takes the same "charge."

It is very likely that resonant vibration plays a most important part in all manifestations of energy in nature. Throughout space all matter is vibrating, and all rates of vibration are represented, from the lowest musical note to the highest pitch of the chemical rays, hence an atom, or complex of atoms, no matter what its period, must find a vibration with which it is in resonance. When we consider the enormous rapidity of the light vibrations, we realize the impossibility of producing such vibrations directly with any apparatus of measurable dimensions, and we are driven to the only possible means of attaining the object of setting up waves of light by electrical means and economically, that is, to affect the molecules or atoms of a gas, to cause them to collide and vibrate. We then must ask ourselves—How can free molecules or atoms be affected?

It is a fact that they can be affected by electrostatic force, as is apparent in many of these experiments. By varying the electrostatic force we can agitate the atoms, and cause them to collide accompanied by evolution of heat and light. It is not demonstrated beyond doubt that we can affect them otherwise. If a luminous discharge is produced in a closed exhausted tube, do the atoms arrange themselves in obedience to any other but to electrostatic force acting in straight lines from atom to atom? Only recently I investigated the mutual action between two circuits with extreme rates of vibration. When a battery of a few jars (*c c c c*, Fig. 196) is discharged through

FIG. 196.

a primary P of low resistance (the connections being as illustrated in Figs. 183a, 183b and 183c), and the frequency of vibration is many millions there are great differences of potential between points on the primary not more than a few inches apart. These differences may be 10,000 volts per inch, if not more, taking the maximum value of the E. M. F. The secondary s is therefore acted upon by electrostatic induction, which is in such extreme cases of much greater importance than the electro-dynamic. To such sudden impulses the primary as well as the secondary are poor conductors, and therefore great differences of potential may be produced by electrostatic induction between adjacent points on the secondary. Then sparks may jump between the wires and streamers become visible in the dark if the light of the discharge through the spark gap *d d* be carefully excluded. If now we substitute a closed vacuum tube for the metallic secondary s, the differences of potential produced in the tube by electrostatic induction from the primary are fully sufficient to excite portions of it; but as the points of certain differences of potential on the primary are not fixed, but are generally constantly changing in position, a luminous band is produced in the tube, apparently not touching the glass, as it should, if the points of maximum and minimum differences of potential were fixed on the primary. I do not exclude the possibility of such a tube being excited only by electro-dynamic induction, for very able physicists hold this view; but in my opinion, there is as yet no positive proof given that atoms of a gas in a closed tube may arrange themselves in chains under the action of an electromotive impulse produced by electro-dynamic induction in the tube. I have been unable so far to produce striae in a tube, however long, and at whatever degree of exhaustion, that is, striae at right angles to the supposed direction of the discharge or the axis of the tube; but I have distinctly observed in a large bulb, in which a wide luminous band was produced by passing a discharge of a battery through a wire surrounding the bulb, a circle of feeble luminosity between two luminous bands, one of which was more intense than the other. Furthermore, with my present experience I do not think that such a gas discharge in a closed tube can vibrate, that is, vibrate as a whole. I am convinced that no discharge through a gas can vibrate. The atoms of a gas behave very curiously in respect to sudden electric impulses. The gas does not seem to possess any appreciable inertia to such impulses, for it is a fact, that the higher the frequency of the impulses, with the greater freedom does the discharge pass through the gas. If the gas possesses no inertia then it cannot vibrate, for some inertia is necessary for the free vibration. I conclude from this that if a lightning discharge occurs between two clouds, there can be no oscillation, such as would be expected, considering the capacity of the clouds. But if the lightning discharge strike the earth, there is always vibration—in the earth, but not in the cloud. In a gas discharge each atom vibrates at its own rate, but there is no vibration of the conducting gaseous mass as a whole. This is an important consideration in the great problem of producing light economically, for it teaches us that to reach this result we must use impulses of very high frequency and necessarily also of high potential. It is a fact that oxygen produces a more intense light in a tube. Is it because oxygen atoms possess some inertia and the vibration does not die out instantly? But then nitrogen should be as good, and chlorine and vapors of many other bodies much better than oxygen, unless the magnetic properties of the latter enter prominently into play. Or, is the process in the tube of an electrolytic nature? Many observations certainly speak for it, the most important being that matter is always carried away from the electrodes and the vacuum in a bulb cannot be permanently maintained. If such process takes place in reality, then again must we take refuge in high frequencies, for, with such, electrolytic action should be reduced to a minimum, if not rendered entirely impossible.

It is an undeniable fact that with very high frequencies, provided the impulses be of harmonic nature, like those obtained from an alternator, there is less deterioration and the vacua are more permanent. With disruptive discharge coils there are sudden rises of potential and the vacua are more quickly impaired, for the electrodes are deteriorated in a very short time. It was observed in some large tubes, which were provided with heavy

FIG. 197.

carbon blocks B B_1, connected to platinum wires w w_1 (as illustrated in Fig. 197), and which were employed in experiments with the disruptive discharge instead of the ordinary air gap, that the carbon particles under the action of the powerful magnetic field in which the tube was placed, were deposited in regular fine lines in the middle of the tube, as illustrated. These lines were attributed to the deflection or distortion of the discharge by the magnetic field, but why the deposit occurred principally where the field was most intense did not appear quite clear. A fact of interest, likewise noted, was that the presence of a strong magnetic field increases the deterioration of the electrodes, probably by reason of the rapid interruptions it produces, whereby there is actually a higher E. M. F. maintained between the electrodes.

Much would remain to be said about the luminous effects produced in gases at low or ordinary pressures. With the present experiences before us we cannot say that the essential nature of these charming phenomena is sufficiently known. But investigations in this direction are being pushed with exceptional ardor. Every line of scientific pursuit has its fascinations, but electrical investigation appears to possess a peculiar attraction, for there is no experiment or observation of any kind in the domain of this wonderful science which would not forcibly appeal to us. Yet to me it seems, that of all the many marvelous things we observe, a vacuum tube, excited by an electric impulse from a distant source, bursting forth out of the darkness and illuminating the room with its beautiful light, is as lovely a phenomenon as can greet our eyes. More interesting still it appears when, reducing the fundamental discharges across the gap to a very small number and waving the tube about we produce all kinds of designs in luminous lines. So by way of amusement I take a straight long tube, or a square one, or a square attached to a straight tube, and by whirling them

Fig. 198.

about in the hand, I imitate the spokes of a wheel, a Gramme winding, a drum winding, an alternate current motor winding, etc. (Fig. 198). Viewed from a distance the effect is weak and much of its beauty is lost, but being near or holding the tube in the hand, one cannot resist its charm.

In presenting these insignificant results I have not attempted to arrange and co-ordinate them, as would be proper in a strictly scientific investigation, in which every succeeding result should be a logical sequence of the preceding, so that it might be guessed in advance by the careful reader or attentive listener. I have preferred to concentrate my energies chiefly upon advancing novel facts or ideas which might serve as suggestions to others, and this may serve as an excuse for the lack of harmony. The explanations of the phenomena have been given in good faith and in the spirit of a student prepared to find that they admit of a better interpretation. There can be no great harm in a student taking an erroneous view, but when great minds err, the world must dearly pay for their mistakes.

CHAPTER XXIX

TESLA ALTERNATING CURRENT GENERATORS FOR HIGH FREQUENCY, IN DETAIL

It has become a common practice to operate arc lamps by alternating or pulsating, as distinguished from continuous, currents; but an objection which has been raised to such systems exists in the fact that the arcs emit a pronounced sound, varying with the rate of the alternations or pulsations of current. This noise is due to the rapidly alternating heating and cooling, and consequent expansion and contraction, of the gaseous matter forming the arc, which corresponds with the periods or impulses of the current. Another disadvantageous feature is found in the difficulty of maintaining an alternating current arc in consequence of the periodical increase in resistance corresponding to the periodical working of the current. This feature entails a further disadvantage, namely, that small arcs are impracticable.

Theoretical considerations have led Mr. Tesla to the belief that these disadvantageous features could be obviated by employing currents of a sufficiently high number of alternations, and his anticipations have been confirmed in practice. These rapidly alternating currents render it possible to maintain small arcs which, besides, possess the advantages of silence and persistency. The latter quality is due to the necessarily rapid alternations, in consequence of which the arc has no time to cool, and is always maintained at a high temperature and low resistance.

At the outset of his experiments Mr. Tesla encountered great difficulties in the construction of high frequency machines. A generator of this kind is described here, which, though constructed quite some time ago, is well worthy of a detailed description. It may be mentioned, in passing, that dynamos of this type have been used by Mr. Tesla in his lighting researches and experiments with currents of high potential and high frequency, and reference to them will be found in his lectures elsewhere printed in this volume.

In the accompanying engravings, Figs. 199 and 200 show the machine, respectively, in side elevation and vertical cross-section; Figs. 201, 202, and 203 showing enlarged details of construction. As will be seen, A is an annular magnetic frame, the interior of which is provided with a large number of pole-pieces D.

FIG. 199.

Owing to the very large number and small size of the poles and the spaces between them, the field coils are applied by winding an insulated conductor F zigzag through the grooves, as shown in Fig. 203, carrying the wire around the annulus to form as many layers as is desired. In this way the pole-pieces D will be energized with alternately opposite polarity around the entire ring.

For the armature, Mr. Tesla employs a spider carrying a ring J, turned down, except at its edges, to form a trough-like receptacle for a mass of fine annealed iron wires K, which are wound in the groove to form the core proper for the armature-coils. Pins L are set in the sides of the ring J and the coils M are wound over the periphery of the armature-structure and around the pins. The coils M are connected together in series, and these terminals N carried through the hollow shaft H to contact-rings P P, from which the currents are taken off by brushes O.

FIGS. 200, 201, 202, and 203.

In this way a machine with a very large number of poles may be constructed. It is easy, for instance, to obtain in this manner three hundred and seventy-five to four hundred poles in a machine that may be safely driven at a speed of fifteen hundred or sixteen hundred revolutions per minute, which will produce ten thousand or eleven thousand alternations of current per second. Arc lamps R R are shown in the diagram as connected up in series with the machine in Fig. 200. If such a current be applied to running arc lamps, the sound produced by or in the arc becomes practically inaudible, for, by increasing the rate of change in the current, and consequently the number of vibrations per unit of time of the gaseous material of the arc up to, or beyond, ten thousand or eleven thousand per second, or to what is regarded as the limit of audition, the sound due to such vibrations will not be audible. The exact number of changes or undulations necessary to produce this result will vary somewhat according to the size of the arc—that is to say, the smaller the arc, the greater the number of changes that will be required to render it inaudible within certain limits. It should also be stated that the arc should not exceed a certain length.

The difficulties encountered in the construction of these machines are of a mechanical as well as an electrical nature. The machines may be designed on two plans: the field may be formed either of alternating poles, or of polar projections of the same polarity. Up to about 15,000 alternations per second in an experimental machine, the former plan may be followed, but a more efficient machine is obtained on the second plan.

In the machine above described, which was capable of running two arcs of normal candle power, the field was composed of a ring of wrought iron 32 inches outside diameter, and about 1 inch thick. The inside diameter was 30 inches. There were 384 polar projections. The wire was wound in zigzag form, but two wires were wound so as to completely envelop the projections. The distance between the projections is about 3/16 inch, and they are a little over 1/16 inch thick. The field magnet was made relatively small so as to adapt the machine for a constant current. There are 384 coils connected in two series. It was found impracticable to use any wire much thicker than No. 26 B. and S. gauge on account of the local effects. In such a machine the clearance should be as small as possible; for this reason the machine was made only 1¼ inch wide, so that the binding wires might be obviated. The armature wires must be wound with great care, as they are apt to fly off in consequence of the great peripheral speed. In various experiments this machine has been run as high as 3,000

Fig. 204.

revolutions per minute. Owing to the great speed it was possible to obtain as high as 10 amperes out of the machine. The electromotive force was regulated by means of an adjustable condenser within very wide limits, the limits being the greater, the greater the speed. This machine was frequently used to run Mr. Tesla's laboratory lights.

FIG. 205.

The machine above described was only one of many such types constructed. It serves well for an experimental machine, but if still higher alternations are required and higher efficiency is necessary, then a machine on a plan shown in Figs. 204 to 207, is preferable. The principal advantage of this type of machine is that there is not much magnetic leakage, and that a field may be produced, varying greatly in intensity in places not much distant from each other.

In these engravings, Figs. 204 and 205 illustrate a machine in which the armature conductor and field coils are stationary, while the field magnet core revolves. Fig. 206 shows a machine embodying the same plan of construction, but having a stationary field magnet and rotary armature.

The conductor in which the currents are induced may be arranged in various ways; but Mr. Tesla prefers the following method: He employs an annular plate of copper D, and by means of a saw cuts in it radial slots from one edge nearly through to the other, beginning alternately from opposite edges. In this way a continuous zigzag conductor is formed. When the polar projections are ⅛ inch wide, the width of the conductor should not, under any circumstances, be more than ¹⁄₃₂ inch wide; even then the eddy effect is considerable.

To the inner edge of this plate are secured two rings of non-magnetic metal E, which are insulated from the copper conductor, but held firmly thereto by means of the bolts F. Within the rings E is then placed an annular coil G, which is the energizing coil for the field magnet. The conductor D and the parts attached thereto are supported by means of the cylindrical shell or casting A A, the two parts of which are brought together and clamped to the outer edge of the conductor D.

The core for the field magnet is built up of two circular parts H H, formed with annular grooves I, which, when the two parts are brought together, form a space for the reception of the energizing coil G. The hubs of the cores are trued off, so as to fit closely against one another, while the outer portions or flanges which form the polar faces J J, are reduced somewhat in thickness to make room for the conductor D, and are serrated on their faces. The number of serrations in the polar faces is arbitrary:

FIG. 206.

but there must exist between them and the radial portions of the conductor D certain relation, which will be understood by reference to Fig. 207 in which N N represent the projections or points on one face of the core of the field, and S S the points of the other face. The conductor D is shown in this figure in section $a\ a'$ designating the radial portions of the conductor, and b the insulating divisions between them. The relative width of the parts $a\ a'$ and the space between any two adjacent points N N or S S is such that when the radial portions a of the conductor are passing between the opposite points N S where the field is strongest, the intermediate radial portions a' are passing through the widest spaces midway between such points and where the field is weakest. Since the core on one side is of opposite polarity to the part facing it, all the projections of one polar face will be of opposite polarity to those of the other face. Hence, although the space between any two adjacent points on the same face may be extremely small, there will be no leakage of the magnetic lines between any two points of the same name, but the lines of force will pass across from one set of points to the other. The construction followed obviates to a great degree the distortion of the magnetic lines by the action of the current in the conductor D, in which it will be observed the current is flowing at any given time from the center toward the periphery in one set of radial parts a and in the opposite direction in the adjacent parts a'.

FIG. 207.

In order to connect the energizing coil G, Fig. 204, with a source of continuous current, Mr. Tesla utilizes two adjacent radial portions of the conductor D for connecting the terminals of the coil G with two binding posts M. For this purpose the plate D is cut entirely through, as shown, and the break thus made is bridged over by a short conductor C. The plate D is cut through to form two terminals d, which are connected to binding posts N. The core H H, when rotated by the driving pulley, generates in the conductors D an alternating current, which is taken off from the binding posts N.

When it is desired to rotate the conductor between the faces of a stationary field magnet, the construction shown in Fig. 206, is adopted. The conductor D in this case is or may be made in substantially the same manner as above described by slotting an annular conducting-plate and supporting it between two heads O, held together by bolts *o* and fixed to the driving-shaft K. The inner edge of the plate or conductor D is preferably flanged to secure a firmer union between it and the heads O. It is insulated from the head. The field-magnet in this case consists of two annular parts H H, provided with annular grooves I for the reception of the coils. The flanges or faces surrounding the annular groove are brought together, while the inner flanges are serrated, as in the previous case, and form the polar faces. The two parts H H are formed with a base R, upon which the machine rests. S S are non-magnetic bushings secured or set in the central opening of the cores. The conductor D is cut entirely through at one point to form terminals, from which insulated conductors T are led through the shaft to collecting-rings V.

In one type of machine of this kind constructed by Mr. Tesla, the field had 480 polar projections on each side, and from this machine it was possible to obtain 30,000 alternations per second. As the polar projections must necessarily be very narrow, very thin wires or sheets must be used to avoid the eddy current effects. Mr. Tesla has thus constructed machines with a stationary armature and rotating field, in which case also the field-coil was supported so that the revolving part consisted only of a wrought iron body devoid of any wire and also machines with a rotating armature and stationary field. The machines may be either drum or disc, but Mr. Tesla's experience shows the latter to be preferable.

* * * * *

In the course of a very interesting article contributed to the *Electrical World* in February, 1891, Mr. Tesla makes some suggestive remarks on these high frequency machines and his experiences with them, as well as with other parts of the high frequency apparatus. Part of it is quoted here and is as follows:

The writer will incidentally mention that anyone who attempts for the first time to construct such a machine will have a tale of woe to tell. He will first start out, as a matter of course, by making an armature with the required number of polar projections. He will then get the satisfaction of having produced an apparatus which is fit to accompany a thoroughly Wagnerian opera. It may besides possess the virtue of converting mechanical energy into heat in a nearly perfect manner. If there is a reversal in the polarity of the projections, he will get heat out of the machine; if there is no reversal, the heating will be less, but the output will be next to nothing. He will then abandon the iron in the armature, and he will get from the Scylla to the Charybdis. He will look for one difficulty and will find another, but, after a few trials, he may get nearly what he wanted.

Among the many experiments which may be performed with such a machine, of not the least interest are those performed with a high-tension induction coil. The character of the discharge is completely changed. The arc is established at much greater distances, and it is so easily affected by the slightest current of air that it often wriggles around in the most singular manner. It usually emits the rhythmical sound peculiar to the alternate current arcs, but the curious point is that the sound may be heard with a number of alternations far above ten thousand per second, which by many is considered to be about the limit of audition. In many respects the coil behaves like a static machine. Points impair considerably the sparking interval, electricity escaping from them freely, and from a wire attached to one of the terminals streams of light issue, as though it were connected to a pole of a powerful Toepler machine. All these phenomena are, of course, mostly due to the enormous differences of potential obtained. As a consequence of the self-induction of the coil and the high frequency, the current is minute while there is a corresponding rise of pressure. A current impulse of some strength started in such a coil should persist to flow no less than four ten-thousandths of a second. As this time is greater than half the period, it occurs that an opposing electromotive force begins to act while the current is still flowing. As a consequence, the pressure rises as in a tube filled with liquid and vibrated rapidly around its axis. The current is so small that, in the opinion and involuntary experience of the writer, the discharge of even a very large coil cannot produce seriously injurious effects, whereas, if the same coil were operated with a current of lower frequency, though the electromotive force would be much smaller, the discharge would be most certainly injurious. This result, however, is due in part to the high frequency. The writer's experiences tend to show that the higher the frequency the greater the amount of electrical energy which may be passed through the body without serious discomfort; whence it seems certain that human tissues act as condensers.

One is not quite prepared for the behavior of the coil when connected to a Leyden jar. One, of course, anticipates that since the frequency is high the capacity of the jar should be small. He therefore takes a very small jar, about the size of a small wine glass, but he finds that even with this jar the coil is practically short-circuited. He then reduces the capacity until he comes to about the capacity of two spheres, say, ten centimeters in diameter and two to four centimeters apart. The discharge then assumes the form of a serrated band exactly like a succession of sparks viewed in a rapidly revolving mirror; the serrations, of course, corresponding to the condenser discharges. In this case one may observe a queer phenomenon. The discharge starts at the nearest points, works gradually up, breaks somewhere near the top of the spheres, begins again at the bottom, and so on. This goes on so fast that several serrated bands are seen at once. One may be puzzled for a few minutes, but the explanation is simple enough. The discharge begins at the nearest points, the air is heated and carries the arc upward until it breaks, when it is re-established at the nearest points, etc. Since the current passes easily through a condenser of even small capacity, it will be found quite natural that connecting only one terminal to a body of the same size, no matter how well insulated, impairs considerably the striking distance of the arc.

Experiments with Geissler tubes are of special interest. An exhausted tube, devoid of electrodes of any kind, will light up at some distance from the coil. If a tube from a vacuum pump is near the coil the whole of the pump is brilliantly lighted. An incandescent lamp approached to the coil lights up and gets perceptibly hot. If a lamp have the terminals connected to one of the binding posts of the coil and the hand is approached to the bulb, a very curious and rather unpleasant discharge from the glass to the hand takes place, and the filament may become incandescent. The discharge resembles to some extent the stream issuing from the plates of a powerful Toepler machine, but is of incomparably greater quantity. The lamp in this case acts as a condenser, the rarefied gas being one coating, the operator's hand the other. By taking the globe of a lamp in the hand, and by bringing the metallic terminals near to or in contact with a conductor connected to the coil, the carbon is brought to bright incandescence and the glass is rapidly heated. With a 100-volt 10 c. p. lamp one may without great discomfort stand as much current as will bring the lamp to a considerable brilliancy; but it can be held in the hand only for a few minutes, as the glass is heated in an incredibly short time. When a tube is lighted by bringing it near to the coil it may be made to go out by interposing a metal plate on the hand between the coil and tube; but if the metal plate be fastened to a glass rod or otherwise insulated, the tube may remain lighted if the plate be interposed, or may even increase in luminosity. The effect depends on the position of the plate and tube relatively to the coil, and may be always easily foretold by *assuming* that conduction takes place from one terminal of the coil to the other. According to the position of the plate, it may either divert from or direct the current to the tube.

In another line of work the writer has in frequent experiments maintained incandescent lamps of 50 or 100 volts burning at any desired candle power with both the terminals of each lamp connected to a stout copper wire of no more than a few feet in length. These experiments seem interesting enough, but they are not more so than the queer experiment of Faraday, which has been revived and made much of by recent investigators, and in which a discharge is made to jump between two points of a bent copper wire. An experiment may be cited here which may seem equally interesting. If a Geissler tube, the terminals of which are joined by a copper wire, be approached to the coil, certainly no one would be prepared to see the tube light up. Curiously enough, it does light up, and, what is more, the wire does not seem to make much difference. Now one is apt to think in the first moment that the impedance of the wire might have something to do with the phenomenon. But this is of course immediately rejected, as for this an enormous frequency would be required. This result, however, seems puzzling only at first; for upon reflection it is quite clear that the wire can make but little difference. It may be explained in more than one way, but it agrees perhaps best with observation to assume that conduction takes place from the terminals of the coil through the space. On this assumption, if the tube with the wire be held in any position, the wire can divert little more than the current which passes through the space occupied by the wire and the metallic terminals of the tube; through the adjacent space the current passes practically undisturbed. For this reason, if the tube be held in any position at right angles to the line joining the binding posts of the coil, the wire makes hardly any difference, but in a position more or less parallel with that line it impairs to a certain extent the brilliancy of the tube and its facility to light up. Numerous other phenomena may be explained on the same assumption. For instance, if the ends of the tube be provided with washers of sufficient size and held in the line joining the terminals of the coil, it will not light up, and then nearly the whole of the current, which would otherwise pass uniformly through the space between the washers, is diverted through the wire. But if the tube be inclined sufficiently to that line, it will light up in spite of the washers. Also, if a metal plate be fastened upon a glass rod and held at right angles to the line joining the binding posts, and nearer to one of them, a tube held more or less parallel with the line will light up instantly when one of the terminals touches the plate, and will go out when separated from the plate. The greater the surface of the plate, up to a certain limit, the easier the tube will light up. When a tube is placed at right angles to the straight line joining the binding posts, and then rotated, its luminosity steadily increases until it is parallel with that line. The writer must state, however, that he does not favor the idea of a leakage or current through the space any more than as a suitable explanation, for he is convinced that all these experiments could not be performed with a static machine yielding a constant difference of potential, and that condenser action is largely concerned in these phenomena.

It is well to take certain precautions when operating a Ruhmkorff coil with very rapidly alternating currents. The primary current should not be turned on too long, else the core may get so hot as to melt the gutta-percha or paraffin, or otherwise injure the insulation, and this may occur in a surprisingly short time, considering the current's strength. The primary current being turned on, the fine wire terminals may be joined without great risk, the impedance being so great that it is difficult to force enough current through the fine wire so as to injure it, and in fact the coil may be on the whole much safer when the terminals of the fine wire are connected than when they are insulated; but special care should be taken when the terminals are connected to the coatings of a Leyden jar, for with anywhere near the critical capacity, which just counteracts the self-induction at the existing frequency, the coil might meet the fate of St. Polycarpus. If an expensive vacuum pump is lighted up by being near to the coil or touched with a wire connected to one of the terminals, the current should be left on no more than a few moments, else the glass will be cracked by the heating of the rarefied gas in one of the narrow passages—in the writer's own experience *quod erat demonstrandum.**

There are a good many other points of interest which may be observed in connection with such a machine. Experiments with the telephone, a conductor in a strong field or with a condenser or arc, seem to afford certain proof that sounds far above the usual accepted limit of hearing would be perceived. A telephone will emit notes of twelve to thirteen thousand vibrations per second; then the inability of the core to follow such rapid alternations begins to tell. If, however, the magnet and core be replaced by a condenser and the terminals connected to the high-tension secondary of a transformer, higher notes may still be heard. If the current be sent around a finely laminated core and a small piece of thin sheet iron be held gently against the core, a sound may be still heard with thirteen to fourteen thousand alternations per second, provided the current is sufficiently strong. A small coil, however, tightly packed between the poles of a powerful magnet, will emit a sound with the above number of alternations, and arcs may be audible with a still higher frequency. The limit of audition is variously estimated. In Sir William Thomson's writings it is stated somewhere that ten thousand per second, or nearly so, is the limit. Other, but less reliable, sources give it as high as twenty-four thousand per second. The above experiments have convinced the writer that notes of an incomparably higher number of vibrations per second would be perceived provided they could be produced with sufficient power. There is no reason why it should not be so. The condensations and rarefactions of the air would necessarily set the diaphragm in a corresponding vibration and some sensation would be produced, whatever—within certain limits—the velocity of

---

* It is thought necessary to remark that, although the induction coil may give quite a good result when operated with such rapidly alternating currents, yet its construction, quite irrespective of the iron core, makes it very unfit for such high frequencies, and to obtain the best results the construction should be greatly modified.

transmission to their nerve centers, though it is probable that for want of exercise the ear would not be able to distinguish any such high note. With the eye it is different; if the sense of vision is based upon some resonance effect, as many believe, no amount of increase in the intensity of the ethereal vibration could extend our range of vision on either side of the visible spectrum.

The limit of audition of an arc depends on its size. The greater the surface by a given heating effect in the arc, the higher the limit of audition. The highest notes are emitted by the high-tension discharges of an induction coil in which the arc is, so to speak, all surface. If R be the resistance of an arc, and c the current, and the linear dimensions be n times increased, then the resistance is R/n, and with the same current density the current would be n^2C; hence the heating effect is n^3 times greater, while the surface is only n^2 times as great. For this reason very large arcs would not emit any rhythmical sound even with a very low frequency. It must be observed, however, that the sound emitted depends to some extent also on the composition of the carbon. If the carbon contain highly refractory material, this, when heated, tends to maintain the temperature of the arc uniform and the sound is lessened; for this reason it would seem that an alternating arc requires such carbons.

With currents of such high frequencies it is possible to obtain noiseless arcs, but the regulation of the lamp is rendered extremely difficult on account of the excessively small attractions or repulsions between conductors conveying these currents.

An interesting feature of the arc produced by these rapidly alternating currents is its persistency. There are two causes for it, one of which is always present, the other sometimes only. One is due to the character of the current and the other to a property of the machine. The first cause is the more important one, and is due directly to the rapidity of the alternations. When an arc is formed by a periodically undulating current, there is a corresponding undulation in the temperature of the gaseous column, and, therefore, a corresponding undulation in the resistance of the arc. But the resistance of the arc varies enormously with the temperature of the gaseous column, being practically infinite when the gas between the electrodes is cold. The persistence of the arc, therefore, depends on the inability of the column to cool. It is for this reason impossible to maintain an arc with the current alternating only a few times a second. On the other hand, with a practically continuous current, the arc is easily maintained, the column being constantly kept at a high temperature and low resistance. The higher the frequency the smaller the time interval during which the arc may cool and increase considerably in resistance. With a frequency of 10,000 per second or more in an arc of equal size excessively small variations of temperature are superimposed upon a steady temperature, like ripples on the surface of a deep sea. The heating effect is practically continuous and the arc behaves like one produced by a continuous current, with the exception, however, that it may not be quite as easily started, and that the electrodes are equally consumed; though the writer has observed some irregularities in this respect.

The second cause alluded to, which possibly may not be present, is due to the tendency of a machine of such high frequency to maintain a practically constant current. When the arc is lengthened, the electromotive force rises in proportion and the arc appears to be more persistent.

Such a machine is eminently adapted to maintain a constant current, but it is very unfit for a constant potential. As a matter of fact, in certain types of such machines a nearly constant current is an almost unavoidable result. As the number of poles or polar projections is greatly increased, the clearance becomes of great importance. One has really to do with a great number of very small machines. Then there is the impedance in the armature, enormously augmented by the high frequency. Then, again, the magnetic leakage is facilitated. If there are three or four hundred alternate poles, the leakage is so great that it is virtually the same as connecting, in a two-pole machine, the poles by a piece of iron. This disadvantage, it is true, may be obviated more or less by using a field throughout of the same polarity, but then one encounters difficulties of a different nature. All these things tend to maintain a constant current in the armature circuit.

In this connection it is interesting to notice that even today engineers are astonished at the performance of a constant current machine, just as, some years ago, they used to consider it an extraordinary performance if a machine was capable of maintaining a constant potential difference between the terminals. Yet one result is just as easily secured as the other. It must only be remembered that in an inductive apparatus of any kind, if constant potential is required, the inductive relation between the primary or exciting and secondary or armature circuit must be the closest possible; whereas, in an apparatus for constant current just the opposite is required. Furthermore, the opposition to the current's flow in the induced circuit must be as small as possible in the former and as great as possible in the latter case. But opposition to a current's flow may be caused in more than one way. It may be caused by ohmic resistance or self-induction. One may make the induced circuit of a dynamo machine or transformer of such high resistance that when operating devices of considerably smaller resistance within very wide limits a nearly constant current is maintained. But such high resistance involves a great loss in power, hence it is not practicable. Not so self-induction. Self-induction does not necessarily mean loss of power. The moral is, use self-induction instead of resistance. There is, however, a circumstance which favors the adoption of this plan, and this is, that a very high self-induction may be obtained cheaply by surrounding a comparatively small length of wire more or less completely with iron, and, furthermore, the effect may be exalted at will by causing a rapid undulation of the current. To sum up, the requirements for constant current are: Weak magnetic connection between the induced and inducing circuits, greatest possible self-induction with the least resistance, greatest practicable rate of change of the current. Constant potential, on the other hand, requires: Closest magnetic connection between the circuits, steady induced current, and, if possible, no reaction. If the latter conditions could be fully satisfied in a constant potential machine, its output would surpass many times that of a machine primarily designed to give constant current. Unfortunately, the type of machine in which these conditions may be satisfied is of little practical value, owing to the small electromotive force obtainable and the difficulties in taking off the current.

With their keen inventor's instinct, the now successful arc-light men have early recognized the desiderata of a constant current machine. Their arc light machines have weak fields, large armatures, with a great length of copper wire and few commutator segments to produce great variations in the current's strength and to bring self-induction into play. Such machines may maintain within considerable limits of variation in the resistance of the circuit a practically constant current. Their output is of course correspondingly diminished, and, perhaps with the object in view not to cut down the output too much, a simple device compensating exceptional variations is employed. The undulation of the current is almost essential to the commercial success of an arc-light system. It introduces in the circuit a steadying element taking the place of a large ohmic resistance, without involving a great loss in power, and, what is more important, it allows the use of simple clutch lamps, which with a current of a certain number of impulses per second, best suitable for each particular lamp, will, if properly attended to, regulate even better than the finest clock-work lamps. This discovery has been made by the writer—several years too late.

It has been asserted by competent English electricians that in a constant-current machine or transformer the regulation is effected by varying the phase of the secondary current. That this view is erroneous may be easily proved by using, instead of lamps, devices each possessing self-induction and capacity or self-induction and resistance—that is, retarding and accelerating components—in such proportions as to not affect materially the phase of the secondary current. Any number of such devices may be inserted or cut out, still it will be found that the regulation occurs, a constant current being maintained, while the electromotive force is varied with the number of the devices. The change of phase of the secondary current is simply a result following from the changes in resistance, and, though secondary reaction is always of more or less importance, yet the real cause of the regulation lies in the existence of the conditions above enumerated. It should be stated, however, that in the case of a machine the above remarks are to be restricted to the cases in which the machine is independently excited. If the excitation be effected by commutating the armature current, then the fixed position of the brushes makes any shifting of the neutral line of the utmost importance, and it may not be thought immodest of the writer to mention that, as far as records go, he seems to have been the first who has successfully regulated machines by providing a bridge connection between a point of the external circuit and the commutator by means of a third brush. The armature and field being properly proportioned and the brushes placed in their determined positions, a constant current or constant potential resulted from the shifting of the diameter of commutation by the varying loads.

In connection with machines of such high frequencies, the condenser affords an especially interesting study. It is easy to raise the electromotive force of such a machine to four or five times the value by simply connecting the condenser to the circuit, and the writer has continually used the condenser for the purposes of regulation, as suggested by Blakesley in his book on alternate currents, in which he has treated the most frequently occurring condenser problems with exquisite simplicity and clearness. The high frequency allows the use of small capacities and renders investigation easy. But, although in most of the experiments the result may be foretold, some phenomena observed seem at first curious. One experiment performed three or four months ago with such a machine and a condenser may serve as an illustration. A machine was used giving about 20,000 alternations per second. Two bare wires about twenty feet long and two millimeters in diameter, in close proximity to each other, were connected to the terminals of the machine at the one end, and to a condenser at the other. A small transformer without an iron core, of course, was used to bring the reading within range of a Cardew voltmeter by connecting the voltmeter to the secondary. On the terminals of the condenser the electromotive force was about 120 volts, and from there inch by inch it gradually fell until at the terminals of the machine it was about 65 volts. It was virtually as though the condenser were a generator, and the line and armature circuit simply a resistance connected to it. The writer looked for a case of resonance, but he was unable to augment the effect by varying the capacity very carefully and gradually or by changing the speed of the machine. A case of pure resonance he was unable to obtain. When a condenser was connected to the terminals of the machine—the self-induction of the armature being first determined in the maximum and minimum position and the mean value taken—the capacity which gave the highest electromotive force corresponded most nearly to that which just counteracted the self-induction with the existing frequency. If the capacity was increased or diminished, the electromotive force fell as expected.

With frequencies as high as the above mentioned, the condenser effects are of enormous importance. The condenser becomes a highly efficient apparatus capable of transferring considerable energy.

* * * * *

In an appendix to this book will be found a description of the Tesla oscillator, which its inventor believes will among other great advantages give him the necessary high frequency conditions, while relieving him of the inconveniences that attach to generators of the type described at the beginning of this chapter.

Chapter XXX

Alternate Current Electrostatic Induction Apparatus[*]

About a year and a half ago while engaged in the study of alternate currents of short period, it occurred to me that such currents could be obtained by rotating charged surfaces in close proximity to conductors. Accordingly I devised various forms of experimental apparatus of which two are illustrated in the accompanying engravings.

In the apparatus shown in Fig. 208, A is a ring of dry shellacked hard wood provided on its inside with two sets of tinfoil coatings, *a* and *b*, all the a coatings and all the *b* coatings being connected together,

Fig. 208.

respectively, but independent from each other. These two sets of coatings are connected to two terminals, T. For the sake of clearness only a few

---

[*]   Article by Mr. Tesla in *The Electrical Engineer,* N.Y., May 6, 1891.

coatings are shown. Inside of the ring A, and in close proximity to it there is arranged to rotate a cylinder B, likewise of dry, shellacked hard wood, and provided with two similar sets of coatings, a^1 and b^1, all the coatings a^1 being connected to one ring and all the others, b^1, to another marked + and –. These two sets, a^1 and b^1 are charged to a high potential by a Holtz or Wimshurst machine, and may be connected to a jar of some capacity. The inside of ring A is coated with mica in order to increase the induction and also to allow higher potentials to be used.

When the cylinder B with the charged coatings is rotated, a circuit connected to the terminals T is traversed by alternating currents. Another

FIG. 209.

form of apparatus is illustrated in Fig. 209. In this apparatus the two sets of tinfoil coatings are glued on a plate of ebonite, and a similar plate which is rotated, and the coatings of which are charged as in Fig. 208, is provided.

The output of such an apparatus is very small, but some of the effects peculiar to alternating currents of short periods may be observed. The effects, however, cannot be compared with those obtainable with an induction coil which is operated by an alternate current machine of high frequency, some of which were described by me a short while ago.

CHAPTER XXXI

"MASSAGE" WITH CURRENTS
OF HIGH FREQUENCY*

I trust that the present brief communication will not be interpreted as an effort on my part to put myself on record as a "patent medicine" man, for a serious worker cannot despise anything more than the misuse and abuse of electricity which we have frequent occasion to witness. My remarks are elicited by the lively interest which prominent medical practitioners evince at every real advance in electrical investigation. The progress in recent years has been so great that every electrician and electrical engineer is confident that electricity will become the means of accomplishing many things that have been heretofore, with our existing knowledge, deemed impossible. No wonder then that progressive physicians also should expect to find in it a powerful tool and help in new curative processes. Since I had the honor to bring before the American Institute of Electrical Engineers some results in utilizing alternating currents of high tension, I have received many letters from noted physicians inquiring as to the physical effects of such currents of high frequency. It may be remembered that I then demonstrated that a body perfectly well insulated in air can be heated by simply connecting it with a source of rapidly alternating high potential. The heating in this case is due in all probability to the bombardment of the body by air, or possibly by some other medium, which is molecular or atomic in construction, and the presence of which has so far escaped our analysis—for according to my ideas, the true ether radiation with such frequencies as even a few millions per second must be very small. This body may be a good conductor or it may be a very poor conductor of electricity with little change in the result. The human body is, in such a case, a fine conductor, and if a person insulated in a room, or no matter where, is brought into contact with such a source of rapidly alternating high potential, the skin is heated by bombardment. It is a mere question of the dimensions and character of the apparatus to produce any degree of heating desired.

---

* Article by Mr. Tesla in *The Electrical Engineer* of Dec. 23d, 1891.

It has occurred to me whether, with such apparatus properly prepared, it would not be possible for a skilled physician to find in it a means for the effective treatment of various types of disease. The heating will, of course, be superficial, that is, on the skin, and would result, whether the person operated on were in bed or walking around a room, whether dressed in thick clothes or whether reduced to nakedness. In fact, to put it broadly, it is conceivable that a person entirely nude at the North Pole might keep himself comfortably warm in this manner.

Without vouching for all the results, which must, of course, be determined by experience and observation, I can at least warrant the fact that heating would occur by the use of this method of subjecting the human body to bombardment by alternating currents of high potential and frequency such as I have long worked with. It is only reasonable to expect that some of the novel effects will be wholly different from those obtainable with the old familiar therapeutic methods generally used. Whether they would all be beneficial or not remains to be proved.

Chapter XXXII

Electric Discharge in Vacuum Tubes*

In *The Electrical Engineer* of June 10 I have noted the description of some experiments of Prof. J. J. Thomson, on the "Electric Discharge in Vacuum Tubes," and in your issue of June 24 Prof. Elihu Thomson describes an experiment of the same kind. The fundamental idea in these experiments is to set up an electromotive force in a vacuum tube— preferably devoid of any electrodes—by means of electro-magnetic induction, and to excite the tube in this manner.

As I view the subject I should think that to any experimenter who had carefully studied the problem confronting us and who attempted to find a solution of it, this idea must present itself as naturally as, for instance, the idea of replacing the tinfoil coatings of a Leyden jar by rarefied gas and exciting luminosity in the condenser thus obtained by repeatedly charging and discharging it. The idea being obvious, whatever merit there is in this line of investigation must depend upon the completeness of the study of the subject and the correctness of the observations. The following lines are not penned with any desire on my part to put myself on record as one who has performed similar experiments, but with a desire to assist other experimenters by pointing out certain peculiarities of the phenomena observed, which, to all appearances, have not been noted by Prof. J. J. Thomson, who, however, seems to have gone about systematically in his investigations, and who has been the first to make his results known. These peculiarities noted by me would seem to be at variance with the views of Prof. J. J. Thomson, and present the phenomena in a different light.

My investigations in this line occupied me principally during the winter and spring of the past year. During this time many different experiments were performed, and in my exchanges of ideas on this subject with Mr. Alfred s. Brown, of the Western Union Telegraph

---

* Article by Mr. Tesla in *The Electrical Engineer*. N.Y., July 1, 1891

FIG. 210.

Company, various different dispositions were suggested which were carried out by me in practice. Fig. 210 may serve as an example of one of the many forms of apparatus used. This consisted of a large glass tube sealed at one end and projecting into an ordinary incandescent lamp-bulb. The primary, usually consisting of a few turns of thick, well-insulated copper sheet was inserted within the tube, the inside space of the bulb furnishing the secondary. This form of apparatus was arrived at after some experimenting, and was used principally with the view of enabling me to place a polished reflecting surface on the inside of the tube, and for this purpose the last turn of the primary was covered with a thin silver sheet. In all forms of apparatus used there was no special difficulty in exciting a luminous circle or cylinder in proximity to the primary.

As to the number of turns, I cannot quite understand why Prof. J. J. Thomson should think that a few turns were "quite sufficient," but lest I should impute to him an opinion he may not have, I will add that I have gained this impression from the reading of the published abstracts of his lecture. Clearly, the number of turns which gives the best result in any case, is dependent on the dimensions of the apparatus, and, were it not for various considerations, one turn would always give the best result.

I have found that it is preferable to use in these experiments an alternate current machine giving a moderate number of alternations per second to excite the induction coil for charging the Leyden jar which discharges through the primary—shown diagrammatically in Fig. 211—as in such case, before the disruptive discharge takes place, the tube or bulb is slightly excited and the formation of the luminous circle is decidedly facilitated. But I have also used a Wimshurst machine in some experiments.

FIG. 211.

Prof. J. J. Thomson's view of the phenomena under consideration seems to be that they are wholly due to electro-magnetic action. I was, at one time, of the same opinion, but upon carefully investigating the subject I was led to the conviction that they are more of an electrostatic nature. It must be remembered that in these experiments we have to deal with primary currents of an enormous frequency or rate of change and of high potential, and that the secondary conductor consists of a rarefied gas, and that under such conditions electrostatic effects must play an important part.

In support of my view I will describe a few experiments made by me. To excite luminosity in the tube it is not absolutely necessary that the conductor should be closed. For instance, if an ordinary exhausted tube (preferably of large diameter) be surrounded by a spiral of thick copper wire serving as the primary, a feebly luminous spiral may be induced in

FIG. 212.

the tube, roughly shown in Fig. 212. In one of these experiments a curious phenomenon was observed; namely, two intensely luminous

FIG. 213.

circles, each of them close to a turn of the primary spiral, were formed inside of the tube, and I attributed this phenomenon to the existence of nodes on the primary. The circles were connected by a faint luminous spiral parallel to the primary and in close proximity to it. To produce this effect I have found it necessary to strain the jar to the utmost. The turns of the spiral tend to close and form circles, but this, of course, would be expected, and does not necessarily indicate an electro-magnetic effect; whereas the fact that a glow can be produced along the primary in the form of an open spiral argues for an electrostatic effect.

In using Dr. Lodge's recoil circuit, the electrostatic action is likewise apparent. The arrangement is illustrated in Fig. 213. In his experiment two hollow exhausted tubes H H were slipped over the wires of the recoil circuit and upon discharging the jar in the usual manner luminosity was excited in the tubes.

Another experiment performed is illustrated in Fig. 214. In this case an ordinary lamp-bulb was surrounded by one or two turns of thick copper wire P and the luminous circle L excited in the bulb by discharging the jar through the primary. The lamp-bulb was provided with a tinfoil coating on the side opposite to the primary and each time the tinfoil coating was connected to the ground or to a large object the luminosity of the circle was considerably increased. This was evidently due to electrostatic action.

In other experiments I have noted that when the primary touches the glass the luminous circle is easier produced and is more sharply defined; but I have not noted that, generally speaking, the circles induced were very sharply defined, as Prof. J. J. Thomson has observed; on the contrary, in my experiments they were broad and often the whole of the bulb or

FIG. 214

tube was illuminated; and in one case I have observed an intensely purplish glow, to which Prof. J. J. Thomson refers. But the circles were always in close proximity to the primary and were considerably easier produced when the latter was very close to the glass, much more so than would be expected assuming the action to be electromagnetic and considering the distance; and these facts speak for an electrostatic effect.

Furthermore I have observed that there is a molecular bombardment in the plane of the luminous circle at right angles to the glass—supposing the circle to be in the plane of the primary—this bombardment being evident from the rapid heating of the glass near the primary. Were the bombardment not at right angles to the glass the heating could not be so rapid. If there is a circumferential movement of the molecules constituting the luminous circle, I have thought that it might be rendered manifest by placing within the tube or bulb, radially to the circle, a thin plate of mica coated with some phosphorescent material and another such plate tangentially to the circle. If the molecules would move circumferentially, the former plate would be rendered more intensely phosphorescent. For want of time I have, however, not been able to perform the experiment.

Another observation made by me was that when the specific inductive capacity of the medium between the primary and secondary is increased, the inductive effect is augmented. This is roughly illustrated in Fig. 215. In this case luminosity was excited in an exhausted tube or bulb B and a glass tube T slipped between the primary and the bulb, when the effect pointed out was noted. Were the action wholly electromagnetic no change could possibly have been observed.

FIG. 215

I have likewise noted that when a bulb is surrounded by a wire closed upon itself and in the plane of the primary, the formation of the luminous circle within the bulb is not prevented. But if instead of the wire a broad strip of tinfoil is glued upon the bulb, the formation of the luminous band was prevented, because then the action was distributed over a greater surface. The effect of the closed tinfoil was no doubt of an electrostatic nature, for it presented a much greater resistance than the closed wire and produced therefore a much smaller electromagnetic effect.

Some of the experiments of Prof. J. J. Thomson also would seem to show some electrostatic action. For instance, in the experiment with the bulb enclosed in a bell jar, I should think that when the latter is exhausted so far that the gas enclosed reaches the maximum conductivity, the formation of the circle in the bulb and jar is prevented because of the space surrounding the primary being highly conducting; when the jar is further exhausted, the conductivity of the space around the primary diminishes and the circles appear necessarily first in the bell jar, as the rarefied gas is nearer to the primary. But were the inductive effect very powerful, they would probably appear in the bulb also. If, however, the bell jar were exhausted to the highest degree they would very likely show themselves in the bulb only, that is, supposing the vacuous space to be non-conducting. On the assumption that in these phenomena electrostatic actions are concerned we find it easily explicable why the introduction of mercury or the heating of the bulb prevents the formation of the luminous band or shortens the after-glow; and also why in some cases a platinum wire may prevent the excitation of the tube. Nevertheless some of the experiments of Prof. J. J. Thomson would seem to indicate an electromagnetic effect. I may add that in one of my experiments in which a vacuum was produced by the Torricellian method, I was unable to produce the luminous band, but this may have been due to the weak exciting current employed.

My principal argument is the following: I have experimentally proved that if the same discharge which is barely sufficient to excite a luminous band in the bulb when passed through the primary circuit be so directed as to exalt the electrostatic inductive effect—namely, by converting upwards—an exhausted tube, devoid of electrodes, may be excited at a distance of several feet.

* * * * *

Some Experiments on the Electric Discharge in Vacuum Tubes*

By Prof. J. J. Thomson, M.A., F.R.S.

The phenomena of vacuum discharges were, Prof. Thomson said, greatly simplified when their path was wholly gaseous, the complication of the dark space surrounding the negative electrode, and the stratifications so commonly observed in ordinary vacuum tubes, being absent. To produce discharges in tubes devoid of electrodes was, however, not easy to accomplish, for the only available means of producing an electromotive force in the discharge circuit was by electro-magnetic induction. Ordinary methods of producing variable induction were valueless, and recourse was had to the oscillatory discharge of a Leyden jar, which combines the two essentials of a current whose maximum value is enormous, and

---

* Abstract of a paper read before Physical Society of London.

whose rapidity of alternation is immensely great. The discharge circuits, which may take the shape of bulbs, or of tubes bent in the form of coils, were placed in close proximity to glass tubes filled with mercury, which formed the path of the oscillatory discharge. The parts thus corresponded to the windings of an induction coil, the vacuum tubes being the

FIG. 216.   FIG. 217.

secondary, and the tubes filled with mercury the primary. In such an apparatus the Leyden jar need not be large, and neither primary nor secondary need have many turns, for this would increase the self-induction of the former, and lengthen the discharge path in the latter. Increasing the self-induction of the primary reduces the E. M. F. induced in the secondary, whilst lengthening the secondary does not increase the E. M. F. per unit length. The two or three turns, as shown in Fig. 216, in each, were found to be quite sufficient, and, on discharging the Leyden jar between two highly polished knobs in the primary circuit, a plain uniform band of light was seen to pass round the secondary. An exhausted bulb, Fig. 217, containing traces of oxygen was placed within a primary spiral of three turns, and, on passing the jar discharge, a circle of light was seen within the bulb in close proximity to the primary circuit, accompanied by a purplish glow, which lasted for a second or more. On heating the bulb, the duration of the glow was greatly diminished, and it could be instantly extinguished by the presence of an electro-magnet. Another exhausted bulb, Fig. 218, surrounded by a primary spiral, was contained in a bell-jar, and when the pressure of air in the jar was about that of the atmosphere, the secondary discharge occurred in the bulb, as is ordinarily the case. On exhausting the jar, however, the luminous discharge grew fainter, and a point was reached at which no secondary discharge was visible. Further exhaustion of the jar caused the secondary discharge to appear outside of the bulb. The fact of obtaining no luminous discharge, either in the bulb or jar, the author could only

Fig. 218.   Fig. 219.

explain on two suppositions, viz.: that under the conditions then existing the specific inductive capacity of the gas was very great, or that a discharge could pass without being luminous. The author had also observed that the conductivity of a vacuum tube without electrodes increased as the pressure diminished, until a certain point was reached, and afterwards diminished again, thus showing that the high resistance of a nearly perfect vacuum is in no way due to the presence of the electrodes. One peculiarity of the discharges was their local nature, the rings of light being much more sharply defined than was to be expected. They were also found to be most easily produced when the chain of molecules in the discharge were all of the same kind. For example, a discharge could be easily sent through a tube many feet long, but the introduction of a small pellet of mercury in the tube stopped the discharge, although the conductivity of the mercury was much greater than that of the vacuum. In some cases he had noticed that a very fine wire placed within a tube, on the side remote from the primary circuit, would prevent a luminous discharge in that tube.

Fig. 219 shows an exhausted secondary coil of one loop containing bulbs; the discharge passed along the inner side of the bulbs, the primary coils being placed within the secondary.

* * * * *

In *The Electrical Engineer* of August 12, I find some remarks of Prof. J. J. Thomson, which appeared originally in the London *Electrician* and which have a bearing upon some experiments described by me in your issue of July 1.*

I did not, as Prof. J. J. Thomson seems to believe, misunderstand his position in regard to the cause of the phenomena considered, but I thought that in his experiments, as well as in my own, electrostatic effects were of great importance. It did not appear, from the meagre description of his experiments, that all possible precautions had been taken to exclude these effects. I did not doubt that luminosity could be excited in a closed tube when electrostatic action is completely excluded. In fact, at the outset, I myself looked for a purely electrodynamic effect and believed that I had obtained it. But many experiments performed at that time proved to me that the electrostatic effects were generally of far greater importance, and admitted of a more satisfactory explanation of most of the phenomena observed.

In using the term *electrostatic* I had reference rather to the nature of the action than to a stationary condition, which is the usual acceptance of the term. To express myself more clearly, I will suppose that near a closed exhausted tube be placed a small sphere charged to a very high potential. The sphere would act inductively upon the tube, and by distributing electricity over the same would undoubtedly produce luminosity (if the potential be sufficiently high), until a permanent condition would be reached. Assuming the tube to be perfectly well insulated, there would be only one instantaneous flash during the act of distribution. This would be due to the electrostatic action simply.

But now, suppose the charged sphere to be moved at short intervals with great speed along the exhausted tube. The tube would now be permanently excited, as the moving sphere would cause a constant redistribution of electricity and collisions of the molecules of the rarefied gas. We would still have to deal with an electrostatic effect, and in addition an electrodynamic effect would be observed. But if it were found that, for instance, the effect produced depended more on the specific inductive capacity than on the magnetic permeability of the

---

* Article by Mr. Tesla in *The Electrical Engineer*, N.Y., August 26, 1891.

medium—which would certainly be the case for speeds incomparably lower than that of light—then I believe I would be justified in saying that the effect produced was more of an electrostatic nature. I do not mean to say, however, that any similar condition prevails in the case of the discharge of a Leyden jar through the primary, but I think that such an action would be desirable.

It is in the spirit of the above example that I used the terms "more of an electrostatic nature," and have investigated the influence of bodies of high specific inductive capacity, and observed, for instance, the importance of the quality of glass of which the tube is made. I also endeavored to ascertain the influence of a medium of high permeability by using oxygen. It appeared from rough estimation that an oxygen tube when excited under similar conditions—that is, as far as could be determined—gives more light; but this, of course, may be due to many causes.

Without doubting in the least that, with the care and precautions taken by Prof. J. J. Thomson, the luminosity excited was due solely to electrodynamic action, I would say that in many experiments I have observed curious instances of the ineffectiveness of the screening, and I have also found that the electrification through the air is often of very great importance, and may, in some cases, determine the excitation of the tube.

In his original communication to the *Electrician*, Prof. J. J. Thomson refers to the fact that the luminosity in a tube near a wire through which a Leyden jar was discharged was noted by Hittorf. I think that the feeble luminous effect referred to has been noted by many experimenters, but in my experiments the effects were much more powerful than those usually noted.

The following is the communication* referred to:

$$* \ * \ * \ * \ *$$

"Mr. Tesla seems to ascribe the effects he observed to electrostatic action, and I have no doubt, from the description he gives of his method of conducting his experiments, that in them electrostatic action plays a very important part. He seems, however, to have misunderstood my position with respect to the cause of these discharges, which is not, as he implies, that luminosity in tubes without electrodes cannot be produced by electrostatic action, but that it can

---

* Note by Prof. J. J. Thompson in the London *Electrician*, July 24, 1891.

also be produced when this action is excluded. As a matter of fact, it is very much easier to get the luminosity when these electrostatic effects are operative than when they are not. As an illustration of this I may mention that the first experiment I tried with the discharge of a Leyden jar produced luminosity in the tube, but it was not until after six weeks' continuous experimenting that I was able to get a discharge in the exhausted tube which I was satisfied was due to what is ordinarily called electrodynamic action. It is advisable to have a clear idea of what we mean by electrostatic action. If, previous to the discharge of the jar, the primary coil is raised to a high potential, it will induce over the glass of the tube a distribution of electricity. When the potential of the primary suddenly falls, this electrification will redistribute itself, and may pass through the rarefied gas and produce luminosity in doing so. Whilst the discharge of the jar is going on, it is difficult, and, from a theoretical point of view, undesirable, to separate the effect into parts, one of which is called electrostatic, the other electromagnetic; what we can prove is that in this case the discharge is not such as would be produced by electromotive forces derived from a potential function. In my experiments the primary coil was connected to earth, and, as a further precaution, the primary was separated from the discharge tube by a screen of blotting paper, moistened with dilute sulphuric acid, and connected to earth. Wet blotting paper is a sufficiently good conductor to screen off a stationary electrostatic effect, though it is not a good enough one to stop waves of alternating electromotive intensity. When showing the experiments to the Physical Society I could not, of course, keep the tubes covered up, but, unless my memory deceives me, I stated the precautions which had been taken against the electrostatic effect. To correct misapprehension I may state that I did not read a formal paper to the Society, my object being to exhibit a few of the most typical experiments. The account of the experiments in the Electrician was from a reporter's note, and was not written, or even read, by me. I have now almost finished writing out, and hope very shortly to publish, an account of these and a large number of allied experiments, including some analogous to those mentioned by Mr. Tesla on the effect of conductors placed near the discharge tube, which I find, in some cases, to produce a diminution, in others an increase, in the brightness of the discharge, as well as some on the effect of the presence of substances of large specific inductive capacity. These seem to me to admit of a satisfactory explanation, for which, however, I must refer to my paper."

* * * * *

Part III

Miscellaneous Inventions and Writings

Chapter XXXIII

Method of Obtaining Direct from Alternating Currents

This method consists in obtaining direct from alternating currents, or in directing the waves of an alternating current so as to produce direct or substantially direct currents by developing or producing in the branches of a circuit including a source of alternating currents, either permanently or periodically, and by electric, electro-magnetic, or magnetic agencies, manifestations of energy, or what may be termed active resistances of opposite electrical character, whereby the currents or current waves of opposite sign will be diverted through different circuits, those of one sign passing over one branch and those of opposite sign over the other.

We may consider herein only the case of a circuit divided into two paths, inasmuch as any further subdivision involves merely an extension of the general principle. Selecting, then, any circuit through which is flowing an alternating current, Mr. Tesla divides such circuit at any desired point into two branches or paths. In one of these paths he inserts some device to create an electromotive force counter to the waves or impulses of current of one sign and a similar device in the other branch which opposes the waves of opposite sign. Assume, for example, that these devices are batteries, primary or secondary, or continuous current dynamo machines. The waves or impulses of opposite direction composing the main current have a natural tendency to divide between the two branches; but by reason of the opposite electrical character or effect of the two branches, one will offer an easy passage to a current of a certain direction, while the other will offer a relatively high resistance to the passage of the same current. The result of this disposition is, that the waves of current of one sign will, partly or wholly, pass over one of the paths or branches, while those of the opposite sign pass over the other. There may thus be obtained from an alternating current two or more direct currents without the employment of any commutator such as it has been heretofore regarded as necessary to use. The current in either branch may be used in the same way and for the same purposes as any other direct current—that is, it may be made to charge secondary batteries, energize electro-magnets, or for any other analogous purpose.

Fig. 220.

Fig. 220 represents a plan of directing the alternating currents by means of devices purely electrical in character. Figs. 221, 222, 223, 224, 225, and 226 are diagrams illustrative of other ways of carrying out the invention.

In Fig. 220, A designates a generator of alternating currents, and B B the main or line circuit therefrom. At any given point in this circuit at or near which it is desired to obtain direct currents, the circuit B is divided into two paths or branches C D. In each of these branches is placed an electrical generator, which for the present we will assume produces direct or continuous currents. The direction of the current thus produced is opposite in one branch to that of the current in the other branch, or, considering the two branches as forming a closed circuit, the generators E F are connected up in series therein, one generator in each part or half of the circuit. The electromotive force of the current sources E and F may be equal to or higher or lower than the electromotive forces in the branches C D, or between the points X and Y of the circuit B B. If equal, it is evident that current waves of one sign will be opposed in one branch and assisted in the other to such an extent that all the waves of one sign will pass over one branch and those of opposite sign over the other. If, on the other hand, the electromotive force of the sources E F be lower than that between X and Y, the currents in both branches will be alternating, but the waves of one sign will preponderate. One of the generators or sources of current E or F may be dispensed with; but it is preferable to employ both, if they offer an appreciable resistance, as the two branches will be thereby better balanced. The translating or other devices to be acted upon by the current are designated by the letters G, and they are inserted in the branches C D in any desired manner; but in order to better preserve an even balance between the branches due regard should, of course, be had to the number and character of the devices.

FIG. 221.

Figs. 221, 222, 223, and 224 illustrate what may termed "electro-magnetic" devices for accomplishing a similar result—that is to say, instead of producing directly by a generator an electromotive force in each branch of the circuit, Mr. Tesla establishes a field or fields of force and leads the branches through the same in such manner that an active opposition of opposite effect or direction will be developed therein by the passage, or tendency to pass, of the alternations of current. In Fig. 221, for example, A is the generator of alternating currents, B B the line circuit, and C D the branches over which the alternating currents are directed. In each branch is included the secondary of a transformer or induction coil, which, since they correspond in their functions to the batteries of the previous figure, are designated by the letters E F. The primaries H H′ of the induction coils or transformers are connected either in parallel or series with a source of direct or continuous currents I, and the number of convolutions is so calculated for the strength of the current from I that the cores J J′ will be saturated. The connections are such that the conditions in the two transformers are of opposite character—

that is to say, the arrangement is such that a current wave or impulse corresponding in direction with that of the direct current in one primary, as H, is of opposite direction to that in the other primary H′. It thus results that while one secondary offers a resistance or opposition to the passage through it of a wave of one sign, the other secondary similarly opposes a wave of opposite sign. In consequence, the waves of one sign will, to a greater or less extent, pass by way of one branch, while those of opposite sign in like manner pass over the other branch.

In lieu of saturating the primaries by a source of continuous current, we may include the primaries in the branches C D, respectively, and periodically short-circuit by any suitable mechanical devices—such as an ordinary revolving commutator—their secondaries. It will be understood, of course, that the rotation and action of the commutator must be in synchronism or in proper accord with the periods of the alternations in order to secure the desired results. Such a disposition is represented diagrammatically in Fig. 222.

Fig. 222.

Corresponding to the previous figures, A is the generator of alternating currents, B B the line, and C D the two branches for the direct currents. In branch C are included two primary coils E E′, and in branch D are two similar primaries F F′. The corresponding secondaries for these coils and which are on the same subdivided cores J or J′, are in circuits the terminals of which connect to opposite segments K K′, and L L′, respectively, of a commutator. Brushes *b b* bear upon the commutator and alternately short-circuit the plates κ and κ′, and ʟ and ʟ′, through a connection *c*. It is obvious that either the magnets and commutator, or the brushes, may revolve.

The operation will be understood from a consideration of the effects of closing or short-circuiting the secondaries. For example, if at the instant when a given wave of current passes, one set of secondaries be short-circuited, nearly all the current flows through the corresponding primaries; but the secondaries of the other branch being open-circuited, the self-induction in the primaries is highest, and hence little or no current will pass through that branch. If, as the current alternates, the secondaries of the two branches are alternately short-circuited, the result will be that the currents of one sign pass over one branch and those of the opposite sign over the other. The disadvantages of this arrangement, which would seem to result from the employment of sliding contacts, are in reality very slight, inasmuch as the electromotive force of the secondaries may be made exceedingly low, so that sparking at the brushes is avoided.

Fig. 223.

Fig. 223 is a diagram, partly in section, of another plan of carrying out the invention. The circuit ʙ in this case is divided, as before, and each branch includes the coils of both the fields and revolving armatures of two induction devices. The armatures o ᴘ are preferably mounted on the same shaft, and are adjusted relatively to one another in such manner that when the self-induction in one branch, as c, is maximum, in the other branch ᴅ it is minimum. The armatures are rotated in synchronism with the alternations from the source ᴀ. The winding or position of the armature coils is such that a current in a given direction passed through both armatures would establish in one, poles similar to those in the adjacent poles of the field, and in the other, poles unlike the adjacent field poles, as indicated by *n n s s* in the diagram. If the like poles are presented, as shown in circuit D, the condition is that of a closed secondary upon a primary, or the position of least inductive resistance; hence a given alternation of current will pass mainly through D. A half revolution of the armatures produces an opposite effect and the succeeding current impulse passes through C. Using this figure as an illustration, it is evident that the fields N M may be permanent magnets or independently excited and the armatures O P driven, as in the present case, so as to produce alternate currents, which will set up alternately impulses of opposite direction in the two branches D C, which in such case would include the armature circuits and translating devices only.

FIG. 224.

In Fig. 224, a plan alternative with that shown in Fig. 222 is illustrated. In the previous case illustrated, each branch C and D contained one or more primary coils, the secondaries of which were periodically short circuited in synchronism with the alternations of current from the main source A, and for this purpose a commutator was employed. The latter may, however, be dispensed with and an armature with a closed coil substituted.

Referring to Fig. 224 in one of the branches, as C, are two coils M′, wound on laminated cores, and in the other branches D are similar coils N′. A subdivided or laminated armature O′, carrying a closed coil R′, is rotatably supported between the coils M′ N′, as shown. In the position shown—that is, with the coil R′ parallel with the convolutions of the primaries N′ M′—practically the whole current will pass through branch D, because the self-induction in coils M′ M′ is maximum. If, therefore, the armature and coil be rotated at a proper speed relatively to the periods or alternations of the source A, the same results are obtained as in the case of Fig. 222.

FIG. 225.

Fig. 225 is an instance of what may be called, in distinction to the others, a "magnetic" means of securing the result. V and W are two strong permanent magnets provided with armatures V′ W′, respectively. The armatures are made of thin laminæ of soft iron or steel, and the amount of magnetic metal which they contain is so calculated that they will be fully or nearly saturated by the magnets. Around the armatures are coils E F, contained, respectively, in the circuits C and D. The connections and electrical conditions in this case are similar to those in Fig. 221, except that the current source of I, Fig. 221, is dispensed with and the saturation of the core of coils E F obtained from the permanent magnets.

The previous illustrations have all shown the two branches or paths containing the translating or induction devices as in derivation one to the other; but this is not always necessary. For example, in Fig. 226, A is an alternating-current generator; B B, the line wires or circuit. At any given point in the circuit let us form two paths, as D D′, and at another point two paths, as C C′. Either pair or group of paths is similar to the previous dispositions with the electrical source or induction device in

FIG. 226.

one branch only, while the two groups taken together form the obvious equivalent of the cases in which an induction device or generator is included in both branches. In one of the paths, as D, are included the devices to be operated by the current. In the other branch, as D′, is an induction device that opposes the current impulses of one direction and directs them through the branch D. So, also, in branch c are translating devices G, and in branch c′ an induction device or its equivalent that diverts through c impulses of opposite direction to those diverted by the device in branch D′. The diagram shows a special form of induction device for this purpose. J J′ are the cores, formed with pole-pieces, upon which are wound the coils M N. Between these pole-pieces are mounted at right angles to one another the magnetic armatures O P, preferably mounted on the same shaft and designed to be rotated in synchronism with the alternations of current. When one of the armatures is in line with the poles or in the position occupied by armature P, the magnetic circuit of the induction device is practically closed; hence there will be the greatest opposition to the passage of a current through coils N N. The alternation will therefore pass by way of branch D. At the same time, the magnetic circuit of the other induction device being broken by the position of the armature O, there will be less opposition to the current in coils M, which will shunt the current from branch c. A reversal of the current being attended by a shifting of the armatures, the opposite effect is produced.

Other modifications of these methods are possible, but need not be pointed out. In all these plans, it will be observed, there is developed in one or all of these branches of a circuit from a source of alternating currents, an active (as distinguished from a dead) resistance or opposition to the currents of one sign, for the purpose of diverting the currents of that sign through the other or another path, but permitting the currents of opposite sign to pass without substantial opposition.

Whether the division of the currents or waves of current of opposite sign be effected with absolute precision or not is immaterial, since it will be sufficient if the waves are only partially diverted or directed, for in such case the preponderating influence in each branch of the circuit of the waves of one sign secures the same practical results in many if not all respects as though the current were direct and continuous.

An alternating and a direct current have been combined so that the waves of one direction or sign were partially or wholly overcome by the direct current; but by this plan only one set of alternations are utilized, whereas by the system just described the entire current is rendered available. By obvious applications of this discovery Mr. Tesla is enabled to produce a self-exciting alternating dynamo, or to operate direct current meters on alternating-current circuits or to run various devices—such as arc lamps—by direct currents in the same circuit with incandescent lamps or other devices operated by alternating currents.

It will be observed that if an intermittent counter or opposing force be developed in the branches of the circuit and of higher electromotive force than that of the generator, an alternating current will result in each branch, with the waves of one sign preponderating, while a constantly or uniformly acting opposition in the branches of higher electromotive force than the generator would produce a pulsating current, which conditions would be, under some circumstances, the equivalent of those described.

CHAPTER XXXIV

CONDENSERS WITH PLATES IN OIL

In experimenting with currents of high frequency and high potential, Mr. Tesla has found that insulating materials such as glass, mica, and in general those bodies which possess the highest specific inductive capacity, are inferior as insulators in such devices when currents of the kind described are employed compared with those possessing high insulating power, together with a smaller specific inductive capacity; and he has also found that it is very desirable to exclude all gaseous matter from the apparatus, or any access of the same to the electrified surfaces, in order to prevent heating by molecular bombardment and the loss or injury consequent thereon. He has therefore devised a method to accomplish these results and produce highly efficient and reliable condensers, by using oil as the dielectric.* The plan admits of a particular construction of condenser, in which the distance between the plates is adjustable, and of which he takes advantage.

In the accompanying illustrations, Fig. 227 is a section of a condenser constructed in accordance with this principle and having stationary plates; and Fig. 228 is a similar view of a condenser with adjustable plates.

Any suitable box or receptacle A may be used to contain the plates or armatures. These latter are designated by B and C and are connected, respectively, to terminals D and E, which pass out through the sides of the case. The plates ordinarily are separated by strips of porous insulating material F, which are used merely for the purpose of maintaining them in position. The space within the can is filled with

---

* Mr. Tesla's experiments, as the careful reader of his three lectures will perceive, have revealed a very important fact which is taken advantage of in invention. Namely, he has shown that in a condenser a considerable amount of energy may be wasted, and the condenser may break down merely because gaseous matter is present between surfaces. A number of experiments are described in the lectures, which bring out this fact forcibly and serve as a guide in the operation of high tension apparatus. But besides bearing upon this point, these experiments also throw a light upon investigations of a purely scientific nature and explain now the lack of harmony among the observations of various investigators. Mr. Tesla shows that in a fluid such as oil the losses are very small as compared with those incurred in a gas.

oil G. Such a condenser will prove highly efficient and will not become heated or permanently injured.

FIG. 227.

FIG. 228.

In many cases it is desirable to vary or adjust the capacity of a condenser, and this is provided for by securing the plates to adjustable supports—as, for example, to rods H—passing through stuffing boxes K in the sides of case A and furnished with nuts L, the ends of the rods being threaded for engagement with the nuts.

It is well known that oils possess insulating properties, and it has been a common practice to interpose a body of oil between two conductors for purposes of insulation; but Mr. Tesla believes he has discovered peculiar properties in oils which render them very valuable in this particular form of device.

CHAPTER XXXV

ELECTROLYTIC REGISTERING METER

An ingenious form of electrolytic meter attributable to Mr. Tesla is one in which a conductor is immersed in a solution, so arranged that metal may be deposited from the solution or taken away in such a manner that the electrical resistance of the conductor is varied in a definite proportion to the strength of the current the energy of which is to be computed, whereby this variation in resistance serves as a measure of the energy and also may actuate registering mechanism, whenever the resistance rises above or falls below certain limits.

In carrying out this idea Mr. Tesla employs an electrolytic cell, through which extend two conductors parallel and in close proximity to each other. These conductors he connects in series through a resistance, but in such manner that there is an equal difference of potential between them throughout their entire extent. The free ends or terminals of the conductors are connected either in series in the circuit supplying the current to the lamps or other devices, or in parallel to a resistance in the circuit and in series with the current consuming devices. Under such circumstances a current passing through the conductors establishes a difference of potential between them which is proportional to the strength of the current, in consequence of which there is a leakage of current from one conductor to the other across the solution. The strength of this leakage current is proportional to the difference of potential, and, therefore, in proportion to the strength of the current passing through the conductors. Moreover, as there is a constant difference of potential between the two conductors throughout the entire extent that is exposed to the solution, the current density through such solution is the same at all corresponding points, and hence the deposit is uniform along the whole of one of the conductors, while the metal is taken away uniformly from the other. The resistance of one conductor is by this means diminished, while that of the other is increased, both in proportion to the strength of the current passing through the conductors. From such variation in the resistance of either or both of the conductors forming the positive and negative electrodes of the cell, the current energy expended may be readily computed. Figs. 229 and 230 illustrate two forms of such a meter.

Fig. 229.

In Fig. 229 G designates a direct-current generator. L L are the conductors of the circuit extending therefrom. A is a tube of glass, the ends of which are sealed, as by means of insulating plugs or caps B B. C C′ are two conductors extending through the tube A, their ends passing out through the plugs B to terminals thereon. These conductors may be corrugated or formed in other proper ways to offer the desired electrical resistance. R is a resistance connected in series with the two conductors C C′, which by their free terminals are connected up in circuit with one of the conductors L.

The method of using this device and computing by means thereof the energy of the current will be readily understood. First, the resistances of the two conductors C C′, respectively, are accurately measured and noted. Then a known current is passed through the instrument for a given time, and by a second measurement the increase and diminution of the resistances of the two conductors are respectively taken. From these data the constant is obtained—that is to say, for example, the increase of resistance of one conductor or the diminution of the resistance of the other per lamp hour. These two measurements evidently serve as a check, since the gain of one conductor should equal the loss of the other. A further check is afforded by measuring both wires in series with the resistance, in which case the resistance of the whole should remain constant.

In Fig. 230 the conductors c c′ are connected in parallel, the current device at x passing in one branch first through a resistance r′ and then through conductor c, while on the other branch it passes first through conductor c′, and then through resistance r″. The resistances r′ r″ are equal, as also are the resistances of the conductors c c′. It is, moreover, preferable that the respective resistances of the conductors c c′ should be a known and convenient fraction of the coils or resistances r′ r″. It will be observed that in the arrangement shown in Fig. 230 there is a constant potential difference between the two conductors c c′ throughout their entire length.

FIG. 230.

It will be seen that in both cases illustrated, the proportionality of the increase or decrease of resistance to the current strength will always be preserved, for what one conductor gains the other loses, and the resistances of the conductors c c′ being small as compared with the resistances in series with them. It will be understood that after each measurement or registration of a given variation of resistance in one or both conductors, the direction of the current should be changed or the instrument reversed, so that the deposit will be taken from the conductor which has gained and added to that which has lost. This principle is capable of many modifications. For instance, since there is a section of the circuit—to wit, the conductor c or c′—that varies in resistance in proportion to the current strength, such variation may be utilized, as is done in many analogous cases, to effect the operation of various automatic devices, such as registers. It is better, however, for the sake of simplicity to compute the energy by measurements of resistance.

The chief advantages of this arrangement are, first, that it is possible to read off directly the amount of the energy expended by means of a properly constructed ohm-meter and without resorting to weighing the deposit; secondly it is not necessary to employ shunts, for the whole of the current to be measured may be passed through the instrument; third, the accuracy of the instrument and correctness of the indications are but slightly affected by changes in temperature. It is also said that such meters have the merit of superior economy and compactness, as well as of cheapness in construction. Electrolytic meters seem to need every auxiliary advantage to make them permanently popular and successful, no matter how much ingenuity may be shown in their design.

CHAPTER XXXVI

THERMO-MAGNETIC MOTORS AND PYRO-MAGNETIC GENERATORS

No electrical inventor of the present day dealing with the problems of light and power considers that he has done himself or his opportunities justice until he has attacked the subject of thermo-magnetism. As far back as the beginning of the seventeenth century it was shown by Dr. William Gilbert, the father of modern electricity, that a loadstone or iron bar when heated to redness loses its magnetism; and since that time the influence of heat on the magnetic metals has been investigated frequently, though not with any material or practical result.

For a man of Mr. Tesla's inventive ability, the problems in this field have naturally had no small fascination, and though he has but glanced at them, it is to be hoped he may find time to pursue the study deeper and further. For such as he, the investigation must undoubtedly bear fruit. Meanwhile he has worked out one or two operative devices worthy of note.* He obtains mechanical power by a reciprocating action resulting from the joint operations of heat, magnetism, and a spring or weight or other force—that is to say he subjects a body magnetized by induction or otherwise to the action of heat until the magnetism is sufficiently neutralized to allow a weight or spring to give motion to the body and lessen the action of the heat, so that the magnetism may be sufficiently restored to move the body in the opposite direction, and again subject the same to the demagnetizing power of the heat.

Use is made of either an electro-magnet or a permanent magnet, and the heat is directed against a body that is magnetized by induction, rather

---

* It will, of course, be inferred from the nature of these devices that the vibration obtained in this manner is very slow owing to the inability of the iron to follow rapid changes in temperature. In an interview with Mr. Tesla on this subject, the compiler learned of an experiment which will interest students. A simple horseshoe magnet is taken and a piece of sheet iron bent in the form of an L is brought in contact with one of the poles and placed in such a position that it is kept in the attraction of the opposite pole delicately suspended. A spirit lamp is placed under the sheet iron piece and when the iron is heated to a certain temperature it is easily set in vibration oscillating as rapidly as 400 to 500 times a minute. The experiment is very easily performed and is interesting principally on account of the very rapid rate of vibration.

than directly against a permanent magnet, thereby avoiding the loss of magnetism that might result in the permanent magnet by the action of heat. Mr. Tesla also provides for lessening the volume of the heat or for intercepting the same during that portion of the reciprocation in which the cooling action takes place.

In the diagrams are shown some of the numerous arrangements that may be made use of in carrying out this idea. In all of these figures the magnet-poles are marked N S, the armature A, the Bunsen burner or other source of heat H, the axis of motion M, and the spring or the equivalent thereof—namely, a weight—is marked W.

FIG. 231.   FIG. 232.   FIG. 233.

In Fig. 231 the permanent magnet N is connected with a frame, F, supporting the axis M, from which the arm P hangs, and at the lower end of which the armature A is supported. The stops 2 and 3 limit the extent of motion, and the spring W tends to draw the armature A away from the magnet N. It will now be understood that the magnetism of N is sufficient to overcome the spring W and draw the armature A toward the magnet N. The heat acting upon the armature A neutralizes its induced magnetism sufficiently for the spring W to draw the armature A away from the magnet N and also from the heat at H. The armature now cools, and the attraction of the magnet N overcomes the spring W and draws the armature A back again above the burner H, so that the same is again heated and the operations are repeated. The reciprocating movements thus obtained are employed as a source of mechanical power in any desired manner. Usually a connecting-rod to a crank upon a fly-wheel shaft would be made use of, as indicated in Fig. 240.

Fig. 232 represents the same parts as before described; but an electromagnet is illustrated in place of a permanent magnet. The operations, however, are the same.

In Fig. 233 are shown the same parts as in Figs. 231 and 232, but they are differently arranged. The armature A, instead of swinging, is stationary and held by arm P′, and the core N S of the electro-magnet is made to swing within the helix Q, the core being suspended by the arm P from the pivot M. A shield, R, is connected with the magnet-core and swings with it, so that after the heat has demagnetized the armature A to such an extent that the spring W draws the core N S away from the armature A, the shield R comes between the flame H and armature A, thereby intercepting the action of the heat and allowing the armature to cool, so that the magnetism, again preponderating, causes the movement of the core N S toward the armature A and the removal of the shield R from above the flame, so that the heat again acts to lessen or neutralize the magnetism. A rotary or other movement may be obtained from this reciprocation.

FIG. 234.   FIG. 235.   FIG. 236.

Fig. 234 corresponds in every respect with Fig. 233, except that a permanent horseshoe-magnet, N S is represented as taking the place of the electro-magnet in Fig. 233.

In Fig. 235 is shown a helix, Q, with an armature adapted to swing toward or from the helix. In this case there may be a soft-iron core in the helix, or the armature may assume the form of a solenoid core, there being no permanent core within the helix.

Fig. 236 is an end view, and Fig. 237 a plan view, illustrating the method as applied to a swinging armature, A, and a stationary permanent magnet, N S. In this instance Mr. Tesla applies the heat to an auxiliary

FIG. 237.

armature or keeper, T, which is adjacent to and preferably in direct contact with the magnet. This armature T, in the form of a plate of sheet-iron, extends across from one pole to the other and is of sufficient section to practically form a keeper for the magnet, so that when the armature T is cool nearly all the lines of force pass over the same and very little free magnetism is exhibited. Then the armature A, which swings freely on the pivots M in front of the poles N S, is very little attracted and the spring W pulls the same way from the poles into the position indicated in the diagram. The heat is directed upon the iron plate T at some distance from the magnet, so as to allow the magnet to keep comparatively cool. This heat is applied beneath the plate by means of the burners H, and there is a connection from the armature A or its pivot to the gas-cock 6, or other device for regulating the heat. The heat acting upon the middle portion of the plate T, the magnetic conductivity of the heated portion is diminished or destroyed, and a great number of the lines of force are deflected over the armature A, which is now powerfully attracted and drawn into line, or nearly so, with the poles N S. In so doing the cock 6 is nearly closed and the plate T cools, the lines of force are again deflected over the same, the attraction exerted upon the armature A is diminished, and the spring W pulls the same away from the magnet into the position shown by full lines, and the operations are repeated. The arrangement shown in Fig. 236 has the advantages that the magnet and armature are kept cool and the strength of the permanent magnet is better preserved, as the magnetic circuit is constantly closed.

In the plan view, Fig. 238, is shown a permanent magnet and keeper plate, T, similar to those in Figs. 236 and 237, with the burners H for the gas beneath the same; but the armature is pivoted at one end to one pole

FIG. 238.   FIG. 239.

of the magnet and the other end swings toward and from the other pole of the magnet. The spring W acts against a lever arm that projects from the armature, and the supply of heat has to be partly cut off by a connection to the swinging armature, so as to lessen the heat acting upon the keeper plate when the armature A has been attracted.

Fig. 239 is similar to Fig. 238, except that the keeper T is not made use of and the armature itself swings into and out of the range of the intense action of the heat from the burner H. Fig. 240 is a diagram similar to Fig. 231, except that in place of using a spring and stops, the armature is shown as connected by a link, to the crank of a fly-wheel, so that the fly-wheel will be revolved as rapidly as the armature can be heated and cooled to the necessary extent. A spring may be used in addition, as in Fig. 231. In Fig. 241 the armatures A A are connected by a link, so that one will be heating while the other is cooling, and the attraction exerted to move the cooled armature is availed of to draw away the heated armature instead of using a spring.

FIG. 240.   FIG. 241.

Mr. Tesla has also devoted his attention to the development of a pyromagnetic generator of electricity* based upon the following laws: First, that electricity or electrical energy is developed in any conducting body by subjecting such body to a varying magnetic influence; and second, that the magnetic properties of iron or other magnetic substance may be partially or entirely destroyed or caused to disappear by raising it to a certain temperature, but restored and caused to reappear by again lowering its temperature to a certain degree. These laws may be applied in the production of electrical currents in many ways, the principle of which is in all cases the same, viz., to subject a conductor to a varying magnetic influence, producing such variations by the application of heat, or, more strictly speaking, by the application or action of a varying temperature upon the source of the magnetism. This principle of operation may be illustrated by a simple experiment: Place end to end, and preferably in actual contact, a permanently magnetized steel bar and a strip or bar of soft iron. Around the end of the iron bar or plate wind a coil of insulated wire. Then apply to the iron between the coil and the steel bar a flame or

---

*   The chief point to be noted is that Mr. Tesla attacked this problem in a way which was, from the standpoint of theory, and that of an engineer, far better than that from which some earlier trials in this direction started. The enlargement of these ideas will be found in Mr. Tesla's work on the pyromagnetic generator, treated in this chapter. The chief effort of the inventor was to economize the heat, which was accomplished by inclosing the iron in a source of heat well insulated, and by cooling the iron by means of steam, utilizing the steam over again. The construction also permits of more rapid magnetic changes per unit of time, meaning larger output.

other source of heat which will be capable of raising that portion of the iron to an orange red, or a temperature of about 600° centigrade. When this condition is reached, the iron somewhat suddenly loses its magnetic properties, if it be very thin, and the same effect is produced as though the iron had been moved away from the magnet or the heated section had been removed. This change of position, however, is accompanied by a shifting of the magnetic lines, or, in other words, by a variation in the magnetic influence to which the coil is exposed, and a current in the coil is the result. Then remove the flame or in any other way reduce the temperature of the iron. The lowering of its temperature is accompanied by a return of its magnetic properties, and another change of magnetic conditions occurs, accompanied by a current in an opposite direction in the coil. The same operation may be repeated indefinitely, the effect upon the coil being similar to that which would follow from moving the magnetized bar to and from the end of the iron bar or plate.

The device illustrated below is a means of obtaining this result, the features of novelty in the invention being, first, the employment of an artificial cooling device, and, second, inclosing the source of heat and that portion of the magnetic circuit exposed to the heat and artificially cooling the heated part.

Fig. 243.

Fig. 242.

These improvements are applicable generally to the generators constructed on the plan above described—that is to say, we may use an artificial cooling device in conjunction with a variable or varied or uniform source of heat.

Fig. 242 is a central vertical longitudinal section of the complete apparatus and Fig. 243 is a cross-section of the magnetic armature-core of the generator.

Let A represent a magnetized core or permanent magnet the poles of which are bridged by an armature-core composed of a casing or shell B inclosing a number of hollow iron tubes C. Around this core are wound the conductors E E′, to form the coils in which the currents are developed. In the circuits of these coils are current-consuming devices, as F F′.

D is a furnace or closed fire-box, through which the central portion of the core B extends. Above the fire is a boiler K, containing water. The flue L from the fire-box may extend up through the boiler.

G is a water-supply pipe, and H is the steam-exhaust pipe, which communicates with all the tubes C in the armature B, so that steam escaping from the boiler will pass through the tubes.

In the steam-exhaust pipe H is a valve V, to which is connected the lever I, by the movement of which the valve is opened or closed. In such a case as this the heat of the fire may be utilized for other purposes after as much of it as may be needed has been applied to heating the core B. There are special advantages in the employment of a cooling device, in that the metal of the core B is not so quickly oxidized. Moreover, the difference between the temperature of the applied heat and of the steam, air, or whatever gas or fluid be applied as the cooling medium, may be increased or decreased at will, whereby the rapidity of the magnetic changes or fluctuations may be regulated.

Chapter XXXVII

Anti-Sparking Dynamo Brush and Commutator

In direct current dynamos of great electromotive force—such, for instance, as those used for arc lighting—when one commutator bar or plate comes out of contact with the collecting-brush a spark is apt to appear on the commutator. This spark may be due to the break of the complete circuit, or to a shunt of low resistance formed by the brush between two or more commutator-bars. In the first case the spark is more apparent, as there is at the moment when the circuit is broken a discharge of the magnets through the field helices, producing a great spark or flash which causes an unsteady current, rapid wear of the commutator bars and brushes, and waste of power. The sparking may be reduced by various devices, such as providing a path for the current at the moment when the commutator segment or bar leaves the brush, by short-circuiting the field-helices, by increasing the number of the commutator-bars, or by other similar means; but all these devices are expensive or not fully available, and seldom attain the object desired.

To prevent this sparking in a simple manner, Mr. Tesla some years ago employed with the commutator-bars and intervening insulating material, mica, asbestos paper or other insulating and incombustible material, arranged to bear on the surface of the commutator, near to and behind the brush.

Fig. 244.   Fig. 245.

In the drawings, Fig. 244 is a section of a commutator with an asbestos insulating device; and Fig. 245 is a similar view, representing two plates of mica upon the back of the brush.

In Fig. 244, c represents the commutator and intervening insulating material; B B, the brushes. *d d* are sheets of asbestos paper or other suitable non-conducting material. *f f* are springs, the pressure of which may be adjusted by means of the screws *g g*.

In Fig. 245 a simple arrangement is shown with two plates of mica or other material. It will be seen that whenever one commutator segment passes out of contact with the brush, the formation of the arc will be prevented by the intervening insulating material coming in contact with the insulating material on the brush.

Asbestos paper or cloth impregnated with zinc-oxide, magnesia, zirconia, or other suitable material, may be used, as the paper and cloth are soft, and serve at the same time to wipe and polish the commutator; but mica or any other suitable material can be employed, provided the material be an insulator or a bad conductor of electricity.

A few years later Mr. Tesla turned his attention again to the same subject, as, perhaps, was very natural in view of the fact that the commutator had always been prominent in his thoughts, and that so much of his work was even aimed at dispensing with it entirely as an objectionable and unnecessary part of dynamos and motors. In these later efforts to remedy commutator troubles, Mr. Tesla constructs a commutator and the collectors therefor in two parts mutually adapted to one another, and, so far as the essential features are concerned, alike in mechanical structure. Selecting as an illustration a commutator of two segments adapted for use with an armature the coils or coil of which have but two free ends, connected respectively to the segments, the bearing-surface is the face of a disc, and is formed of two metallic quadrant segments and two insulating segments of the same dimensions, and the face of the disc is smoothed off, so that the metal and insulating segments are flush. The part which takes the place of the usual brushes, or the "collector," is a disc of the same character as the commutator and has a surface similarly formed with two insulating and two metallic segments. These two parts are mounted with their faces in contact and in such manner that the rotation of the armature causes the commutator to turn upon the collector, whereby the currents induced in the coils are taken off by the collector segments and thence conveyed off by suitable conductors leading from the collector segments. This is the general plan of the construction adopted. Aside from certain adjuncts, the nature and functions of which are set forth later, this means of commutation will be seen to possess many important advantages. In the first place the short-circuiting and the breaking of the armature coil connected to the commutator-segments occur at the same instant, and from the nature of the construction this will be done with the greatest precision; secondly, the duration of both the break and of the short circuit will be reduced to a minimum. The first results in a reduction which amounts practically to a suppression of the spark, since the break and the short circuit produce opposite effects in the armature-coil. The second has the effect of diminishing the destructive effect of a spark, since this would be in a measure proportional to the duration of the spark; while lessening the duration of the short circuit obviously increases the efficiency of the machine.

Fig. 246.   Fig. 247.

The mechanical advantages will be better understood by referring to the accompanying diagrams, in which Fig. 246 is a central longitudinal section of the end of a shaft with the improved commutator carried thereon. Fig. 247 is a view of the inner or bearing face of the collector. Fig. 248 is an end view from the armature side of a modified form of commutator. Figs. 249 and 250 are views of details of Fig. 248. Fig. 251 is a longitudinal central section of another modification, and Fig. 252 is a sectional view of the same. A is the end of the armature-shaft of a dynamo-electric machine or motor. A′ is a sleeve of insulating material around the shaft, secured in place by a screw, *a*′.

Fig. 249.   Fig. 250.

Fig. 248.

The commutator proper is in the form of a disc which is made up of four segments D D′ G G′, similar to those shown in Fig. 248. Two of these segments, as D D′, are of metal and are in electrical connection with the ends of the coils on the armature. The other two segments are of insulating material. The segments are held in place by a band, B, of insulating material. The disc is held in place by friction or by screws, *g*′ *g*′, Fig. 248, which secure the disc firmly to the sleeve A′.

The collector is made in the same form as the commutator. It is composed of the two metallic segments E E′ and the two insulating segments F F′, bound together by a band, C. The metallic segments E E′ are of the same or practically the same width or extent as the insulating segments or spaces of the commutator. The collector is secured to a sleeve, B′, by screws *g g*, and the sleeve is arranged to turn freely on the shaft A. The end of the sleeve B′ is closed by a plate, *f*, upon which presses a pivot-pointed screw, *h*, adjustable in a spring, H, which acts to maintain the collector in close contact with the commutator and to compensate for the play of the shaft. The collector is so fixed that it cannot turn with the shaft. For example, the diagram shows a slotted plate, K, which is designed to be attached to a stationary support, and an arm extending from the collector and carrying a clamping screw, L, by which the collector may be adjusted and set to the desired position.

Mr. Tesla prefers the form shown in Figs. 246 and 247 to fit the insulating segments of both commutator and collector loosely and to provide some means—as, for example, light springs, *e e*, secured to the bands A´ B´, respectively, and bearing against the segments—to exert a light pressure upon them and keep them in close contact and to compensate for wear. The metal segments of the commutator may be moved forward by loosening the screw a´.

The line wires are fed from the metal segments of the collector, being secured thereto in any convenient manner, the plan of connections being shown as applied to a modified form of the commutator in Fig. 251. The commutator and the collector in thus presenting two flat and smooth bearing-surfaces prevent most effectually by mechanical action the occurrence of sparks.

The insulating segments are made of some hard material capable of being polished and formed with sharp edges. Such materials as glass, marble, or soapstone may be advantageously used. The metal segments are preferably of copper or brass; but they may have a facing or edge of durable material—such as platinum or the like—where the sparks are liable to occur.

In Fig. 248 a somewhat modified form of the invention is shown, a form designed to facilitate the construction and replacing of the parts. In this modification the commutator and collector are made in substantially the same manner as previously described, except that the bands B c are omitted. The four segments of each part, however, are secured to their respective sleeves by screws g´ g´, and one edge of each segment is cut away, so that small plates a b may be slipped into the spaces thus formed. Of these plates a a are of metal, and are in contact with the metal segments D D´, respectively. The other two, b b, are of glass or marble, and they are all better square, as shown in Figs. 249 and 250, so that they may be turned to present new edges should any edge become worn by use. Light springs d bear upon these plates and press those in the commutator toward those in the collector, and insulating strips c c are secured to the periphery of the discs to prevent the blocks from being thrown out by centrifugal action. These plates are, of course, useful at those edges of the segments only where sparks are liable to occur, and, as they are easily replaced, they are of great advantage. It is considered best to coat them with platinum or silver.

In Figs. 251 and 252 is shown a construction where, instead of solid segments, a fluid is employed. In this case the commutator and collector are made of two insulating discs, s T, and in lieu of the metal segments a space is cut out of each part, as at R R′, corresponding in shape and size to a metal segment. The two parts are fitted smoothly and the collector T held by the screw *h* and spring H against the commutator S. As in the other cases, the commutator revolves while the collector remains stationary. The ends of the coils are connected to binding-posts *s s*, which are in electrical connection with metal plates *t t* within the recesses in the two parts S T. These chambers or recesses are filled with mercury, and in the collector part are tubes w w. with screws *w w*, carrying springs x and pistons x′, which compensate for the expansion and contraction of the mercury under varying temperatures, but which are sufficiently strong not to yield to the pressure of the fluid due to centrifugal action, and which serve as binding-posts.

Fig. 251.

Fig. 252.

In all the above cases the commutators are adapted for a single coil, and the device is particularly suited to such purposes. The number of segments may be increased, however, or more than one commutator used with a single armature. Although the bearing-surfaces are shown as planes at right angles to the shaft or axis, it is evident that in this particular the construction may be very greatly modified.

Chapter XXXVIII

Auxiliary Brush Regulation of Direct Current Dynamos

An interesting method devised by Mr. Tesla for the regulation of direct current dynamos, is that which has come to be known as the "third brush" method. In machines of this type, devised by him as far back as 1885, he makes use of two main brushes to which the ends of the field magnet coils are connected, an auxiliary brush, and a branch or shunt connection from an intermediate point of the field wire to the auxiliary brush.*

The relative positions of the respective brushes are varied, either automatically or by hand, so that the shunt becomes inoperative when the auxiliary brush has a certain position upon the commutator; but when the auxiliary brush is moved in its relation to the main brushes, or the latter are moved in their relation to the auxiliary brush, the electric condition is disturbed and more or less of the current through the field-helices is diverted through the shunt or a current is passed over the shunt to the field-helices. By varying the relative position upon the commutator of the respective brushes automatically in proportion to the varying electrical conditions of the working-circuit, the current developed can be regulated in proportion to the demands in the working-circuit.

Fig. 253 is a diagram illustrating the invention, showing one core of the field-magnets with one helix wound in the same direction throughout. Figs. 254 and 255 are diagrams showing one core of the field-magnets with a portion of the helices wound in opposite directions. Figs. 256 and 257 are diagrams illustrating the electric devices that may be employed for automatically adjusting the brushes, and Fig. 258 is a

---

* The compiler has learned partially from statements made on several occasions in journals and partially by personal inquiry of Mr. Tesla, that a great deal of work in this interesting line is unpublished. In these inventions as will be seen, the brushes are automatically shifted, but in the broad method barely suggested here the regulation is effected without any change in the position of the brushes. This auxiliary brush invention, it will be remembered, was very much discussed a few years ago, and it may be of interest that this work of Mr. Tesla, then unknown in this field, is now brought to light.

diagram illustrating the positions of the brushes when the machine is being energized at the start.

a and *b* are the positive and negative brushes of the main or working-circuit, and *c* the auxiliary brush. The working-circuit D extends from the brushes *a* and *b*, as usual, and contains electric lamps or other devices, D′, either in series or in multiple arc.

M M′ represent the field-helices, the ends of which are connected to the main brushes *a* and *b*. The branch or shunt wire *c*′ extends from the auxiliary brush *c* to the circuit of the field-helices, and is connected to the same at an intermediate point, *x*.

H represents the commutator, with the plates of ordinary construction. When the auxiliary brush *c* occupies such a position upon the commutator that the electro-motive force between the brushes *a* and *c* is to the electro-motive force between the brushes *c* and *b* as the resistance of the circuit *a* M *c*′ *c* A is to the resistance of the circuit *b* M′ *c*′ *c* B, the potentials of the points *x* and Y will be equal, and no current will flow

Fig. 253.

over the auxiliary brush; but when the brush *c* occupies a different position the potentials of the points *x* and Y will be different, and a current will flow over the auxiliary brush to and from the commutator, according to the relative position of the brushes. If, for instance, the commutator-space between the brushes *a* and *c*, when the latter is at the neutral point, is diminished, a current will flow from the point Y over the shunt *c* to the brush *b*, thus strengthening the current in the part M′, and partly neutralizing the current in part M; but if the space between the brushes *a* and *c* is increased, the current will flow over the auxiliary brush in an opposite direction, and the current in M will be strengthened, and in M′, partly neutralized.

FIG. 254.

By combining with the brushes *a*, *b*, and *c* any usual automatic regulating mechanism, the current developed can be regulated in proportion to the demands in the working circuit. The parts M and M′ of the field wire may be wound in the same direction. In this case they are arranged as shown in Fig. 253; or the part M may be wound in the opposite direction, as shown in Figs. 254 and 255.

FIG. 255.

It will be apparent that the respective cores of the field-magnets are subjected to neutralizing or intensifying effects of the current in the shunt through *c′*, and the magnetism of the cores will be partially neutralized, or the points of greatest magnetism shifted, so that it will be more or less remote from or approaching to the armature, and hence the aggregate energizing actions of the field magnets on the armature will be correspondingly varied.

In the form indicated in Fig. 253 the regulation is effected by shifting the point of greatest magnetism, and in Figs. 254 and 255 the same effect is produced by the action of the current in the shunt passing through the neutralizing helix.

The relative positions of the respective brushes may be varied by moving the auxiliary brush, or the brush *c* may remain stationary and the core P be connected to the main-brush holder A, so as to adjust the brushes *a b* in their relation to the brush *c*. If, however, an adjustment is applied to all the brushes, as seen in Fig. 257, the solenoid should be connected to both *a* and *c*, so as to move them toward or away from each other.

There are several known devices for giving motion in proportion to an electric current. In Figs. 256 and 257 the moving cores are shown as convenient devices for obtaining the required extent of motion with very slight changes in the current passing through the helices. It is understood that the adjustment of the main brushes causes variations in the strength of the current independently of the relative position of those brushes to the auxiliary brush. In all cases the adjustment should be such that no current flows over the auxiliary brush when the dynamo is running with its normal load.

FIG. 256.   FIG. 257.

In Figs. 256 and 257 A A indicate the main-brush holder, carrying the main brushes, and c the auxiliary-brush holder, carrying the auxiliary brush. These brush-holders are movable in arcs concentric with the center of the commutator-shaft. An iron piston, P, of the solenoid S, Fig. 256, is attached to the auxiliary-brush holder c. The adjustment is effected by means of a spring and screw or tightener.

In Fig. 257 instead of a solenoid, an iron tube inclosing a coil is shown. The piston of the coil is attached to both brush-holders A A and C. When the brushes are moved directly by electrical devices, as shown in Figs. 256 and 257, these are so constructed that the force exerted for adjusting is practically uniform through the whole length of motion.

It is true that auxiliary brushes have been used in connection with the helices of the field-wire; but in these instances the helices receive the entire current through the auxiliary brush or brushes, and these brushes could not be taken off without breaking the circuit through the field. These brushes cause, moreover, heavy sparking at the commutator. In the present case the auxiliary brush causes very little or no sparking, and can be taken off without breaking the circuit through the field-helices. The arrangement has, besides, the advantage of facilitating the self-excitation of the machine in all cases where the resistance of the field-wire is very great comparatively to the resistance of the main circuit at the start—for instance, on arc-light machines. In this case the auxiliary

FIG. 258.

brush c is placed near to, or better still in contact with, the brush b, as shown in Fig. 258. In this manner the part M′ is completely cut out, and as the part M has a considerably smaller resistance than the whole length of the field-wire the machine excites itself, whereupon the auxiliary brush is shifted automatically to its normal position.

In a further method devised by Mr. Tesla, one or more auxiliary brushes are employed, by means of which a portion or the whole of the field coils is shunted. According to the relative position upon the commutator of the respective brushes more or less current is caused to pass through the helices of the field, and the current developed by the machine can be varied at will by varying the relative positions of the brushes.

FIG. 259.

FIG. 260.

FIG. 261.

FIG. 262.   FIG. 263.

In Fig. 259, a and b are the positive and negative brushes of the main circuit, and c an auxiliary brush. The main circuit D extends from the brushes a and b, as usual, and contains the helices M of the field wire and the electric lamps or other working devices. The auxiliary brush c is connected to the point x of the main circuit by means of the wire c'. H is a commutator of ordinary construction. It will have been seen from what was said already that when the electro-motive force between the brushes a and c is to the electromotive force between the brushes c and b as the resistance of the circuit a M c' c A is to the resistance of the circuit b C B c c' D, the potentials of the points x and y will be equal, and no current will pass over the auxiliary brush c; but if that brush occupies a different position relatively to the main brushes the electric condition is disturbed, and current will flow either from y to x or from x to y, according to the relative position of the brushes. In the first case the current through the field-helices will be partly neutralized and the magnetism of the field magnets will be diminished. In the second case the current will be increased and the magnets gain strength. By combining with the brushes at a b c any automatic regulating mechanism, the current developed can be regulated automatically in proportion to the demands of the working circuit.

In Figs. 264 and 265 some of the automatic means are represented that may be used for moving the brushes. The core P, Fig. 264, of the solenoid-helix S is connected with the brush c to move the same, and in Fig. 265 the core P is shown as within the helix S, and connected with brushes a and c, so as to move the same toward or from each other, according to the strength of the current in the helix, the helix being within an iron tube, S', that becomes magnetized and increases the action of the solenoid.

FIG. 264. FIG. 265.

In practice it is sufficient to move only the auxiliary brush, as shown in Fig. 264, as the regulation is very sensitive to the slightest changes; but the relative position of the auxiliary brush to the main brushes may be varied by moving the main brushes, or both main and auxiliary brushes may be moved, as illustrated in Fig. 265. In the latter two cases, it will be understood, the motion of the main brushes relatively to the neutral line of the machine causes variations in the strength of the current independently of their relative position to the auxiliary brush. In all cases the adjustment may be such that when the machine is running with the ordinary load, no current flows over the auxiliary brush.

The field helices may be connected, as shown in Fig. 259, or a part of the field helices may be in the outgoing and the other part in the return circuit, and two auxiliary brushes may be employed as shown in Figs. 261 and 262. Instead of shunting the whole of the field helices, a portion only of such helices may be shunted, as shown in Figs. 260 and 262.

The arrangement shown in Fig. 262 is advantageous, as it diminishes the sparking upon the commutator, the main circuit being closed through the auxiliary brushes at the moment of the break of the circuit at the main brushes.

The field helices may be wound in the same direction, or a part may be wound in opposite directions.

The connection between the helices and the auxiliary brush or brushes may be made by a wire of small resistance, or a resistance may be interposed (r, Fig. 263) between the point x and the auxiliary brush or brushes to divide the sensitiveness when the brushes are adjusted.

The accompanying sketches also illustrate improvements made by Mr. Tesla in the mechanical devices used to effect the shifting of the brushes, in the use of an auxiliary brush. Fig. 266 is an elevation of the regulator with the frame partly in section; and Fig. 267 is a section at the line $x\,x$, Fig. 266. c is the commutator; B and B′, the brush-holders, B carrying the main brushes $a\,a'$, and B′ the auxiliary or shunt brushes $b\,b$. The axis of the brush-holder B is supported by two pivot-screws, $p\,p$. The other brush-holder, B′, has a sleeve, d, and is movable around the axis of the brush-holder B. In this way both brush-holders can turn very freely, the friction of the parts being reduced to a minimum. Over the brush-holders is mounted the solenoid s, which rests upon a forked column, c. This column also affords a support for the pivots $p\,p$, and is fastened upon a solid bracket or projection, P, which extends from the base of the machine, and is cast in one piece with the same. The brush-holders B B′ are connected by means of the links $e\,e$ and the cross-piece F to the iron core I, which slides freely in the tube T of the solenoid. The iron core I

FIG. 266.   FIG. 267.

has a screw, *s*, by means of which it can be raised and adjusted in its position relatively to the solenoid, so that the pull exerted upon it by the solenoid is practically uniform through the whole length of motion which is required to effect the regulation. In order to effect the adjustment with greater precision, the core I is provided with a small iron screw, *s'*. The core being first brought very nearly in the required position relatively to the solenoid by means of the screw *s*, the small screw *s'* is then adjusted until the magnetic attraction upon the core is the same when the core is in any position. A convenient stop, *t*, serves to limit the upward movement of the iron core.

To check somewhat the movement of the core I, a dash-pot, K, is used. The piston L of the dash-pot is provided with a valve, V, which opens by a downward pressure and allows an easy downward movement of the iron core I, but closes and checks the movement of the core when it is pulled up by the action of the solenoid.

To balance the opposing forces, the weight of the moving parts, and the pull exerted by the solenoid upon the iron core, the weights w w may be used. The adjustment is such that when the solenoid is traversed by the normal current it is just strong enough to balance the downward pull of the parts.

The electrical circuit-connections are substantially the same as indicated in the previous diagrams, the solenoid being in series with the circuit when the translating devices are in series, and in shunt when the devices are in multiple arc. The operation of the device is as follows: When upon a decrease of the resistance of the circuit or for some other reason, the current is increased, the solenoid s gains in strength and pulls up the iron core ι, thus shifting the main brushes in the direction of rotation and the auxiliary brushes in the opposite way. This diminishes the strength of the current until the opposing forces are balanced and the solenoid is traversed by the normal current; but if from any cause the current in the circuit is diminished, then the weight of the moving parts overcomes the pull of the solenoid, the iron core ι descends, thus shifting the brushes the opposite way and increasing the current to the normal strength. The dash-pot connected to the iron core ι may be of ordinary construction; but it is better, especially in machines for arc lights, to provide the piston of the dash-pot with a valve, as indicated in the diagrams. This valve permits a comparatively easy downward movement of the iron core, but checks its movement when it is drawn up by the solenoid. Such an arrangement has the advantage that a great number of lights may be put on without diminishing the light-power of the lamps in the circuit, as the brushes assume at once the proper position. When lights are cut out, the dash-pot acts to retard the movement; but if the current is considerably increased the solenoid gets abnormally strong and the brushes are shifted instantly. The regulator being properly adjusted, lights or other devices may be put on or out with scarcely any perceptible difference. It is obvious that instead of the dash-pot any other retarding device may be used.

Chapter XXXIX

Improvement in the Construction of Dynamos and Motors

This invention of Mr. Tesla is an improvement in the construction of dynamo or magneto electric machines or motors, consisting in a novel form of frame and field magnet which renders the machine more solid and compact as a structure, which requires fewer parts, and which involves less trouble and expense in its manufacture. It is applicable to generators and motors generally, not only to those which have independent circuits adapted for use in the Tesla alternating current system, but to other continuous or alternating current machines of the ordinary type generally used.

Fig. 268.

Fig. 268 shows the machine in side elevation. Fig. 269 is a vertical sectional view of the field magnets and frame and an end view of the armature; and Fig. 270 is a plan view of one of the parts of the frame and the armature, a portion of the latter being cut away.

The field magnets and frame are cast in two parts. These parts are identical in size and shape, and each consists of the solid plates or ends A B, from which project inwardly the cores C D and the side bars or bridge pieces, E F. The precise shape of these parts is largely a matter of

FIG. 269.

FIG. 270.

choice—that is to say, each casting, as shown, forms an approximately rectangular frame; but it might obviously be more or less oval, round, or square, without departure from the invention. It is also desirable to reduce the width of the side bars, E F, at the center and to so proportion the parts that when the frame is put together the spaces between the pole pieces will be practically equal to the arcs which the surfaces of the poles occupy.

The bearings G for the armature shaft are cast in the side bars E F. The field coils are either wound on the pole pieces or on a form and then slipped on over the ends of the pole pieces. The lower part or casting is secured to the base after being finished off. The armature K on its shaft is then mounted in the bearings of the lower casting and the other part of the frame placed in position, dowel pins L or any other means being used to secure the two parts in proper position.

In order to secure an easier fit, the side bars E F, and end pieces, A B, are so cast that slots M are formed when the two parts are put together.

This machine possesses several advantages. For example, if we magnetize the cores alternately, as indicated by the characters N S, it will be seen that the magnetic circuit between the poles of each part of a casting is completed through the solid iron side bars. The bearings for the shaft are located at the neutral points of the field, so that the armature core is not affected by the magnetic condition of the field.

The improvement is not restricted to the use of four pole pieces, as it is evident that each pole piece could be divided or more than four formed by the shape of the casting.

CHAPTER XL

TESLA DIRECT CURRENT ARC LIGHTING SYSTEM

At one time, soon after his arrival in America, Mr. Tesla was greatly interested in the subject of arc lighting, which then occupied public attention and readily enlisted the support of capital. He therefore worked out a system which was confided to a company formed for its exploitation, and then proceeded to devote his energies to the perfection of the details of his more celebrated "rotary field" motor system. The Tesla arc lighting apparatus appeared at a time when a great many other lamps and machines were in the market, but it commanded notice by its ingenuity. Its chief purpose was to lessen the manufacturing cost and simplify the processes of operation.

FIG. 271.

We will take up the dynamo first. Fig. 271 is a longitudinal section, and Fig. 272 a cross-section of the machine. Fig. 273 is a top view, and Fig. 274 a side view of the magnetic frame. Fig. 275 is an end view of the commutator bars, and Fig. 276 is a section of the shaft and commutator bars. Fig. 277 is a diagram illustrating the coils of the armature and the connections to the commutator plates.

The cores $c\ c\ c\ c$ of the field-magnets are tapering in both directions, as shown, for the purposes of concentrating the magnetism upon the middle of the pole-pieces.

FIG. 272.   FIG. 273.

The connecting-frame F F of the field-magnets is in the form indicated in the side view, Fig. 274, the lower part being provided with the spreading curved cast legs *e e*, so that the machine will rest firmly upon two base-bars, *r r.*

FIG. 274.

To the lower pole, S, of the field-magnet M is fastened, by means of babbitt or other fusible diamagnetic material, the base B, which is provided with bearings *b* for the armature-shaft H. The base B has a projection, P, which supports the brush-holders and the regulating devices, which are of a special character devised by Mr. Tesla.

The armature is constructed with the view to reduce to a minimum the loss of power due to Foucault currents and to the change of polarity, and also to shorten as much as possible the length of the inactive wire wound upon the armature core.

FIG. 275.   FIG. 276.

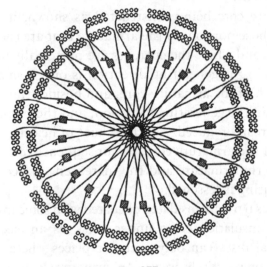

FIG. 277.

It is well known that when the armature is revolved between the poles of the field-magnets, currents are generated in the iron body of the armature which develop heat, and consequently cause a waste of power. Owing to the mutual action of the lines of force, the magnetic properties of iron, and the speed of the different portions of the armature core, these currents are generated principally on and near the surface of the armature core, diminishing in strength gradually toward the center of the core. Their quantity is under some conditions proportional to the length of the iron body in the direction in which these currents are generated. By subdividing the iron core electrically in this direction, the generation of these currents can be reduced to a great extent. For instance, if the length of the armature-core is twelve inches, and by a suitable construction it is subdivided electrically, so that there are in the generating direction six inches of iron and six inches of intervening air-spaces or insulating material, the waste currents will be reduced to fifty percent.

As shown in the diagrams, the armature is constructed of thin iron discs D D D, of various diameters, fastened upon the armature-shaft in a suitable manner and arranged according to their sizes, so that a series of iron bodies, *i i i*, is formed, each of which diminishes in thickness from the center toward the periphery. At both ends of the armature the inwardly curved discs *d d*, of cast iron, are fastened to the armature shaft.

The armature core being constructed as shown, it will be easily seen that on those portions of the armature that are the most remote from the axis, and where the currents are principally developed, the length of iron in the generating direction is only a small fraction of the total length of the armature core, and besides this the iron body is subdivided in the generating direction, and therefore the Foucault currents are greatly reduced. Another cause of heating is the shifting of the poles of the armature core. In consequence of the subdivision of the iron in the armature and the increased surface for radiation, the risk of heating is lessened.

The iron discs D D D are insulated or coated with some insulating-paint, a very careful insulation being unnecessary, as an electrical contact between several discs can only occur at places where the generated currents are comparatively weak. An armature core constructed in the manner described may be revolved between the poles of the field magnets without showing the slightest increase of temperature.

The end discs, *d d*, which are of sufficient thickness and, for the sake of cheapness, of cast-iron, are curved inwardly, as indicated in the drawings. The extent of the curve is dependent on the amount of wire to be wound upon the armatures. In this machine the wire is wound upon the armature in two superimposed parts, and the curve of the end discs, *d d*, is so calculated that the first part—that is, practically half of the wire—just fills up the hollow space to the line *x x*; or, if the wire is wound in any other manner, the curve is such that when the whole of the wire is wound, the outside mass of wires, *w*, and the inside mass of wires, w', are equal at each side of the plane $x\,x$. In this case the passive or electrically-inactive wires are of the smallest length practicable. The arrangement has further the advantage that the total lengths of the crossing wires at the two sides of the plane x x are practically equal.

To equalize further the armature coils at both sides of the plate that are in contact with the brushes, the winding and connecting up is effected in the following manner: The whole wire is wound upon the armature-core in two superimposed parts, which are thoroughly insulated from each other. Each of these two parts is composed of three separated groups of coils. The first group of coils of the first part of wire being wound and connected to the commutator-bars in the usual manner, this group is insulated and the second group wound; but the coils of this second group, instead of being connected to the next following commutator bars, are connected to the directly opposite bars of the commutator. The second group is then insulated and the third group wound, the coils of this group being connected to those bars to which they would be connected in the usual way. The wires are then thoroughly insulated and the second part of wire is wound and connected in the same manner.

Suppose, for instance, that there are twenty-four coils—that is, twelve in each part—and consequently twenty-four commutator plates. There will be in each part three groups, each containing four coils, and the coils will be connected as follows:

	Groups.	Commutator Bars.
First part of wire	First	1—5
	Second	17—21
	Third	9—13
Second part of wire	First	13—17
	Second	5—9
	Third	21—1

In constructing the armature core and winding and connecting the coils in the manner indicated, the passive or electrically inactive wire is reduced to a minimum, and the coils at each side of the plates that are in contact with the brushes are practically equal. In this way the electrical efficiency of the machine is increased.

The commutator plates *t* are shown as outside the bearing *b* of the armature shaft. The shaft H is tubular and split at the end portion, and the wires are carried through the same in the usual manner and connected to the respective commutator plates. The commutator plates are upon a cylinder, *u*, and insulated, and this cylinder is properly placed and then secured by expanding the split end of the shaft by a tapering screw plug, *v*.

The arc lamps invented by Mr. Tesla for use on the circuits from the above described dynamo are those in which the separation and feed of the carbon electrodes or their equivalents is accomplished by means of electro-magnets or solenoids in connection with suitable clutch mechanism, and were designed for the purpose of remedying certain faults common to arc lamps.

He proposed to prevent the frequent vibrations of the movable carbon "point" and flickering of the light arising therefrom; to prevent the falling into contact of the carbons; to dispense with the dash pot, clock work, or gearing and similar devices; to render the lamp extremely sensitive, and to feed the carbon almost imperceptibly, and thereby obtain a very steady and uniform light.

In that class of lamps where the regulation of the arc is effected by forces acting in opposition on a free, movable rod or lever directly connected with the electrode, all or some of the forces being dependent on the strength of the current, any change in the electrical condition of the circuit causes a vibration and a corresponding flicker in the light. This difficulty is most apparent when there are only a few lamps in circuit. To lessen this difficulty lamps have been constructed in which the lever or armature, after the establishing of the arc, is kept in a fixed position and cannot vibrate during the feed operation, the feed mechanism acting independently; but in these lamps, when a clamp is employed, it frequently occurs that the carbons come into contact and the light is momentarily extinguished, and frequently parts of the circuit are injured. In both these classes of lamps it has been customary to use dash pot, clock work, or equivalent retarding devices; but these are often unreliable and objectionable, and increase the cost of construction.

Mr. Tesla combines two electro-magnets—one of low resistance in the main or lamp circuit, and the other of comparatively high resistance in a shunt around the arc—a movable armature lever, and a special feed mechanism, the parts being arranged so that in the normal working position of the armature lever the same is kept almost rigidly in one position, and is not affected even by considerable changes in the electric circuit; but if the carbons fall into contact the armature will be actuated by the magnets so as to move the lever and start the arc, and hold the carbons until the arc lengthens and the armature lever returns to the normal position. After this the carbon rod holder is released by the action of the feed mechanism, so as to feed the carbon and restore the arc to its normal length.

FIG. 278.

FIG. 282.

FIG. 279.

FIG. 281.

FIG. 280.

Fig. 278 is an elevation of the mechanism made use of in this arc lamp. Fig. 279 is a plan view. Fig. 280 is an elevation of the balancing lever and spring; Fig. 281 is a detached plan view of the pole pieces and armatures upon the friction clamp, and Fig. 282 is a section of the clamping tube.

M is a helix of coarse wire in a circuit from the lower carbon holder to the negative binding screw –. N is a helix of fine wire in a shunt between the positive binding screw + and the negative binding screw –. The upper carbon holder s is a parallel rod sliding through the plates s' s² of the frame of the lamp, and hence the electric current passes from the positive binding post + through the plate s², carbon holder s, and upper carbon to the lower carbon, and thence by the holder and a metallic connection to the helix M.

The carbon holders are of the usual character, and to insure electric connections the springs *l* are made use of to grasp the upper carbon holding rod s, but to allow the rod to slide freely through the same. These springs *l* may be adjusted in their pressure by the screw *m*, and the spring *l* may be sustained upon any suitable support. They are shown as connected with the upper end of the core of the magnet N.

Around the carbon-holding rod s, between the plates s' s², there is a tube, R, which forms a clamp. This tube is counter-bored, as seen in the section Fig. 282, so that it bears upon the rod s at its upper end and near the middle, and at the lower end of this tubular clamp R there are armature segments *r* of soft iron. A frame or arm, *n*, extending, preferably, from the core N², supports the lever A by a fulcrum-pin, *o*. This lever A has a hole, through which the upper end of the tubular clamp R passes freely, and from the lever A is a link, *q*, to the lever *t*, which lever is pivoted at *y* to a ring upon one of the columns s³. This lever *t* has an opening or bow surrounding the tubular clamp R, and there are pins or pivotal connections *w* between the lever *t* and this clamp R, and a spring, *r²*, serves to support or suspend the weight of the parts and balance them, or nearly so. This spring is adjustable.

At one end of the lever A is a soft-iron armature block, *a*, over the core M' of the helix M, and there is a limiting screw, *c*, passing through this armature block *a*, and at the other end of the lever A is a soft iron armature block, *b*, with the end tapering or wedge shaped, and the same comes close to and in line with the lateral projection *e* on the core N². The lower ends of the cores M' N² are made with laterally projecting pole-pieces

M³ N³, respectively, and these pole-pieces are concave at their outer ends, and are at opposite sides of the armature segments *r* at the lower end of the tubular clamp R.

The operation of these devices is as follows: In the condition of inaction, the upper carbon rests upon the lower one, and when the electric current is turned on it passes freely, by the frame and spring *l*, through the rods and carbons to the coarse wire and helix M, and to the negative binding post V and the core M′ thereby is energized. The pole piece M³ attracts the armature *r*, and by the lateral pressure causes the clamp R to grasp the rod s′, and the lever A is simultaneously moved from the position shown by dotted lines, Fig. 278, to the normal position shown in full lines, and in so doing the link *q* and lever *t* are raised, lifting the clamp R and s, separating the carbons and forming the arc. The magnetism of the pole piece *e* tends to hold the lever A level, or nearly so, the core N² being energized by the current in the shunt which contains the helix N. In this position the lever A is not moved by any ordinary variation in the current, because the armature *b* is strongly attracted by the magnetism of *e*, and these parts are close to each other, and the magnetism of *e* acts at right angles to the magnetism of the core M′. If, now, the arc becomes too long, the current through the helix M is lessened, and the magnetism of the core N³ is increased by the greater current passing through the shunt, and this core N³, attracting the segmental armature *r*, lessens the hold of the clamp R upon the rod s, allowing the latter to slide and lessen the length of the arc, which instantly restores the magnetic equilibrium and causes the clamp R to hold the rod s. If it happens that the carbons fall into contact, then the magnetism of N² is lessened so much that the attraction of the magnet M will be sufficient to move the armature a and lever A so that the armature *b* passes above the normal position, so as to separate the carbons instantly; but when the carbons burn away, a greater amount of current will pass through the shunt until the attraction of the core N² will overcome the attraction of the core M′ and bring the armature lever A again into the normal horizontal position, and this occurs before the feed can take place. The segmental armature pieces *r* are shown as nearly semicircular. They are square or of any other desired shape, the ends of the pole pieces M³, N³ being made to correspond in shape.

In a modification of this lamp, Mr. Tesla provided means for automatically withdrawing a lamp from the circuit, or cutting it out when, from a failure of the feed, the arc reached an abnormal length; and also means for automatically reinserting such lamp in the circuit when the rod drops and the carbons come into contact.

Fig. 283 is an elevation of the lamp with the case in section. Fig. 284 is a sectional plan at the line *x x*. Fig. 285 is an elevation, partly in section, of the lamp at right angles to Fig. 283. Fig. 286 is a sectional plan at the line *y y* of Fig. 283. Fig. 287 is a section of the clamp in about full size. Fig. 288 is a detached section illustrating the connection of the spring to the lever that carries the pivots of the clamp, and Fig. 289 is a diagram showing the circuit-connections of the lamp.

In Fig. 283, M represents the main and N the shunt magnet, both securely fastened to the base A, which with its side columns, S S, are cast in one piece of brass or other diamagnetic material. To the magnets are soldered or otherwise fastened the brass washers or discs *a a a a*. Similar washers, *b b*, of fibre or other insulating material, serve to insulate the wires from the brass washers.

The magnets M and N are made very flat, so that their width exceeds three times their thickness, or even more. In this way a comparatively small number of convolutions is sufficient to produce the required magnetism, while a greater surface is offered for cooling off the wires.

The upper pole pieces, *m n*, of the magnets are curved, as indicated in the drawings, Fig. 283. The lower pole pieces *m′ n′*, are brought near together, tapering toward the armature *g*, as shown in Figs. 284 and 286. The object of this taper is to concentrate the greatest amount of the developed magnetism upon the armature, and also to allow the pull to be exerted always upon the middle of the armature *g*. This armature *g* is a piece of iron in the shape of a hollow cylinder, having on each side a segment cut away, the width of which is equal to the width of the pole pieces *m′ n′*.

The armature is soldered or otherwise fastened to the clamp *r*, which is formed of a brass tube, provided with gripping-jaws *e e*, Fig. 287. These jaws are arcs of a circle of the diameter of the rod R, and are made of hardened German silver. The guides *f f*, through which the carbon-holding rod R slides, are made of the same material. This has the advantage of reducing greatly the wear and corrosion of the parts coming in frictional contact with the rod, which frequently causes trouble. The jaws *e e* are fastened to the inside of the tube *r*, so that one is a little lower than the other. The object of this is to provide a greater opening for the passage of

FIG. 286.　　　FIG. 283.　　　　　　　FIG. 285.

FIG. 284.　　　　　　　FIG. 287.

FIG. 288.

the rod when the same is released by the clamp. The clamp *r* is supported
on bearings *w w*, Figs. 283, 285 and 287, which are just in the middle
between the jaws *e e.* The bearings *w w* are carried by a lever, *t*, one end
of which rests upon an adjustable support, *q*, of the side columns, s, the
other end being connected by means of the link *e′* to the armature-lever
L. The armature-lever L is a flat piece of iron in N shape, having its ends
curved so as to correspond to the form of the upper pole-pieces of the
magnets M and N. It is hung upon the pivots *v v*, Fig. 284, which are in

the jaw x of the top plate B. This plate B, with the jaw, is cast in one piece and screwed to the side columns, S S, that extend up from the base A. To partly balance the overweight of the moving parts, a spring, s', Figs. 284 and 288, is fastened to the top plate, B, and hooked to the lever t. The hook o is toward one side of the lever or bent a little sidewise, as seen in Fig. 288. By this means a slight tendency is given to swing the armature toward the pole-piece m' of the main magnet.

The binding-posts K K' are screwed to the base A. A manual switch, for short-circuiting the lamp when the carbons are renewed, is also fastened to the base. This switch is of ordinary character, and is not shown in the drawings.

The rod R is electrically connected to the lamp-frame by means of a flexible conductor or otherwise. The lamp-case receives a removable cover, s^2, to inclose the parts.

FIG. 289.

The electrical connections are as indicated diagrammatically in Fig. 289. The wire in the main magnet consists of two parts, x' and p'. These two parts may be in two separated coils or in one single helix, as shown in the drawings. The part x' being normally in circuit, is, with the fine wire upon the shunt-magnet, wound and traversed by the current in the same direction, so as to tend to produce similar poles, N N or S S, on the corresponding pole-pieces of the magnets M and N. The part p' is only in circuit when the lamp is cut out, and then the current being in the opposite direction produces in the main magnet, magnetism of the opposite polarity.

The operation is as follows: At the start the carbons are to be in contact, and the current passes from the positive binding-post K to the lamp-frame, carbon-holder, upper and lower carbon, insulated return-wire in one of the side rods, and from there through the part x' of the wire on the main magnet to the negative binding-post. Upon the passage of the current the main magnet is energized and attracts the clamping-armature g, swinging the clamp and gripping the rod by means of the gripping jaws $e\ e$. At the same time the armature lever L is pulled down and the carbons are separated. In pulling down the armature lever L the main magnet is assisted by the shunt-magnet N, the latter being magnetized by magnetic induction from the magnet M.

It will be seen that the armatures L and g are practically the keepers for the magnets M and N, and owing to this fact both magnets with either one of the armatures L and g may be considered as one horseshoe magnet, which we might term a "compound magnet." The whole of the soft-iron parts M, m', g, n', N and L form a compound magnet.

The carbons being separated, the fine wire receives a portion of the current. Now, the magnetic induction from the magnet M is such as to produce opposite poles on the corresponding ends of the magnet N; but the current traversing the helices tends to produce similar poles on the corresponding ends of both magnets, and therefore as soon as the fine wire is traversed by sufficient current the magnetism of the whole compound magnet is diminished.

With regard to the armature g and the operation of the lamp, the pole m' may be considered as the "clamping" and the pole n' as the "releasing" pole.

As the carbons burn away, the fine wire receives more current and the magnetism diminishes in proportion. This causes the armature lever L to swing and the armature g to descend gradually under the weight of the moving parts until the end p, Fig. 283, strikes a stop on the top plate, B. The adjustment is such that when this takes place the rod R is yet gripped securely by the jaws *e e*. The further downward movement of the armature lever being prevented, the arc becomes longer as the carbons are consumed, and the compound magnet is weakened more and more until the clamping armature *g* releases the hold of the gripping-jaws *e e* upon the rod R, and the rod is allowed to drop a little, thus shortening the arc. The fine wire now receiving less current, the magnetism increases, and the rod is clamped again and slightly raised, if necessary. This clamping and releasing of the rod continues until the carbons are consumed. In practice the feed is so sensitive that for the greatest part of the time the movement of the rod cannot be detected without some actual measurement. During the normal operation of the lamp the armature lever L remains practically stationary, in the position shown in Fig. 283.

Should it happen that, owing to an imperfection in it, the rod and the carbons drop too far, so as to make the arc too short, or even bring the carbons in contact, a very small amount of current passes through the fine wire, and the compound magnet becomes sufficiently strong to act as at the start in pulling the armature lever L down and separating the carbons to a greater distance.

It occurs often in practical work that the rod sticks in the guides. In this case the arc reaches a great length, until it finally breaks. Then the light goes out, and frequently the fine wire is injured. To prevent such an accident Mr. Tesla provides this lamp with an automatic cut-out which operates as follows: When, upon a failure of the feed, the arc reaches a certain predetermined length, such an amount of current is diverted through the fine wire that the polarity of the compound magnet is reversed. The clamping armature *g* is now moved against the shunt magnet N until it strikes the releasing pole *n'*. As soon as the contact is established, the current passes from the positive binding post over the clamp *r*, armature *g*, insulated shunt magnet, and the helix *p'* upon the main magnet M to the negative binding post. In this case the current passes in the opposite direction and changes the polarity of the magnet M, at the same time maintaining by magnetic induction in the core of the shunt magnet the required magnetism without reversal of polarity, and the armature *g* remains against the shunt magnet pole *n'*. The lamp is thus cut out as long as the carbons are separated. The cut out may be used in this form without any further improvement; but Mr. Tesla arranges it so that if the rod drops and the carbons come in contact the arc is started again. For this purpose he proportions the resistance of part p' and the number of the convolutions of the wire upon the main magnet so that when the carbons come in contact a sufficient amount of current is diverted through the carbons and the part x' to destroy or neutralize the magnetism of the compound magnet. Then the armature g, having a slight tendency to approach to the clamping pole m', comes out of contact with the releasing pole n'. As soon as this happens, the current through the part p' is interrupted, and the whole current passes through the part x. The magnet M is now strongly magnetized, the armature g is attracted, and the rod clamped. At the same time the armature lever L is pulled down out of its normal position and the arc started. In this way the lamp cuts itself out automatically when the arc gets too long, and reinserts itself automatically in the circuit if the carbons drop together.

Chapter XLI

Improvement in "Unipolar" Generators

Another interesting class of apparatus to which Mr. Tesla has directed his attention, is that of "unipolar" generators, in which a disc or a cylindrical conductor is mounted between magnetic poles adapted to produce an approximately uniform field. In the disc armature machines the currents induced in the rotating conductor flow from the center to the periphery, or conversely, according to the direction of rotation or the lines of force as determined by the signs of the magnetic poles, and these currents are taken off usually by connections or brushes applied to the disc at points on its periphery and near its center. In the case of the cylindrical armature machine, the currents developed in the cylinder are taken off by brushes applied to the sides of the cylinder at its ends.

In order to develop economically an electromotive force available for practicable purposes, it is necessary either to rotate the conductor at a very high rate of speed or to use a disc of large diameter or a cylinder of great length; but in either case it becomes difficult to secure and maintain a good electrical connection between the collecting brushes and the conductor, owing to the high peripheral speed.

It has been proposed to couple two or more discs together in series, with the object of obtaining a higher electro-motive force; but with the connections heretofore used and using other conditions of speed and dimension of disc necessary to securing good practicable results, this difficulty is still felt to be a serious obstacle to the use of this kind of generator. These objections Mr. Tesla has sought to avoid by constructing a machine with two fields, each having a rotary conductor mounted between its poles. The same principle is involved in the case of both forms of machine above described, but the description now given is confined to the disc type, which Mr. Tesla is inclined to favor for that machine. The discs are formed with flanges, after the manner of pulleys, and are connected together by flexible conducting bands or belts.

The machine is built in such manner that the direction of magnetism or order of the poles in one field of force is opposite to that in the other, so that rotation of the discs in the same direction develops a current in one from center to circumference and in the other from circumference to center. Contacts applied therefore to the shafts upon which the discs are mounted form the terminals of a circuit the electro-motive force in which is the sum of the electro-motive forces of the two discs.

It will be obvious that if the direction of magnetism in both fields be the same, the same result as above will be obtained by driving the discs in opposite directions and crossing the connecting belts. In this way the difficulty of securing and maintaining good contact with the peripheries of the discs is avoided and a cheap and durable machine made which is useful for many purposes—such as for an exciter for alternating current generators, for a motor, and for any other purpose for which dynamo machines are used.

FIG. 290.   FIG. 291.

Fig. 290 is a side view, partly in section, of this machine. Fig. 291 is a vertical section of the same at right angles to the shafts.

In order to form a frame with two fields of force, a support, A, is cast with two pole pieces B B′ integral with it. To this are joined by bolts E a casting D, with two similar and corresponding pole pieces C C′. The pole pieces B B′ are wound and connected to produce a field of force of given polarity, and the pole pieces C C′ are wound so as to produce a field of opposite polarity. The driving shafts F G pass through the poles and are journaled in insulating bearings in the casting A D, as shown.

H K are the discs or generating conductors. They are composed of copper, brass, or iron and are keyed or secured to their respective shafts. They are provided with broad peripheral flanges J. It is of course obvious that the discs may be insulated from their shafts, if so desired. A flexible metallic belt L is passed over the flanges of the two discs, and, if desired, may be used to drive one of the discs. It is better, however, to use this belt merely as a conductor, and for this purpose sheet steel, copper, or other suitable metal is used. Each shaft is provided with a driving pulley M, by which power is imparted from a driving shaft.

N N are the terminals. For the sake of clearness they are shown as provided with springs P, that bear upon the ends of the shafts. This machine, if self-exciting, would have copper bands around its poles; or conductors of any kind—such as wires shown in the drawings—may be used.

* * * * *

It is thought appropriate by the compiler to append here some notes on unipolar dynamos, written by Mr. Tesla, on a recent occasion.

* * * * *

Notes on a Unipolar Dynamo[*]

It is characteristic of fundamental discoveries, of great achievements of intellect, that they retain an undiminished power upon the imagination of the thinker. The memorable experiment of Faraday with a disc rotating between the two poles of a magnet, which has borne such magnificent fruit, has long passed into everyday experience; yet there are certain features about this embryo of the present dynamos and

---

[*] Article by Mr. Tesla, contributed to *The Electrical Engineer*, N. Y., Sept. 2, 1891.

motors which even today appear to us striking, and are worthy of the most careful study.

Consider, for instance, the case of a disc of iron or other metal revolving between the two opposite poles of a magnet, and the polar surfaces completely covering both sides of the disc, and assume the current to be taken off or conveyed to the same by contacts uniformly from all points of the periphery of the disc. Take first the case of a motor. In all ordinary motors the operation is dependent upon some shifting or change of the resultant of the magnetic attraction exerted upon the armature, this process being effected either by some mechanical contrivance on the motor or by the action of currents of the proper character. We may explain the operation of such a motor just as we can that of a water-wheel. But in the above example of the disc surrounded completely by the polar surfaces, there is no shifting of the magnetic action, no change whatever, as far as we know, and yet rotation ensues. Here, then, ordinary considerations do not apply; we cannot even give a superficial explanation, as in ordinary motors, and the operation will be clear to us only when we shall have recognized the very nature of the forces concerned, and fathomed the mystery of the invisible connecting mechanism.

Considered as a dynamo machine, the disc is an equally interesting object of study. In addition to its peculiarity of giving currents of one direction without the employment of commutating devices, such a machine differs from ordinary dynamos in that there is no reaction between armature and field. The armature current tends to set up a magnetization at right angles to that of the field current, but since the current is taken off uniformly from all points of the periphery, and since, to be exact, the external circuit may also be arranged perfectly symmetrical to the field magnet, no reaction can occur. This, however, is true only as long as the magnets are weakly energized, for when the magnets are more or less saturated, both magnetizations at right angles seemingly interfere with each other.

For the above reason alone it would appear that the output of such a machine should, for the same weight, be much greater than that of any other machine in which the armature current tends to demagnetize the field. The extraordinary output of the Forbes unipolar dynamo and the experience of the writer confirm this view.

Again, the facility with which such a machine may be made to excite itself is striking, but this may be due—besides to the absence of armature reaction—to the perfect smoothness of the current and non-existence of self-induction.

If the poles do not cover the disc completely on both sides, then, of course, unless the disc be properly subdivided, the machine will be very inefficient. Again, in this case there are points worthy of notice. If the disc be rotated and the field current interrupted, the current through the armature will continue to flow and the field magnets will lose their strength comparatively slowly. The reason for this will at once appear when we consider the direction of the currents set up in the disc.

FIG. 292.

Referring to the diagram Fig. 292, *d* represents the disc with the sliding contacts B B′ on the shaft and periphery. N and s represent the two poles of a magnet. If the pole N be above, as indicated in the diagram, the disc being supposed to be in the plane of the paper, and rotating in the direction of the arrow D, the current set up in the disc will flow from the center to the periphery, as indicated by the arrow A. Since the magnetic action is more or less confined to the space between the poles N s, the other portions of the disc may be considered inactive. The current set up will therefore not wholly pass through the external circuit F, but will close through the disc itself, and generally, if the disposition be in any way similar to the one illustrated, by far the greater portion of the current generated will not appear externally, as the circuit F is practically short-circuited by the inactive portions of the disc. The direction of the resulting currents in the latter may be assumed to be as indicated by the dotted lines and arrows *m* and *n*; and the direction of the energizing field current being indicated by the arrows *a b c d*, an inspection of the figure shows that one of the two branches of the eddy current, that is, A B′ *m* B, will tend to demagnetize the field, while the other branch, that is, A B′ *n* B, will have the opposite effect. Therefore, the branch A B′ *m* B, that is, the one which is *approaching* the field, will repel the lines of the same, while branch A B′ *n* B, that is, the one *leaving* the field, will gather the lines of force upon itself.

In consequence of this there will be a constant tendency to reduce the current flow in the path A B′ *m* B, while on the other hand no such opposition will exist in path A B′ *n* B, and the effect of the latter branch or path will be more or less preponderating over that of the former. The joint effect of both the assumed branch currents might be represented by that of one single current of the same direction as that energizing the field. In other words, the eddy currents circulating in the disc will energize the field magnet. This is a result quite contrary to what we might be led to suppose at first, for we would naturally expect that the resulting effect of the armature currents would be such as to oppose the field current, as generally occurs when a primary and secondary conductor are placed in inductive relations to each other. But it must be remembered that this results from the peculiar disposition in this case, namely, two paths being afforded to the current, and the latter selecting that path which offers the least opposition to its flow. From this we see that the eddy currents flowing in the disc partly energize the field, and for this reason when the field current is interrupted the currents in the disc will continue to flow, and the field magnet will lose its strength with comparative slowness and may even retain a certain strength as long as the rotation of the disc is continued.

The result will, of course, largely depend on the resistance and geometrical dimensions of the path of the resulting eddy current and on the speed of rotation; these elements, namely, determine the retardation of this current and its position relative to the field. For a certain speed there would be a maximum energizing action; then at higher speeds, it would gradually fall off to zero and finally reverse, that is, the resultant eddy current effect would be to weaken the field. The reaction would be best demonstrated experimentally by arranging the fields N S, N′ S′, freely movable on an axis concentric with the shaft of the disc. If the latter were rotated as before in the direction of the arrow D, the field would be dragged in the same direction with a torque, which, up to a certain point, would go on increasing with the speed of rotation, then fall off, and, passing through zero, finally become negative; that is, the field would begin to rotate in opposite direction to the disc. In experiments with alternate current motors in which the field was shifted by currents of differing phase, this interesting result was observed. For very low speeds of rotation of the field the motor would show a torque of 900 lbs. or more, measured on a pulley 12 inches in diameter. When the speed of rotation of the poles was increased, the torque would diminish, would finally go down to zero, become negative, and then the armature would begin to rotate in opposite direction to the field.

To return to the principal subject; assume the conditions to be such that the eddy currents generated by the rotation of the disc strengthen the field, and suppose the latter gradually removed while the disc is kept rotating at an increased rate. The current, once started, may then be sufficient to maintain itself and even increase in strength, and then we have the case of Sir William Thomson's "current accumulator." But from the above considerations it would seem that for the success of the experiment the employment of a disc *not subdivided** would be essential, for if there should be a radial subdivision, the eddy currents could not form and the self-exciting action would cease. If such a radially subdivided disc were used it would be necessary to connect the spokes by a conducting rim or in any proper manner so as to form a symmetrical system of closed circuits.

The action of the eddy currents may be utilized to excite a machine of any construction. For instance, in Figs. 293 and 294 an arrangement is shown by which a machine with a disc armature might be excited. Here a number of magnets, N S, N S, are placed radially on each side of a metal disc D carrying on its rim a set of insulated coils, C C. The magnets form two separate fields, an internal and an external one, the solid disc rotating in the field nearest the axis, and the coils in the field further from it. Assume the magnets slightly energized at the start;

---

* Mr. Tesla here refers to an interesting article which appeared in July, 1865, in the *Phil. Magazine*, by Sir W. Thomson, in which Sir William, speaking of his "uniform electric current accumulator," assumes that for self-excitation it is desirable to subdivide the disc into an infinite number of infinitely thin spokes, in order to prevent diffusion of the current. Mr. Tesla shows that diffusion is absolutely necessary for the excitation and that when the disc is subdivided no excitation can occur.

FIG. 293. FIG. 294.

they could be strengthened by the action of the eddy currents in the solid disc so as to afford a stronger field for the peripheral coils. Although there is no doubt that under proper conditions a machine might be excited in this or a similar manner, there being sufficient experimental evidence to warrant such an assertion, such a mode of excitation would be wasteful.

But a unipolar dynamo or motor, such as shown in Fig. 292, may be excited in an efficient manner by simply properly subdividing the disc or cylinder in which the currents are set up, and it is practicable to do away with the field coils which are usually employed. Such a plan is illustrated in Fig. 295. The disc or cylinder D is supposed to be arranged to rotate between the two poles N and S of a magnet, which completely cover it on both sides, the contours of the disc and poles being represented by the circles d and d^1 respectively, the upper pole being omitted for the sake of clearness. The cores of the magnet are supposed to be hollow, the shaft C of the disc passing through them. If the unmarked pole be below, and the disc be rotated screw fashion, the current will be, as before, from the center to the periphery, and may be taken off by suitable sliding contacts, B B', on the shaft and periphery respectively. In this arrangement the current flowing through the disc and external circuit will have no appreciable effect on the field magnet.

But let us now suppose the disc to be subdivided spirally, as indicated by the full or dotted lines, Fig. 295. The difference of potential between

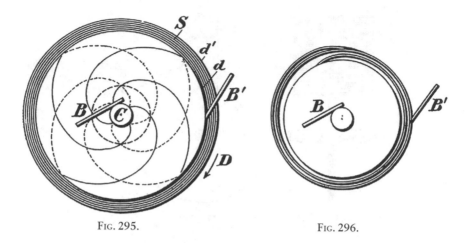

FIG. 295.   FIG. 296.

a point on the shaft and a point on the periphery will remain unchanged, in sign as well as in amount. The only difference will be that the resistance of the disc will be augmented and that there will be a greater fall of potential from a point on the shaft to a point on the periphery when the same current is traversing the external circuit. But since the current is forced to follow the lines of subdivision, we see that it will tend either to energize or de-energize the field, and this will depend, other things being equal, upon the direction of the lines of subdivision. If the subdivision be as indicated by the full lines in Fig. 295, it is evident that if the current is of the same direction as before, that is, from center to periphery, its effect will be to strengthen the field magnet; whereas, if the subdivision be as indicated by the dotted lines, the current generated will tend to weaken the magnet. In the former case the machine will be capable of exciting itself when the disc is rotated in the direction of arrow D; in the latter case the direction of rotation must be reversed. Two such discs may be combined, however, as indicated, the two discs rotating in opposite fields, and in the same or opposite direction.

Similar disposition may, of course, be made in a type of machine in which, instead of a disc, a cylinder is rotated. In such unipolar machines, in the manner indicated, the usual field coils and poles may be omitted and the machine may be made to consist only of a cylinder or of two discs enveloped by a metal casting.

Instead of subdividing the disc or cylinder spirally, as indicated in Fig. 295, it is more convenient to interpose one or more turns between the disc and the contact ring on the periphery, as illustrated in Fig. 296.

A Forbes dynamo may, for instance, be excited in such a manner. In the experience of the writer it has been found that instead of taking the current from two such discs by sliding contacts, as usual, a flexible conducting belt may be employed to advantage. The discs are in such case provided with large flanges, affording a very great contact surface. The belt should be made to bear on the flanges with spring pressure to take up the expansion. Several machines with belt contact were constructed by the writer two years ago, and worked satisfactorily; but for want of time the work in that direction has been temporarily suspended. A number of features pointed out above have also been used by the writer in connection with some types of alternating current motors.

Part IV

Appendix:
Early Phase Motors and
the Tesla Mechanical
and Electrical Oscillator

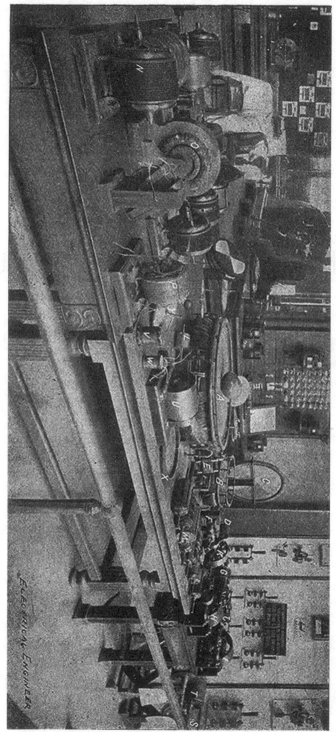

Fig. 297.

Chapter XLII

Mr. Tesla's Personal Exhibit
at the World's Fair

While the exhibits of firms engaged in the manufacture of electrical apparatus of every description at the Chicago World's Fair, afforded the visitor ample opportunity for gaining an excellent knowledge of the state of the art, there were also numbers of exhibits which brought out in strong relief the work of the individual inventor, which lies at the foundation of much, if not all, industrial or mechanical achievement. Prominent among such personal exhibits was that of Mr. Tesla, whose apparatus occupied part of the space of the Westinghouse Company, in Electricity Building.

This apparatus represented the results of work and thought covering a period of ten years. It embraced a large number of different alternating motors and Mr. Tesla's earlier high frequency apparatus. The motor exhibit consisted of a variety of fields and armatures for two, three and multiphase circuits, and gave a fair idea of the gradual evolution of the fundamental idea of the rotating magnetic field. The high frequency exhibit included Mr. Tesla's earlier machines and disruptive discharge coils and high frequency transformers, which he used in his investigations and some of which are referred to in his papers printed in this volume.

Fig. 297 shows a view of part of the exhibits containing the motor apparatus. Among these is shown at A a large ring intended to exhibit the phenomena of the rotating magnetic field. The field produced was very powerful and exhibited striking effects, revolving copper balls and eggs and bodies of various shapes at considerable distances and at great speeds. This ring was wound for two-phase circuits, and the winding was so distributed that a practically uniform field was obtained. This ring was prepared for Mr. Tesla's exhibit by Mr. C. F. Scott, electrician of the Westinghouse Electric and Manufacturing Company.

A smaller ring, shown at B, was arranged like the one exhibited at A but designed especially to exhibit the rotation of an armature in a rotating field. In connection with these two rings there was an interesting exhibit shown by Mr. Tesla which consisted of a magnet with a coil, the magnet being arranged to rotate in bearings. With this magnet he first demonstrated the identity between a rotating field and a rotating magnet; the latter, when rotating, exhibited the same phenomena as the rings when they were energized by currents of differing phase. Another prominent exhibit was a model illustrated at c which is a two-phase motor, as well as an induction motor and transformer. It consists of a large outer ring of laminated iron wound with two superimposed, separated windings which can be connected in a variety of ways. This is one of the first models used by Mr. Tesla as an induction motor and rotating transformer. The armature was either a steel or wrought iron disc with a closed coil. When the motor was operated from a two phase generator the windings were connected in two groups, as usual. When used as an induction motor, the current induced in one of the windings of the ring was passed through the other winding on the ring and so the motor operated with only two wires. When used as a transformer the outer winding served, for instance, as a secondary and the inner as a primary. The model shown at d is one of the earliest rotating field motors, consisting of a thin iron ring wound with two sets of coils and an armature consisting of a series of steel discs partly cut away and arranged on a small arbor.

At e is shown one of the first rotating field or induction motors used for the regulation of an arc lamp and for other purposes. It comprises a ring of discs with two sets of coils having different self-inductions, one set being of German silver and the other of copper wire. The armature is wound with two closed-circuited coils at right angles to each other. To the armature shaft are fastened levers and other devices to effect the regulation. At f is shown a model of a magnetic lag motor; this embodies a casting with pole projections protruding from two coils between which is arranged to rotate a smooth iron body. When an alternating current is sent through the two coils the pole projections of the field and armature within it are similarly magnetized, and upon the cessation or reversal of the current the armature and field repel each other and rotation is produced in this way. Another interesting exhibit, shown at g, is an early model of a two field motor energized by currents of different phase. There are two independent fields of laminated iron joined by brass bolts; in each field is mounted an armature, both armatures being on the same shaft. The armatures were originally so arranged as to be placed in any position relatively to each other, and the fields also were arranged to be connected in a number of ways. The motor has served for the exhibition of a number of features; among other things, it has been used as a dynamo for the production of currents of any frequency between wide limits. In this case the field, instead of being energized by direct current, was energized by currents differing in phase, which produced a rotation of the field; the armature was then rotated in the same or in opposite direction to the movement of the field; and so any number of alternations of the currents induced in the armature, from a small to a high number, determined by the frequency of the energizing field coils and the speed of the armature, was obtained.

The models H, I, J, represent a variety of rotating field, synchronous motors which are of special value in long distance transmission work. The principle embodied in these motors was enunciated by Mr. Tesla in his lecture before the American Institute of Electrical Engineers, in May, 1888.* It involves the production of the rotating field in one of the elements of the motor by currents differing in phase and energizing the other element by direct currents. The armatures are of the two and three phase

FIG. 298.

type. K is a model of a motor shown in an enlarged view in Fig. 298. This machine, together with that shown in Fig. 299, was exhibited at the same

---

* See Part I, Chap. III, page 247.

lecture, in May, 1888. They were the first rotating field motors which were independently tested, having for that purpose been placed in the hands of Prof. Anthony in the winter of 1887–88. From these tests it was shown that the efficiency and output of these motors was quite satisfactory in every respect.

FIG. 299.

It was intended to exhibit the model shown in Fig. 299, but it was unavailable for that purpose owing to the fact that it was some time ago handed over to the care of Prof. Ayrton in England. This model was originally provided with twelve independent coils; this number, as Mr. Tesla pointed out in his first lecture, being divisible by two and three, was selected in order to make various connections for two and three-phase operations, and during Mr. Tesla's experiments was used in many ways with from two to six phases. The model, Fig. 298, consists of a magnetic frame of laminated iron with four polar projections between which an armature is supported on brass bolts passing through the frame. A great variety of armatures was used in connection with these two and other fields. Some of the armatures are shown in front on the table, Fig. 297, and several are also shown enlarged in Figs. 300 to 310. An interesting exhibit is that shown at L, Fig. 297. This is an armature of hardened steel which was used in a demonstration before the Society of Arts in Boston, by Prof. Anthony. Another curious exhibit is shown enlarged in Fig. 301. This consists of thick discs of wrought iron placed lengthwise, with a mass of copper cast around them. The discs were arranged longitudinally to afford an easier starting by reason of the induced current formed in the iron discs, which differed in phase from those in the copper. This armature would start with a single circuit and run in synchronism, and represents one of the earliest types of such an armature. Fig. 305 is another striking exhibit. This is one of the earliest types of an armature with holes beneath the periphery, in which copper conductors are imbedded. The armature has eight closed circuits and was used in many different ways. Fig. 304 is a type of synchronous armature consisting of a block of soft steel wound with a coil closed upon itself. This armature was used in connection with the field shown in Fig. 298 and gave excellent results.

Fig. 300.

Fig. 301.

Fig. 302.

Fig. 303.

Fig. 304.

Fig. 305.

Fig. 306.

Fig. 307.

Fig. 308.

Fig. 309.

Fig. 310.

Fig. 302 represents a synchronous armature with a large coil around a body of iron. There is another very small coil at right angles to the first. This small coil was used for the purpose of increasing the starting torque and was found very effective in this connection. Figs. 306 and 308 show a favorite construction of armature; the iron body is made up of two sets of discs cut away and placed at right angles to each other, the interstices being wound with coils. The one shown in Fig. 308 is provided with an additional groove on each of the projections formed by the discs, for the purpose of increasing the starting torque by a wire wound in these projections. Fig. 307 is a form of armature similarly constructed, but with four independent coils wound upon the four projections. This armature was used to reduce the speed of the motor with reference to that of the generator. Fig. 300 is still another armature with a great number of independent circuits closed upon themselves, so that all the dead points on the armature are done away with, and the armature has a large starting torque. Fig. 303 is another type of armature for a four-pole motor but with coils wound upon a smooth surface. A number of these armatures have hollow shafts, as they have been used in many ways. Figs. 309 and 310 represent armatures to which either alternating or direct current was conveyed by means of sliding rings. Fig. 309 consists of a soft iron body with a single coil wound around it, the ends of the coil being connected to two sliding rings to which, usually, direct current was conveyed. The armature shown in Fig. 310 has three insulated rings on a shaft and was used in connection with two or three phase circuits.

All these models shown represent early work, and the enlarged engravings are made from photographs taken early in 1888. There is a great number of other models which were exhibited, but which are not brought out sharply in the engraving, Fig. 297. For example at M is a model of a motor comprising an armature with a hollow shaft wound with two or three coils for two or three-phase circuits; the armature was arranged to be stationary and the generating circuits were connected directly to the generator. Around the armature is arranged to rotate on its shaft a casting forming six closed circuits. On the outside this casting was turned smooth and the belt was placed on it for driving with any desired appliance. This also is a very early model.

On the left side of the table there are seen a large variety of models, N, O, P, etc., with fields of various shapes. Each of these models involves some distinct idea and they all represent gradual development chiefly interesting as showing Mr. Tesla's efforts to adapt his system to the existing high frequencies.

On the right side of the table, at S, T, are shown, on separate supports, larger and more perfected armatures of commercial motors, and in the space around the table a variety of motors and generators supplying currents to them was exhibited.

The high frequency exhibit embraced Mr. Tesla's first original apparatus used in his investigations. There was exhibited a glass tube with one layer of silk-covered wire wound at the top and a copper ribbon on the inside. This was the first disruptive discharge coil constructed by him. At U is shown the disruptive discharge coil exhibited by him in his lecture before the American Institute of Electrical Engineers, in May, 1891.* At V and W are shown some of the first high frequency transformers. A number of various fields and armatures of small models of high frequency apparatus as shown at X and Y, and others not visible in the picture, were exhibited. In the annexed space the dynamo then used by Mr. Tesla at Columbia College was exhibited; also another form of high frequency dynamo used.

In this space also was arranged a battery of Leyden jars and his large disruptive discharge coil which was used for exhibiting the light phenomena in the adjoining dark room. The coil was operated at only a small fraction of its capacity, as the necessary condensers and transformers could not be had and as Mr. Tesla's stay was limited to one week; notwithstanding, the phenomena were of a striking character. In the room were arranged two large plates placed at a distance of about eighteen feet from each other. Between them were placed two long tables with all sorts of phosphorescent bulbs and tubes; many of these were prepared with great care and marked legibly with the names which would shine with phosphorescent glow. Among them were some with the names of Helmholtz, Faraday, Maxwell, Henry, Franklin, etc. Mr. Tesla had also not forgotten the greatest living poet of his own country, Zmaj Jovan; two or three were prepared with inscriptions, like "Welcome, Electricians," and produced a beautiful effect. Each represented

---

* See Part II, Chap. XXVI., page 393.

some phase of this work and stood for some individual experiment of importance. Outside the room was the small battery seen in Fig. 311, for the exhibition of some of the impedance and other phenomena of interest. Thus, for instance, a thick copper bar bent in arched form was provided with clamps for the attachment of lamps, and a number of lamps were kept at incandescence on the bar; there was also a little motor shown on the table operated by the disruptive discharge.

FIG. 311.

As will be remembered by those who visited the Exposition, the Westinghouse Company made a fine exhibit of the various commercial motors of the Tesla system, while the twelve generators in Machinery Hall were of the two-phase type constructed for distributing light and power. Mr. Tesla also exhibited some models of his oscillators.

Chapter XLIII

The Tesla Mechanical and Electrical Oscillators

On the evening of Friday, August 25, 1893, Mr. Tesla delivered a lecture on his mechanical and electrical oscillators, before the members of the Electrical Congress, in the hall adjoining the Agricultural Building, at the World's Fair, Chicago. Besides the apparatus in the room, he employed an air compressor, which was driven by an electric motor.

Mr. Tesla was introduced by Dr. Elisha Gray, and began by stating that the problem he had set out to solve was to construct, first, a mechanism which would produce oscillations of a perfectly constant period independent of the pressure of steam or air applied, within the widest limits, and also independent of frictional losses and load. Secondly, to produce electric currents of a perfectly constant period independently of the working conditions, and to produce these currents with mechanism which should be reliable and positive in its action without resorting to spark gaps and breaks. This he successfully accomplished in his apparatus, and with this apparatus, now, scientific men will be provided with the necessaries for carrying on investigations with alternating currents with great precision. These two inventions Mr. Tesla called, quite appropriately, a mechanical and an electrical oscillator, respectively.

The former is substantially constructed in the following way. There is a piston in a cylinder made to reciprocate automatically by proper dispositions of parts, similar to a reciprocating tool. Mr. Tesla pointed out that he had done a great deal of work in perfecting his apparatus so that it would work efficiently at such high frequency of reciprocation as he contemplated, but he did not dwell on the many difficulties encountered. He exhibited, however, the pieces of a steel arbor which had been actually torn apart while vibrating against a minute air cushion.

With the piston above referred to there is associated in one of his models in an independent chamber an air spring, or dash pot, or else he obtains the spring within the chambers of the oscillator itself. To appreciate the beauty of this it is only necessary to say that in that disposition, as he showed it, no matter what the rigidity of the spring and no matter what the weight of the moving parts, in other words, no matter what the period of vibrations, the vibrations of the spring are always isochronous with the applied pressure. Owing to this, the results obtained with these vibrations are truly wonderful. Mr. Tesla provides for an air spring of tremendous rigidity, and he is enabled to vibrate big weights at an enormous rate, considering the inertia, owing to the recoil of the spring. Thus, for instance, in one of these experiments, he vibrates a weight of approximately 20 pounds at the rate of about 80 per second and with a stroke of about $7/8$ inch, but by shortening the stroke the weight could be vibrated many hundred times, and has been, in other experiments.

To start the vibrations, a powerful blow is struck, but the adjustment can be so made that only a minute effort is required to start, and, even without any special provision it will start by merely turning on the pressure suddenly. The vibration being, of course, isochronous, any change of pressure merely produces a shortening or lengthening of the stroke. Mr. Tesla showed a number of very clear drawings, illustrating the construction of the apparatus from which its working was plainly discernible. Special provisions are made so as to equalize the pressure within the dash pot and the outer atmosphere. For this purpose the inside chambers of the dash pot are arranged to communicate with the outer atmosphere so that no matter how the temperature of the enclosed air might vary, it still retains the same mean density as the outer atmosphere, and by this means a spring of constant rigidity is obtained. Now, of course, the pressure of the atmosphere may vary, and this would vary the rigidity of the spring, and consequently the period of vibration, and this feature constitutes one of the great beauties of the apparatus; for, as Mr. Tesla pointed out, this mechanical system acts exactly like a string tightly stretched between two points, and with fixed nodes, so that slight changes of the tension do not in the least alter the period of oscillation.

The applications of such an apparatus are, of course, numerous and obvious. The first is, of course, to produce electric currents, and by a number of models and apparatus on the lecture platform, Mr. Tesla showed how this could be carried out in practice by combining an electric generator with his oscillator. He pointed out what conditions must be observed in order that the period of vibration of the electrical system might not disturb the mechanical oscillation in such a way as to alter the periodicity, but merely to shorten the stroke. He combines a condenser with a self-induction, and gives to the electrical system the same period as that at which the machine itself oscillates, so that both together then fall in step and electrical and mechanical resonance is obtained, and maintained absolutely unvaried.

Next he showed a model of a motor with delicate wheelwork, which was driven by these currents at a constant speed, no matter what the air pressure applied was, so that this motor could be employed as a clock. He also showed a clock so constructed that it could be attached to one of the oscillators, and would keep absolutely correct time. Another curious and interesting feature which Mr. Tesla pointed out was that, instead of controlling the motion of the reciprocating piston by means of a spring, so as to obtain isochronous vibration, he was actually able to control the mechanical motion by the natural vibration of the electro-magnetic system, and he said that the case was a very simple one, and was quite analogous to that of a pendulum. Thus, supposing we had a pendulum of great weight, preferably, which would be maintained in vibration by force, periodically applied; now that force, no matter how it might vary, although it would oscillate the pendulum, would have no control over its period.

Mr. Tesla also described a very interesting phenomenon which he illustrated by an experiment. By means of this new apparatus, he is able to produce an alternating current in which the E. M. F. of the impulses in one direction preponderates over that of those in the other, so that there is produced the effect of a direct current. In fact he expressed the hope that these currents would be capable of application in many instances, serving as direct currents. The principle involved in this preponderating E. M. F. he explains in this way: Suppose a conductor is moved into the magnetic field and then suddenly withdrawn. If the current is not retarded, then the work performed will be a mere fractional one; but if the current is retarded, then the magnetic field acts as a spring. Imagine that the motion of the conductor is arrested by the current generated, and that at the instant when it stops to move into the field, there is still the maximum current flowing in the conductor; then this current will, according to Lenz's law, drive the conductor out of the field again, and if the conductor has no resistance, then it would leave the field with the velocity it entered it. Now it is clear that if, instead of simply depending on the current to drive the conductor out of the field, the mechanically applied force is so timed that it helps the conductor to get out of the field, then it might leave the field with higher velocity than it entered it, and thus one impulse is made to preponderate in E. M. F. over the other.

With a current of this nature, Mr. Tesla energized magnets strongly, and performed many interesting experiments bearing out the fact that one of the current impulses preponderates. Among them was one in which he attached to his oscillator a ring magnet with a small air gap between the poles. This magnet was oscillated up and down 80 times a second. A copper disc, when inserted within the air gap of the ring magnet, was brought into rapid rotation. Mr. Tesla remarked that this experiment also seemed to demonstrate that the lines of flow of current through a metallic mass are disturbed by the presence of a magnet in a manner quite independently of the so-called Hall effect. He showed also a very interesting method of making a connection with the oscillating magnet. This was accomplished by attaching to the magnet small insulated steel rods, and connecting to these rods the ends of the energizing coil. As the magnet was vibrated, stationary nodes were produced in the steel rods, and at these points the terminals of a direct current source were attached. Mr. Tesla also pointed out that one of the uses of currents, such as those produced in his apparatus, would be to select any given one of a number of devices connected to the same circuit by picking out the vibration by resonance. There is indeed little doubt that with Mr. Tesla's devices, harmonic and synchronous telegraphy will receive a fresh impetus, and vast possibilities are again opened up.

Mr. Tesla was very much elated over his latest achievements, and said that he hoped that in the hands of practical, as well as scientific men, the devices described by him would yield important results. He laid special stress on the facility now afforded for investigating the effect of mechanical vibration in all directions, and also showed that he had observed a number of facts in connection with iron cores.

The engraving, Fig. 312, shows, in perspective, one of the forms of apparatus used by Mr. Tesla in his earlier investigations in this field of work, and its interior construction is made plain by the sectional view shown in Fig. 313. It will be noted that the piston P is fitted into the hollow of a cylindercwhich is provided with channel ports o o, and ı, extending all around the inside surface. In this particular apparatus there are two channels o o for the outlet of the working fluid and one, ı, for the inlet. The piston P is provided with two slots s s′ at a carefully determined distance, one from the other. The tubes T T which are screwed into the holes drilled into the piston, establish communication between the slots s s′ and chambers on each side of the piston, each of these chambers connecting with the slot which is remote from it. The piston P is screwed tightly on a shaft A which passes through fitting boxes at the end of the cylinder c. The boxes project to a carefully determined distance into the hollow of the cylinder c, thus determining the length of the stroke.

FIG. 312.

Surrounding the whole is a jacket ı. This jacket acts chiefly to diminish the sound produced by the oscillator and as a jacket when the oscillator is driven by steam, in which case a somewhat different arrangement of the magnets is employed. The apparatus here illustrated was intended for demonstration purposes, air being used as most convenient for this purpose.

A magnetic frame M M is fastened so as to closely surround the oscillator and is provided with energizing coils which establish two strong magnetic fields on opposite sides. The magnetic frame is made up of thin sheet iron. In the intensely concentrated field thus produced, there are arranged two pairs of coils H H supported in metallic frames which are screwed on the shaft A of the piston and have additional bearings in the boxes B B on each side. The whole is mounted on a metallic base resting on two wooden blocks.

The operation of the device is as follows: The working fluid being admitted through an inlet pipe to the slot I and the piston being supposed to be in the position indicated, it is sufficient, though not necessary, to give a gentle tap on one of the shaft ends protruding from the boxes B. Assume that the motion imparted be such as to move the piston to the left (when looking at the diagram) then the air rushes through the slot s′ and tube T into the chamber to the left. The pressure now drives the piston towards the right and, owing to its inertia, it overshoots the position of equilibrium and allows the air to rush through the slot s and tube T into the chamber to the right, while the communication to the left hand chamber is cut off, the air of the latter chamber escaping through the outlet O on the left. On the return stroke a similar operation takes place on the right hand side. This oscillation is maintained continuously and the apparatus performs vibrations from a scarcely perceptible quiver amounting to no more than ¼ of an inch, up to vibrations of a little over ⅜ of an inch, according to the air pressure and load. It is indeed interesting to see how an incandescent lamp is kept burning with the apparatus showing a scarcely perceptible quiver.

To perfect the mechanical part of the apparatus so that oscillations are maintained economically was one thing, and Mr. Tesla hinted in his lecture at the great difficulties he had first encountered to accomplish this. But to produce oscillations which would be of constant period was another task of no mean proportions. As already pointed out, Mr. Tesla obtains the constancy of period in three distinct ways. Thus, he provides properly calculated chambers, as in the case illustrated, in the oscillator itself; or he associates with the oscillator an air spring of constant resilience. But the most interesting of all, perhaps, is the maintenance of the constancy of oscillation by the reaction of the electromagnetic part of the combination. Mr. Tesla winds his coils, by preference, for high tension and associates with them a condenser, making the natural period of the combination fairly approximating to the average period at which the piston would oscillate without any particular provision being made for the constancy of period under varying pressure and load. As the piston with the coils is perfectly free to move, it is extremely susceptible to the influence of the natural vibration set up in the circuits of the coils H H. The mechanical efficiency of the apparatus is very high owing to the fact that friction is reduced to a minimum and the weights which are moved are small; the output of the oscillator is therefore a very large one.

Fig. 313.

Theoretically considered, when the various advantages which Mr. Tesla holds out are examined, it is surprising, considering the simplicity of the arrangement, that nothing was done in this direction before. No doubt many inventors, at one time or other, have entertained the idea of generating currents by attaching a coil or a magnetic core to the piston of a steam engine, or generating currents by the vibrations of a tuning fork, or similar devices, but the disadvantages of such arrangements from an engineering standpoint must be obvious. Mr. Tesla, however, in the introductory remarks of his lecture, pointed out how by a series of conclusions he was driven to take up this new line of work by the necessity of producing currents of constant period and as a result of his endeavors to maintain electrical oscillation in the most simple and economical manner.